POLITICAL SCIENCE
RESEARCH METHODS

POLITICAL SCIENCE RESEARCH METHODS

SIXTH EDITION

Janet Buttolph Johnson
University of Delaware

H.T. Reynolds
University of Delaware

with
Jason D. Mycoff
University of Delaware

CQ PRESS

A Division of SAGE

Washington, D.C.

CQ Press
2300 N Street, NW, Suite 800
Washington, DC 20037

Phone: 202-729-1900; toll-free, 1-866-4CQ-PRESS (1-866-427-7737)

Web: www.cqpress.com

Cover design by McGaughy Design, Centreville, VA
Typesetting by BMWW, Baltimore, MD

☺ The paper used in this publication exceeds the requirements of the American National Standard for Information Sciences—Permanence of Paper for Printed Library Materials, ANSI Z39.48-1992.

Printed and bound in the United States of America

12   11   10   09   08   2   3   4   5   6

Library of Congress Cataloging-in-Publication Data

Johnson, Janet Buttolph
    Political science research methods / Janet Buttolph Johnson, H.T. Reynolds with Jason D. Mycoff. — 6th ed.
        p.   cm.
    Includes bibliographical references and index.
    ISBN 978-0-87289-442-6 (pbk. : alk. paper)   1. Political science—Methodology.
I. Reynolds, H.T. (Henry T.)   II. Mycoff, Jason D.   III. Title

    JA71.J55 2008
    320.072—dc22

                                                                              2007046524

To the instructors and students
who have used this book over the years
J. B. J.

To my friends and family
H. T. R.

Contents

Figures

Preface

Political Science Research Methods, now in its sixth edition, continues to hold true to the three primary objectives that have guided us since the book's inception. Our first goal is to illustrate important aspects of the research process and to demonstrate that political scientists produce worthwhile knowledge about significant political phenomena using the methods we describe in this book. To illustrate this as vividly as possible, we begin again with several case studies of political science research drawn from different areas of the discipline that address key issues and controversies in the study of politics. For this edition, we updated and extended the example on income inequality and redistribution in OECD (Organization for Economic Co-operation and Development) countries, which links nicely to the example on voter turnout. We expanded the case on judicial decision making to include a recent study on the effects of judicial decision making that makes use of a panel study research design. The example of executive branch control of the bureaucracy now mentions research on the design of new agencies by Congress with the intention to insulate them from presidential control. Finally, we changed the example on public opinion about U.S. military involvement in foreign affairs using research on the tolerance for casualties in the Iraq war.

We also made changes to fulfill our other two objectives: to give readers the tools necessary to conduct their own empirical research projects and evaluate others' research, and to help students with limited mathematical backgrounds understand the statistical calculations that are part of social science research. Throughout the book we reduced or eliminated lengthy sections on methods and techniques that students were unlikely to use or encounter in published research, and relocated some of the equations from the body of the text to a new feature called "How It's Done." To reflect the ever-increasing interest in Web-based research and resources, we've expanded our discussion of how to locate and properly use Internet sources. The book now also makes an extraordinary effort to encourage students to understand and think about the practical and theoretical implications of statistical results. To this end, we have significantly reworked several chapters to provide students with more intuitive introductions to some of the field's most important methods. For instance, Chapter 11, on statistics, now begins with "the first steps," a friendlier and more accessible introduction to the topic with better discussions of data preparation, description, inference, interpretation, and the communication of results. We hope that by meeting these goals, this book will continue to satisfy the needs of our undergraduate and graduate students as they embark on their studies in the field.

This book is organized to show that research starts with ideas—we call these hypotheses—and then follows a series of logical steps. Chapter 1 introduces the case studies that are integrated into our discussion of the research process in the subsequent chapters. We chose these cases, which form the backbone of the book, to demonstrate a wide range of research topics within the discipline of political science: American politics, public administration, the courts, international relations, comparative politics, and public policy. We refer to these cases throughout the book to demonstrate the issues, choices, decisions, and obstacles that political scientists typically confront while doing research. We want to show what takes place behind the scenes in the production of research, and the best way to do this is to refer to actual articles. The advantage to this approach, which we feel has been borne out by the book's success over the years, is that it helps students relate substance to methods. Chapter 2 examines the definition of scientific research and the development of empirical political science. We discuss the role of theory in the research process and review some of the debates in modern and contemporary political science.

Chapters 3 and 4 address the building blocks of social scientific research: hypotheses, core concepts, variables, and measurements. Chapter 5 covers research design with an expanded discussion of small-N studies and includes a handy table that summarizes and compares the features of alternative research designs. In Chapter 6 we turn to the topic of conducting a literature review. We think that once students have some idea of the initial steps of investigating research propositions, they should learn how to locate resources to help select and refine research topics and investigate the literature related to topics of interest. This chapter has been modified to reflect changes to the ways researchers conduct Internet-based searches. We give examples of using Web of Science and other sources to find refereed articles. For the first time, we include a discussion of how to write a literature review and analyze an example of one. Chapter 7, on sampling, precedes the chapters on data collection on the reasoning that sampling is not a topic solely of interest to those conducting survey research. Chapters 8–10 discuss data collection, with an emphasis on the research methods that political scientists frequently employ and that students are likely to find useful in conducting or evaluating empirical research. We consider the principles of ethical research and the role of human subject review boards and note the ethical issues related to methods of data collection. We examine observation in Chapter 8 and document analysis and the use of aggregate data in Chapter 9. Chapter 10, on survey research, includes a discussion of face-to-face interviewing. In Chapters 9 and 10 we include updated and Web-based sources of aggregate statistics and survey questions and data.

Chapters 11–13 focus on data analysis: How do we interpret data and present them to others? All three chapters contain updated examples and discussions and are supplemented by a host of new figures and tables designed to illustrate the various techniques in as friendly and intuitive a way possible. We also strengthened the discussion of tests of statistical significance. Our goals are to make the logic of the tests more comprehensible and to stress the differences among statistical, theoretical, and practical significance. Chapter 13 includes material on logistic regression, an increasingly important statistical tool in social research. In all of this, we attempted to be as rigorous as possible without overwhelming readers with theoretical fine points or computational details. The content is still accessible to anyone with a basic understanding of high school algebra. Our goal, as always, is to provide an intuitive understanding of these sometimes intimidating topics without distorting the concepts or misleading our readers.

Finally, in Chapter 14, we present a new research report, using a published journal article that investigates whether there is bias in newspaper photograph selection of political candidates. We strongly suggest that instructors who assign a research paper have their students consult the example in this chapter and use it to pattern their own writing.

In addition to the new "How It's Done" feature, the "Helpful Hints" boxes, highlighted by the light bulb icon, continue to give students practical tips, and there are several new ones in this edition, particularly in the later chapters, where they're most needed. Each chapter has updated suggested reading lists and lists of terms introduced. A glossary at the end of the book, with more than 250 definitions, lists important terms and provides a convenient study guide.

Jason D. Mycoff joins us for this edition, bringing considerable talent and valuable classroom experience to bear with revisions to a handful of chapters in the text. He is also the brains behind much of the material on the book's new companion Web site. For the first time, the site houses review and study materials for students and a host of password-protected instructor resources as well. Both can be found at http://psrm.cqpress.com. Students will find useful chapter summaries to help review and study for exams, a set of multiple-choice and true/false quiz questions, flashcards with all of the key terms from the book for self-testing, crossword puzzles that make use of the vocabulary terms, and a set of annotated Web links to help with further research. Instructors have access to a testbank with 350 fill-in-the-blank and long-answer style questions, a set of 250 PowerPoint lecture slides covering key topics in each of the chapters, and all of the tables and figures from the book in .jpg and PowerPoint format for use in lectures.

In addition to revising the text, we have substantially revised the accompanying workbook, *Working with Political Science Research Methods,* Second Edition, providing many new exercises and retaining the ones we feel worked well in the previous edition. We have also taken care to revise and create clear, straightforward instructions for each of the exercises, and we made sure that they build logically over the course of each chapter. Each workbook chapter briefly reviews key concepts covered by the corresponding chapter in the text. Students and instructors will find data sets and other documents and materials used in the workbook exercises at http://psrm.cqpress.com. The data sets, available in a variety of platforms, may also be used for additional exercises and test items developed by instructors. Instructors may want to add on to the data sets or have their students do so as part of a research project. A solutions manual for the workbook with suggestions for class work is available to instructors who adopt the workbook.

In closing, we would like to make a comment on statistical software. Instructors remain divided over the extent to which computers should be part of an introductory research course and what particular programs to require. We believe many members of the discipline are wedded to a suite of programs such as SPSS. Others prefer that students take a more hands-on approach and use an "environment," such as R, which offers step-by-step control, interactive data analysis, enriched graphics, and simulation tools, as well as the familiar statistical routines. (We also encourage instructors and students to begin exploring the many other online statistical resources such as SDA, ICPSR, Athens, American Factfinder, Rice Virtual Statistics Lab, and Vassarstats. There are many others as well.) We do, however, explicitly mention SPSS on occasion, not because we prefer or endorse it, but because it is widely adopted in the introductory courses we are trying to reach. At the same time, we have tried to keep our discussions and descriptions of "results" sufficiently general to be applicable to almost any system.

Acknowledgments

We would like to thank our careful reviewers who helped us shape this new edition: Lawrence Butler, Rowan University; Robin M. Lauermann, Messiah College; Greg Petrow, University of Nebraska–Omaha; and Emmanuel Uwalaka, St. Louis University. Each of these reviewers has helped make the sixth edition stronger than ever, and we are grateful for their assistance.

We would like to thank several people who have contributed to this edition: Elise Frasier, our editor at CQ Press; Amy Marks, our copy editor; Kerry Kern, our production editor; and Steve Pazdan, managing editor of CQ Press's College Division. We also thank Jim Taylor and Leroy Stirewalt, our compositors at BMWW. A book with as many bells and whistles as this one needs many sets of eyes to watch over it. We are glad to have had so many good ones.

Janet Buttolph Johnson
H. T. Reynolds

CHAPTER 1

Introduction

Political scientists are interested in learning about and understanding a variety of important political phenomena. Some of us are interested, for example, in the conditions that lead to stable and secure political regimes without civil unrest, rebellion, or government repression. Some are interested in the relationships and interactions between nations and how some nations exercise power over others. Other political scientists are more interested in the relationship between the populace and public officials in democratic countries and, in particular, whether or not public opinion influences the policy decisions of public officials. Still others are concerned with how particular political institutions function. They conduct research on questions such as the following: Does Congress serve the interests of organized groups rather than of the general populace? Do judicial decisions depend upon the personal values of individual judges and the group dynamics of judicial groups or on the relative power of the litigants? To what extent can American presidents influence the behavior of members of the federal bureaucracy? Does the use of nonprofit service organizations to deliver public services change government control of and accountability for those services? Do political parties enhance or retard democratic processes? How much do the policy outputs of states vary and why do they vary?

This book is an introduction to the process and methods of using **empirical research**—research based on the actual, "objective" observation of phenomena—to achieve scientific knowledge about political phenomena. Scientific knowledge, which is discussed in more detail in Chapter 2, differs from other types of knowledge, such as intuition, common sense, superstition, or mystical knowledge. One difference stems from the way in which scientific knowledge is acquired. In conducting empirical research the researcher adheres to certain well-defined principles for collecting, analyzing, and evaluating information. **Political science,** then, is simply the application of these principles to the study of phenomena that are political in nature.

Students should learn about how political scientists conduct empirical research for two major reasons. First, citizens in contemporary American society are often called upon to evaluate empirical research about political

Research on Winners and Losers in Politics

In 1936 Harold Lasswell published *Politics: Who Gets What, When, How.*[2] Ever since then, political scientists have liked this title because it succinctly states an important truth: politics is about winning and losing. No political system, not even a perfectly democratic one, can always be all things to all people. Inevitably, policies favor some and disadvantage others. So important is this observation that one of political science's main tasks is discovering precisely which individuals and groups benefit the most from political struggle and why.

As one might expect, efforts to explain political outcomes have taken widely different forms. According to one approach, called power resources theory, the social strata or classes that win depends heavily on their political resources, not the legitimacy of their claims or fair treatment by public servants. In this view, a democratic political system can be thought of as an arena in which struggles for material and symbolic rewards take place. On any given issue the "team" with the greatest strength—as measured by, say, organizational skill and size, access to expertise, wealth, and the like—will fare far better than those with less. The way the field is set up (that is, the structure of the government) plays a role because it may favor one side over another. But it is really the players' strengths and weaknesses that determine the outcome.[3] This view is at one and the same time optimistic and pessimistic because it holds that although "average" citizens *can* participate in policymaking, in fact they seldom do, except, for example, as members of powerful organizations like labor unions.

Other social scientists have objected to this theory because it portrays politicians and bureaucrats as more or less passive bystanders who simply do the bidding of one side or the other. So another school of thought, often called state-centered theory or institutionalism, argues that even members of democratic governments (sometimes called state managers) have their own interests, which they can frequently impose on those outside the political system. Far from being an arena, the state itself is best thought of as one of several participants (a sort of team) in the struggle over political gains and losses.[4]

As one might surmise, many other kinds of answers to the question "Who gets what?" have been proposed. Which of these theories is right? Given the importance of the debate, it should come as no surprise that an enormous amount of research and thought has gone into the argument. Our purpose in this book, however, is not to decide this particular controversy but rather to describe some of the methods used to study the matter.

A common and effective approach is to derive specific predictions from each of the theories and then to look for information, or "data," that supports one side and undercuts the others. Presumably the theory with the most empirically confirmed propositions would be judged to be in some sense the best. An interesting example of this kind of research is found in the work of David Bradley and his colleagues. Their article "Distribution and Redistribution in Postindustrial Democracies" illustrates the steps in the research process and provides many of the examples for this book.[5]

After posing the question of what explains political outcomes—in their words, what determines "distributive and redistributive processes"—Bradley and colleagues conducted a thorough review of the literature, a subject we take up in Chapter 6. Using this background information, they then stated a series of hypotheses or statements whose validity were to be checked empirically. The power resource theory, for instance, states that a group or class that can through democratic means such as elections gain control of a government for a sustained period of time will win concrete benefits for itself. More precisely, to the extent that political parties representing workers and unions dominate the legislative and executive branches of government they will enact economic and employment policies favorable to the lower classes. In contrast, if the lower classes are poorly or weakly organized and parties representing the interests of those higher up the social and economic scale are in power, government policies will tend to favor business and the wealthy. Since parties of the working class are often called leftist or even socialist, one specific test of the power resources theory is to see if a relationship exists between the magnitude and length of left-wing control of government and distributive and redistributive outcomes.

Bradley and his colleagues formalized their analysis by defining what are called dependent variables (see Chapter 3). One of these dependent variables, "pre-tax and transfer inequality," measures how evenly or unevenly income is distributed among a nation's households before taxes are levied or payments from government programs are received. (The numerical indicator of this variable is the gini coefficient, which varies between 0 and 1. If all the households in a country have exactly the same income, there is no inequality and the coefficient equals 0. If, however, one family has all the income and the remainder none, the coefficient is 1.0, meaning complete inequality.) Since the definition means that values close to 1 imply inequality, while those near 0 suggest the opposite, this measure provides a systematic quantitative way to compare income distributions across nations. The other dependent variable in the analysis is the (percentage) reduction in inequality after government taxes and transfers have been taken into account.[6] Since it too provides a precise numerical indicator, specific predictions can be assessed empirically. Bradley and colleagues' so-called explanatory or independent variables include measures of leftist political party control of government. (The difference between types of variables and the incorporation of variables in hypotheses are discussed in detail in Chapter 3.)

The idea behind the Bradley group's research is that if the power resources theory holds, then in nations where the lower and working classes are able to mobilize through left-wing parties and control government long enough, there will be relatively large reductions in inequality as the result of government action. If, instead, the strength of leftist parties is (to take one case) counterbalanced by the power of state managers, then there will be little or no relationship between the partisanship of governments and the reduction in inequality. Of course, innumerable factors affect the distribution of income besides the nature of the political system. So to analyze the theories fully the researchers had to measure and control for several additional variables, such as nations' positions in the global economy, their level of female participation in the workforce, and their degree of industrialization.

Data on these dependent, explanatory, and control variables were collected for about a dozen industrial democracies, including the United States, Australia, Canada, and many European countries, and analyzed by a variation of a statistical technique called regression analysis (see Chapter 13). This technique allows investigators to see how and how closely two variables are related to each other when still other factors have been held constant.

The findings of this analysis strongly supported the authors' main hypothesis. They concluded, "Taken together, the results of our study are a resounding vindication of power resources theory."[7] They found a strong relationship between reduction in inequality and the ability of subordinate classes to mobilize and support leftist governments. In the United States, for instance, the leftist, or liberal, party, the Democratic Party, has at best had a tenuous hold on government, and as expected the reduction in inequality has been small compared with nations such as Sweden and Norway that have much stronger leftist parties.

In a more recent study, Lane Kenworthy and Jonas Pontusson analyzed trends in the distribution of gross market income—the distribution of income before taxes and government transfers—for affluent OECD (Organization for Economic Co-operation and Development) countries using data from the Luxembourg Income Study, the same source of data used by Bradley and his colleagues.[8] Kenworthy and Pontusson were interested in whether inequality in market income has increased and to what extent government policies have responded to changes in market income inequality. In particular, they were interested in testing the median-voter model developed by Allan H. Meltzer and Scott F. Richard.[9]

According to the median-voter model, support for government redistributive spending depends on the distance between the income of the median voter and the average income of all voters. The greater this distance, the greater the income inequality and, thus, the greater the demand from lower-income voters for government spending. Countries with the greatest market inequalities should have more government spending. One way to test the median-voter model is to see whether changes in redistribution are related to changes in market inequality. One would expect that larger changes in market inequality would cause larger changes in redistribution. The authors found this to be the case, although the United States, Germany, and the United Kingdom do not fit the pattern very well. In further analyses in which they look at country by country responsiveness to market inequality over several decades, Kenworthy and Pontusson found that most OECD countries are responsive to market income inequalities, although to varying degrees, and that the United States is the least responsive.

Perhaps, Kenworthy and Pontusson suggest, government responsiveness to market inequality is related to voter turnout. If one assumes that lower-income voters are less likely to turn out to vote than are higher-income voters, then one would expect that the lower the turnout, the less likely governments would be pressured to respond to income inequality. The median-voter model still would apply, but in countries with low voter turnout, the median voter would be less likely to represent lower-income households. Kenworthy and Pontusson used regression analysis and a scatterplot (Figure 1-1) to show that the higher the voter turnout, the more responsive a country is to market income inequality. The results provide an explanation for why the United States is less responsive to changes in market inequality than are other nations: the United States has the lowest turnout rate among the nations included in the analysis.

These research findings are more interesting in view of another body of research that we consider shortly: the possibility that since the mid-1950s the bottom classes in America have been increasingly dropping out of electoral politics. This means the poor, skilled and unskilled workers, and others are not able to mobilize through leftist parties to advance their interests in the political arena.

The point of these examples is not to make a statement about the

FIGURE 1-1

Redistribution Coefficients by Average Voter Turnout

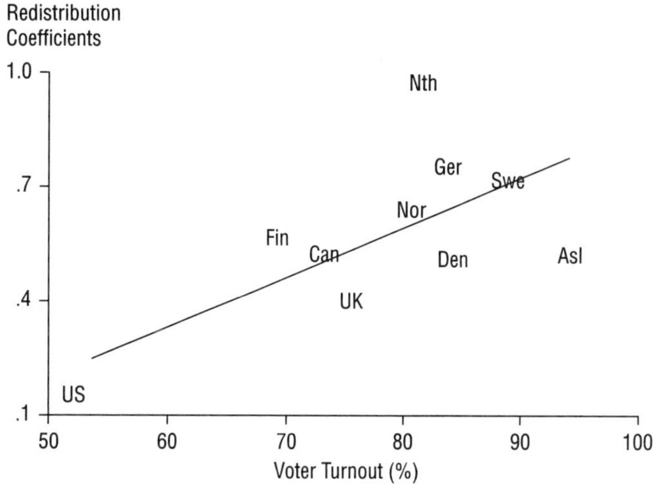

Redistribution Coefficients

Source: Adapted from Figure 9 in Lane Kenworthy and Jonas Pontusson, "Rising Inequality and the Politics of Redistribution in Affluent Countries," *Perspectives on Politics* 3 (September 2005): 462.

Note: Asl = Australia; Can = Canada; Den = Denmark; Fin = Finland; Ger = Germany; Nth = Netherlands; Swe = Sweden; UK = United Kingdom; US = United States. Presidential elections for the United States; general parliamentary elections for the other countries. Redistribution data are for working-age households only.

value of particular ideologies or parties. Instead, we want to stress that important questions—what could be more crucial than knowing who gets what from a political system?—can be answered systematically and objectively, even if tentatively, with careful thought and analysis. Moreover, we hope to show that the techniques used in these debates are not beyond the understanding of students of the social sciences.

Who Votes, Who Doesn't?

The previous example of research showed the importance of group power in determining political winners and losers. Political participation is a major factor: those individuals who make themselves heard in politics "do better" than those who are apathetic. So a natural question is, Why do some people participate more than others?

A good place to start looking for the answer is with the decision to vote. Except for new research that we review briefly later in this chapter, most political scientists accept two generalizations about voting in the United States. First, voting varies by socioeconomic class. Members of the lower classes participate less frequently than do more affluent and better-educated citizens. There is, in short, a "class gap" in turnout rates.[10] The second finding is that since the 1950s fewer and fewer people are going to the polls. Voting rates dropped more or less steadily until the 2000 presidential election—a closely contested race—when barely more than half of the eligible citizens took part. The voting rate has been even lower in recent congressional or "off-year" elections and in the South.

The political scientist Walter Dean Burnham combined these findings into an argument that has come to be known as "selective class demobilization."[11] In a nutshell, Burnham's thesis is that the decline in turnout is especially pronounced in the lower and working classes, those with relatively little education and income and working in manual, routine service and unskilled occupations. Those higher up the ladder, so to speak, have voted at more or less the same rates since the 1950s. In Burnham's words, "[T]he attrition rate among various working-class categories is more than three times as high as in the professional and technical category and well over twice as high as for the middle class as a whole."[12] In other words, for every upper-class nonvoter, there are now two or three lower-class nonvoters. It appears to Burnham and others that the lower classes are effectively abandoning electoral politics. And as a consequence and in keeping with the research on winners and losers, it appears that political rewards may increasingly favor the middle and upper classes.

In the tradition of modern political science, Burnham supports his case by using hard empirical data—measured turnout rates for various social strata. (He relies heavily on census data and graphical and tabular displays to make his points.) But even more important, he supplies a theory that explains this apparent selective demobilization. He contends that political parties in America, never very strong to begin with, have become even weaker in the post–World War II years as a result of many factors, including the rise of candidate-centered campaigns and the increased use of primary elections in party nominations. The weakening of party organization has been especially pronounced in the Democratic Party.[13] The decline of parties places an especially onerous burden on the working and lower classes. Why? Because these groups, having less education and information about government, rely more heavily on cues and motivation supplied by political parties, and without this guidance these citizens lose their way in politics and frequently drop out.[14]

So selective class demobilization has a cause—the decline of parties—and a consequence—the loss of political influence. If true, Burnham's analysis would have enormous implications for the understanding of American politics. Stated bluntly, public policy will have an upper-class bias. Being so provocative, Burnham's thesis naturally sparked considerable comment and controversy, a fact that illustrates an important aspect of scientific research.

As discussed in Chapter 2, science demands independent verification of findings. Conclusions such as Burnham's are not accepted at face value but must be verified by others working separately. In this case, additional research has produced mixed results. Some investigators agree with Burnham that the decline in turnout has been concentrated disproportionately among lower socioeconomic classes.[15] But others, using alternative measures of class and other data sets, come to a different conclusion. Jan E. Leighley and Jonathan Nagler, for instance, found "that the class bias [in nonvoting] has not increased since 1964."[16]

Complicating matters even further, recent research calls into question even the basic belief that voter turnout in general has been declining. These newer investigations say the apparent decrease in the rate of electoral participation stems from an artifact in how turnout is measured. The voting rate has typically been measured as the number of votes cast divided by the number of eligible voters. This procedure may seem straightforward, but a problem arises. How should the eligible voting population be defined? The Census Bureau uses the so-called voting-age population (VAP) as its measure of the eligible electorate. But, as Michael P. McDonald and Samuel L. Popkin maintain, this approach "includes people who are ineligible to vote, such as noncitizens, felons, and the mentally incompetent, and fails to include [Americans] living overseas but otherwise eligible."[17] They developed an alternative measure of the pool of eligible voters and showed that when it is used in the denominator of voting-rate calculations "nationally and outside the South there are virtually no identifiable turnout trends from 1972 onward, and within the South there is a clear trend of *increasing* turnout rates."[18]

Adjusting the VAP to exclude felons raises another series of questions investigated by political scientists: To what extent does the disenfranchisement of nonincarcerated offenders (a practice that in the United States results in the disenfranchisement of large numbers of citizens) alter the outcome of elections? Why is the United States alone among democratic countries in this regard? And, what accounts for differences in restricting access to the ballot among the American states?[19]

Finally, here is another curious twist in research on voter turnout. Some investigators approach the study of political phenomena by building what are known as formal models. Although we discuss this methodology in Chapter 5, we can say here that modelers begin with a set of a priori assumptions and propositions and use logic to deduce further statements from them. Up to this point, the truth of the conclusions depends simply on logic; no empirical verification is required. But the conclusions are sometimes translated into operational terms and tested by using data from sample surveys and other sources.

In the case of voter turnout, the modeling approach begins with the assumption that citizens are rational, in the sense that they try to maximize their utility (the things that they value) at the least cost to themselves. So a potential voter will think about the personal benefits of going to the polls and weigh these against the costs of doing so (for example, taking the time to become informed, registering, and finding and driving to the polling place). Surprisingly, many models lead to the conclusion that a rational person—one who wants to maximize utility at least cost—will decide that voting is not worth the effort and simply abstain.[20] The reason for this conclusion: one single individual's participation has an exceedingly small probability of affecting the outcome of an election. And so, according to the deduction, the small chance of bringing benefits by voting is easily outweighed by the costs, however low. Consequently the formal model predicts that hardly anyone will vote. But in point of fact millions of Americans do vote, which seems to belie the model's conclusions. This situation, which has been called the "paradox of voting," has sparked an enormous amount of discussion and controversy since the 1950s.[21]

This is not the place to sort out these arguments and counterarguments. Instead we have used studies of voter turnout to illustrate some features of research that are described in more detail in the following chapters, including the derivation of hypotheses from existing theory, measurement of concepts, and the use of objective standards to adjudicate among competing ideas. The major point, perhaps, is that if one's procedures are stated clearly, others can pick up the thread of analysis and independently investigate the problem. In this sense empirical political research is, like all science, a cumulative process. Usually no one person or group can discover a definitive answer to a complicated phenomenon like voting or nonvoting. Rather the answers come, if they do at all, from the gradual accumulation of findings from numerous investigators working independently of one another to validate or invalidate each other's claims.

Repression of Human Rights

In recent years increased public and scholarly interest has focused on the human rights practices of governments. Several organizations (Amnesty International, the U.S. Department of State, and Freedom House, for example) publish annual reports on the human rights performance of nations worldwide. Researchers Stephen C. Poe and C. Neal Tate investigated the causes of state terrorism, which involves violations of personal integrity rights and includes such acts as murder, torture, forced disappearance, and imprisonment of persons for their political views.[22] Poe and Tate sought to explain variation of government or regime performance in human rights in 153 countries during the 1980s. Most of the previous studies on human rights abuses compared the practices of a more limited set of nations at a single point in time. Poe and Tate built upon the earlier research by examining change that occurred within a given country across a specified period of time.

Poe and Tate considered the following as possible explanations of variations in state terrorism:

■ Presence of democratic procedures and protections (democracy). Because they offer an alternative method of settling conflicts, and because they provide citizens with the tools to oust from office potentially abusive leaders, democratic procedures and protections minimized serious threats to human rights. In fact, some definitions of human rights include wording that notes the need for access to democratic procedures and protections. (By focusing on variation in state coercion, the authors made sure that their measures of democracy were distinct from their measures of human rights abuses.)

■ Population size and growth. A large population increases the number of opportunities for government coercion and creates stress on available resources. Rapid population growth creates even greater stress. In addition, rapid population growth results in an increase in the proportion of the population that is young, an age cohort that is more likely to engage in criminal behavior and threaten public order.

■ Level of economic development and economic growth. Rapid economic growth creates resentment within those classes that are not benefiting from the new wealth. Such resentment could destabilize a regime and thereby promote repression. However, higher levels of economic development will result in less repression as highly developed countries are able to meet the needs of their people.

- Establishment of leftist regime. States governed by a socialist party or coalition that does not permit effective electoral competition with non-socialist opposition and removal from office of abusive leaders are predicted to have higher levels of state terrorism.

- Establishment of military regime. Military regimes are expected to be more coercive than other types of regimes, but this hypothesis needs to be tested because some military regimes take power claiming that the previous regime was violating the rights of citizens.

- British cultural influence. A colonial tie to British political and cultural influences is more favorable to protection of human rights than is a tie to other colonial influences.

- International and civil war experience. A state's experiences with both international and civil war influence a government's use of repression to control its citizens.

Poe and Tate found that governments with established democratic procedures and protections were unlikely to engage in human rights abuses. They also found that governments in more populous countries were more likely to engage in human rights abuses than were governments in less populous countries. However, population growth in and of itself did not affect levels of government repression. Neither did a history of British cultural influence nor the presence of an established military regime. Economic development had only a weak impact on reducing human rights abuses, and the impact of leftist governments was mixed. Poe and Tate found that national experience in international and civil wars had "statistically significant and substantively important impacts on national respect for the personal integrity of citizens . . . with civil war participation having a somewhat larger impact than participation in international war."[23] Experience with civil war appeared less likely to provoke human rights abuses in countries with democratic governments. Poe and Tate concluded that "basic rights can be enhanced by actors who would encourage countries to solve their political conflicts short of war, and use whatever means are at their disposal to assist them in doing so."[24]

A Look into Judicial Decision Making and Its Effects

When the decisions of public officials clearly and visibly affect the lives of the populace, political scientists are interested in the process by which those decisions are reached. This is as true when the public officials are judges as when they are legislators or executives. As one legal scholar states in his review of the development of empirical research on judicial decision making:

"Given the often critical role judges play in our constitutional, political, and social lives, it is axiomatic that we need to better understand how and why judges reach the decisions they do in the course of discharging their judicial roles."[25] The decision-making behavior of the nine justices of the U.S. Supreme Court is especially intriguing because they are not elected officials, their deliberations are secret, they serve for life, and their decisions constrain other judges and public officials. As a result, political scientists have been curious for some time about how Supreme Court justices reach their decisions.

Researchers have approached the study of judicial decision making from several different perspectives. Early studies investigated the influences of a judge's background (for example, as a prosecutor or defense attorney) and personal attributes such as race or gender. The results have been mixed, with little evidence to support the influence of these factors.[26] One school of thought concerning judicial decision making holds that decisions are shaped primarily by legal doctrine and precedent. Because most Supreme Court judges have spent many years rendering judicial decisions while serving on lower courts, and because judges in general are thought to respect the decisions made by previous courts, this approach posits that the decisions of Supreme Court justices depend on a search for, and discovery of, relevant legal precedent.

Another view of judicial decision making proposes that judges, like other politicians, make decisions in part based on personal political beliefs and values. Furthermore, because Supreme Court judges are not elected, serve for life, seldom seek any other office, and are not expected to justify their decisions to the public, they are in an ideal position to act in accord with their personal value systems.[27]

One of the obstacles to discovering the relationship between the personal attitudes of justices and the decisions handed down by the Court is the difficulty of measuring judicial attitudes. Supreme Court justices do not often consent to give interviews to researchers while they are on the bench, nor do they fill out attitudinal surveys. Their deliberations are secret, they seldom make public speeches during their terms, and their written publications consist mainly of their case decisions. Consequently, about all we can observe of the political attitudes of Supreme Court justices during their terms are the written decisions they offer, which are precisely what researchers are seeking to explain. Some researchers use political party and the appointing president as indicators of judicial attitudes, although these are less than satisfactory measures.

An inventive attempt to overcome this obstacle is contained within Jeffrey A. Segal and Albert D. Cover's article "Ideological Values and the Votes of U.S. Supreme Court Justices."[28] Segal and Cover decided that an appropriate way in which to measure the attitudes of judges, independent of the decisions they make, would be to analyze the editorial columns written about them in four major U.S. daily newspapers after their nomination by the president but before their confirmation by the Senate. This data source, the researchers argue, provides a comparable measure of attitudes for all justices studied, independent of the judicial decisions rendered and free of systematic errors. Here, too, though, the researchers had to accept a measure that was not ideal, for the editorial columns reflected journalists' perceptions of judicial attitudes rather than the attitudes themselves.

Despite this limitation, the editorial columns did provide an independent measure of the attitudes of the eighteen Supreme Court justices who served between 1953 and 1987. Segal and Cover found a strong relationship between the justices' decisions on cases dealing with civil liberties and the justices' personal attitudes as evinced in editorial columns. Those justices who were perceived to be liberal *before* their term on the Supreme Court voted in a manner consistent with this perception once they got on the Court. Judicial attitudes, then, do seem to be an important component of judicial decision making.

Other researchers have investigated the influence of so-called extra-legal factors on the decisions of Supreme Court justices. Are there factors in addition to ideology but outside of legal precedent that influence judicial decision making? Do judges behave strategically to increase their prestige or influence vis-à-vis other judges and other branches of government?[29] Are they subject to influence by other judges and governmental actors? Among the possibilities are congressional influence (given the ability of Congress to pass legislation that overrides Court decisions and to initiate constitutional amendments, among others actions), presidential influence, and public opinion.[30]

The presidential election in 2000 brought into sharp relief for many Americans the importance of Supreme Court decisions to American politics. Some people felt that the high regard that Americans have for the Supreme Court brought closure to the highly contentious election and that support for the Supreme Court as an institution helped people to accept its decision in *Bush v. Gore* (2000). Others argued that general support and respect for the Supreme Court was undermined among those disappointed by the decision. Interestingly, political scientist Valerie J. Hoekstra was already busy investigating the two general questions raised so vividly by the 2000 decision: (1) How does the content of Supreme Court decisions affect support for the Court? That is, does respect for the Court decline among people who disagree with a decision? (2) Do Supreme Court decisions have any effect on public opinion? In other words, does the public change its mind about public policy issues once the Supreme Court has spoken?[31]

Hoekstra's work demonstrates how the choice of a research design (the topic of Chapter 5) affects a researcher's ability to answer research questions with confidence. Hoekstra noted that public opinion polls generally show that the Supreme Court enjoys higher and more stable levels of public support than Congress or presidents, but that stability of aggregate-level measures such as public opinion polls does not mean that the opinions of individuals have not changed.[32] She argued that a panel study, one in which the same individuals are interviewed before and after a Supreme Court decision, is best for examining how support for the Supreme Court changes and whether individuals change their views about an issue in response to the Court's decision in a case. She also argues that it is important to interview individuals who are aware of the case to be decided by the Court. One cannot expect a decision of the Supreme Court to influence how people feel about an issue, if people are not aware of its decision. Most Supreme Court decisions do not have the national significance and high level of public awareness as in *Bush v. Gore*. Therefore, Hoekstra selected four cases and interviewed people in the communities from which the cases originated.

Hoekstra hypothesized that people who are more supportive of the Supreme Court are more likely to change their view of an issue in the direction of the Court's decision and that people who have strong opinions about an issue are less likely to change their views than are people whose opinions are not as strong. In two of the four cases, Hoekstra found that public opinion shifted in the direction of the Court's decision, but initial levels of support for the Court did not have an effect on the amount of change.[33] She did find that people who paid more attention to politics, and presumably were more aware of the issue, were more likely to change opinion in direction of the Court's decision.[34] Overall, she found limited support for the persuasive effect of Supreme Court decisions.

In terms of the effect of Supreme Court decisions on the public's support of the Court, Hoekstra found that people who were pleased with the Court's decision became more confident in and supportive of the Court, whereas those who were disappointed with the decision became less supportive. These changes were affected by how strongly a person felt about the issue: those who cared strongly about an issue tended to change their views of the Court more than those who did not care as much about the issue.[35]

Influencing Bureaucracies

Political control of the bureaucracy is an ongoing topic of discussion and investigation by political scientists. A variety of theories about political influence on bureaucratic activities has ascended, only to be superseded by new theories based on yet more research. Theories have evolved from the politics-administration dichotomy, which strictly separates politics and administration, to the iron triangle, or capture, theory, which views agencies as responsive to a narrow range of advantaged and special interests assisted by a few strategically located members of Congress, leaving the president with relatively little influence. A more recent theory, the agency theory, suggests that presidents and Congress do have ways to control bureaucratic activities. According to this theory, policymakers use rewards or sanctions to bring agency activities back in line when they stray too far from the policy preferences of elected politicians. Control mechanisms include budgeting, political appointments, structure and reorganization, personnel power, and oversight.[36]

Research shows that agency outputs vary with political changes. The emergence of a new presidential administration, the seating of new personnel on the courts, and change in the ideological stances of congressional oversight committees all influence agency outputs. B. Dan Wood and Richard W. Waterman tried to find out more about political control of bureaucracies. They studied a broad range of agencies to identify how agencies are controlled and to assess the relative effectiveness of the different control mechanisms.[37] They were also interested in determining whether Congress or the president is more effective at influencing bureaucratic outputs.

Wood and Waterman selected seven federal agencies, each representing a different organizational design. Using archives and interviews of high-ranking agency officials, they identified events that should have caused change in bureaucratic outputs. They then gathered information on agency outputs of regulatory enforcement activities, such as litigations, sanctions, and administrative decisions. In contrast to previous researchers, they obtained information on outputs on a monthly or quarterly basis. They then looked to see whether agency outputs had changed in the ways predicted based on the changes that had occurred at the political level. Wood and Waterman found evidence that political controls did cause changes in outputs in all seven agencies. Political appointment had a very important impact on political control; reorganization, congressional oversight, and budgeting were also important factors in accounting for change in agency activities. These findings, indicating that agency outputs do respond to political manipulation, led Wood and Waterman to suggest that policy monitoring should become routine for federal agencies. Information on outputs could make politicians and bureaucrats more accountable and informed and help all participants in the policy process have access to information. The information could also aid scholars in further research on the behavior and decisions of public bureaucracies.

More recent research investigated the issue of political control of public agencies at their inception and illustrates that politicians believe that they can influence them. David E. Lewis observed that presidents and Congress compete over the control of agencies and that agencies vary in the extent to which they are designed to be insulated from presidential control.[38] He hypothesized that divided government and the size of the majority in Congress affects the probability of Congress creating an insulated agency. For example, when the same party controls the presidency and Congress, and the size of the majority in Congress is large, Congress will create an agency less insulated from the president. If, however, the party control of the executive and legislative branches is divided and the majority in Congress is large, Congress will want to create an agency that is insulated from the president's control, although expectations about the presidency changing party hands also play a role in Congress's strategy. To test these hypotheses, Lewis collected data on agencies created between 1946 and 1997 measuring the extent to which they were designed with insulating features. He found that the percentage of new agencies with insulating characteristics correlated with periods of divided government, and that during periods of divided government with large majorities in Congress, the percentage of insulated agencies was higher than when majorities were small.[39] Higher presidential approval ratings were associated with less agency insulation during periods of unified government and with more agency insulation during periods of divided government.[40] Lewis concluded that, in general, weak presidents are less able than strong presidents to resist congressional desires to insulate agencies. Lewis's research would seem to suggest that Wood and Waterman's research should be expanded to include more agencies and to take into account whether an agency was designed to be more responsive to Congress or the president.

Effects of Campaign Advertising on Voters

Enormous sums of money are spent on campaign advertising by candidates vying for political office. Political scientists have long been interested in the effects of campaign advertising on voters. Some have argued that advertising has little effect due to the public's ability to screen out messages conflicting with their existing views. Others have suggested that campaign activity, including advertising, stimulates voter interest and increases turnout. More recent conventional wisdom suggests that negative campaign advertising, particularly television advertisements, has harmful effects on the democratic process: negative campaign ads are thought to increase cynicism about politics and to cause the electorate to turn away from elections in disgust, a phenomenon called demobilization.

A 1994 study on so-called attack advertising by Stephen D. Ansolabehere, Shanto Iyengar, Adam Simon, and Nicholas Valentino is widely recognized as establishing support for the demobilization theory. Noting that "[m]ore often than not, candidates criticize, discredit, or belittle their opponents rather than providing their own ideas," they hypothesized that, rather than stimulating voter turnout, such campaigns would depress turnout.[41]

and Brians's study by noting that survey recall data are prone to inaccuracies: recall is a poor measure of actual exposure, and people who are likely to vote are more likely to recall seeing a political ad.[46] They analyzed the survey data for the 1992 and 1996 elections, making adjustments for exposure to campaign ads that Wattenberg and Brians did not. They used data measuring the volume of ads in the different senatorial elections, noting that higher-volume campaigns have disproportionately more negative ads. They also noted that the tone of campaigns becomes more negative as elections approach. Thus respondents surveyed earlier in an election will have been exposed to less negative campaigning than those interviewed later in an election. Their analysis showed that recall of negative ads was significantly higher in states with higher levels of advertising and in the latter stages of the campaign and that intentions to vote were lower in states with more television advertising and in the latter stages of campaigns. Thus they concluded that the survey data show that negative advertising has a negative impact on voter turnout. They also replicated their analyses of the Senate races using official FEC data (previously they had used data obtained directly from the election officers in each state) and concluded that, on average, turnout in positive campaigns is nearly 5 percentage points higher than turnout in negative campaigns.

In closing, we should point out that other political scientists have been actively investigating an important related question: Do attack ads work? The authors of an analysis of research on this topic concluded that negative ads have not been shown to be more effective than positive political ads in a statistically significant way, but their effect could be "politically significant or even decisive" in some campaigns.[47] As long as the conventional wisdom that attack ads work persists, campaign managers and candidates are unlikely to abandon them.

In Chapter 5 we discuss some ways to design research to investigate the effects of advertising on political behavior. We simply note for now that this issue will surely continue to preoccupy researchers and illustrates some of the complexities and excitement of the empirical study of politics.[48]

Research on Public Support for U.S. Foreign Involvement

The ongoing conflicts in Iraq and Afghanistan highlight the relevance of public support for U.S. military involvement in foreign affairs and the extent to which the American public judges the president's performance based on foreign policy issues. Researchers have investigated a wide range of factors associated with public support for U.S. military involvement, some of which focus on attributes of individuals, including attitudes toward the use of military force and U.S. involvement in world affairs in general, education, knowledge of foreign affairs, and others that focus on situational factors such as the primary purpose or objective of U.S. military involvement, the relative power of the U.S. vis-à-vis an adversary, the costs of involvement (particularly U.S. military casualties), the extent of elite consensus over whether the United States should be involved, and multilateral support for involvement.[49] Let's take a look at one recent example of such work.

In an article entitled "Success Matters: Casualty Sensitivity and the War in Iraq," Christopher Gelpi, Peter D. Feaver, and Jason Reifler investigated the claim that the U.S. public will support military operations only if combat deaths are minimal, the so-called casualty-phobia thesis.[50] Their work responded to a debate over whether the American public reflexively withdraws its support for U.S. military missions as casualties mount or whether casualties are but one factor in a more complicated cost-benefit analysis affecting support for military missions. And, if the latter case is true, they ask, how important are casualties in the analysis? They also are interested in how individuals' assessments of the likelihood of success and other beliefs affect their tolerance for combat deaths.

In the first part of their investigation, they collected weekly data on U.S. combat deaths in Iraq from January 1, 2003, through November 1, 2004, and used public opinion data to measure public support for the war. Several polling questions, which can be used to measure public support, have been asked routinely over the course of the war in Iraq: approval of the president in general, approval of the president's handling of the situation in Iraq, and whether the Iraq war has been "worth it." They found that the poll results to the three questions were closely correlated, so that their results were not affected by which question was used to measure support for the war. Therefore, in their article they reported the results from their analyses for just one of the questions, presidential approval. They looked at the relationship between the average weekly approval ratings for President George W. Bush from March 2003 to November 2004 and cumulative U.S. military deaths in Iraq for three time periods: the major combat phase, the occupation period, and the period subsequent to the transfer of sovereignty to an Iraqi authority.

The researchers were interested especially in the extent to which the public's expectation of a successful outcome and its belief in the rightness of the war explain support for the war in Iraq, even in the face of mounting casualties. To begin to explore the impact of the public perception of success, they calculated the impact of casualties on presidential approval for each of the three phases of the Iraq war. For the major combat period, they found a positive relationship between casualties and presidential approval (as casualties went up, so did approval of the president). During this period, despite the casualties, one would expect a "rally round the flag" effect and perception of success was assumed to be high. This relationship changed during the second period, U.S. occupation, when success was less certain: mounting casualties were associated with a decline in presidential approval. In the third period, in which Iraqis assumed sovereignty, which could be interpreted as an indication of a likely successful outcome to the war, casualties had no impact on presidential approval even though media coverage was intense and the number of casualties approached one thousand.

As Gelpi, Feaver, and Reifler point out, using aggregate data to explore the impact of several individual attitudes and beliefs and their interaction is problematic because such data do not allow us to link individuals' answers to different polling questions. Thus we do not know for sure whether individuals who believe that the mission in Iraq will be successful are more willing to tolerate mounting casualties than those who doubt that the mission will be successful. We also do not know how beliefs about the rightness of the war in Iraq and the likelihood of its success interact to affect tolerance for casualties.

To address this limitation of aggregate data, the researchers conducted a series of surveys in which they asked respondents questions about their confidence in victory, whether they believed that President Bush did the right thing in attacking Iraq, and their tolerance for military deaths. They also asked survey respondents questions about the primary policy objective of the war in Iraq, their perceptions of an elite consensus, and their attitudes toward multilateralism, in order to test competing explanations for tolerance of casualties. They found that these other factors do influence tolerance of casualties but that beliefs in success and rightness have a greater impact on the dependent variable.

Figure 1-2 shows the impact of a respondent's beliefs in success and rightness of the war (the latter indicated in the figure as approval) on the probability that the respondent was willing to tolerate 1,500 U.S. casualties. As one goes from left to right, that is, from lower beliefs in success to higher, the groups of columns increase in height. Looking at each group of columns, one can see that as approval of the war increases, so does tolerance for casualties, although not very much for those respondents who think that success is not very likely. The figure shows that beliefs about the expected success and rightness of the war work together to explain tolerance for casualties, but the columns in the figure clearly indicate that beliefs about success have a greater impact on tolerance for military deaths than beliefs about the rightness of going to war; in general the columns tend to increase in height more rapidly as one goes by group to group from left to right rather than within groups. In addition, a respondent who believes that success is not at all likely, but who strongly approves of the rightness of the war has only a 23 percent chance of tolerating 1,500 deaths whereas a respondent who believes that success is very likely, but strongly disapproves of the rightness of the war (right-front corner) has a 50 percent chance of tolerating 1,500 casualties.

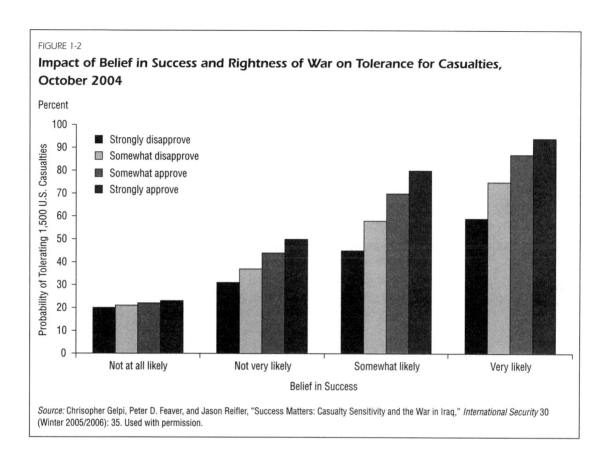

FIGURE 1-2

Impact of Belief in Success and Rightness of War on Tolerance for Casualties, October 2004

Percent

Legend:
- Strongly disapprove
- Somewhat disapprove
- Somewhat approve
- Strongly approve

Y-axis: Probability of Tolerating 1,500 U.S. Casualties

X-axis (Belief in Success): Not at all likely, Not very likely, Somewhat likely, Very likely

Source: Chrisopher Gelpi, Peter D. Feaver, and Jason Reifler, "Success Matters: Casualty Sensitivity and the War in Iraq," *International Security* 30 (Winter 2005/2006): 35. Used with permission.

Clearly, both citizens and politicians have quite a bit to learn from recent political science research on the conditions under which the public will support the use of military force and foreign policies advocated by national political leaders. It is exciting for researchers to investigate these issues and to pursue greater understanding of these and related questions.

Conclusion

Political scientists are continually adding to and revising our understanding of politics and government. As the several examples in this chapter illustrate, empirical research in political science is useful for satisfying intellectual curiosity and for evaluating real-world political conditions. New ways of designing investigations, the availability of new types of data, and new statistical techniques contribute to the ever-changing body of political science knowledge. Conducting empirical research is not a simple process, however. The information a researcher chooses to use, the method that he or she follows to investigate a research question, and the statistics used to report research findings may affect the conclusions drawn. For instance, some of these examples used sample surveys to measure important phenomena such as public opinion on a variety of public policy issues. Yet surveys are not always an accurate reflection of people's beliefs and attitudes. In addition, how a researcher measures the phenomena of interest can affect the conclusions reached. Finally, some researchers conducted experiments in which they were able to control the application of the experimental or test factor, whereas others compared naturally occurring cases in which the factors of interest varied.

Sometimes researchers are unable to measure political phenomena themselves and have to rely on information collected by others, particularly government agencies. Can we always find readily available data to investigate a topic? If not, do we choose a different topic or collect our own data? How do we collect data firsthand? When we are trying to measure cause and effect in the real world of politics, rather than in a carefully controlled laboratory setting, how can we be sure that we have identified all the factors that could affect the phenomena we are trying to explain? Finally, do research findings based on the study of particular people, agencies, courts, communities, or countries have general applications to all people, agencies, courts, communities, or countries? To develop answers to these questions we need to understand the process of scientific research, the subject of this book.

Notes

1. *Recreational research* is a term used by W. Phillips Shively in *The Craft of Political Research,* 2d ed. (Englewood Cliffs, N.J.: Prentice-Hall, 1980), chap. 1.

2. Harold Lasswell, *Politics: Who Gets What, When, How* (New York: Hittlesey House, 1936). A more recent statement of the idea is found in Benjamin I. Page, *Who Gets What from Government* (Berkeley: University of California Press, 1983).

3. See, for example, Gosta Esping-Andersen and Walter Korpi, "Social Policy as Class Politics in Postwar Capitalism: Scandinavia, Austria, and Germany," in John Goldthorpe, ed., *Order and Conflict in Contemporary Capitalism: Studies in the Political Economy of Western European Nations* (New York: Oxford University Press, 1984), esp. 181.

4. For example, Theda Skocpol, ed., *Bringing the State Back In* (New York: Cambridge University Press, 1985).

5. David Bradley, Evelyne Huber, Stephanie Moller, Françoise Nielsen, and John D. Stephens, "Distribution and Redistribution in Postindustrial Democracies," *World Politics* 55 (January 2003): 193–228.

6. In other words, this variable indicates how much state programs reduced the gini coefficient.

7. Bradley et al., "Distribution and Redistribution," 227.

8. Lane Kenworthy and Jonas Pontusson, "Rising Inequality and the Politics of Redistribution in Affluent Countries," *Perspectives on Politics* 3 (September 2005): 449–471.

9. Allan H. Meltzer and Scott F. Richard, "A Rational Theory of the Size of Government," *Journal of Political Economy* 89 (October 1981): 914–927.

10. Thomas E. Patterson, *The Vanishing Voter* (New York: Vintage Books, 2003), 44–46.

11. Walter Dean Burnham, "The Turnout Problem," in A. James Richley, ed., *Elections American Style* (Washington D.C.: Brookings Institution, 1987).

12. Ibid., 125.

13. Burnham writes, "While no one doubts that the Republican party suffers from some internal divisions and even occasional bouts of selective abstention among its supporters . . . the GOP remains much closer to being a true party in the comparative sense than do today's Democrats." Ibid., 124. This remark is as true in the early twenty-first century as it was in the mid-1980s when Burnham wrote it.

14. Ibid., 123–124.

15. Stephen E. Bennett, "Left Behind: Exploring Declining Turnout among Noncollege Young Whites, 1964–1988," *Social Science Quarterly* 72 (1991): 314–333, and Patterson, *The Vanishing Voter,* chap. 2.

16. Jan E. Leighley and Jonathan Nagler, "Socioeconomic Class Bias in Turnout, 1964–1988: The Voters Remain the Same," *American Political Science Review* 86 (September 1992): 734. Also see Ruy A. Teixeria, *Why Americans Don't Vote: Turnout Decline in the United States, 1960–1984* (Westport, Conn.: Greenwood, 1987).

17. Michael P. McDonald and Samuel L. Popkin, "The Myth of the Vanishing Voter," *American Political Science Review* 95 (December 2001): 963.

18. Ibid., 968 (emphasis added). Also see Michael P. McDonald, "On the Overreport Bias of the National Election Study Turnout Rate," *Political Analysis* 11 (Spring 2003): 180–186.

19. Jeff Manza and Christopher Uggen, "Punishment and Democracy: Disenfranchisement of Nonincarcerated Felons in the United States," *Perspectives on Politics* 2 (September 2004): 491–505.

20. One of the first to arrive at this conclusion was the economist Anthony Downs, whose seminal book *An Economic Theory of Democracy* (New York: Harper and Row, 1957) sparked a generation of research into the seeming irrationality of voting.

21. See Donald P. Green and Ian Shapiro, *The Pathologies of Rational Choice Theory: A Critique of Applications in the Social Sciences* (New Haven: Yale University Press, 1994); and Jeffrey Friedman, ed., *The Rational Choice Controversy: Economic Models of Politics Reconsidered* (New Haven: Yale University Press, 1996).

22. Stephen C. Poe and C. Neal Tate, "Repression of Human Rights to Personal Integrity in the 1980s: A Global Analysis," *American Political Science Review* 88 (December 1994): 853–872.

23. Ibid., 866.

24. Ibid.

25. Michael Heise, "The Past, Present, and Future of Empirical Legal Scholarship: Judicial Decision Making and the New Empiricism," *University of Illinois Law Review* 4 (2002): 832.

26. Ibid., 834–835.

27. For an example of research that considers both precedent and values, see Youngsik Lim, "An Empirical Analysis of Supreme Court Justices' Decision Making," *Journal of Legal Studies* 29 (June 2000): 721–752.

28. Jeffrey A. Segal and Albert D. Cover, "Ideological Values and the Votes of U.S. Supreme Court Justices," *American Political Science Review* 83 (June 1989): 557–565.

29. For an example of an investigation of strategic considerations, see Forrest Maltzman and Paul J. Wahlbeck, "Strategic Policy Considerations and Voting Fluidity on the Burger Court," *American Political Science Review* 90 (September 1996): 581–592.

30. See Thomas G. Hansford and David F. Damore, "Congressional Preferences, Perceptions of Threat, and Supreme Court Decision Making," *American Politics Quarterly* 28 (October 2000): 490–510; and Jeff Yates and Andrew Whitford, "Presidential Power and the United States Supreme Court," *Political Research Quarterly* 51 (June 1998): 539–550.

31. Valerie J. Hoekstra, *Public Reaction to Supreme Court Decisions* (New York: Cambridge University Press, 2003).

32. Ibid., 13.

33. Ibid., 113.

34. Ibid., 114.

35. Ibid., 137.

note only phenomena that reinforce their beliefs while ignoring or dismissing those that do not. Thus their knowledge is based on selective and biased experience and observation. Superstitious people are often fearful of empirically testing their superstitions and resist doing so.

Some philosophers of science, in fact, insist that a key characteristic of scientific claims is **falsifiability,** meaning the statements or hypotheses can in principle be rejected in the face of contravening empirical evidence. A claim not refutable by any conceivable observation or experiment is nonscientific. In this sense, the findings of science are usually considered tentative: they are "champion" only so long as competing ideas do not upend them. Indeed, the philosopher Karl Popper argues that scientists should think solely in terms of attempting to refute or falsify theories, not prove them.[8]

In any event, note that commonsense knowledge as well as knowledge derived from casual observation may be valid. Yet they do not constitute scientific knowledge until they have been empirically verified in a systematic and unbiased way. Alan Isaak notes that commonsense knowledge is often accepted "without question, as a matter of faith," which means that facts are accepted without being established by commonly accepted rules and procedures of science.[9]

In view of the importance of verification and falsification, scientists must always remain open to alterations and improvements of their research. To say that scientific knowledge is provisional does not mean that the evidence accumulated to date can be ignored or is worthless. It does mean, however, that future research could always significantly alter what we currently believe. In a word, scientific knowledge is tentative. Often when people think of science and scientific knowledge, they think of scientific "laws." A scientific law is a "generalization that was tested and confirmed through empirical verification."[10] But these laws often have to be modified or discarded in light of new evidence. So even though political scientists strive to develop law-like generalizations, they understand and accept the fact that such statements are subject to revision.[11]

Sometimes efforts to investigate commonsense knowledge have surprising results. For example, given America's high levels of literacy, the emergence of mass communications, modern transportation networks, and the steady expansion of voting rights for the last two hundred years, we might assume that participation in national elections would be high and even increase as time goes by. But, as the example in Chapter 1 suggested, neither of these conditions holds. Lots of evidence indicates that half or more of eligible Americans regularly skip voting and that the number doing so may be increasing despite all the economic and civic progress that has been made. In the studies described in Chapter 1, all of the researchers subjected their claims and explanations to empirical verification. They observed the phenomena they were trying to understand, recorded instances of the occurrence (and nonoccurrence) of these phenomena, and looked for patterns in their observations that were consistent with their expectations. In other words, they accumulated a body of evidence that gave other social scientists a basis for further study of the phenomena.

Scientific knowledge is supposedly "value free." Empiricism addresses what is, what might be in the future, and why. It does not typically address whether or not the existence of something is good or bad, although it may be useful in making these types of determinations. Political scientists use the words *normative* and *nonnormative* to express the distinction. Knowledge that is evaluative, value laden, and concerned with prescribing what ought to be is known as **normative knowledge.** Knowledge that is concerned not with evaluation or prescription but with factual or objective determinations is known as **nonnormative knowledge.** Most scientists would agree that science is (or should attempt to be) a nonnormative enterprise.

This is not to say that empirical research operates in a valueless vacuum. A researcher's values and interests, which are indeed subjective, affect the selection of research topics, time periods, populations, and the like. A criminologist, for example, may feel that crime is a serious problem and that long prison sentences for those who commit crimes deter would-be criminals. He or she may therefore advocate stiff mandatory sentences as a way to reduce crime. But a test of the proposition that stiff penalties reduce the crime rate should be conducted in such a way that the researcher's values and predilections do not bias the results of the study. And it is the responsibility of other social scientists to evaluate whether or not the research meets the criteria of empirical verification. Scientific principles and methods of observation thus help both researchers and those who must evaluate and use their findings. Note, however, that within the discipline of political science, as well as in other disciplines, the relationship between values and scientific research is frequently debated. We have more to say about this subject later in this chapter.

Even though political scientists may strive to minimize the impact of biases on their work, it is difficult, if not impossible, to achieve total objectivity. An additional characteristic of scientific knowledge helps to identify and weed out prejudices (inadvertent or otherwise) that may creep into research activities.[12] Scientific knowledge

DISTINGUISHING EMPIRICAL FROM NORMATIVE CLAIMS

It is sometimes tricky to tell an empirical statement from a normative one. The key is to infer the author's intention: Is he or she asserting that something is simply the way it is, no matter what anyone's preference may be? Or is the person stating an opinion or a desire or an aspiration? Sometimes normative arguments contain auxiliary verbs, such as *should* or *ought,* which express an obligation or wanting and thus suggest a normative position. Empirical arguments, by contrast, often use variations of *to be* or direct verbs to convey the idea that "this is the way it really is in the world." Naturally, people occasionally believe that their values are matters of fact, but scientists must be careful to keep the types of claims separate.

then explained by the conjunction of the condition and the proposition. The goal of explanation is, sometimes, to account for a particular event—the demise of the Soviet Union, for example—but more often it is to explain general classes of phenomena such as wars or revolutions or voting behavior.

Explanation, then, answers "why" and "how" kinds of questions. The questions may be specific, as, for instance, "Why did a particular event take place at a particular time?" or more general, as, for example, "Why do upper-class people vote more regularly than, say, blue-collar workers?" Observing and describing facts is, of course, important. But most political scientists want more than mere facts. They are usually interested in identifying the factors that account for or explain human behavior. Studies of turnout are valuable because they do more than simply describe particular election results; they offer an explanation of political behavior in general.

An especially important kind of explanation for science is that which asserts *causality* between two events or trends. A causal relation means that in some sense the emergence or presence of one condition or event will always (or with high probability) bring about another. Causation implies more than one thing follows another: instead, it means one necessarily follows the other. It is one thing to say that economic status is somehow related to the level of political participation. It is quite another to assert that economics determines or causes behavior. Statements asserting cause and effect are generally considered more informative and perhaps useful than ones simply stating an unexplained connection exists. After all, there may be a relationship between the birthrate in countries and the size of their stork populations. But this connection is purely coincidental. We discuss causality in more detail in Chapter 5.

In this vein, explanatory knowledge is also important because it can be predictive by offering systematic, reasoned anticipation of future events. Note that prediction based on explanation is not the same as forecasting or soothsaying or astrology, which do not rest on empirically verified explanations. An explanation gives scientific reasons or justifications—for why a certain outcome is to be expected. In fact, many scientists consider the ultimate test of an explanation to be its usefulness in prediction. Prediction is an extremely valuable type of knowledge, since it may be used to avoid undesirable and costly events and to achieve desired outcomes. Of course, whether or not a prediction is "useful" is a normative question. Consider, for example, a government that uses scientific research to predict the outbreak of domestic violence but uses the knowledge not to alleviate the underlying conditions but to suppress the discontented with force.

In political science, explanations rarely account for all the variation observed in attributes or behavior. So exactly how accurate, then, do scientific explanations have to be? Do they have to account for or predict phenomena 100 percent of the time? Most political scientists, like scientists in other disciplines, accept **probabilistic explanation,** in which it is not necessary to explain or predict a phenomenon with 100 percent accuracy.

At this point we should acknowledge that many explanations and predictions in political science are weak or even false. Indeed some have so many counterinstances that they do not seem worthy of the designation *scientific,* and many critics rightfully point out that the social sciences have never come close to the rigor and precision of the natural sciences. For this reason, philosophers and methodologists maintain that social scientists cannot achieve the exactitude and precision of the natural sciences and that instead they should attempt not to explain behavior but to understand it.[17] Needless to say, we do not entirely agree with this view; but later in the chapter we acknowledge that this position has merits.

Scientists also recognize another characteristic of scientific knowledge, **parsimony,** or simplicity. Suppose, for instance, two researchers have developed explanations of why some people trust and follow authoritarian leaders. The first account mentions only the immediate personal social and economic situation of the individuals, whereas the second account accepts these factors but also adds deep-seated psychological states stemming from traumatic childhood experiences. And imagine that both provide equally compelling accounts and predictions of behavior. Yet, since the first relies on fewer explanatory factors than the second, it will generally be the preferred explanation, all other things being equal. This is the principle of Ockham's razor, which might be summed up as "keep explanations as simple as possible."

The Importance of Theory

The accumulation of related explanations sometimes leads to the creation of a **theory**—that is, a body of statements that systematize knowledge of, and explain, phenomena. Stated differently, theories help "organize, systematize, and coordinate existing knowledge" in a unified explanatory framework.[18] A theory about a subject such as war or voting or bureaucracy consists of several components: a set of "primitive" terms (words and concepts whose meanings are taken for granted); assumptions or axioms about some of the subject matter; explicit definitions of key concepts; a commitment to a particular set of empirical tools such as survey research (that is, polling) or document analysis; and, most important, general, verifiable statements that explain the subject matter. Two crucial aspects of empirical theory are (1) that it leads to specific, testable predictions and (2) that the more observations there are to support these predictions, the more the theory is confirmed.

To clarify some of these matters, let us take a quick look at an example. The "proximity theory of electoral choice" provides a concise explanation for why voters choose parties and candidates.[19] Superficially the theory may seem simplistic. Its simplicity can be deceiving, however, for it rests on

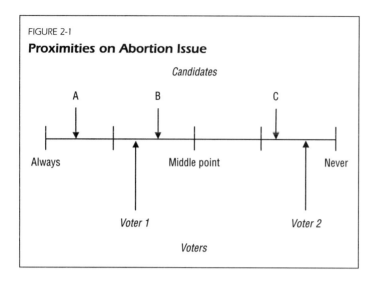

FIGURE 2-1

Proximities on Abortion Issue

Candidates

A   B   C

Always   Middle point   Never

Voter 1   Voter 2

Voters

many years of multidisciplinary research[20] and involves considerable sophisticated thinking.[21] But essentially the theory boils down to the assertion that people support parties and candidates who are "closest" to them on policy issues.

Take a particularly simple case. Suppose we consider the abortion debate. Positions on this issue might be arrayed along a single continuum running from, say, "Abortion should always be allowed" to "Abortion should never be permitted" (see Figure 2-1). Proximity theorists believe that both voters and candidates (or parties) can be placed or located on this scale and, consequently, that the distances or proximities between them (voters and candidates) can be compared. The theory's prediction is straightforward: an individual votes for the candidate to whom he or she lies closest.

To expand a bit, theorists in this camp argue (1) that analysts using proper measurement techniques can position both issues and candidates on scales that show how "close" they are to each other and to other objects and (2) that voters vote for candidates who are closest (most proximate) to themselves. People choose nearby candidates out of their desire to maximize utility, or the value that results from one choice over another. Knowing this fact, candidates adjust their behavior to maximize the votes they receive. Adjusting behavior means not only taking or moving to positions as close as possible to the average or typical voter (the so-called median voter) but also, if and when necessary, obscuring one's true position (that is, following a strategy of ambiguity).[22] Figure 2-1, for instance, shows that Voter 1's position is closest to Candidate B's; therefore, Voter 1 would presumably vote for that candidate. Similarly Voter 2 would prefer Candidate C. Note also that Candidate A could attract Voter 1's support by moving closer to the middle, perhaps by campaigning on an "abortion-only-in-certain-circumstances" platform.

The proximity theory has many of the characteristics of an empirical theory. It explains why things happen as they do, and it offers specific and testable predictions. It is also an implicitly causal theory in that it hypothesizes that the desire to maximize utility "causes" voters to support specific candidates. It is general since it claims to apply to any election in any place at any time. As such, it provides a much more sweeping explanation of voting than a theory that uses time- and place-bounded terms such as "the 2006 gubernatorial election in Pennsylvania." In addition, it provides a parsimonious or relatively simple account of candidate choice. It does not invoke additional explanatory factors such as psychological or mental states, social class membership, or current economic conditions to describe the voting act. Most important, although the proximity theory rests on considerable formal (and abstract) economic and decision-making reasoning, it puts itself on the line by making specific empirical predictions.

As a theory, it incorporates or uses numerous primitive or undefined terms such as *issue, candidate,* and *utility.* These words and concepts may have well-accepted dictionary meanings, but the theory itself takes their common understanding for granted. When a theory is challenged, part of the dispute might involve slightly divergent interpretations of these terms. At the same time the theory makes explicit various other assumptions. It assumes among other things that a researcher can place individuals on issue dimensions, that people occupy these positions for reasonably long time periods, that voters are rational in that they maximize utility, and that candidates have objective positions on these issues.[23] Moreover, by assumption, certain possibilities are not considered. The theory does not delve into the question of whether or not a person holds a "correct" position on the scale, given his or her objective interests. Finally, to test the proximity or spatial idea, researchers assume that one can assign individuals meaningful spatial positions by asking certain kinds of questions on surveys or polls.[24] This may be a perfectly reasonable assumption (we touch on that matter in Chapter 10), but it is an assumption nevertheless.

Still, spatial modelers, as they are called, go to lengths to define and explain key concepts. How *distance* is defined is a serious matter because different definitions can lead to different substantive conclusions.[25] And, as we noted earlier, the theory establishes clear hypotheses. Consider, for example, Voter 1 in Figure 2-1. The theory predicts that this person will vote for Candidate B, not A or C, because that candidate is closest. Voter 2, on the other hand, is closest to C and will vote for that candidate. All of these predictions can be checked with appropriate survey data.

No theory rests entirely on "facts" because it invariably contains unproven or unexamined definitions and statements. These assumptions may be based on previous usage and research, but the theory does not address them directly, except possibly to acknowledge their existence. For example, a theory of war might assert that one nation will attack another one if conditions X, Y, and Z hold. In making this argument, however, the theory may use words and ideas (e.g., *aggression, nation state, balance of power*) that go undefined; or it may assume that the best way to see if the conditions apply is to use certain historical documents. Hence, if the theory's main assertion—if conditions X, Y, and Z occur, an armed conflict follows—may fail to explain or predict the occurrence of a specific war because (1) the theory itself is just wrong, (2) one or more of its underlying concepts or assumptions is incorrect or ambiguous, or (3) both (1) and (2) are correct (the most likely case). This characteristic means that scientific theories are provisional, that is, always subject to revision and change. In fact, according to the philosopher Thomas S. Kuhn, most "normal" activity in science involves checking the adequacy and implications of existing theories.[26] Thus new observations, more accurate measurements, improved research design, and the testing of alternative explanations may reveal the limitations or empirical inadequacies of a theory. In this case, the theory will have to be modified or rejected.[27] One of the excitements of reading scholarly literature is to witness the battle of clashing theories.[28]

Theories are sometimes described by their explanatory range, or the breadth of the phenomena they purport to explain. Usually one does not have a theory of "why George Bush won the 2004 presidential election." (It is, of course possible to find several theories that account for this particular outcome. But note that 2004 election results are an instance, or "token," of the kind of event with which these theories deal.) Instead, a good theory of electoral outcomes presumably pertains to more than simply the 2004 contest but also to other elections in other times and places. In the social sciences, so-called narrow-gauge or middle-range theories pertain to limited classes of events or behaviors such as a theory of voting behavior or a theory about the role of revolution in political development.[29] Thus a theory of voting may explain voter turnout by proposing factors that affect people's perceptions of the costs and benefits of voting: socioeconomic class, degree of partisanship, the ease of registration and voting laws, choices among candidates, availability of election news in the media, and so forth.[30] Global or broad-range theories, by contrast, claim to describe and account for an entire body of human behavior. Hence, we find theories of "international relations" or "the rise and fall of civilizations."[31] In short, theories play a prominent role in natural and social sciences because they provide general accounts of phenomena.[32] Indeed, other things being equal, the broader the range of the things to be explained, the more valuable the theory.

We can see the utility of theory building in the work of Bruce A. Williams and Albert R. Matheny, who evaluated several competing theories in their examination of variations in state regulation of hazardous waste disposal.[33] Regulation of hazardous waste disposal is an example of social regulation, regulation that imposes costs on a specific group to benefit the public or some segment of it. Improper waste disposal imposes costs on the environment and human health. These costs, known as negative externalities, are not reflected in the price of a product. Put somewhat differently, the people who produce and use the products that generate hazardous waste do not pay the costs that arise when improper storage of that waste threatens the community. Avoiding or preventing negative externalities by requiring safe disposal of hazardous waste means imposing substantial costs on industry. There are at least three theories to explain and predict the amount or nature of waste disposal regulation enacted by the states.

According to economic theory, negative externalities are a type of market failure, because the market fails to deal with the problem in that the price of goods does not reflect their true cost. When this happens, government regulation of the market becomes justified and necessary. The market failure theory of government regulation predicts that social regulation is related to the severity of the market failure and that the costs to regulated industries should be equal to the costs or harms created by unsafe hazardous waste disposal.

Others argue that social regulation corresponds to more than just the presence or magnitude of market failures. They claim that social regulation is the result of political behavior and political influence. This theory predicts that exaggerated claims about dangers imposed by market failures must be made in order to generate public awareness, "moral outrage," and thence support for regulation. Consequently, the resulting regulations impose unnecessarily high costs on industry. The flip side of this theory has industry opposing regulation and dominating the regulatory process by threatening economic slowdown, unemployment, even change of location. The political strength of an industry is related to its importance to the economy and to the level of government considering the regulation. Threats to relocate have a greater impact at the local or state level than at the national level. Thus regulation may be related to conditions of industry dominance, not the extent of market failure or actual pollution.

A third theory states that, although industry dominates the regulatory process, it does not necessarily oppose all regulation. According to this view, industry supports regulation as long as the costs of regulation can be shifted to government and away from industry. This regulatory outcome is called the socialization of the costs of production and is predicted by neo-Marxists, who maintain that many private industries could not make a profit without evading actual production costs. They also argue that effective regulation and avoidance of negative externalities is not possible without fundamental institutional reform of both government and the economy. We have, then, three quite different theories of why and how much government regulation of hazardous waste disposal occurs. Each theory has something different to say about the power of public interest groups and industry groups and the outcome of social regulatory efforts. The conflicting beliefs about the politics they represent fuel many a debate about environmental regulations and the performance of government and the economy. Researchers investigating examples of social regulation may be far more interested in determining which of these theories seems to fit with the observed data than in the actual amount and consequence of a specific regulatory program. In fact, researchers may become quite attached to a particular theory and be convinced that it is the correct theory.

But theory builders must not forget the basic standards for judging scientific theories: Are assumptions and axioms clearly separated from substantive propositions and hypotheses? Can the claims be verified or falsified? Are they empirical, not normative? Do they provide general explanations and add to existing knowledge? Are their statements transmissible to others? And, are they parsimonious?

Acquiring Empirical Knowledge: The Scientific Method

What produces scientific knowledge? What, in other words, is the type of thinking that leads to knowledge with the authoritative label *scientific*? Is there one valid path to scientific truth or can it be reached from different starting points? In reality, no scientist in the field or laboratory adheres to a prescribed set of steps like someone following a script. They rely on not just formal procedures, but intuition, imagination, and even luck at times. Nevertheless, we may conceptualize what they do by identifying the underlying logic of their activities. Here is a brief reconstruction of an ideal research program:

- Develop an idea to investigate or a problem to solve. A scientist gets topics from any number of sources including literature about a subject, general observation, intuition (or hunch), the existence of conflicts or anomalies in reported research findings, or the implications of an established theory. For example, newspaper accounts suggest that evangelical Christians tend to support conservative candidates because of "moral values." Several research questions are raised by these accounts: Do evangelicals behave in politics differently than do other religious groups? Do evangelicals turn out to vote more in elections where there are distinct differences between candidates on moral issues than in elections where the differences are small?

- Hypothesis formation. After selecting a topic, an investigator tries to translate the idea or problem into a series of specific hypotheses. As we see in Chapter 3, hypotheses are tentative statements that, if confirmed, show how and why one thing is related to another or why a condition comes into existence. These statements have to be worded unambiguously and in a way that their specific claims can be evaluated by commonly accepted procedures. After all, one of the requirements of science is for others to be able to independently corroborate a discovery. If assertions are not completely transparent, how can someone else verify them? In the preceding example, we might hypothesize that evangelical Christians are more likely than others to base their vote on candidates' positions on moral issues or that evangelical Christians are more likely than other voters to vote for conservative candidates.

■ Research. This is where the rubber meets the road: the essence of science comes in the empirical testing of hypotheses through the collection and analysis of data. We need to define in operational and understandable terms the concepts *moral values, conservative,* and *evangelical Christian.* We might, for instance, tentatively identify evangelicals as people who attend certain churches, moral values as attitudes toward abortion and gay marriage, and support for conservatives as voting for Republican candidates for state and national office. It would be possible (but not necessarily easy) to write a series of questions to be administered in a survey or poll to elicit this information. If this operational hypothesis holds water, we would expect certain responses (for example, opposition to gay marriage) to be associated with certain behaviors (for example, voting for Republicans).

■ Decision. The logical next step is to see whether or not the observed results are consistent with the hypotheses. Simple in principle, judging how well data support scientific hypotheses is usually not an easy matter. Suppose, for example, we find that 75 percent of evangelical Christians opposed gay marriage and 90 percent of these individuals voted for a Republican House candidate in 2006. So far, so good. But suppose, in addition, that 70 percent of non-evangelicals also oppose gay marriage and that more than 90 percent of these people also voted Republican in the same election. It appears that attitudes might be affecting voting, but this does not necessarily establish a connection between religious preference and political behavior. When it comes to weighing quantitative or statistical evidence, this step requires expertise, practice, and knowledge of the subject matter plus good judgment and is often difficult to teach. Still, chapters in this book are devoted to showing ways to make valid inferences about tenability of empirical hypotheses.

■ Modification and extension. Depending on the outcome of the test one can tentatively accept, abandon, or modify the hypotheses. If the results are favorable, it might be possible to derive new predictions to investigate. If, however, the data do not or only very weakly support the hypotheses, it will be necessary to modify or discard them. Let us stress here that negative results—that is, those that do not support a particular hypothesis—can still be interesting or helpful.[34] As we suggested earlier, some scholars such as Popper believe that science advances by disproving claims, not by accepting them. Consequently, a valuable contribution to science can come from disconfirming widely held beliefs, and the only way to do that is by replicating or reinvestigating the research upon which the beliefs rest. The key is not so much the result of a hypothesis test as how substantively important the hypothesis is to begin with.

In essence the scientific method entails using quantitative or qualitative data to test specific propositions. But exactly how does one use evidence to establish a hypothesis? What kind of thinking is involved?

Deduction and Induction

Most people probably believe that scientists prove their results. But the term *prove* may be too strong because it suggests that a conclusion cannot possibly be wrong. Of course, in some areas of science, such as mathematics, the proper application of logic guarantees the truthfulness of a proposition. This type of reasoning is called **deduction.** A valid deductive argument is one in which, if the premises are true, the conclusion must necessarily be true as well. The classic example is the syllogism:

> All men are mortal.
> Socrates is a man.
> _____
> Socrates is mortal.

The conclusion (the sentence below the line) must be true if the premises (the statements above the line) are true. In this example, if all men are mortal and Socrates is a man, how could he *not* be mortal? Whether or not the premises are true is immaterial to the validity of the reasoning. In a valid deduction, it is the structure of the argument that counts: if the conditions ("All men. . .") are true and the argument is stated correctly, then the conclusion must be true.

As noted earlier, a common application of deductive reasoning is found in mathematics, in which theorems are deduced from a set of premises assumed to be or having been established as true. Deductive arguments find their way into political science, too. Social scientists have attempted to develop many axiomatic or deductive accounts of voting, group and coalition behavior, decision making, or the outbreak of war. For example, voters are motivated to vote on the basis of the costs and benefits to them of the policies espoused by the candidates in an election. If this premise is true and large policy differences exist between candidates in an election, then turnout in the election will be high because voter motivation will be higher than it is in elections in which there are small policy differences.

A more common type of reasoning is **induction.** Induction refers to the process of drawing an inference from a set of premises and observations. This type of reasoning differs from deduction because the premises do not guarantee the conclusion but instead lend support to it. An inductive argument, in other words, does not rely on formal proof but rather gives us (more or less solid) reasons for believing in the conclusion's truthfulness. A common type of inductive argument in the social sciences is one that makes a generalization on the basis of a sample. An argument based on sampling, for instance, has the following general form:[35]

In a particular sample, X percent of A's are B's.

X percent of *all* A's are B's.

One might argue, for instance, that 75 (X) percent of those people in a sample of Americans who attend church more than once a week (A) think that the Bible is the literal word of God (B). Then, you would have some reason to believe (but would not have proved deductively) that in the population as a whole 75 percent of frequent church attendees will regard the Bible as the actual word of God. This is, in effect, the kind of argumentation used by pollsters who want to make a generalization about a population based on the results of a sample. For us to accept the argument, we must have confidence in the sampling and analysis procedures. But even if we do, there is no assurance that the conclusion is true. It might seem probable or likely, but we have not demonstrated it conclusively.

Another type of inductive argument is the use of analogy or similarity to establish a point.[36] Here's an example. Imagine that you have made the following three observations. First, the Bemba of south-central Africa live a life of marginal subsistence consisting of nine months of abundance and three months of hunger. Despite deplorable conditions, there is no outbreak of violence or protest within the tribe during the three-month hunger period.[37] Second, the income of African Americans compared with that of whites of equal education rose rapidly during the 1940s and early 1950s but then declined precipitously so that half the relative gains were lost by 1960. Subsequently, violence broke out among blacks living in U.S. urban areas in the 1960s.[38] Third, political violence in Europe occurred during the growth of industrial and commercial centers, even though alternatives to the peasant's hard life emerged at the same time.[39]

In the first and second case studies, the objective well-being of the population declined, but only in the second instance did violence break out. In the third case study, there was no decline in the objective well-being of the population, yet violence occurred. Let's assume that in seeking an explanation in the first case, you reason that the cycle of the seasons and its ensuing periods of feast and famine had been experienced for many years and was unlikely to change. In the second case you reason that African Americans expected to maintain the economic gains they had made in the previous decade. And in the third case you reason that during the period of emerging industrialization, all people expected to improve their living conditions, yet some members of society gained much more than others from the increased industry and commerce. Based on this reasoning, you could conclude that the second and third cases were similar because a discrepancy existed between expected and actual conditions, whereas in the first case there was no discrepancy. From this you might conclude that a large discrepancy or gap between expected and actual economic gains causes discontent, which in turn leads to violence. Thus you might develop a theory of relative deprivation from a few observations of specific cases of deprivation and violence.

The Scientific Method at Work

Generally speaking, it is difficult to point to examples of pure induction, since often a researcher starts with a hunch and then collects information that he or she expects will show certain patterns in line with that hunch. While not a full-blown theory, a hunch places the researcher farther along in his or her investigation than observation alone. In practice, science is an iterative process that typically employs several kinds of reasoning and arguments. Thus a researcher may start with a well-established theory and deduce certain phenomena that he or she will attempt to observe. If the observations are not quite what were expected, some modification of the theory will be made and the revised theory subjected to further testing. Sometimes the theory may have to be discarded and, on the basis of observations, a new theory induced.

A good example is found in the work of two researchers studying news coverage and social trust.[40] For some time psychologists Stephen Holloway and Harvey A. Hornstein had been studying social trust by observing the rate at which people returned wallets dropped on New York City streets to the addresses of the owners identified inside. The researchers would periodically drop wallets in various locations and wait and see how many were returned. Typically, half the wallets dropped were eventually returned. However, one day something happened that had never happened before: none of the wallets was returned. This unexpected result led Holloway and Hornstein to search for a plausible explanation. They set out to develop an explanation based on an observation—that is, they proceeded to the process of induction.

It so happened that on this particular day in June 1968 Robert Kennedy, a senator from New York and candidate for the Democratic presidential nomination, was assassinated. The investigators wondered if Kennedy's assassination could have something to do with the failure to return any of the wallets. Perhaps the news coverage of the event made people upset, mistrustful of strangers, and unwilling to help people they did not know or had not seen. Holloway and Hornstein hypothesized that exposure to "bad" news makes people less socially trusting and cooperative.

To test this hypothesis the researchers devised a series of experiments in which people were divided into two groups and were subtly exposed to "bad" or "good" news broadcasts. Then they were asked to reveal their attitudes toward other people and to play a game with other people that allowed observation of their degree of cooperation. Holloway and Hornstein were testing a general theory with research designed to measure the occurrence of certain predicted observations—that is, they were using deduction.

The experiments demonstrated that those exposed to bad news were, indeed, less socially trusting and cooperative, confirming the researchers' hypothesis. Both induction and deduction had been involved in accumulating an empirical, verifiable, transmissible, explanatory, general (yet provisional) body of evidence regarding an important social phenomenon.

Applying an existing theory to new situations, deciding which phenomena to observe and how to measure them, and developing a theory that explains many more things than the specific observations that led to its discovery are all creative enterprises. Unfortunately, it is difficult to teach creativity. But being aware of the general tenets of science will help make your own evaluation and conduct of research more worthwhile.

Is Political Science Really "Science"?

We have implied throughout this chapter that politics can and should be studied scientifically. Some people question this position, however, because the discipline involves the study of human political behavior, and studying people—as opposed to material objects such as atoms or stars—raises all sorts of complexities. As a consequence, compared with the natural sciences, progress in developing and testing political empirical theories has been agonizingly slow. It is no surprise, then, that scientists and nonscientists alike often observe that both the methods and the content of political science have not come even close to the exactitude and depth of sciences such as biology or physics, and consequently nowhere can we find empirical generalizations with the level of precision and confirmation enjoyed by, say, Einstein's theories of relativity. Moreover, if political science is a science in the same way that the natural sciences are, behavior must ultimately be describable by contingent causal laws.[41] Yet if human beings do not act predictably, or if their actions are not susceptible to description by general laws, political scientists, acting as scientists, encounter serious problems.

We can identify two objections to treating political science as a subdivision of science in general. The first we might term logistical or practical. The other is more philosophical.[42]

Practical Objections

The search for regularities in behavior assumes that men and women act consistently and in a discoverable manner. Nonetheless, even if we accept that individuals are generally predictable, some persons may deliberately act in unpredictable or misleading ways. This problem is occasionally encountered among subjects "cooperating" in a research project. For example, a subject may figure out that he or she is part of an experiment to test a theory about how people behave when put in a difficult or stressful or confusing situation. He or she may then act in a way not predicted by, or in conflict with, the theory. Or the subject may try to conform to what he or she thinks the researcher is looking for. Similarly, people may never reveal what is really on their minds or what they have done in the past or would do in the future. In other words, our ability to accurately observe the attributes of people can at times be severely limited. It is, for instance, frequently difficult to measure and explain illegal or socially unacceptable behaviors such as drug use.

Measurement problems also arise because the concepts of interest to many political scientists are abstract and value laden. Chapter 1 showed that a phrase as seemingly straightforward as "the number of eligible voters" can present problems that affect our substantive conclusions about how civic minded Americans are. Or consider unemployment, a seemingly unambiguous concept. One measure of unemployment takes into account persons who are out of work but actively seeking employment. An argument may be made that such a measure greatly underestimates unemployment because it does not include those who are so discouraged by their failure to find a job that they are no longer actively seeking work. Finding an adequate definition of poverty can be just as difficult, because people live in different types of households and have available different kinds of support beyond just their observed income. What one scholar may feel constitutes poverty another may see as nothing more than acceptable hardship.

Furthermore, political scientists must face the fact that consistent and rational human behavior is complex, perhaps even more complex than the subject matter of other sciences (genes, subatomic particles, insects, and so on). Complexity has been a significant obstacle to the discovery of general theories that accurately explain and predict almost every kind of behavior. After all, developing a theory with broad applicability requires the identification and specification of innumerable variables and the linkages among them. Consequently, when a broad theory is proposed, it can be attacked on the grounds that it is too simple or that too many exceptions to it exist. Certainly to date no empirically verified generalizations in political science match the simplicity and explanatory power of Einstein's famous equation, $E = mc^2$.[43]

There are other practical obstacles.[44] The data needed to test explanations and theories may be extremely hard to obtain. People with the needed information may not want to release it for political or personal reasons. Or they may not want to answer potentially embarrassing or threatening questions honestly or completely. Pollsters, for instance, find refusal to answer certain questions, such as attitudes toward ethnic groups, to be a major problem in gauging public opinion. Similarly, some experiments require manipulation of people. But since humans are the subjects, the researchers must contend with ethical considerations that might preclude them from obtaining all the information they want. Asking certain questions can interfere with privacy rights, and exposing subjects to certain stimuli might put the participants at physical or emotional risk. Tempting someone to commit a crime, to take an obvious case, might tell a social scientist a lot about adherence to the law but would be unacceptable nevertheless.

All these claims about the difficulty of studying political behavior scientifically have merit. Yet they can be overstated. Consider, for example, that scientists studying natural phenomena encounter many of the same problems. Paleontologists must attempt to explain events that occurred millions or even billions of years ago. Astronomers and geologists cannot mount repeated experiments on most of the phenomena of greatest interest to them. They certainly cannot visit many of the places they study most intensively, like other planets or the center of the earth. And what can be more complex than organisms and their components, which consist of thousands of compounds and chemical interactions? Stated quite simply, it is in no way clear that severe practical problems distinguish political science from any of the other sciences.

Philosophical Objections

Before moving on, we want to emphasize again that the scientific method is not the only path to knowledge. In fact, some scholars believe that because the social sciences attempt to explain human **actions**—that is, behavior that is done for reasons—and not mere physical movement, they face challenges not encountered in the natural sciences. Opponents of the empirical approach claim that scientific methods do not explain nearly as much about behavior as their practitioners think. The problem, one set of critics argues, is that to understand human behavior one must try to see the world the way individuals do. (These are the interpretationists, mentioned earlier.) And doing so requires empathy, or the ability to identify and in some sense experience the subjective moods or feelings or thoughts of those being studied. Instead of acting as outside, objective observers, we need to "see" how individuals themselves view their actions. For only by reaching this level of understanding can we hope to answer "why" questions such as "why did John vote Democratic in the last election" or "why did this group revolt against the duly elected government?" The answers require the interpretation of behavior, not its scientific explanation in terms of general laws. In short, interpretation means decoding verbal and physical actions, which is a much different task than proposing and testing hypotheses.

Indeed, some objections to the application of science to the study of human behavior go even further for they raise the question of what constitutes scientific knowledge. Empiricists, as we have suggested, take reality pretty much as a given. That is, the objects they study—elections, wars, constitutions, government agencies—have an existence independent of observers and can be studied more or less objectively. But an alternative perspective that since the 1970s has taken root in political science and international relations theory is called the social construction of reality or **constructionism**.[45] Constructivists, as supporters of the theory are called, claim that humans do not simply discover knowledge of the real world through a neutral process like the scientific method but rather create it. In other words, instead of knowing reality directly in its unvarnished or pure form, our perceptions, understandings, and beliefs about many "facts" stem largely, if not entirely, from human cultural and historical experiences and practices. We put *facts* in quotation marks in this context to stress the constructionist belief that what people often assume to be pure facts are conditioned by the observers' perceptions, experiences, opinions, and similar mental states. This position is perhaps another way of saying, "facts do not speak for themselves but are always interpreted or constructed by humans in specific historical times and settings." One version of this position admits that entities (for example, molecules, planets) exist separately from anyone's thoughts about them, but it also insists that much of what people take for granted as being "real" or "true" of the world is built from learning and interaction with others and does not have an existence apart from human thought.[46] Consider the term *Democratic Party*. Instead of having an independent, material existence like an electron or a strand of DNA, a political party exists only because citizens behave as if it did exist. This means that two individuals who come from different social, historical, and cultural backgrounds may not comprehend and respond to the term in the same way. What is important in studying, say, the individuals' responses to Democratic candidates is fathoming their personal beliefs and attitudes about the party.

Constructionist thinking now plays a strong role in international relations theory, where a concept such as anarchy is not considered a "given and immutable" cause of the behavior of states (for example, their desire for security through power politics). Rather, terms like this one have to be understood as what actors (individuals, states) make of them.[47]

The constructionist viewpoint, which comes in innumerable varieties, challenges the idea of an objective epistemology, or theory of knowledge. Such ideas, however, are of a deeply methodological nature and raise deep philosophical issues that go well beyond the task of describing the empirical methods used in the discipline.[48] We thus acknowledge that the scientific study of politics is controversial but nevertheless maintain that the procedures we describe in the chapters that follow are widely accepted and can in many circumstances lead to valuable understandings of political processes and behavior. Moreover, they have greatly shaped the research agenda and teaching of the discipline, as can be seen by looking at the evolution of the field in the twentieth century.

A Brief History of Political Science as a Discipline

Steven B. Smith wrote, "From its very beginnings political science has been a complex disciple torn in conflicting directions."[49] The history of the field can be divided into roughly four periods: the traditional era, the behavioral or empirical "revolution," the reaction to the rise of the behavioral perspective, and the contemporary period of accommodation.

The Era of Traditional Political Science

Traditional political science, which grew out of the study of law, institutions, and ethics, flourished until the early 1960s. It emphasized historical, legalistic, and institutional subjects.[50] The historical emphasis produced detailed descriptions of the developments leading to political events and practices. Legalism, in contrast, involved the study of constitutions and legal codes, and the concentration on institutions included studies of the powers and functions of political institutions such as legislatures, bureaucracies, and courts. In general, traditional political science focused on formal governments and their legally defined powers. Legal and historical documents, including laws, constitutions, proclamations, and treaties, were studied to trace the development of international organizations and key concepts such as sovereignty, the state, federalism, and imperialism. Informal political processes—the exercise of informal power and the internal dynamics of institutions, for example— were frequently ignored.

In the heyday of the traditional approach (roughly 1930 to 1960), the study of politics was usually taught in the history and philosophy departments of colleges and universities. Political theories concerning human nature and politics, the purpose and most desirable form of government, and the philosophy of law were the province of philosophy departments. When separate departments did appear, they were frequently called departments of government, reflecting the emphasis on formal structures rather than on political processes and behavior. In fact, some universities still have government departments.

Since scientific methodology did not inform most of the study of government, traditional political science was primarily descriptive rather than explanatory. Most of its practitioners did not feel a need to conduct research that had the characteristics of the so-called hard sciences, which were often deemed inapplicable to social behavior and institutions. Critics were later to charge that the traditional school lacked rigor and generality and that, although theorists occasionally came up with intriguing and well-reasoned verbal theories, these discoveries were usually not subjected to rigorous and extensive empirical verification.

The Empirical Revolution

The emergence of the scientific study of politics in the United States after World War II, and especially in the late 1950s, can be attributed to several developments.[51] First, many of the European social scientists and theorists who emigrated to the United States in the 1930s were skilled in the use of new, scientific research methods.[52] Second, war-related social research in the following decade promoted the exchange of ideas among scientifically minded persons from the disciplines of political science, sociology, psychology, and economics. In fact, considerable evidence indicates that the U.S. government looked to colleges and universities for scientific social science research that would be of use in fighting the cold war against the Soviet Union.[53]

In addition, systematic research was aided by two related developments: the collection of large amounts of empirical data and the development of computers to store and process this information. For example, beginning in the late 1930s Paul F. Lazarsfeld pioneered the use of large-scale sample surveys or polls to study voting behavior and continued to refine the technique while working for the federal government during World War II. After the war he applied survey research methods to his study of the 1948 and 1952 presidential elections.[54] In these endeavors Lazarsfeld and others were influenced by the developing field of market research. This development makes sense if one thinks of voters as consumers who must choose among competing products (that is, candidates). Once the use of survey research got under way, the field exploded and is now perhaps the most common source of knowledge about politics in the United States and abroad.[55]

Moreover, as the empirical school ascended, sister disciplines in the social sciences began to investigate political problems and ultimately helped shape the content and methods of political science. A classic example is Anthony Downs's *An Economic Theory of Democracy*. As his title suggests, Downs (an economist) applied many concepts from economic theory to develop a formal model of voting and party behavior. His reasoning depended on concepts such as "utility maximization" and "indifference." (One surprising conclusion was that under most circumstances it is irrational for individuals to vote in any but the very smallest elections. Similarly, the statistical methods used by economists, called econometrics, also played a critical role in the newer political science. Indeed, statistics is now deemed to be so essential to the field that practically every graduate student must take at least a one-year course in applied statistics. In the same vein, sociology, psychology, and anthropology have all influenced political scientists. Indeed, looking at the sway of other disciplines makes one appreciate the multidisciplinary nature of the field.

The important point, however, is that unlike the traditional school, the newer approach consciously embraces the spirit of scientific inquiry, as illustrated by David B. Easton's influential 1967 article, "The Current Meaning of 'Behavioralism.' "

> *There are discoverable uniformities in political behavior. These can be expressed in generalizations or theories with explanatory and predictive value. Means for acquiring and interpreting data . . . need to be examined self-consciously, refined, and validated. Precision in the recording of data and the statement of findings requires measurement and quantification. Ethical evaluation and empirical explanation involve two different kinds of propositions that, for the sake of clarity, should be kept analytically distinct. Research ought to be systematic.*[56]

Behavioral political science assumes and advocates the search for fundamental units of analysis that can provide a common base for the investigation of human behavior by all social scientists. Some political scientists, for instance, suggest that groups are an important unit on which to focus, while others are more interested in decision making and decisions.[57] Whatever the case, the hope was that units of analysis would be found and examined in much the same way that physicists and chemists focus on atoms, molecules, and the like.

Reaction to Empiricism

From the very beginning of the empirical movement, critics appeared. They pointed to the trivial nature of some of the supposedly scientific findings and applications. Common sense would have told us the same thing, they argued. But, as we explained earlier, there is a difference between intuition and scientific knowledge. To build a solid base for further research and accumulation of scientific knowledge in politics, commonsense knowledge must be verified empirically and, as is frequently the case, discarded when wrong.

Some political scientists were also concerned about the prominence of non-political factors in explanations of political behavior. Psychological explanations of political behavior stress the effect of personality on political behavior, whereas economic explanations attempt to show how costs and benefits affect people's actions. These competing approaches to understanding political behavior sometimes disturbed those used to studying political institutions or political philosophies. To them it looked as though politics was being taken out of the study of politics.

A more serious criticism of the scientific study of politics is that it leads to a failure to focus enough scholarly research attention on important social issues and problems. Some critics contend that, in the effort to be scientific and precise, political science overlooks the moral and policy issues that make the discipline relevant to the real world. The implications of research findings for important public policy choices or political reform are rarely addressed. In other words, the quest for scientific knowledge of politics has led to a focus on topics that are quantifiable and relatively easy to verify empirically but that are not related to significant, practical, and relevant societal concerns.[58] These worries led to, among other things, the emergence of new or revisionist approaches to political science.

Even as early as the late 1960s the president of the American Political Science Association, David Easton, offered a critique of behavioral or empirical political research. He argued, among other things, that to be more relevant to current issues, political scientists acting as scientists should consider these points:[59]

- Substance should determine technique. For example, the widespread availability of quantitative techniques such as market research tools should not govern the choice of research topics. As the interpretationists maintain, sometimes it is as important to understand attitudes as to measure them.

- The scientific study of politics may conceal a "conservative" bias because it studies institutions and practices as they are, not as they should be or would be under different circumstances. Scientists, as noted earlier, strive to be value free. But is this a proper stance for someone studying despotism or revolution or poverty? Isn't it appropriate to suggest alternatives to the status quo based on values?

- Research must not "lose touch with reality." Anyone who skims articles in *The Journal of American Political Science* might wonder if they concern current government and politics or are a form of higher mathematics. One of the commonest complaints about behavioral research is that it has little relevance to "practical politics," and many observers lament the failure of government or society to benefit from the knowledge and perspectives of political science.[60]

■ Even natural scientists have an obligation to think about and improve their communities and nations, especially in times of crises or turmoil. After all, to say, "I'm only concerned with facts," may be to turn a blind eye to injustice.

Political Science Today: Peaceful Coexistence?

During the growth of empirical political science and the reaction it produced, political science became extremely self-conscious about its methods and methodology. Innumerable journal articles and books debated the relative merits of attempting to study politics scientifically. Many of the debates became acrimonious, with the participants charging each other with misunderstanding and misstating each other's positions. Departments at many colleges and universities became bitterly divided between "behavioralists" and "anti-behavioralists." Being associated with one camp or the other could jeopardize someone's job or chances for tenure, and many scholars charged that the major journals in the field—the most important venues for getting published and advancing careers—were being "taken over" by methodological purists. (If you wanted to publish an article on Plato's philosophy of justice, you might be out of luck finding a leading journal that would publish a paper without an empirical, preferably quantitative, slant.) Indeed, this level of discord would surprise those students who believe that scholarship is a calm, dispassionate activity.

Fortunately, although deep differences remain, the field of political science entered a period of truce beginning in the early 1980s. (What is more, the fights are not as public or bitter.) We might, then, think of the current era as eclectic, meaning that although the discipline continues to be divided by empiricism versus interpretative and constructionist schools, the sides seem to live in relative harmony.

On one hand, the empirical or behavioral approach dominates certain subfields such as the study of electoral behavior, public opinion, decision making, policy, and political economy. The research coming from this side has become increasingly technical, especially in its use of mathematics, statistics, and deductive logic; hence, the need for courses and texts in research methods.

On the other hand, the hostile reaction to the emergence and domination of the empirical perspective has brought about renewed interest in normative philosophical questions of "what ought to be" rather than "what is."[61] Beyond this stance, part of the discipline has become receptive to variations of **critical theory,** or the belief that a proper goal of social science is to critique and change society as a whole rather than merely understand or explain it. They feel, in other words, that by simply analyzing a polity as it is amounts to a tacit endorsement of its institutions and the distribution of power. Many critical theorists argue that proposing and working for reforms are legitimate activities for the social sciences. They therefore analyze institutions, practices, ideologies, and beliefs not only for their surface characteristics but also for their "hidden meanings" and implications for behavior.

Take, for example, the statement "I'm just not interested in politics."[62] An empirical political scientist might take this simply as a cut-and-dried case of apathy. He or she might then look for variables (for example, age, gender, ethnicity) associated with "not interested" responses on questionnaires. Its meaning is simply taken for granted. A critical theorist, by contrast, might ask, "Does this person really have *no* interest in current events? After all, isn't everyone affected by most political outcomes like decisions about taxes, war and peace, and the environment and thus in fact *have* an interest in politics? So, perhaps we have a case of, say, 'false consciousness,' and it is crucial to uncover the reasons for lack of awareness of one's 'real' stake in politics. Is the indifference a matter of choice, or does it stem from the (adverse) effects of the educational system, the mass media, modern campaigning, or some other source?"

Here is another case in point. An important challenge to research in political science (as well as in other social science disciplines such as sociology) has come from feminist scholars. Among the criticisms raised is that "the nature of political action and the scope of political research have been defined in ways that, in particular, exclude *women as women* from politics" (emphasis added).[63] Accordingly, "What a feminist political science must do is develop a new vocabulary of politics so that it can express the specific and different ways in which women have wielded power, been in authority, practiced citizenship, and understood freedom."[64] Even short of arguing that political science concepts and theories have been developed from a male-only perspective, it is all too easy to point to examples of gender bias in political science research. Examples of such bias include failing to focus on policy issues of importance to women, assuming that findings apply to everyone when the population studied was predominantly male, and using biased wording in survey questions.[65]

A related complaint is that political science in the past ignored the needs, interests, and views of the poor, the lower class, and the powerless and served mainly to reinforce the belief that existing institutions were as good as they could be. Concerns about the proper scope and direction of political science have not abated, although nearly all researchers and teachers accept the need to balance the scientific approach with consideration of practical problems and moral issues.[66]

In sum, the widespread acceptance of empiricism has certainly not silenced critical reflection on political science as a discipline.[67] Nor has it prevented the acceptance of alternative methodologies such as interpretation and constructionism. For the most part, however, both empiricists and nonempiricists live with each other peacefully.

Conclusion

In this chapter we described the characteristics of scientific knowledge and the scientific method. We presented reasons why political scientists are attempting to become more scientific in their research and discussed some of the difficulties associated with empirical political science. We also touched on questions about the value of the scientific approach to the study of politics. Despite these difficulties and uncertainties, the empirical approach is widely embraced, and students of politics need to be familiar with it. In Chapter 3 we begin to examine how to develop a strategy for investigating a general topic or question about some political phenomenon scientifically.

Notes

1. Boston *Globe,* "Not Fit for the Court," January 14, 2006. Retrieved January 5, 2007, from www.boston.com/news/globe/editorial_opinion/editorials/articles/2006/01/14/not_fit_for _the_court/.

2. *Washington Post,* "Confirm Judge Alito," January 15, 2006, B06.

3. Jeffrey A. Segal and Albert D. Cover, "Ideological Values and the Votes of U.S. Supreme Court Justices," *American Political Science Review* 83 (June 1989): 557–565.

4. We hasten to add that there is not one, definitive definition or interpretation of science and the scientific method. Philosophers, scientists, and social scientists have argued long and hard about core ideas and propositions. Our listing of the characteristics of scientific knowledge, however, includes widely accepted attributes, even if other writers describe them in different terms.

5. Whether or not political science or any social science can find causal laws is very much a contentious issue in philosophy. See, for instance, Alexander Rosenberg, *Sociobiology and the Preemption of Social Science* (Baltimore: Johns Hopkins University Press, 1980).

6. Alan C. Isaak, *Scope and Methods of Political Science,* 4th ed. (Homewood, Ill.: Dorsey, 1985), 106.

7. Ibid., 107.

8. The most ardent proponent of the idea that science really amounts to an effort to falsify (not prove) hypotheses and theories is Karl Popper. See for example, *The Logic of Scientific Discovery* (New York: Basic Books, 1959).

9. Isaak, *Scope and Methods,* 66; see also 67.

10. Ibid., 297.

11. Remembering that scientific explanations are tentative can help clarify certain current controversies. You hear, for example, the claim that Darwinian evolution is not a fact but merely a theory. This statement is correct on both accounts. Nevertheless, the ideas and predictions of evolutionary theory have been repeatedly tested and confirmed by scientific methods and standards. So nearly every scientist accepts it as the more or less valid account of, say, human ancestry. But they will gladly abandon Darwinism if and when a better *scientific* explanation comes along.

12. Isaak, *Scope and Methods,* 30.

13. Ibid., 31.

14. The studies are reported in H. J. Eysenck and D. K. B. Nias, *Sex, Violence, and the Media* (London: Temple Smith, 1978), 103–104.

15. It may be tempting to think that historians are interested in describing and explaining only unique, one-time events, such as the outbreak of a particular war. This is not the case, however. Many historians search for generalizations that account for several specific events. Some even claim to have discovered the "laws of history."

16. Isaak, *Scope and Methods,* 103.

17. These are the interpretationists mentioned earlier. See, for example, R. G. Collingwood, *The Idea of History* (Oxford: Oxford University Press, 1946). For a good introduction to the distinction between understanding behavior and explaining it, see Martin Hollis, *The Philosophy of Social Science: An Introduction* (Cambridge: Cambridge University Press, 1994), chap. 7.

18. Ibid., 167, 169.

19. Many varieties of this theory exist, but they share the components presented here.

20. Anthony Downs, an economist, provided one of the first explications of the theory in *An Economic Theory of Democracy* (New York: Harper & Row, 1957). His ideas in turn flowed from earlier economic analysis. See, for example, Harold Hotelling, "Stability in Competition," *Economic Journal* 39 (March 1929): 41–57.

21. See James Enelow and Melvin Hinch, *The Spatial Theory of Voting: An Introduction* (New York: Cambridge University Press, 1984).

22. Kenneth Shepsle, "The Strategy of Ambiguity: Uncertainty and Electoral Competition," *American Political Science Review* 66 (June 1972): 555–568.

23. As an example, see Anders Westholm, "Distance Versus Direction: The Illusory Defeat of the Proximity Theory of Electoral Choice," *American Political Science Review* 91 (December 1997): 870.

24. Here is an example: "Please look at . . . the booklet. Some people believe that we should spend much less money for defense. Suppose these people are at one end of a scale, at point 1. Others feel that defense spending should be greatly increased. Suppose these people are at the other end, at point 7. And, of course, some other people have opinions somewhere in between, at points 2, 3, 4, 5 or 6." *American National Election Study (ANES) 2004 Codebook.* Available at Survey Documentation and Analysis, University of California, Berkeley. Retrieved March 12, 2007, from http://sda.berkeley.edu/D3/NES2004public/Doc/nes0.htm.

25. The conceptualization of distance and other matters related to the proximity theory are debated in Westholm, "Distance Versus Direction," 865–873 and Stuart Elaine MacDonald, George Rabinowitz, and Olga Listhaug, "On Attempting to Rehabilitate the Proximity Model: Sometimes the Patient Just Can't Be Helped," *Journal of Politics* (August 1998): 653–690.

26. Thomas S. Kuhn, *The Structure of Scientific Revolutions,* 2d ed. (Chicago: University of Chicago Press, 1971).

27. Ibid.

28. The discussion of voter turnout presented in Chapter 1 provides a clear and important example.

29. A good example is Theda Skocpol *States and Social Revolutions: A Comparative Analysis of France, Russia and China* (New York: Cambridge University Press, 1979).

30. See Raymond E. Wolfinger and Steven J. Rosenstone, *Who Votes?* (New Haven: Yale University Press, 1980).

31. An excellent example of the latter is Jared Diamond's study of the demise of the Mayan, Anasazi, and other societies. See his *Collapse: How Societies Choose to Fail or Succeed* (New York: Viking, 2005), especially part 2; and *Guns, Germs, and Steel: The Fates of Human Societies* (New York: Norton, 1999).

32. Isaak, *Scope and Methods,* 167.

33. Bruce A. Williams and Albert R. Matheny, "Testing Theories of Social Regulation: Hazardous Waste Regulation in the American States," *Journal of Politics* 46 (May 1984): 428–458.

34. An often remarked on characteristic of scholarly journals is that they tend to report mostly positive findings. An article that shows "X is related to Y" may be more likely to be accepted for publication than one that asserts "X is *not* related to Y." Whether or not this practice makes sense depends on the theoretical significance of the findings. If the X-Y relationship is trivial, it probably does not matter much if it is confirmed or disconfirmed.

35. Merrilee H. Salmon, *Introduction to Logic and Critical Thinking,* 2d. ed. (San Diego: Harcourt Brace Jovanovich, 1989), 88–97.

36. Ibid., 80–85.

37. Ted Robert Gurr, *Why Men Rebel* (Princeton: Princeton University Press, 1970), 57.

38. Ibid., 54.

39. Ibid., 51.

40. The wallet-dropping episode is described in Stephen Holloway and Harvey A. Hornstein, "How Good News Makes Us Good," *Psychology Today,* December 1976, 76–78. The results of the subsequent experiments are discussed in Stephen Holloway, Lyle Tucker, and Harvey A. Hornstein, "The Effects of Social and Nonsocial Information on Interpersonal Behavior of Males: The News Makes News," *Journal of Personality and Social Psychology* 35 (July 1977): 514–522; and in Harvey A. Hornstein, Elizabeth Lakind, Gladys Frankel, and Stella Manne, "Effects of Knowledge about Remote Social Events on Prosocial Behavior, Social Conception, and Mood," *Journal of Personality and Social Psychology* 32 (December 1975): 1038–1046.

41. See Alexander Rosenberg, *The Philosophy of Social Science,* 2d ed. (Boulder: Westview, 1998).

42. In fact, there are myriad concerns about the epistemological status of political science, but to simplify matters we use these two broad categories.

43. For further discussion of complete and partial explanations, see Isaak, *Scope and Methods,* 143.

44. See Charles A. McCoy and John Playford, eds., *Apolitical Politics: A Critique of Behavioralism* (New York: Thomas Y. Crowell, 1967).

45. The term *constructionism* encompasses an enormous variety of philosophical perspectives, the description of which goes far beyond the purposes of this book. The seminal work that brought the ideas into sociology and from there into political science is Peter L. Berger and Thomas Luckmann, *The Social Construction of Reality* (New York: Doubleday, 1966). An excellent but challenging analysis of constructionism is Ian Hacking, *The Social Construction of What?* (Cambridge: Harvard University Press, 1999). Equally important, members of this school have widely varying opinions about the place of empiricism in social research. Many constructivists feel their position is perfectly consistent with the scientific study of politics; others do not.

46. See John R. Searle, *The Construction of Social Reality* (New York: Free Press, 1995).

47. Alexander Wendt, "Anarchy Is What States Make of It: The Social Construction of Power Politics," *International Organization* 46 (Spring 1992): 391–425.

48. For an excellent collection of articles about the pros and cons of studying human behavior scientifically, see Michael Martin and Lee C. Anderson, eds., *Readings in the Philosophy of Social Science* (Cambridge: MIT Press, 1996).

49. Steven B. Smith, "Political Science and Political Philosophy: An Uneasy Relationship," *PS: Political Science and Politics* 33 (June 2000): 189.

50. Isaak, *Scope and Methods,* 34–38.

51. Ibid., 38–39. For a history of the development of survey research, see also Earl F. Babbie, *Survey Research Methods* (Belmont, Calif.: Wadsworth, 1973), 42–45.

52. For early American sources of behavioralism, see Charles E. Merriam, *New Aspects of Politics* (Chicago: University of Chicago Press, 1924).

53. See, for example, the excellent collection of articles entitled "Science and the Cold War: A Roundtable," in *Diplomatic History* 24 (Winter 2000). The essay by Jefferson P. Marquis, "Social Science and Nation Building in Vietnam," 79–105, is especially relevant.

54. Paul F. Lazarsfeld, Bernard Berelson, and Hazel Gaudet, *The People's Choice* (New York: Duell, Sloane and Pearce, 1944).

55. It is interesting to note that survey research (polling) is used to study attitudes and behavior in many authoritarian and unstable nations.

56. David B. Easton, "The Current Meaning of 'Behavioralism'," in James C. Charlesworth, ed., *Contemporary Political Analysis* (New York: Free Press, 1967), 16–17.

57. David B. Truman, *The Governmental Process* (New York: Knopf, 1951); and Robert A. Dahl, *Who Governs? Democracy and Power in an American City* (New Haven: Yale University Press, 1961).

58. See McCoy and Playford, *Apolitical Politics.*

59. David Easton, "The New Revolution in Political Science," *American Political Science Review* 63 (December 1969): 1051.

60. See Richard P. Nathan, *Social Science in Government: Uses and Misuses* (New York: Basic Books, 1988).

61. Isaak, *Scope and Methods,* 45.

62. This example is based on an article by Issac Balbus, "The Concept of Interest in Pluralist and Marxian Analysis," *Politics & Society* (February 1971): 151–177.

63. Kathleen B. Jones and Anna G. Jonasdottir, "Introduction: Gender as an Analytic Category in Political Science," in Kathleen B. Jones and Anna G. Jonasdottir, eds., *The Political Interests of Gender* (Beverly Hills, Calif.: Sage Publications, 1988), 2.

64. Kathleen B. Jones, "Towards the Revision of Politics," in Jones and Jonasdottir, *The Political Interests of Gender,* 25.

65. Margrit Eichler, *Nonsexist Research Methods: A Practical Guide* (Boston: Allen and Unwin, 1987).

66. See the symposium "Special to PS: Political Science and Political Philosophy" in *PS: Political Science and Politics* 33 (June 2000): 189–197.

67. For example, see David M. Ricci, *The Tragedy of Political Science: Politics, Scholarship, and Democracy* (New Haven: Yale University Press, 1984).

Terms Introduced

ACTIONS. Physical human movement or behavior done for a reason.

CONSTRUCTIONISM. An approach to knowledge that asserts humans actually construct—through their social interactions and cultural and historical practices—many of the facts they take for granted as having an independent, objective, or material reality.

CRITICAL THEORY. The philosophical stance that disciplines such as political science should assess critically and change society, not merely study it objectively.

CUMULATIVE. Characteristic of scientific knowledge; new substantive findings and research techniques are built upon those of previous studies.

DEDUCTION. A process of reasoning from a theory to specific observations.

EMPIRICAL GENERALIZATION. A statement that summarizes the relationship between individual facts and that communicates general knowledge.

EMPIRICAL VERIFICATION. Characteristic of scientific knowledge; demonstration by means of objective observation that a statement is true.

EXPLANATORY. Characteristic of scientific knowledge; signifying that a conclusion can be derived from a set of general propositions and specific initial considerations; providing a systematic, empirically verified understanding of why a phenomenon occurs as it does.

FALSIFIABILITY. A property of a statement or hypothesis such that it can (in principle, at least) be rejected in the face of contravening evidence.

GENERAL. Characteristic of scientific knowledge; applicable to many rather than to a few cases.

INDUCTION. Induction is the process of drawing an inference from a set of premises and observations. The premises of an inductive argument support its conclusion but do not prove it.

INTERPRETATION. Philosophical approach to the study of human behavior that claims that one must understand the way individuals see their world in order to understand truly their behavior or actions; philosophical objection to the empirical approach to political science.

NONNORMATIVE KNOWLEDGE. Knowledge concerned not with evaluation or prescription but with factual or objective determinations.

NORMATIVE KNOWLEDGE. Knowledge that is evaluative, value laden, and concerned with prescribing what ought to be.

PARSIMONY. The principle that among explanations or theories with equal degrees of confirmation, the simplest—the one based on the fewest assumptions and explanatory factors—is to be preferred. (Sometimes known as Ockham's razor.)

PROBABILISTIC EXPLANATION. An explanation that does not explain or predict events with 100 percent accuracy.

THEORY. A statement or series of statements that organize, explain, and predict phenomena.

TRANSMISSIBLE. Characteristic of scientific knowledge; indicates that the methods used in making scientific discoveries are made explicit.

Suggested Readings

Eichler, Margrit. *Nonsexist Research Methods: A Practical Guide.* Boston: Allen and Unwin, 1987.

Elster, Jon. *Nuts and Bolts for the Social Sciences.* Cambridge: Cambridge University Press, 1990.

Heil, John. *Philosophy of the Mind: A Contemporary Introduction.* London: Routledge, 1998.

Isaak, Alan C. *Scope and Methods of Political Science,* 4th ed. Homewood, Ill.: Dorsey, 1985.

Kuhn, Thomas. *The Structure of Scientific Revolutions,* 2d ed. Chicago: University of Chicago Press, 1971.

Martin, Michael, and Lee C. McIntyre, eds. *Readings in the Philosophy of the Social Sciences.* Cambridge: MIT Press, 1994.

McCoy, Charles A., and John Playford, eds. *Apolitical Politics: A Critique of Behavioralism.* New York: Thomas Y. Crowell, 1967.

Nielsen, Joyce McCarl, ed. *Feminist Research Methods: Exemplary Readings in the Social Sciences.* Boulder, Colo.: Westview, 1990.

Rosenberg, Alexander. *The Philosophy of Social Science,* 2d ed. Boulder, Colo.: Westview, 1998.

CHAPTER 3

The Building Blocks of Social Scientific Research:

Hypotheses, Concepts, and Variables

In Chapters 1 and 2 we discussed what it means to acquire scientific knowledge and presented examples of political science research intended to produce this type of knowledge. In this chapter we consider the initial steps in an empirical research project. We emphasize explaining or exploring relationships between political phenomena. These steps require us to (1) specify the question or problem with which the research is concerned; (2) propose a suitable explanation for the phenomena under study; (3) formulate testable hypotheses; and (4) define the concepts identified in the hypotheses. Although we discuss these steps as if they represent a logical sequence, the actual order may vary. All the steps must be taken eventually, however, before such a research project can be completed successfully. The sooner the issues and decisions involved in each of the steps are addressed, the sooner the other portions of the research project can be completed.

Specifying the Research Question

One of the most important purposes of social scientific research is to answer questions about social phenomena. The research projects summarized in Chapter 1, for example, attempt to answer questions about some important political attitudes or behaviors: Why is wealth distributed more equally among the population in some countries than in others? Why do some people vote in elections while others do not? Why do Supreme Court justices reach the decisions they do on the cases before them? Do Supreme Court decisions affect people's opinions on issues and people's support of the Supreme Court? Under what circumstances are people most likely to support U.S. involvement in foreign affairs? How sensitive is the American public to combat casualties, and does the number of casualties affect public support for the war in Iraq? Does negative campaign advertising have any impact on the electorate? Do partisan divisions in Congress and between Congress and the White House affect the design of new federal agencies and do variations in the structure of agencies affect the ability of Congress and the White House to influence them? In each case the researchers identified a political phenomenon that interested them and tried to answer questions about that phenomenon.

The phenomena investigated by political scientists are diverse and are limited only by whether they are significant (that is, would advance our understanding of politics and government), observable, and political. Political scientists attempt to answer questions about the political behavior of individuals (voters, citizens, residents of a particular area, Supreme Court justices, members of Congress, presidents), groups (political parties, interest groups, labor unions, international organizations), institutions (state legislatures, city councils, bureaucracies, district courts), and political jurisdictions (cities, states, nations).

Most students, when confronting a research project for the first time, will start by saying, "I'm interested in X," where X may be the Supreme Court, media coverage of the war in Iraq, campaign finance policy, the response to Hurricane Katrina, or some other political phenomenon. Thus the first major task in a research effort often is to translate a general topic into a research question or series of questions or propositions with which the research is concerned. The framing of an engaging and appropriate research question will get a research project off to a good start by limiting the scope of the investigation and determining what information has to be collected. A poorly specified question inevitably leads to wasted time and energy. Any of the following questions would probably lead to a politically significant and informative research project:

Why is the voter turnout for local elections higher in some cities than in others?

Why is the rate of recycling higher in some communities than in others?

Why did some members of Congress vote for legislation creating a prescription drug benefit under Medicare, whereas others opposed it?

Why do some states have laws strongly regulating the activities of lobbyists, while other states do not?

Why does the amount spent per pupil by school districts in the state of Pennsylvania vary?

Why has public support for the war in Iraq declined since the start of the war?

Does public support for war generally or always decline over time?

Why are some judges more protective of the rights of the accused than others?

Why do some nations have higher levels of human rights abuses than others?

Why does the cost of medical malpractice insurance vary among the states?

Why do some nations support setting specific targets for limiting carbon dioxide emissions, while others do not?

A research project will get off on the wrong foot if the question that shapes it fails to address a political phenomenon, is unduly concerned with discrete facts, or is focused on reaching normative conclusions. Although the definition of political phenomena is vague, it does not include the study of all human characteristics or behavior.

Research questions, if they dwell on discrete or narrow factual issues, may limit the significance of a research project. Although important, facts alone are not enough to yield scientific explanations. What is missing is a **relationship**— that is, the association, dependence, or covariance of the values of one variable with the values of another. Researchers are generally interested in how to advance and test generalizations relating one phenomenon to another. In the absence of such generalizations, factual knowledge of the type called for by the following research questions will be fundamentally limited in scope:

How many seats in the most recent state legislative elections in your state were uncontested (had only one contestant)?

How many states passed budgets last year that were more than 10 percent lower than the previous year's?

How many members of Congress had favorable environmental voting records in the last session of Congress?

How many trade disputes have been referred to the World Trade Organization (WTO) for resolution in the past five years?

What percentage of registered voters voted in the most recent U.S. Senate elections?

How many cabinet members have been replaced in each of the past three presidential administrations?

Who were the ten largest contributors to the Democratic presidential primary candidates prior to the Iowa caucus? How much did they contribute?

How many people are opposed to affirmative action?

Factual information, however, may lead a researcher to ask "why?" questions. For example, if a researcher has information about the number of uncontested seats and notes that this number varies substantially from state to state, the research question "Why are legislative elections competitive in some states and not in others?" forms the basis of an interesting research project. Alternatively, if one had data from just one state, one could investigate the question "Why do some districts have competitive elections and not others?" This would involve identifying characteristics of districts and elections that might explain the difference.

Or someone might notice that the number of trade disputes referred to the WTO has varied from year to year. What explains this situation? In collecting data on the number of disputes, it might be noticed that the complaints originate in many different countries. It would be interesting then to find out how the disputes are resolved. Is there any pattern to their resolution in regard to which countries benefit or the principles and arguments underlying the decisions? Why? Similarly, the environmental voting records of members of Congress differ. Why? Is political party a likely explanation? Is ideology? Or is some other factor responsible?

Sometimes important research contributions come from descriptive or factual research because the factual information being sought is difficult to obtain or, as we discuss later in this chapter and in Chapter 4, disagreement exists over which information or facts should be used to measure a concept. In this situation a research effort will entail showing how different ways of measuring a concept have important consequences for establishing the facts.

Questions calling for normative conclusions also are inconsistent with the research methods discussed in this book. (Refer to Chapter 2 for the distinction between normative and empirical statements.) For example, questions such as "Should the United States give preference in reconstruction contracts to those nations that supported going to war in Iraq?" or "Should a new federal agency be placed within the Executive Office of the President?" or "Should states give tax breaks to new businesses willing to locate within their borders?" are important and suitable for the attention of political scientists (indeed, for any citizen), but they, too, are inappropriate as framed here. They ask for a normative response, seeking an indication of what is good or of what should be done. Although scientific knowledge may be helpful in answering questions like these, it cannot provide the answers without regard for an individual's personal values or preferences. What someone ultimately likes or dislikes, values or rejects, is involved in the answers to these questions.

Normative questions, however, may lead you to develop an empirical research question. For example, a student of one of the authors felt that Pennsylvania's method of selecting judges using partisan elections was not a good way to choose judges. To contribute to an informed discussion of this issue, she collected data on the amount of money raised and spent by judicial candidates, the amount of money spent per vote cast in judicial races compared with other state elections, and the voter turnout rate in judicial races compared with other races. This information spoke to some of the arguments raised against partisan judicial elections. She discovered that it was very difficult to collect empirical evidence to answer the interesting question of whether reliance on campaign contributions jeopardized the independence and impartiality of judges.

Students sometimes have difficulty formulating interesting and appropriate questions. What constitutes an appropriate research topic will vary, depending on the circumstances. Often the choices will be constrained by the content of a course for which a research paper is required. Choosing an appropriate research topic requires the investment of some time to familiarize oneself with the scope and substance of previous research. You should be prepared to cast a fairly broad net in looking for a topic; although some effort will be spent learning about topics that will not be chosen, the time is not wasted, and the reward is being able to select a topic that is closest to your interests.

In general, it is useful to submit your research question to the "so what?" test: Will the answer to it make a significant contribution to the accumulation of our understanding of and knowledge about political phenomena? Will it be useful for practitioners and policymakers? Will it provide an interesting test of a theory?

Where do the research questions of political scientists originate? There are many answers to this question. Some researchers become interested in a topic because of personal observation or experience. For example, a researcher who works for a candidate who loses a political campaign may wonder what factors are responsible for electoral success, and a researcher who fled her country of birth during a period of civil unrest may be drawn to conducting research on the causes of political disorder. Some researchers are drawn to a topic because of the research and writing of others. A scholar familiar with studies of congressional decision making may want to investigate the reasons for the success and failure of different public policy proposals. Still others select a research topic because of their interest in some broader social theory, as the researcher whose fascination with theories of rational decision making prompted the study of federal bureaucrats' behavior. Simi-

HOW TO COME UP WITH A RESEARCH TOPIC
1. Get started early.
2. Pose a "how many" question. Where possible, collect data for more than one time (e.g., year, election) or for more than one case (e.g., more than one city, state, nation, primary election). Do any patterns emerge? What might explain these patterns? Is it difficult to find information to answer your question? Why? Do you think that the ways in which other researchers have measured what you are interested in are adequate? Are there any validity or reliability problems with the measures? (Measurement validity and reliability are discussed in Chapter 4).
3. Find an assertion or statement in the popular press or a conclusion in a research article that you believe to be incorrect. Look for empirical evidence so that you can assess the statement or examine the evidence used by the author to see if any mistakes were made that could have affected the conclusion.
4. Find two studies that reach conflicting conclusions. Explain or try to reconcile the conflict.
5. Same as No. 1.

Note: We wish to thank one of our anonymous reviewers for suggesting that we include tips for coming up with paper topics and for suggesting these tips.

larly, researchers concerned in general with democratic theory often conduct research on what causes people to participate in politics. Finally, researchers select research topics for practical reasons: because grant money for a particular subject is available or because demonstrating expertise in a particular area will advance their professional career objectives.

Proposing Explanations

Once a researcher has developed a suitable research question or topic, the next step is to propose an explanation for the phenomenon the researcher is interested in understanding. Proposing an explanation involves identifying other phenomena that we think will help us account for the object of our research and then specifying how and why these two (or more) phenomena are related.

In the examples referred to in Chapter 1, the researchers proposed explanations for the political phenomena they were studying. David Bradley and his coauthors thought that the distribution of income among households in a nation would be affected by whether or not a leftist political party was in control of the government. Stephen C. Poe and C. Neal Tate investigated whether governments' violation of their citizens' human rights was related to rapid population growth, military regimes, colonial history, and level of economic development. B. Dan Wood and Richard W. Waterman investigated the activities of federal agencies to see if they changed in response to attempts by presidents and Congress to influence them. And Stephen D. Ansolabehere and his colleagues thought that voter turnout would be affected by the tone of campaign advertising.

A phenomenon that we think will help us explain the political characteristics or behavior that interests us is called an **independent variable.** Independent variables are the measurements of the phenomena that are thought to influence, affect, or cause some other phenomenon. A **dependent variable** is thought to be caused, to depend upon, or to be a function of an independent variable. Thus, if a researcher has hypothesized that acquiring more formal education will lead to increased income later on (in other words, that income may be explained by education), then years of formal education would be the independent variable and income would be the dependent variable. As the word *variable* connotes, we expect the value of the concepts we identify as variables to vary or change. A concept that does not change in value is called a constant and will not make a suitable phenomenon to investigate as part of the research process we focus on in this book. Unfortunately, sometimes a concept is expected to vary and thus be suitable for inclusion in a research project, only for a researcher to discover later on that the concept does not vary. For example, a student working on a survey to be distributed to her classmates wanted to see if students having served in the military or having a family member in the military had different attitudes toward the war in Iraq than students without military service connections. She discovered that none of the students had any military service connections: having military service connections was a constant.

Proposed explanations for political phenomena are often more complicated than the simple identification of one independent variable that is thought to explain a dependent variable. More than one phenomenon is usually needed to account adequately for most political behavior. For example, suppose a researcher proposes the following relationship between state efforts to regulate pollution and the severity of potential harm from pollution: the higher the threat of pollution (independent variable), the greater the effort to regulate pollution (dependent variable). The insightful researcher would realize the possibility that another phenomenon, such as the wealth of a state, might also affect a state's regulatory effort. The proposed explanation for state regulatory effort, then, would involve an alternative variable (wealth) in addition to the original independent variable. As another example, remember from Chapter 1 that Lane Kenworthy and Jonas Pontusson thought that larger changes in market inequality would cause larger changes in redistribution, but that changes in redistribution would also be affected by turnout rates in national elections. It is frequently desirable to compare the effect of each independent variable on the dependent variable. This is done by "controlling for" or holding constant one of the independent variables so that the effect of the other may be observed. This process is discussed in more detail in Chapters 12 and 13.

Sometimes researchers are also able to propose explanations for how the independent variables are related to each other. In particular, we might want to determine which independent variables come before other independent variables and indicate which ones have a more direct, as opposed to indirect, effect on the phenomenon we are trying to explain (the dependent variable). A variable that occurs prior to all other variables and that may affect other independent variables is called an **antecedent variable.** A variable that occurs closer in time to the dependent variable and is itself affected by other independent variables is called an **intervening variable.** The roles of antecedent and intervening variables in the explanation of the dependent variable differ significantly. Consider these examples.

Suppose a researcher hypothesizes that a person who favored national health insurance was more likely to have voted for John Kerry in 2004 than a person who did not favor such extensive coverage. In this case the attitude toward national health insurance would be the independent variable and the presidential vote the dependent variable. The researcher might wonder what causes the attitude toward national health insurance and might propose that those people who have inadequate medical insurance are more apt to favor national health insurance. This new variable (adequacy of a person's present medical insurance) would then be an antecedent variable, since it comes before and affects (we think) the independent variable. Thinking about antecedent variables pushes our explanatory scheme further back in time and, we hope, will lead to a more complete understanding of a particular phenomenon (in this case, presidential voting). Notice how the independent variable in the original hypothesis (attitude toward national health insurance) becomes the dependent variable in the hypothesis involving the antecedent variable (adequacy of health insurance). Also notice that in this example adequacy of health insurance is thought to exert an indirect effect on the dependent variable (presidential voting) via its impact on attitudes toward national health insurance.

Now consider a second example. Suppose a researcher hypothesizes that a voter's years of formal education affect her or his propensity to vote. In this case, education would be the independent variable and voter turnout the dependent variable. If the researcher then begins to think about what it is about education that has this effect, he or she has begun to identify the intervening variables between education and turnout. For example, the researcher might hypothesize that formal education creates or causes a sense of civic duty, which in turn encourages voter turnout, or that formal education causes an ability to understand the different issue positions of the candidates, which in turn causes voter turnout. Intervening variables come between an independent variable and a dependent variable and help explain the process by which one influences the other.

Explanatory schemes that involve numerous independent, alternative, antecedent, and intervening variables can become quite complex. An **arrow diagram** is a handy device for presenting and keeping track of such complicated explanations. The arrow diagram specifies the phenomena of interest; indicates which variables are independent, alternative, antecedent, intervening, and dependent; and shows which variables are thought to affect which other ones. In Figure 3-1 we present arrow diagrams for the two examples we just considered.

In both diagrams the dependent variable is placed at the end of the time line, with the independent, alternative, intervening, and antecedent variables placed in their appropriate locations to indicate which ones come first. Arrows indicate that one variable is thought to explain or be related to another; the direction of the arrow indicates which variable is independent and which is dependent in that proposed relationship.

Figure 3-2 shows two examples of arrow diagrams that have been proposed and tested by political scientists. Both diagrams are thought to explain presidential voting behavior. In the first diagram the ultimate dependent variable, Vote, is thought to be explained by Candidate Evaluations and Party

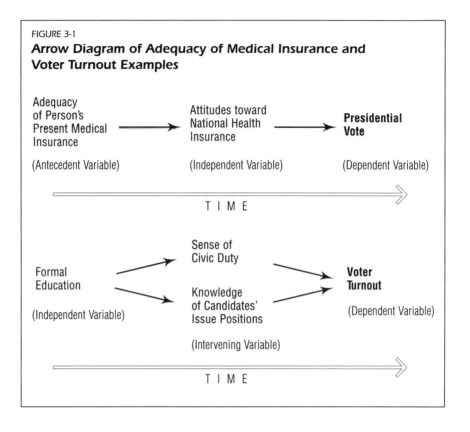

FIGURE 3-1

Arrow Diagram of Adequacy of Medical Insurance and Voter Turnout Examples

Identification. The Candidate Evaluations variable, in turn, is explained by the Issue Losses, Party Identification, and Perceived Candidate Personalities variables. These, in turn, are explained by other concepts in the diagram. The variables at the top of the diagram tend to be antecedent variables (the subscript $t - 1$ denotes that these variables precede variables with subscript t, where t indicates time); the ones in the center tend to be intervening variables. Nine independent variables of one sort or another figure in the explanation of the vote.

The second diagram also has Vote as the ultimate dependent variable, which is explained directly by only one independent variable, Comparative Candidate Evaluations. The latter variable, in turn, is dependent upon six independent variables: Personal Qualities Evaluations, Comparative Policy Distances, Current Party Attachment, Region, Religion, and Partisan Voting History. In this diagram sixteen variables figure, either indirectly or directly, in the explanation of the Vote variable, with the antecedent variables located around the perimeter of the diagram and the intervening variables closer to the center. Both of these diagrams clearly represent complicated and extensive attempts to explain a dependent variable.

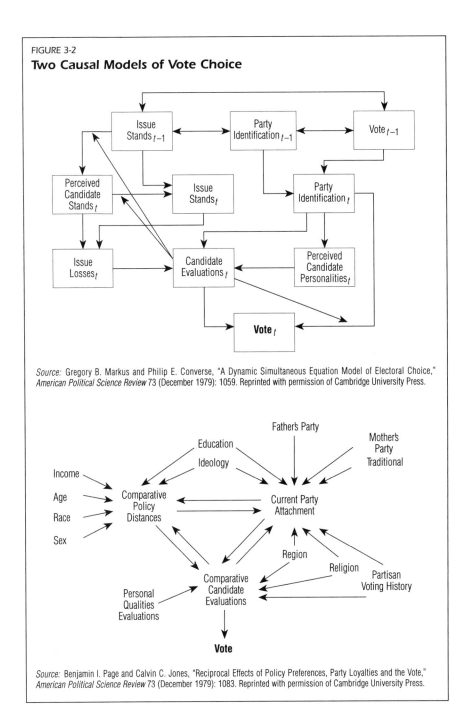

FIGURE 3-2

Two Causal Models of Vote Choice

Source: Gregory B. Markus and Philip E. Converse, "A Dynamic Simultaneous Equation Model of Electoral Choice," *American Political Science Review* 73 (December 1979): 1059. Reprinted with permission of Cambridge University Press.

Source: Benjamin I. Page and Calvin C. Jones, "Reciprocal Effects of Policy Preferences, Party Loyalties and the Vote," *American Political Science Review* 73 (December 1979): 1083. Reprinted with permission of Cambridge University Press.

Note that arrow diagrams show hypothesized causal relationships. A one-headed arrow connecting two variables is a shorthand way of expressing the proposition "*X* directly causes *Y.*" If arrows do not directly link two variables, they may be associated or correlated, but the relationship is indirect, not causal. As we discuss in greater depth in Chapter 5, when we assert *X* causes *Y,* we are in effect making three claims. One is that *X* and *Y* covary—a change in one variable is associated with a change in the other. Second, we are claiming that a change in the independent variable (*X*) *precedes* the change in the dependent variable (*Y*). Finally, we are stating that the covariation between *X* and *Y* is not simply a coincidence or spurious—that is, due to change in some other variable—but is direct.

We have discussed the first two steps in the research process—asking a question and then proposing an explanation—as occurring in this order, but quite often this is not the case. In Chapter 2 we pointed out that researchers might start out with a theory and make deductions based on it. Thus researchers often start with an explanation and look for an appropriate research question that the theory might answer. Theory is an important aspect of explanation, for in order to be able to argue effectively that something causes something else, we need to be able to supply a reason or, to use words from the natural sciences, to identify the *mechanism* behind the relationship. This is the role of theory.

Formulating Hypotheses

Thus far we have discussed two stages in the research process: identifying the research question and proposing explanations for the phenomena of interest. By this point, then, the researcher is ready to state what his or her hypotheses are. A **hypothesis** is an explicit statement that indicates how a researcher thinks the phenomena of interest are related. A hypothesis is a guess (but of an educated nature) that represents the proposed explanation for some phenomenon and that indicates how an independent variable is thought to affect, influence, or alter a dependent variable. Since hypotheses are proposed relationships, they may turn out to be incorrect.

Characteristics of Good Hypotheses

For a hypothesis to be tested adequately and persuasively, it must be stated properly. It is important to start a research project with a clearly stated hypothesis because it provides the foundation for subsequent decisions and steps in the research process. A poorly formulated hypothesis often indicates confusion about the relationship to be tested or can lead to mistakes that will limit the value or meaning of any findings. Many students find it quite challenging to write a hypothesis that precisely states the relationship to be tested: it takes practice to write consistently well-worded hypotheses. A good hypothesis has six characteristics: (1) It is an empirical statement, (2) it is stated as a generality, (3) it is plausible, (4) it is specific, (5) it is stated in a manner that corresponds to the way in which the researcher intends to test it, and (6) it is testable. The following discussion of these six characteristics will alert students to some common mistakes to avoid.

First, hypotheses should be empirical statements. They should be educated guesses about relationships that exist in the real world, not statements about what ought to be true or about what a researcher believes should be the case. Consider someone who is interested in democracy. If the researcher hypothesizes "Democracy is the best form of government," he or she has formulated a normative, nonempirical statement that cannot be tested. The statement communicates the preference of the researcher; it does not explain a phenomenon. By now, this researcher ought to have defined the central concept—in this case, democracy—and those concepts thought to be related to democracy (such as literacy, size of population, geographical isolation, and economic development). Therefore, to produce an acceptable hypothesis, the researcher ought to make an educated guess about the relationship between democracy and another of these concepts, for example: "Democracy is more likely to be found in countries with high literacy than in countries with low literacy." This hypothesis now proposes an explanation for a phenomenon that can be observed empirically. Or, one might think that democracy is preferable to other systems because it produces higher standards of living. We cannot prove that one thing is preferable to another, but we could certainly compare countries on numerous measures of well-being, such as health status. The conclusion might then be, "Compared with people living under dictatorships, citizens of democracies have higher life expectancies." Whether the hypothesis is confirmed empirically is not necessarily related to whether the researcher thinks the phenomenon (in this case, democracy) is good or bad.

In some cases, empirical knowledge can be relevant for normative inquiry. Often, people reach normative conclusions based on their evaluation of empirical relationships. Someone might reason, for example, that negative campaign ads cause voters to become disgusted with politics and not vote in elections, and because low turnout is bad, negative campaign ads are bad as well. The first part of the assertion is an empirical statement, which could be investigated using the techniques developed in this book, whereas the next two (low turnout and negative ads being bad) are normative statements.

Normative thinking is useful because it forces an individual to clarify his or her values, and it encourages research on significant empirical questions. For example, a normative distaste for crime encourages empirical research on the causes of crime or on the effectiveness of particular sentencing policies. Consequently, the two modes of inquiry—normative and empirical—should be viewed as complementary rather than contradictory.

A second characteristic of a good hypothesis is generality. It should explain a general phenomenon rather than one particular occurrence of the phenomenon. For example, one might hypothesize that the cause of World War II was economic upheaval in Germany. If the hypothesis were confirmed, what would be the extent of our knowledge? We would know the cause of one war. This knowledge is valuable, but it would be more useful to know if economic upheaval *in general* causes wars. That would be knowledge pertaining to many occurrences of a phenomenon (in this case, many wars), rather than knowledge about just one occurrence. A more general hypothesis, then, might be, "Countries experiencing economic upheaval are more likely to become involved in a war than countries not experiencing economic upheaval." Knowledge about the causes of particular occurrences of a phenomenon could be helpful in formulating more general guesses about the relationships between concepts, but with a general hypothesis we attempt to expand the scope of our knowledge beyond individual cases. Stating hypotheses in the plural form, rather than the singular, makes it clear that testing the hypothesis will involve more than one case.

The four hypotheses in the left column below are too narrow, whereas the four hypotheses in the right column are more general and more acceptable as research propositions:

Senator X voted for a bill because it is the president's bill and they are both Democrats.	Senators are more likely to vote for bills sponsored by the president if they belong to the same political party as the president.
The United States is a democracy because its population is affluent.	Countries with high levels of affluence are more likely to be democracies than countries with low levels of affluence.
The United States has more murders than other countries because so many people own guns there.	Countries with more guns per capita will experience more murders per capita than countries with fewer guns.
Joe is a liberal because his mother is one, too.	People tend to adopt political viewpoints similar to those of their parents.

A third characteristic of a good hypothesis is that it should be plausible. There should be some logical reason for thinking that it might be confirmed. Of course, since a hypothesis is a guess about a relationship, whether it will be confirmed cannot be known for certain. Any number of hypotheses could be thought of and tested, but many fewer are plausible ones. For example, if a researcher hypothesized that "People who eat dry cereal for breakfast are more likely to be liberal than people who eat eggs," we would question his or her logic even though the form of the hypothesis may be perfectly acceptable. It is difficult to imagine why this hypothesis would be confirmed.

But how do we make sure that a hypothesis has a good chance of being confirmed? Sometimes the justification is provided by specific instances in which the hypothesis was supported (going from specific to general knowledge in the manner discussed in Chapter 2—that is, using induction). For example, a researcher may have observed a particular election in which a hotly contested primary campaign damaged the eventual nominee's chances of winning the general election. The researcher may then have concluded that "The more difficult it is for candidates to secure their party's nomination, the more poorly those candidates will do in the general election."

And, as we pointed out earlier in our discussion of proposing explanations, a hypothesis also may be justified through the process of deduction. A researcher may deduce from more general theories that a particular hypothesis is sensible. For example, there is a general psychological theory that frustration leads to aggression. Some political scientists have adapted this general theory to the study of political violence or civil unrest and hypothesized that civil unrest occurs when a civilian population is frustrated. A population may feel frustrated when many people believe that they are economically or politically worse off than they should be, than they used to be, or than other people like themselves are. This feeling, as we now know, is called relative deprivation, and it has figured prominently in hypotheses seeking to explain civil unrest. In this way the general frustration-aggression theory led to a more specialized, deduced hypothesis for the occurrence of civil unrest.

Formulating plausible hypotheses is one of the reasons why researchers conduct a literature review early in their research projects. Literature reviews (discussed in more detail in Chapter 6) can acquaint researchers both with general theories and with specific hypotheses advanced by others. In either case, reading the literature on a subject can improve the chances that a hypothesis will be confirmed. There are no hard and fast rules to ensure plausibility, however. After all, people used to think that "germs cause diseases" was an implausible hypothesis and that "dirt may be turned into gold" was a plausible one.

The fourth characteristic of a good hypothesis is that it is specific. The researcher should be able to state a **directional hypothesis**—that is, he or she should be able to specify the expected relationship between two or more variables. Following are examples of directional hypotheses that specify the nature of the relationship between concepts:

> *Median family income is higher in urban counties than in rural counties.*

> *States that are characterized by a "moralistic" political culture will have higher levels of voter turnout than will states with an "individualistic" or "traditionalistic" political culture.*

The first hypothesis indicates which relative values of median family income are related to which type or category of county. Similarly, the second hypothesis predicts a particular relationship between specific types of political culture (the independent variable) and voter turnout (the dependent variable).

The direction of the relationship between concepts is referred to as a **positive relationship** if the concepts are predicted to increase in size together or decrease in size together. The following are examples of hypotheses that predict positive relationships:

> *The more education a person has, the higher his or her income.*

> *As the percentage of a country's population that is literate increases, the country's political process becomes more democratic.*

> *The older people become, the more likely they are to be conservative.*

> *People who read the newspaper more are more informed about current events than are people who read the newspaper less.*

> *The lower a state's per-capita income, the less money the state spends per pupil on education.*

If, however, the researcher thinks that as one concept increases in size or amount, another one will decrease in size or amount, then a **negative relationship** is suggested, as in the following examples:

> *Older people are less tolerant of social protest than are younger people.*

> *The more income a person has, the less concerned about mass transit the person will become.*

> *More affluent countries have less property crime than poorer countries.*

In addition, the concepts used in a hypothesis should be defined carefully. For example, a hypothesis that suggests "There is a relationship between personality and political attitudes" is far too ambiguous. What is meant by personality? Which political attitudes? A more specific reformulation of this hypothesis might be, "The more self-esteem a person has, the less likely the person is to be an isolationist." Now personality has been narrowed to self-esteem, and the political attitude has been defined as isolationism, both more precise concepts, although not precise enough. Eventually even these two terms must be given more precise definitions when it comes to measuring them. (We return to the problem of measuring concepts in Chapter 4.) As the concepts become more clearly defined, the researcher is better able to specify the direction of the hypothesized relationship.

Following are four examples of ambiguous hypotheses that have been made more specific:

How a person votes for president depends on the information he or she is exposed to.	The more information favoring candidate X a person is exposed to during a political campaign, the more likely that person is to vote for candidate X.
A country's geographical location matters for the type of political system it develops.	The more borders a country shares with other countries, the more likely that country is to have a non-democratic political process.
A person's capabilities affect his or her political attitudes.	The more intelligent a person is, the more likely he or she is to support civil liberties.
Guns do not cause crime.	People who own guns are less likely to be the victims of crimes than are persons who do not own guns.

A fifth characteristic of a good hypothesis is that it is stated in a manner that corresponds to the way in which the researcher intends to test it—that is, it should be "consistent with the data."[1] For example, although the hypothesis "Higher levels of literacy are associated with higher levels of democracy" does state how the concepts are related, it does not indicate how the researcher plans to test the hypothesis. In contrast, the hypothesis "As the percentage of a country's population that is literate increases, the country's political process becomes more democratic" suggests that the researcher is proposing to use a time series design by measuring the literacy rate and the amount of democracy for a country or countries at several different times to see if increases in democracy are associated with increases in literacy (that is, if changes in one concept lead to changes in another). If, however, the researcher plans to test the hypothesis by measuring the literacy rates and levels of democracy for many countries at one point in time to see if those with higher literacy rates also have higher levels of democracy, it would be better to rephrase the hypothesis as "Countries with higher literacy rates tend to be more democratic than countries with lower literacy rates." This way of phrasing the hypothesis reflects that the researcher is planning to use a cross-sectional research design to compare the levels of democracy in countries with different literacy rates. This differs from comparing a country's level of democracy at more than one point in time to see if it changes in concert with changes in literacy.

Finally, a good hypothesis is testable. It must be possible and feasible to obtain data that will indicate whether the hypothesis is defensible. Hypotheses for which either confirming or disconfirming evidence is impossible to gather are not subject to testing and hence are unusable for empirical purposes.

Consider this example of a promising yet untestable hypothesis: "The more a child is supportive of political authorities, the less likely that child will be to engage in political dissent as an adult." This hypothesis is general, plausible, fairly specific, and empirical, but in its current form it cannot be tested because no data exist to verify the proposition. The hypothesis requires data that measure a set of attitudes for individuals when they are children and a set of behaviors when they are adults. Survey data do exist that include the political attitudes and behavior of seventeen- and eighteen-year-olds and their parents in 1965 and many of the same people in 1973.[2] These data lack childhood measures for the parents, however, and for the others there are only late adolescent and early adulthood (mid-twenties) measures. Consequently, a frustrating practical barrier prevents the testing of an otherwise acceptable hypothesis. Students in one-semester college courses on research methods often run up against this constraint. A semester is not usually long enough to collect and analyze data, and some data may be too expensive to acquire. Many interesting hypotheses go untested simply because researchers do not have the resources to collect the data necessary to test them.

Hypotheses stated in tautological form are also untestable. A **tautology** is a statement linking two concepts that mean essentially the same thing: for example, "The less support there is for a country's political institutions, the more tenuous the stability of that country's political system." This hypothesis would be difficult to disconfirm because the two concepts—support for political institutions and stability of a political system—are so similar. To provide a fair test one would have to measure independently—in different ways—the support for the political institutions and the stability of the political system.

Poe and Tate's study of government maltreatment of citizens defined human rights abuses as coercive activities (such as murder, torture, forced disappearance, and imprisonment of persons for their political views) designed to induce compliance.[3] Other researchers have included lack of democratic processes and poor economic conditions in their definitions of human rights abuses, but Poe and Tate did not include these concepts because they wanted to use democratic rights and economic conditions as independent variables explaining variation in human rights abuses by governments.

There are many hypotheses, then, that are not formulated in a way that permits an informative test of them with empirical research. Readers of empirical research in political science, as well as researchers themselves, should take care that research hypotheses are empirical, general, plausible, specific, consistent with the data, and testable. Hypotheses that do not share these characteristics are likely to cause difficulty for the researcher and reader alike and make a minimal contribution to scientific knowledge.

Specifying Units of Analysis

In addition to proposing a relationship between two or more variables, a hypothesis also specifies the types or levels of political actor to which the hypothesis is thought to apply. This is called the **unit of analysis** of the hypothesis, and it also must be selected thoughtfully.

As noted in Chapter 2, political scientists are interested in understanding the behavior or properties of all sorts of political actors and events: individuals, groups, states, government agencies, organizations, regions, nations, elections, wars, conflicts. The particular type of actor whose political behavior is named in a hypothesis is the unit of analysis for the research project. In a legislative behavior study, for example, the individual members of the House of Representatives might be the units of analysis in the following hypothesis:

> *Members of the House who belong to the same party as the president are more likely to vote for legislation desired by the president than are members who belong to a different party.*

In the following hypothesis, a city is the unit of analysis, since attributes of cities are being explored:

> *Northeastern cities are more likely to have mayors, while western cities have city managers.*

Civil wars are the units of analysis in this hypothesis:

> *Civil wars that are halted by negotiated peace agreements are less likely to re-erupt than are those that cease due to the military superiority of one of the parties to the conflict.*

Elections are the unit of analysis in this example:

> *Elections in which the contestants spend the same amount of money tend to be decided by closer margins of victory than elections in which one candidate spends a lot more than the other candidate(s).*

Finally, consider this proposition:

> *The more affluent a country is, the more likely it is to have democratic political institutions.*

Here the unit of analysis is the country. It is the measurement of national characteristics—affluence (the independent variable) and democratic political institutions (the dependent variable)—that are relevant to testing this hypothesis. In sum, the research hypothesis indicates the researcher's unit of analysis and the behavior or attributes that must be measured for that unit.

Cross-level Analysis: Ecological Inference and Ecological Fallacy

Sometimes researchers conduct what is called **cross-level analysis.** In this type of analysis, researchers use data collected for one unit of analysis to make inferences about another unit of analysis. Christopher H. Achen and W. Phillips Shively point out that "[f]or reasons of cost or availability, theories and descriptions referring to one level of aggregation are frequently testable only with data from another level."[4] A discrepancy between the unit of analysis specified in a hypothesis and the entities whose behavior is empirically observed can cause problems, however.

A frequent goal of cross-level analysis is making **ecological inference,** the use of aggregate data to study the behavior of individuals.[5] Data of many kinds are collected for school districts, voting districts, counties, states, nations, or other aggregates in order to make inferences about individuals. The relationship between schools' average test scores and the percentage of children receiving subsidized lunches, national poverty and child mortality rates, air pollution indexes and the incidence of disease in cities, and the severity of state criminal penalties and crime rates are examples of relationships explored using aggregate data. The underlying hypotheses of such studies are that children who receive subsidized lunches score lower on standardized tests, that poor children are more likely to die of childhood diseases, that individuals' health problems are due to their exposure to air pollutants, and that harsh penalties deter individuals from committing crimes. Yet, if a relationship is found between group indicators or characteristics, it does not necessarily mean that a relationship exists between the characteristics for individuals in the group. Using information that shows a relationship for groups to infer that the same relationship exists for individuals when in fact there is no such relationship at the individual level is called an **ecological fallacy.**

Let's take a look at an example to see how an ecological fallacy might be committed as a result of failing to be clear about the unit of analysis. Suppose a researcher wants to test the hypothesis "African Americans are more likely to support female candidates than are Italian Americans." Individuals are the unit of analysis in this hypothesis. If the researcher selects an election with a female candidate and obtains the voting returns as well as data on the proportions of African Americans and Italian Americans in each election precinct, the data are aggregate data, not data on individual voters. If it is found that female candidates received more votes in precincts with a higher proportion of African Americans than in the precincts with a higher proportion of Italian Americans, the researcher might take this as evidence in support of the hypothesis. There is a fundamental problem with this conclusion, however. Unless a district is 100 percent African American or 100 percent Italian American, the researcher cannot necessarily draw such a conclusion about the behavior of individuals from the behavior of election districts. It could be that a female candidate's support in a district with a high proportion of African American voters came mostly from non–African Americans, and that most of the female candidate's votes in the Italian American districts came from Italian Americans. If this is the case, then the researcher would have committed an ecological fallacy. What is true at the aggregate level is not true at the individual level.

Let us take two hypothetical election precincts to illustrate how this fallacy could occur. Suppose we have Precinct 1, classified as an "African American" district, and Precinct 2, an "Italian American" district. If the African American district voted 67 percent to 33 percent in favor of the female candidate, and the Italian American district voted 53 percent to 47 percent in favor of the female candidate, we might be tempted to conclude that African Americans as individuals voted more heavily for the female candidate than did Italian Americans.

But imagine we peek inside each of the election precincts to see how individuals of different ethnicities behaved; that is, suppose we obtain information about individuals within the districts. The data in Table 3-1 show that in the African American district, African Americans split 25–25 for the woman, Italian Americans voted 18–2 for her, and others voted 24–6 for her. This resulted in the 67–33 percent edge for the woman in Precinct 1. In the Italian American district, Precinct 2, African Americans voted 16–24 against the woman, Italian Americans split 30–20 for her, and others voted 7–3 in her favor. This resulted in the 53–47 percent margin for the woman in Precinct 2. When we compare the percentage of African Americans, Italian Americans, and others voting for the female candidate, the difference in the voting behavior of the ethnic groups becomes clearer. In both precincts, the percentage of African Americans voting for the female candidate was lower than that of the two other groups of voters. In Precinct 1, 50 percent of the African Americans voted for the female candidate, compared with 90 percent of the Italian American voters and 80 percent of the

TABLE 3-1

Voting by African Americans, Italian Americans, and Others for a Female Candidate

Ethnicity	Raw Vote			Percent Vote	
	Number	For Male	For Female	For Male	For Female
Precinct 1					
African Americans	50	25	25	50.0	50.0
Italian Americans	20	2	18	10.0	90.0
Other	30	6	24	20.0	80.0
Total	100	33	67	33.0	67.0
Precinct 2					
African Americans	40	24	16	60.0	40.0
Italian Americans	50	20	30	40.0	60.0
Other	10	3	7	30.0	70.0
Total	100	47	53	47.0	53.0
Voting of Individuals					
African Americans	90	49	41	54.4	45.6
Italian Americans	70	22	48	31.4	68.6
Other	40	9	31	22.5	77.5
Total	200	80	120	40.0	60.0

Note: Hypothetical data.

others. In Precinct 2, only 40 percent of the African Americans voted for the female candidate, compared with 60 percent of the Italian Americans and 70 percent of the other voters. In other words, Italian Americans as individuals were more likely to have voted for the woman candidate than were African Americans as individuals in both precincts. Knowing only the precinct-level totals gave the opposite impression. When the results for both districts are combined and broken down by ethnicity, we see that, overall, 68.6 percent of Italian Americans and 45.6 percent of African Americans voted for the female candidate.

In the research by Ansolabehere and his colleagues reported on in Chapter 1, the tone of campaign advertising and the roll-off rates were measured in thirty-four Senate races, and states with races characterized by a negative tone had higher roll-off rates than states with positive campaigns. The inference is that those individuals exposed to negative campaign ads are less likely to vote than are those exposed to positive campaign ads. But the researchers lacked data that showed the relationship between actual exposure to campaign ads of individuals and their voting behavior in the Senate elections. Remember, however, that the researchers examined and reported on individual-level data obtained from experiments, so they did not rely just on aggregate data to test their hypotheses about individuals. Use of aggregate data to examine hypotheses that pertain to individuals may be unavoidable in some situations because individual-level data are lacking. Achen and Shively point out that before the development of survey research, aggregate data generally were the only data available and were used routinely by political scientists.[6] Several statistical methods have been developed to try to adjust inferences from aggregate-level data, although a discussion of these is beyond the scope of this book.[7]

Another mistake researchers sometimes make is to mix different units of analysis in the same hypothesis. "The more education a person has, the more democratic his country is" doesn't make much sense because it mixes the individual and country as units of analysis. However, "The smaller a government agency, the happier its workers" concerns an attribute of an agency and an attribute of individuals, but in a way that makes sense. The size of the agency in which individuals work may be an important aspect of the context or environment in which the individual phenomenon occurs and may influence the individual attribute. In this case the unit of analysis is clearly the individual, but a phenomenon that is experienced by many cases is used to explain the behavior of individuals, some of whom may well be identically situated.

In short, a researcher must be careful about the unit of analysis specified in a hypothesis and its correspondence with the unit measured. In general, a researcher should not mix units of analysis within a hypothesis.

Defining Concepts

Clear definitions of the concepts of interest to us are important if we are to develop specific hypotheses and avoid tautologies. Clear definitions also are important so that the knowledge we acquire from testing our hypotheses is transmissible and empirical.

Political scientists are interested in why people or social groupings (organizations, political parties, legislatures, states, countries) behave in a certain way or have particular attributes or properties. The words that we choose to describe these behaviors or attributes are called concepts. Concepts should be accurate, precise, and informative.

In our daily life we use concepts frequently to name and describe features of our environment. For example, we describe some snakes as poisonous and others as nonpoisonous, some politicians as liberal and others as conservative, some friends as shy and others as extroverted. These attributes, or concepts, are useful to us because they help us observe and understand aspects of our environment, and they help us communicate with others.

Concepts also contribute to the identification and delineation of the scientific disciplines within which research is conducted. In fact, to a large extent a discipline maintains its identity because different researchers within it share a concern for the same concepts. Physics, for example, is concerned with the concepts of gravity and mass (among others); sociology, with social class and social mobility; psychology, with personality and deviance. By contrast, political science is concerned with concepts such as democracy, power, representation, justice, and equality. The boundaries of disciplines are not well defined or rigid, however. Political scientists, developmental psychologists, sociologists, and anthropologists all share an interest in how new members of a society are socialized into the norms and beliefs of that society, for example. Nonetheless, because a particular discipline has some minimal level of shared consensus concerning its significant concepts, researchers can usually communicate more readily with other researchers in the same discipline than with researchers in other disciplines.

A shared consensus over those concepts thought to be significant is related directly to the development of theories. Thus a theory of politics will identify significant concepts and suggest why they are central to an understanding of political phenomena. Concepts are developed through a process by which some human group (tribe, nation, culture, profession) agrees to give a phenomenon or property a particular name. The process is ongoing and somewhat arbitrary and does not ensure that all peoples everywhere will give the same phenomena the same names. In some areas of the United States, for example, a *soda* is a carbonated beverage, while in other areas it is a drink with ice cream in it. Likewise, the English language has only one word for *love,* whereas the Greeks have three words to distinguish between romantic love, familial love, and generalized feelings of affection.[8] Concepts disappear from a group's language when they are no longer needed, and new ones are invented as new phenomena are noticed that require names (for example, computer *programs* and *software, cultural imperialism,* and *hyperkinetic* behavior).

Some concepts—such as *car, chair,* and *vote*—are fairly precise because there is considerable agreement about their meaning. Others are more abstract and lend themselves to differing definitions—for example, *liberalism, crime, democracy, equal opportunity, human rights, social mobility,* and *alienation.* A similar concept is *orange.* Although there is considerable agreement about it (orange is not usually confused with purple), the agreement is less than total (whether a particular object is orange or red is not always clear). Or when orange blends into red and ceases to be orange.

Many interesting concepts that political scientists deal with are abstract and lack a completely precise, shared meaning. This hinders communication concerning research and creates uncertainty regarding the measurement of a phenomenon. Consequently, a researcher must explain what is meant by the concept so that a measurement strategy may be developed and so that those reading and evaluating the research can decide if the meaning accords with their own understanding of the term. Although some concepts that political scientists use—such as *amount of formal education, presidential vote,* and *amount of foreign trade*—are not particularly abstract, other concepts—such as *partisan realignment, political integration,* and *regime support*—are far more abstract and need more careful consideration and definition.

Suppose, for example, that a researcher is interested in the kinds of political systems that different countries have and, in particular, why some countries are more democratic than others. *Democracy* is consequently a key concept and one that needs definition and measurement. The word contains meaning for most of us; that is, we have some idea what is democratic and what is not. But once we begin thinking about the concept, we quickly realize that it is not as clear as we thought originally. To some, a country is democratic if it has "competing political parties, operating in free elections, with some reasonable level of popular participation in the process."[9] To others, a country is democratic only if legal guarantees protect free speech, the press, religion, and the like. To others, a country is democratic if the political leaders make decisions that are acceptable to the populace. And to still others, democracy implies equality of economic opportunity among the citizenry. If a country has all these attributes, it would be called a democracy by any of the criteria and there would be no problem classifying the country. But if a country possesses only one of these attributes, its classification would be uncertain, since by some definitions it would be democratic but by others it would not be. Different definitions require different measurements and may result in different research findings. Hence, defining one's concepts is important, particularly when the concept is so abstract as to make shared agreement difficult.

Concept definitions have a direct impact on the quality of knowledge produced by research studies. Suppose, for example, that a researcher is interested in the connection between economic development and democracy, the working hypothesis being that countries with a high level of economic development will be more likely to have democratic forms of government. And suppose that there are two definitions of economic development and two definitions of democracy that might be used in the research. Finally, suppose that the researcher has data on twelve countries (A–L) included in the study. In Table 3-2 we show that the definition selected for each concept has a direct bearing on how different countries are categorized on each attribute. By definition 1, countries A, B, C, D, E, and F are economically developed; however, by definition 2, countries A, B, C, G, H, and I are. By definition 1, countries A, B, C, D, E, and F are democracies; by definition 2, countries D, E, F, J, K, and L are.

This is only the beginning of our troubles, however. When we look for a pattern involving the economic development and democracy of countries, we find that our answer depends mightily on how we have defined the two concepts. If we use the first definitions of the two concepts, we find that all economically

TABLE 3-2

Concept Development: The Relationship between Economic Development and Democracy

Is the country economically developed?

		By definition 1:	
		Yes	No
By definition 2:	Yes	A,B,C	G,H,I
	No	D,E,F	J,K,L

Is the country a democracy?

		By definition 1:	
		Yes	No
By definition 2:	Yes	D,E,F	J,K,L
	No	A,B,C	G,H,I

developed countries are also democracies (A, B, C, D, E, F), which supports our hypothesis. If we use the first definition for economic development and the second for democracy (or vice versa), half of the economically developed nations are democracies and half are not. If we use the second definitions of both concepts, none of the economically developed countries is a democracy, whereas all of the undeveloped countries are (D, E, F, J, K, L). In other words, because of our inability to formulate a precise definition of the two concepts, and because the two definitions of each concept yield quite different categorizations of the twelve countries, our hypothesis could be either confirmed or disconfirmed by the data at hand. Our conceptual confusion has put us in a difficult position.

Consider another example. Suppose a researcher is interested in why some people are liberal and some are not. In this case we need to define what is meant by liberal so that those who are liberal can be identified. *Liberal* is a frequently used term, but it has many different meanings: one who favors change, one who favors redistributive income or social welfare policies, one who favors increased government spending and taxation, or one who opposes government interference in the political activities of its citizens. If a person possesses all these attributes, there is no problem deciding whether or not he or she is a liberal. A problem arises, however, when a person possesses some of these attributes but not others.

The examples here illustrate the elusive nature of concepts and the need to define them. The empirical researcher's responsibility to define terms is a necessary and challenging one. Unfortunately, many of the concepts used by political science researchers are fairly abstract and require careful thought and extensive elaboration.

Researchers can clarify the concept definitions they use simply by making the meanings of key concepts explicit. This requires researchers to think carefully about the concepts used in their research and to share their meanings with others. Other researchers often challenge concept definitions, requiring researchers to elaborate upon and justify their meanings.

Another way in which researchers get help defining concepts is by reviewing and borrowing (possibly with modification) definitions developed by others in the field. This is one of the reasons why researchers conduct literature reviews of pertinent research, a task we take up in detail in Chapter 6. For example, a researcher interested in the political attitudes and behavior of the American public would find the following definitions of key concepts in the existing literature:

> *Political participation.* Those activities by private citizens that are more or less directly aimed at influencing the selection of government personnel, the actions they take, or both.[10]

> *Political violence.* All collective attacks within a political community against the political regime, its actors—including competing political groups as well as incumbents—or its policies.[11]

> *Political efficacy.* The feeling that individual political action does have, or can have, an impact upon the political processes—that it is worthwhile to perform one's civic duties.[12]

> *Belief system.* A configuration of ideas and attitudes in which the elements are bound together by some form of constraint or functional interdependence.[13]

Each of these concepts is somewhat vague and lacks complete shared agreement about its meaning. Furthermore, it is possible to raise questions about each of these concept definitions. Notice, for example, that the definition of *political participation* excludes the possibility that government employees (presumably "nonprivate" citizens) engage in political activities, and that the definition of *political efficacy* excludes the impact of collective political action on political processes. Consequently, we may find these and other concept definitions inadequate and revise them to capture more accurately what we mean by the terms.

Over time a discipline cannot proceed very far unless some minimal agreement is reached about the meanings of the concepts with which scientific research is concerned. Researchers must take care to think about the phenomena named in a research project and make explicit the meanings of any problematic concepts.

Conclusion

In this chapter we discussed the beginning stages of a scientific research project. A research project must provide—to both the producer and the consumer of social scientific knowledge—the answers to these important questions: What phenomenon is the researcher trying to understand and explain? What explanation has the researcher proposed for the political behavior or attributes in question? What are the meanings of the concepts used in this explanation? What specific hypothesis relating two or more variables will be tested? What is the unit of analysis for the observations? If these questions are answered adequately, then the research will have a firm foundation.

Notes

1. This term is used by Susan Ann Kay, *Introduction to the Analysis of Political Data* (Englewood Cliffs, N.J.: Prentice-Hall, 1991), 6.
2. For a description of this data set, see M. Kent Jennings and Richard G. Niemi, *Generations and Politics* (Princeton: Princeton University Press, 1981).
3. Steven C. Poe and C. Neal Tate, "Repression of Human Rights to Personal Integrity in the 1980s: A Global Analysis," *American Political Science Review* 88 (December 1994): 853–872.
4. Christopher H. Achen and W. Phillips Shively, *Cross-Level Inference* (Chicago: University of Chicago Press, 1995), 4.
5. Ibid.
6. Ibid., 5–10.
7. For example, see Gary King, *A Solution to the Ecological Inference Problem* (Princeton: Princeton University Press, 1997); Achen and Shively, *Cross-Level Inference;* and Barry C. Burden and David C. Kimball, *Why Americans Split Their Tickets: Campaigns, Competition, and Divided Government* (Ann Arbor: University of Michigan Press, 2002), chap. 3.
8. Kenneth R. Hoover, *The Elements of Social Scientific Thinking* (New York: St. Martin's, 1980), 18–19.
9. W. Phillips Shively, *The Craft of Political Research* (Englewood Cliffs, N.J.: Prentice-Hall, 1980), 33.
10. Sidney Verba and Norman H. Nie, *Participation in America* (New York: Harper and Row, 1972), 2.
11. Ted Robert Gurr, *Why Men Rebel* (Princeton: Princeton University Press, 1970), 3–4.
12. Angus Campbell, Gerald Gurin, and Warren E. Miller, *The Voter Decides* (Evanston, Ill.: Row, Peterson, 1954), 187.
13. Philip E. Converse, "The Nature of Belief Systems in Mass Publics," in David E. Apter, ed., *Ideology and Discontent* (New York: Free Press, 1964), 207.

Terms Introduced

ANTECEDENT VARIABLE. An independent variable that precedes other independent variables in time.

ARROW DIAGRAM. A pictorial representation of a researcher's explanatory scheme.

CROSS-LEVEL ANALYSIS. The use of data at one level of aggregation to make inferences at another level of aggregation.

DEPENDENT VARIABLE. The phenomenon thought to be influenced, affected, or caused by some other phenomenon.

DIRECTIONAL HYPOTHESIS. A hypothesis that specifies the expected relationship between two or more variables.

ECOLOGICAL FALLACY. The fallacy of deducing a false relationship between the attributes or behavior of individuals based on observing that relationship for groups to which the individuals belong.

ECOLOGICAL INFERENCE. The process of inferring a relationship between characteristics of individuals based on group or aggregate data.

HYPOTHESIS. A tentative or provisional or unconfirmed statement that can (in principle) be verified.

INDEPENDENT VARIABLE. The phenomenon thought to influence, affect, or cause some other phenomenon.

INTERVENING VARIABLE. A variable coming between an independent variable and a dependent variable in an explanatory scheme.

NEGATIVE RELATIONSHIP. A relationship in which the values of one variable increase as the values of another variable decrease.

POSITIVE RELATIONSHIP. A relationship in which the values of one variable increase (or decrease) as the values of another variable increase (or decrease).

RELATIONSHIP. The association, dependence, or covariance of the values of one variable with the values of another variable.

TAUTOLOGY. A hypothesis in which the independent and dependent variables are identical, making it impossible to disconfirm.

UNIT OF ANALYSIS. The type of actor (individual, group, institution, nation) specified in a researcher's hypothesis.

Suggested Readings

Achen, Christopher H., and W. Phillips Shively. *Cross-Level Inference.* Chicago: University of Chicago Press, 1995.

King, Gary. *A Solution to the Ecological Inference Problem.* Princeton: Princeton University Press, 1997.

CHAPTER 4

The Building Blocks of Social Scientific Research:

Measurement

In the previous chapters we discussed the beginning stages of political science research projects: the choice of research topics, the formulation of scientific explanations, the development of testable hypotheses, and the definition of concepts. In this chapter we take the next step toward testing empirically the hypotheses we have advanced. This entails understanding some issues involving the **measurement,** or systematic observation and representation by scores or numerals, of the variables we have decided to investigate.

In Chapter 2 we said that scientific knowledge is based on empirical research. In this chapter we confront the implications of this fact. If we are to test empirically the accuracy and utility of a scientific explanation for a political phenomenon, we will have to observe and measure the presence of the concepts we are using to understand that phenomenon. Furthermore, if this test is to be adequate, our measurements of the political phenomenon must be as accurate and precise as possible. The process of measurement is important because it provides the bridge between our proposed explanations and the empirical world they are supposed to explain. How researchers measure their concepts can have a significant impact on their findings, even leading to totally different conclusions.

Lane Kenworthy and Jonas Pontusson's investigation of income inequality in affluent countries illustrates well the impact on research findings of how a concept is measured.[1] One way to measure income is to look at the earnings of full-time employed individuals and to compare the incomes of those at the top and the bottom of the earnings distribution. Kenworthy and Pontusson argue that it is more appropriate to compare the incomes of households rather than individuals. The unemployed are excluded from the calculations of individual earnings inequality, but households will include the unemployed. Also, low-income workers disproportionately drop out of the employed labor force. Using working-age household income will reflect changes in employment among household members. Kenworthy and Pontusson found that when individual income was used as a basis for measuring inequality, inequality had increased the most in the United States, New Zealand, and the United Kingdom, all liberal market economies, and significantly more than in Europe's social market economies and Japan. When household income was used, the data indicated that inequality had increased in all countries with the exception of the Netherlands.

Another example involves the measurement of turnout rates (discussed in Chapter 1). Political scientists have investigated whether turnout rates in the United States have declined in recent decades.[2] The answer may depend on how the number of eligible voters is measured. Should it be the number of all citizens of voting age, or should this number be adjusted to take into account those who are not eligible to vote, or should the turnout rate be calculated using just the number of registered voters as the potential voting population?

The researchers discussed in Chapter 1 measured a variety of political phenomena, some of which posed greater challenges than others. Steven C. Poe and C. Neal Tate measured the presence of democratic procedures and protections, leftist or military regimes, population and economic growth, and international and civil war experience and the incidence of state terrorism.[3] Investigating the relationship between the party in control of government and income redistribution, David Bradley and colleagues measured these and other concepts for fourteen postindustrial democratic countries.[4]

Jeffrey A. Segal and Albert D. Cover measured both the political ideologies and the written opinions of Supreme Court justices in cases involving civil rights and liberties.[5] Valerie J. Hoekstra measured peoples' opinions about issues connected to U.S. Supreme Court cases and their opinions about the Court.[6] B. Dan Wood and Richard W. Waterman measured the decisions of several bureaucratic agencies to determine if they were influenced by presidential and congressional intervention.[7] David Lewis measured party control of the presidency and Congress, the size of the majority in Congress, and the design of new agencies.[8] And Stephen Ansolabehere, Shanto Iyengar, Adam Simon, and Nicholas Valentino measured the intention to vote reported by study participants to see if it was affected by exposure to negative campaign advertising.[9] In each case, some political behavior or attribute was measured so that a scientific explanation could be tested. All of these researchers made important choices regarding their measurements.

Devising Measurement Strategies

As we pointed out in Chapter 3, researchers must define the concepts they use in their hypotheses. They also must decide how to measure the presence, absence, or amount of these concepts in the real world. Political scientists refer to this process as providing an **operational definition** of their concepts—in other words, deciding what kinds of empirical observations should be made to measure the occurrence of an attribute or a behavior.

Let us return, for example, to the researcher trying to explain the existence of democracy in different nations. If the researcher were to hypothesize that higher rates of literacy make democracy more likely, then a definition of two concepts—literacy and democracy—would be necessary. The researcher could then develop a strategy, based on these two definitions, for measuring the existence and amount of both attributes in nations.

Suppose literacy was defined as "the completion of six years of formal education" and democracy was defined as "a system of government in which public officials are selected in competitive elections." These definitions would then be used to develop operational definitions of the two concepts. These operational definitions would indicate what should be observed empirically to measure both literacy and democracy, and they would indicate specifically what data should be collected to test the researcher's hypothesis. In this example, the operational definition of literacy might be "those nations in which at least 50 percent of the population has had six years of formal education, as indicated in a publication of the United Nations," and the operational definition of democracy might be "those countries in which the second-place finisher in elections for the chief executive office has received at least 25 percent of the vote at least once in the past eight years."

When a researcher specifies a concept's operational definition, the concept's precise meaning in a particular research study becomes clear. In the preceding example, we now know exactly what the researcher means by literacy and democracy. Since different people often mean different things by the same concept, operational definitions are especially important. Someone might argue that defining literacy in terms of formal education ignores the possibility that people who complete six years of formal education might still be unable to read or write well. Similarly, it might be argued that defining democracy in terms of competitive elections ignores other important features of democracy, such as freedom of expression and citizen involvement in government actions. In addition, the operational definition of competitive elections is clearly debatable. Is the "competitiveness" of elections based on the number of competing candidates, the size of the margin of victory, or the number of consecutive victories by a single party in a series of elections? Unfortunately, operational definitions are seldom absolutely correct or absolutely incorrect; rather, they are evaluated according to how well they correspond to the concepts they are meant to measure.

It is useful to think of the operational definition as the last stage in the process of defining a concept precisely. We often begin with an abstract concept (such as democracy), then attempt to define it in a meaningful way, and finally decide in specific terms how we are going to measure it. At the end of this process we hope to attain a definition that is sensible, close to our meaning of the concept, and exact in what it tells us about how to go about measuring the concept.

Let us consider another example: the researcher interested in why some individuals are more liberal than others. The concept of liberalism might be defined as "believing that government ought to pursue policies that provide benefits for the less well-off." The task then is to develop an operational definition that can be used to measure whether particular individuals are liberal or not. The following question from the General Social Survey might be used to operationalize the concept:

> 73A. *Some people think that the government in Washington ought to reduce the income differences between the rich and the poor, perhaps by raising the taxes of wealthy families or by giving income assistance to the poor. Others think that the government should not concern itself with reducing this income difference between the rich and the poor.*
>
> *Here is a card with a scale from 1 to 7. Think of a score of 1 as meaning that the government ought to reduce the income differences between rich and poor, and a score of 7 as meaning that the government should not concern itself with reducing income differences. What score between 1 and 7 comes closest to the way you feel? (CIRCLE ONE)*[10]

An abstract concept, liberalism has now been given an operational definition that can be used to measure the concept for individuals. This definition is also related to the original definition of the concept, and it indicates precisely what observations need to be made. It is not, however, the only operational definition possible. Others might suggest that questions regarding affirmative action, school vouchers, the death penalty, welfare benefits, and pornography could be used to measure liberalism. The important thing is to think carefully about the operational definition you choose and try to ensure that the definition coincides closely with the meaning of the original concept. How a concept is operationalized affects how generalizations are made and interpreted. For example, general statements about liberals or conservatives apply to liberals or conservatives only as they have been operationally defined, in this case by this one question regarding government involvement in reducing income differences. As a consumer of research, you should familiarize yourself with the operational definitions used by researchers so that you are better able to interpret and generalize research results.

Examples of Political Measurements: Getting to Operationalization

Let us take a closer look at some operational definitions used by political science researchers referred to in Chapter 1. Wood and Waterman set out to measure "mechanisms of political control" and "bureaucratic responses." They identified events that would represent "political signals that should produce 'anticipative responses.' "[11] The events that represented mechanisms of control included political appointments, resignations, budget increases and decreases, congressional oversight hearings, administrative reorganizations, and legislation. Measures of bureaucratic responses or agency outputs included litigations, sanctions, and administrative decisions. They chose to focus on regulatory agencies because the outputs of regulatory programs are easy to measure. Thus the meaning of "mechanisms of political control" and "bureaucratic responses" became clearer as the researchers specified how they planned to measure these concepts.

The research conducted by Segal and Cover on the behavior of U.S. Supreme Court justices is a good example of an attempt to overcome a serious measurement problem to test a scientific hypothesis.[12] Recall that Segal and Cover were interested, as many others have been before them, in the extent to which the votes cast by Supreme Court justices were dependent on their own personal political attitudes. Measuring the justices' votes on the cases decided by the Supreme Court is no problem; the votes are public information. But measuring the personal political attitudes of judges, *independent of their votes,* is a problem (remember the discussion in Chapter 3 on avoiding tautologies, or statements that link two concepts that mean essentially the same thing). Many of the judges whose behavior is of interest have died, and it is difficult to get living Supreme Court justices to reveal their political attitudes through personal interviews or questionnaires. Furthermore, ideally one would like a measure of attitudes that is comparable across many judges and that measures attitudes related to the cases decided by the Court.

Segal and Cover limited their inquiry to votes on civil liberties cases between 1953 and 1987, so they needed a measure of related political attitudes for the judges serving on the Supreme Court over that same time period. They decided to infer the judges' attitudes from the newspaper editorials written about them in four major daily newspapers from the time each justice was appointed by the president until the justice's confirmation vote by the Senate. Trained analysts read the editorials and coded each paragraph for whether it asserted that a justice designate was liberal, moderate, or conservative (or if the paragraph was inapplicable) regarding "support for the rights of defendants in criminal cases, women and racial minorities in equality cases, and the individual against the government in privacy and First Amendment cases."[13]

They selected the editorials appearing in two liberal papers and in two conservative papers to produce a more accurate measure of judicial attitudes.

Because of practical barriers to ideal measurement, then, Segal and Cover had to rely on an indirect measure of judicial attitudes *as perceived by four newspapers* rather than on a measure of the attitudes themselves. Although this approach *may* have resulted in flawed measures, it also permitted the test of an interesting hypothesis about the behavior of Supreme Court justices that had not been tested previously. If the measures that resulted were both accurate and precise, then this measurement strategy would permit the empirical verification of an important hypothesis. Without such measurements, the hypothesis could not have been tested.

Next, let us consider the research conducted by Bradley and his colleagues on the relationship between party control of government and the distribution and redistribution of wealth. Despite the importance of this research question, a lack of comparable data on an adequate number of cases had limited the exploration of this topic. For their investigation the researchers relied on improvements to the Luxembourg Income Study (LIS) database that provides cross-national income data over time in OECD (Organization for Economic Co-operation and Development) countries.[14] They decided, however, to make adjustments to published LIS data on income inequality that included pensioners. Because some countries make comprehensive provisions for retirees, retirees in these countries make little provision on their own for their retirement. Thus many of these people would be counted as poor before any government transfers. Including pensioners would inflate the pretransfer poverty level as well as the extent of income transfer for these countries. Bradley and his colleagues limited their analyses to households with a head aged twenty-five to fifty-nine (thus excluding the student-age population as well) and calculated their own measures of income inequality from the LIS data. They argued that their data would measure redistribution across income groups, not life-cycle redistributions of income, such as transfers to students and retired persons. Income was defined as income from wages and salaries, self-employment income, property income, and private pension income. They also made adjustments for household size using an equivalence scale, which adjusts the number of persons in a household to an equivalent number of adults. The equivalence scale takes into account the economies of scale resulting from sharing household expenses.

Researchers investigating the impact of exposure to negative campaign ads on elections used a variety of strategies to measure the tone of campaigns and campaign ads. Kim Fridkin Kahn and Patrick J. Kenney measured the tone of campaigns in Senate elections three ways: one was based on a sample of TV commercials from candidates' campaigns, the second on a sample of newspaper articles selected from the largest circulating newspaper in each state in which there was a senate election, and the third by contacting campaign managers and asking them to characterize the level of mudslinging in the campaigns. Each commercial was rated on a three-point scale, and a negativity score, representing the proportion of negative to positive messages associated with each race, was computed. This negativity score was multiplied by the amount of money spent during the race, with the assumption that the more money spent, the more people were exposed to the tone of the ads.[15] Martin P. Wattenberg and Craig Leonard Brians measured exposure by responses to a survey question that asked respondents if they recalled a campaign ad and whether or not it was negative or positive in tone.[16] Finally, Ansolabehere and his colleagues measured exposure to negative campaign ads in the 1990 Senate elections by accessing newspaper and magazine articles about the campaigns and determining how the tone of the campaigns was described in these articles.[17]

The cases discussed here are good examples of researchers' attempts to measure important political phenomena (behaviors or attributes) in the real world. Whether the phenomenon in question was judges' political attitudes, income inequality, the tone of campaign advertising, political control, or bureaucratic response, the researchers devised measurement strategies that could detect and measure the presence and amount of the concept in question. These observations were then generally used as the basis for an empirical test of the researchers' hypotheses.

To be useful in providing scientific explanations for political behavior, measurements of political phenomena must correspond closely to the original meaning of a researcher's concepts. They must also provide the researcher with enough information to make valuable comparisons and contrasts. Hence the quality of measurements is judged in regard to both their *accuracy* and their *precision*.

The Accuracy of Measurements

Because we are going to use our measurements to test whether or not our explanations for political phenomena are valid, those measurements must be as accurate as possible. Inaccurate measurements may lead to erroneous conclusions, since they will interfere with our ability to observe the actual relationship between two or more variables.

There are two major threats to the accuracy of measurements. Measures may be inaccurate because they are *unreliable* or because they are *invalid*.

Reliability

Reliability "concerns the extent to which an experiment, test, or any measuring procedure yields the same results on repeated trials. . . . The more consistent the results given by repeated measurements, the higher the reliability of the measuring procedure; conversely, the less consistent the results, the lower the reliability."[18]

Suppose, for example, you are given the responsibility of counting a stack of 1,000 paper ballots for some public office. The first time you count them, you obtain a particular result. But as you were counting the ballots you might have been interrupted, two or more ballots might have stuck together, some might have been blown onto the floor, or you might have written down the totals incorrectly. As a precaution, then, you count them five more times and get four other people to count them as well. The similarity of the results of all ten counts would be an indication of the reliability of the measure.

Or suppose you design a series of questions to measure how cynical people are and ask a group of people those questions. If a few days later you ask the same questions of the same group of people, the correspondence between the two measures would indicate the reliability of that particular measure of cynicism (assuming that the amount of cynicism has not changed). Similarly, suppose you wanted to test the hypothesis that the *New York Times* is more critical of the federal government than is the *Wall Street Journal*. This would require you to measure the level of criticism found in articles in the two papers. You would need to develop criteria or instructions for identifying or measuring criticism. The reliability of your measuring scheme could be assessed by having two people read all the articles, independently rate the level of criticism in them according to your instructions, and then compare their results. Reliability would be demonstrated if both people reached similar conclusions regarding the content of the articles in question.

The reliability of political science measures can be calculated in many different ways. We describe three methods here that are often associated with written test items or survey questions, but the ideas may be applied in other research contexts. The **test-retest method** involves applying the same "test" to the same observations after a period of time and then comparing the results of the different measurements. For example, if a series of questions measuring liberalism is asked of a group of respondents on two different days, a comparison of their scores at both times could be used as an indication of the reliability of the measure of liberalism. We frequently engage in test-retest behavior in our everyday lives. How often have you stepped on the bathroom scale twice in a matter of seconds?

The test-retest method of measuring reliability may be both difficult and problematic, since one must measure the phenomenon at two different points. It is possible that two different results may be obtained because what is being measured has changed, not because the measure is unreliable. For example, if your bathroom scale gives you two different weights within a few seconds, the scale is unreliable as your weight cannot have changed. However, if you weigh yourself once a week for a month and find that you get different results each time, is the scale unreliable or has your weight changed? A further problem with the test-retest check for reliability is that the administration of the first measure may affect the second measure's results. For instance, the difference between Scholastic Aptitude Test scores the first and second times that individuals take the test may not be assumed to be a measure of the reliability of the test, since test takers might alter their behavior the second time as a result of taking the test the first time.

The **alternative-form method** of measuring reliability also involves measuring the same attribute more than once, but it uses two different measures of the same concept rather than the same measure. For example, a researcher could devise two different sets of questions to measure the concept of liberalism, ask the two sets of questions of the same respondents at two different times, and compare the respondents' scores. Using two different forms of the measure prevents the second scores from being influenced by the first measure, but it still requires the phenomenon to be measured twice, and, depending on the length of time between the two measurements, what is being measured may change.

Going back to our bathroom scale example, if you weigh yourself on your home scale, go to the gym, weigh yourself again, and get the same number, you may conclude that your home scale is reliable. But what if you get two different numbers? Assuming your weight has not changed, what is the problem? If you go back home immediately and step back on your bathroom scale and find that it gives you a measurement that is different from the first, you could conclude that your scale has a faulty mechanism, is inconsistent, and therefore, unreliable. However, what if your bathroom scale gives you the same weight as the first time? It would appear to be reliable. Maybe the gym scale is unreliable. You could test this out by going back to the gym and reweighing yourself. If the gym scale gives a reading different from that given the first time, then it is unreliable. But, what if the gym scale gives consistent readings? Clearly one (or both) of the scales is inaccurate, and you have a measurement problem other than unreliability that needs to be resolved. Before we address this situation, let us mention one more test for reliability.

The **split-halves method** of measuring reliability involves two measures of the same concept, with both measures applied at the same time. The results of the two measures are then compared. This method avoids the problem of the change in the concept being measured. The split-halves method is often used when a multi-item measure can be split into two equivalent halves. For example, a researcher may devise a measure of liberalism consisting of the responses to ten questions on a public opinion survey. Half of these questions could be selected to represent one measure of liberalism and the other half selected to represent a second measure of liberalism. If individual scores on the two measures of liberalism are similar, then the ten-item measure may be said to be reliable by the split-halves approach.

The test-retest, alternative-form, and split-halves methods provide a basis for calculating the similarity of results of two or more applications of the same or equivalent measures. The less consistent the results are, the less reliable the measure is. Reliability of the measures used by political scientists is a serious problem. Survey researchers are often concerned about the reliability of the answers they receive. For example, respondents' answers to survey questions often vary considerably when given at two different times.[19] If respondents are not concentrating or taking the survey seriously, the answers they provide may as well have been pulled out of a hat.

Now, let us return to the problem of a measure that yields consistent results, as in the case of the two scales that differed consistently in how much you weighed. Each scale appears to be reliable (the scales are not giving you different weights at random), but at least one of them is giving you a wrong measurement (that is, not giving you your correct weight). This is the type of problem confronted in trying to assess whether or not measures are valid.

Validity

Essentially, a valid measure is one that measures what it is supposed to measure. Unlike reliability, which depends on whether repeated applications of the same or equivalent measures yield the same result, **validity** refers to the degree of correspondence between the measure and the concept it is thought to measure.

Let us consider first some examples of invalid measures. Suppose a researcher hypothesizes that the larger a city's police force is, the less crime that city will have. This requires the measurement of crime rates in different cities. Now also assume that some police departments systematically overrepresent the number of crimes in their cities to persuade public officials that crime is a serious problem and that the local police need more resources. Some police departments in other cities may systematically underreport crime to make their cities appear safe. If the researcher relied on official, reported measures of crime, the measures would be invalid because they would not correspond closely to the actual amount of crime in some cities. Consider also that not all crimes are reported to the police. A more valid measure of crime might be "victimization" surveys, which ask respondents whether they have been a victim of a crime.

Many studies look into the factors that affect voter turnout. These studies require an accurate measurement of voter turnout. One way of measuring voter turnout is to ask people if they voted in the last election. However, given the social desirability of voting in the United States, will all the people who did not vote in the previous election admit that to an interviewer? More people might say that they voted than actually did, resulting in an invalid measure of voter turnout. In fact, this is what usually happens. Voter surveys commonly overestimate turnout by several percentage points.[20]

A measure's validity is more difficult to demonstrate empirically than is its reliability because validity involves the relationship between the measurement of a concept and the actual presence or amount of the concept itself. Information regarding the correspondence is seldom abundant. Nonetheless, there are ways of evaluating the validity of any particular measure.

Face validity may be asserted (not empirically demonstrated) when the measurement instrument appears to measure the concept it is supposed to measure. To assess the face validity of a measure, we need to know the meaning of the concept being measured and whether the information being collected is "germane to that concept."[21] For example, suppose you want to measure an individual's political ideology, whether someone is conservative, moderate, or liberal. It may be tempting to use an individual's responses to a question on party identification (assuming that all Democrats are liberal, Republicans conservative, and independents moderate). Yet, because some Democrats hold views that are considered to be conservative and some Republicans hold liberal ones, partisan identification does not correspond exactly to the concept of political ideology. Similarly, some observers have argued that the results of many standard IQ tests measure intelligence *and* exposure to middle-class white culture, thus making the test results a less valid measure of intelligence.

In general, measures lack face validity when there are good reasons to question the correspondence of the measure to the concept in question. In other words, assessing face validity is essentially a matter of judgment. If no consensus exists about the meaning of the concept to be measured, the face validity of the measure is bound to be problematic.

Content validity is similar to face validity but involves determining the full domain or meaning of a particular concept and then making sure that measures of all portions of this domain are included in the measurement technique. For example, suppose you wanted to design a measure of the extent to which a nation's political system is democratic. As noted earlier, democracy means many things to many people. Raymond D. Gastil constructed a measure of democracy that included two dimensions, political rights and civil liberties. His checklists for each dimension consisted of eleven items.[22] Political scientists are often interested in concepts with multiple dimensions or complex domains and spend quite a bit of time discussing and justifying the content of their measures.

A third way to evaluate the validity of a measure is by empirically demonstrating **construct validity.** When a measure of a concept is related to a measure of another concept with which the original concept is thought to be related, construct validity is demonstrated. In other words, a researcher may specify, on theoretical grounds, that two concepts ought to be related (say, political efficacy with political participation, or education with income). The researcher then develops a measure of each of the concepts and examines the relationship between them. If the measures are related, then one measure has construct validity for the other measure. If the measures are unrelated, construct validity is absent. In that case the theoretical relationship is in error, at least one of the measures is not an accurate representation of the concept, or the procedure used to test the relationship is inappropriate. The absence of a hypothesized relationship does not mean a measure is invalid, but the presence of a relationship gives some assurance of the measure's validity.

A good example of an attempt to demonstrate construct validity may be found in the Educational Testing Services (ETS) booklet describing the Graduate Record Exam (GRE), a standardized test required for admission to most graduate schools. Since GRE test scores are supposed to measure a person's aptitude for graduate study, presumably construct validity could be demonstrated if the scores did, in fact, accurately predict the person's performance in graduate school. Over the years ETS has tested the relationships between GRE scores and first-year graduate school grade point averages (GPAs). The results, shown in Table 4-1, appear to indicate that GRE scores are not very strong predictors of this measure of graduate school performance and therefore do not have construct validity. In fact, this issue, and the role that GRE scores should play in admissions decisions, have been discussed a great deal in recent years. But this is a good example of a situation in which the absence of a strong relationship does not necessarily mean the measure lacks construct validity. Because persons with low GRE scores are generally not admitted to graduate school, we lack performance measures for them. Thus the people for whom we can test the relationship between scores and performance may be of similar ability and may not exhibit meaningful variation in their graduate school performance. Hence the test scores may be valid indicators of ability and may in fact show a stronger relationship to performance for a less selective sample of test takers (one that would include people who were not admitted to graduate school). The lack of a relationship in Table 4-1 undercuts claims of test score validity, but it does not necessarily disprove such claims.

A fourth way to demonstrate validity is through **interitem association.** This is the type of validity test most often used by political scientists. It relies on the similarity of outcomes of more than one measure of a concept to demonstrate the validity of the entire measurement scheme.

Let us return to the researcher who wants to develop a valid measure of liberalism. First, the researcher might measure people's attitudes toward (1) welfare, (2) military spending, (3) abortion, (4) Social Security benefit levels, (5) affirmative action, (6) a progressive income tax, (7) school vouchers, and

TABLE 4-1

Construct Validity of Graduate Record Examination (GRE) Test Scores

Average Estimated Correlations of GRE General Test Scores and Undergraduate Grade Point Average (GPA) with Graduate First-Year GPA by Department Type

Type of Department	Number of Departments	Number of Examinees	Predictors					
			V	Q	A	U	VQA	VQAU
All Departments	1,038	12,013	.30	.29	.28	.37	.34	.46
Natural Sciences	384	4,420	.28	.27	.26	.36	.31	.44
Engineering	87	1,066	.27	.22	.24	.38	.30	.44
Social Sciences	352	4,211	.33	.32	.30	.38	.37	.48
Humanities & Arts	115	1,219	.30	.33	.27	.37	.34	.46
Education	86	901	.31	.30	.29	.35	.36	.47
Business	14	196	.28	.28	.25	.39	.31	.47

V = GRE Verbal, Q = GRE Quantitative, A = GRE Analytical, U = Undergraduate grade point average.
The departments included in these analyses participated in the GRE Validity Study Service between 1986 and 1990. A minimum of 10 departments and 100 examinees in any departmental grouping were required for inclusion in the table.

Source: GRE materials selected from GRE® 2003–2004 Guide to the Use of Scores, 2003. Reprinted by permission of Educational Testing Service, the copyright owner. Permission to reprint GRE materials does not constitute review or endorsement by Educational Testing Service of this publication as a whole or of any other testing information it may contain.

Note: The numbers in this table are product-moment correlations—numbers that can vary between −1.0 and +1.0 and that indicate the extent to which one variable is associated with another. The closer the correlation is to ±1, the stronger the relationship between the two variables; the closer the correlation is to 0.0, the weaker the relationship. Since the correlations between VQA and graduate first-year GPA in this table are between .30 and .40, the relationships are not very strong. Notice also that undergraduate GPA is also not a very strong predictor of graduate first-year GPA, but that together GRE scores and undergraduate GPA improve predictions.

(8) protection of the rights of the accused. Then the researcher could determine how the responses to each question relate to the responses to each of the other questions. The validity of the measurement scheme would be demonstrated if strong relationships existed among people's responses across the eight questions.

The results of such interitem association tests are often displayed in a **correlation matrix** (Table 4-2). Such a display shows how strongly related each of the items in the measurement scheme is to all the other items. In the hypothetical data shown in Table 4-2, we can see that people's responses to six of the eight measures were strongly related to each other, whereas responses to the questions on protection of the rights of the accused and school vouchers were not part of the general pattern. Thus the researcher would probably conclude that the first six items all measure a dimension of liberalism and that, taken together, they are a valid measurement of liberalism.

Such a procedure was used by Ada W. Finifter and Ellen Mickiewicz in their study of popular attitudes toward political change in the Soviet Union.[23]

TABLE 4-2

Interitem Association Validity Test of a Measure of Liberalism

	Welfare	Military Spending	Abortion	Social Security	Affirmative Action	Income Tax	School Vouchers	Rights of Accused
Welfare	x							
Military Spending	.56	x						
Abortion	.71	.60	x					
Social Security	.80	.51	.83	x				
Affirmative Action	.63	.38	.59	.69	x			
Income Tax	.48	.67	.75	.39	.51	x		
School Vouchers	.28	.08	.19	.03	.30	−.07	x	
Rights of Accused	−.01	.14	−.12	.10	.23	.18	.45	x

Note: Hypothetical data. The figures in this table are product-moment correlations, explained in the note to Table 4-1. A high correlation indicates a strong relationship between how people answered the two different questions designed to measure liberalism. The figures in the last two rows are considerably closer to 0.0 than are the other entries, indicating that people's answers to the school vouchers and rights of the accused questions did not follow the same pattern as their answers to the other questions. Therefore, it looks like school vouchers and rights of the accused are not connected to the same concept of liberalism as measured by other questions.

They designed six survey questions to measure attitudes toward aspects of political change: the pace of political change, the perceived locus of responsibility (individual or collective) for individuals' social well-being, the acceptability of differences in individual incomes and standards of living, the acceptability of unconventional forms of political expression, the importance of free speech, and the level of support for competitive elections. The questions are as follows:

1. *(Rapid vs. slow change)* "*Some people think that to solve our most pressing problems it is necessary to make decisive and rapid changes, since any delay threatens to make things worse. Others, on the other hand, think that changes should be cautious and slow, since you can never be sure that they won't cause more harm than good. Which of these points of view are you more likely to agree with?*

   "Below are some widespread but contradictory statements relating to problems of the development of our society. Which would you be most likely to agree with?"

2. *(Individual vs. state responsibility)* "*The state and government should be mainly responsible for the success and well-being of people*" or "*People should look out for themselves, decide for themselves what to do for success in life.*"

3. *(Income differences) "The state should provide an opportunity for everyone to earn as much as he can, even if it leads to essential differences in people's standard of living and income" or "The state should do everything to reduce differences in people's standard of living and income, even if they won't try to work harder and earn more."*

4. *(Protest vs. traditional methods) "Strikes, spontaneous demonstrations, political meetings and other forms of social protest are completely acceptable methods of mass conduct and an effective means for solving social problems" or "These forms of protest are undesirable for society; they should be avoided in favor of more peaceful, traditional and organized methods of solving social conflicts."*

5. *(Free speech vs. order) "To improve things in our country people should be given the opportunity to say what they want, even if it can lead to public disorder" or "Keeping the peace in society should be the main effort, even if it requires limiting freedom of expression."*

6. *(Competitive elections) "In the coming elections for local soviets, should we elect deputies from among several candidates, as we mostly did in the spring elections?" or "Is it better to avoid the conflicts these elections generated and go back to the old system of voting?"*[24]

Finifter and Mickiewicz found a pattern in Soviet citizen responses to five of the six questions. Attitudes toward the locus of responsibility for individual well-being were only weakly related to the other questions about political change. Therefore the researchers combined only the answers to the other five measures into an attitude scale of "support for political change" as a result of the observed interitem associations among them.

Content and face validity are difficult to assess when agreement is lacking on the meaning of a concept, and construct validity, which requires a well-developed theoretical perspective, usually yields a less-than-definitive result. The interitem association test requires multiple measures of the same concept. Although these validity "tests" provide important evidence, none of them is likely to support an unequivocal decision concerning the validity of particular measures.

Problems with Reliability and Validity in Political Science Measurement

An example of research performed at the Survey Research Center at the University of Michigan illustrates the numerous threats to the reliability and validity of political science measures. In 1980 the center conducted interviews with a national sample of eligible voters and measured their income levels with the following question:

Please look at this page and tell me the letter of the income group that includes the income of all members of your family living here in 1979 before taxes. This figure should include salaries, wages, pensions, dividends, interest, and all other income.

Respondents were given the following choices.

A.	None or less than $2,000	N.	$12,000–$12,999
B.	$2,000–$2,999	P.	$13,000–$13,999
C.	$3,000–$3,999	Q.	$14,000–$14,999
D.	$4,000–$4,999	R.	$15,000–$16,999
E.	$5,000–$5,999	S.	$17,000–$19,999
F.	$6,000–$6,999	T.	$20,000–$22,999
G.	$7,000–$7,999	U.	$23,000–$24,999
H.	$8,000–$8,999	V.	$25,000–$29,999
J.	$9,000–$9,999	W.	$30,000–$34,999
K.	$10,000–$10,999	X.	$35,000–$49,999
M.	$11,000–$11,999	Y.	$50,000 and over

Both the reliability and the validity of this method of measuring income are questionable. Threats to the reliability of the measure include the following:

1. Respondents may not know how much money they make and therefore incorrectly guess their income.

2. Respondents may not know how much money other family members make and guess incorrectly.

3. Respondents may know how much they make but carelessly select the wrong categories.

4. Interviewers may circle the wrong categories when listening to the selections of the respondents.

5. Data entry personnel may touch the wrong numbers when entering the answers into the computer.

6. Dishonest interviewers may incorrectly guess the income of a respondent who does not complete the interview.

7. Respondents may not know which family members to include in the income total; some respondents may include only a few family members while others may include even distant relations.

8. Respondents whose income is on the border between two categories may not know which one to pick. Some pick the higher category, some the lower one.

Each of these problems may introduce some error into the measurement of income, resulting in inaccurate measures that are too high for some respondents and too low for others. Therefore, if this measure were applied to the same people at two different times we could expect the results to vary.

In addition to these threats to reliability, there are numerous threats to the validity of this measure:

1. Respondents may have illegal income they do not want to reveal and therefore may systematically underestimate their income.

2. Respondents may try to impress the interviewer, or themselves, by systematically overestimating their income.

3. Respondents may systematically underestimate their before-tax income because they think of their take-home pay and underestimate how much money is being withheld from their paychecks.

This long list of problems with both the reliability and the validity of this fairly straightforward measure of a relatively concrete concept is worrisome. Imagine how much more difficult it is to develop reliable and valid measures when the concept is abstract (for example, intelligence, self-esteem, or liberalism) and the measurement scheme is more complicated.

The reliability and validity of the measures used by political scientists are seldom demonstrated to everyone's satisfaction. Most measures of political phenomena are neither completely invalid or valid nor thoroughly unreliable or reliable but, rather, are partially accurate. Therefore, researchers generally present the rationale and evidence available in support of their measures and attempt to persuade their audience that their measures are at least as accurate as alternative measures would be. Nonetheless, a skeptical stance on the part of the reader toward the reliability and validity of political science measures is often warranted.

Reliability and validity are not the same thing. A measure may be reliable without being valid. One may devise a series of questions to measure liberalism, for example, that yields the same result for the same people every time but that misidentifies individuals. A valid measure, however, will also be reliable: if it accurately measures the concept in question, then it will do so consistently across measurements. It is more important, then, to demonstrate validity than reliability, but reliability is usually more easily and precisely tested.

The Precision of Measurements

Measurements should be not only accurate but also precise; that is, measurements should contain as much information as possible about the attribute or behavior being measured. The more precise our measures, the more complete and informative can be our test of the relationships between two or more variables.

Suppose, for example, that we wanted to measure the height of political candidates to see if taller candidates usually win elections. Height could be measured in many different ways. We could have two categories of the variable "height"—tall and short—and assign different candidates to the two categories based on whether they were of above-average or below-average height. Or we could compare the heights of candidates running for the same office and measure which candidate was the tallest, which the next tallest, and so on. Or we could take a tape measure and measure each candidate's height in inches and record that measure. The last method of measurement captures the most information about each candidate's height and is, therefore, the most precise measure of the attribute.

Levels of Measurement

When we consider the precision of our measurements, we refer to the **level of measurement.** The level of measurement involves the type of information that we think our measurements contain and the type of comparisons that can be made across a number of observations on the same variable. The level of measurement also refers to the claim we are willing to make when we assign numbers to our measurements.

There are four different levels of measurement: nominal, ordinal, interval, and ratio. Few concepts used in political science research inherently require a particular level of measurement, so the level used in any specific research project is a function of the imagination and resources of the researcher and the decisions made when the method of measuring each of the variables is developed.

A **nominal measurement** is involved whenever the values assigned to a variable represent only different categories or classifications for that variable. In such a case, no category is more or less than another category; they are simply different. For example, suppose we measure the religion of individuals by asking them to indicate whether they are Christian, Jewish, Muslim, or "other." Since the four categories or values for the variable "religion" are simply different, the measurement is at a nominal level. Other common examples of nominal-level measures are gender, marital status, and state of residence. A nominal measure of partisan affiliation might have the following categories: Democrat, Republican, Green, Libertarian, other, and none. Numbers will be assigned to the categories when the data are coded for statistical analysis, but these numbers do not represent mathematical differences between the categories.

Nominal-level measures ought to consist of categories that are exhaustive and mutually exclusive; that is, the categories should include all the possibilities for the measure, and every observation should fit in one and only one category. If we attempted to measure "types of political systems" with the categories democratic, socialist, authoritarian, undeveloped, traditional, capitalist, and monarchical, however, the categories would be neither exhaustive nor mutually exclusive. (In which one category would Japan, Great Britain, and India belong?) The difficulty of deciding the category into which many countries should be put would hinder the very measurement process the variable was intended to further. Notice that the categories of the measure of religion are exhaustive because of the "other" category as well as mutually exclusive (since presumably an individual cannot be of more than one religion). Researchers use "other" when they are unable to specify all alternatives, or when they expect very few of their observations to fall into the "other" category, but want to provide an option for respondents who do not fall into one of the labeled categories (respondents may fail to complete surveys with questions that don't apply to them). If using "other" as a category, you should check your data to make sure that only a relatively few observations fall into it. Otherwise subsequent data analysis will not be very meaningful.

An **ordinal measurement** assumes that a comparison can be made on which observations have more or less of a particular attribute. For example, we could create an ordinal measure of formal education completed with the following categories: "eighth grade or less," "some high school," "high school graduate," "some college," and "college degree or more." Here we are concerned not with the exact difference between the categories of education but only with whether one category is more or less than another. When coding this variable, we would assign higher numbers to higher categories of education. The intervals between the numbers have no meaning; all that matters is that the higher numbers represent more of the attribute than do the lower numbers. An ordinal variable measuring partisan affiliation with the categories "strong Republican," "weak Republican," "neither leaning Republican nor Democrat," "weak Democrat," and "strong Democrat" could be assigned codes 1, 2, 3, 4, 5 or 1, 2, 5, 8, 9 or any other combination of numbers as long as they were in ascending or descending order.

Dichotomous nominal-level measures, that is, nominal-level variables with only two categories are frequently treated as ordinal-level measures. For example, gender is a measure with two categories, male and female. One could interpret being male as being less female than a female. To give another example, a person who did not vote in the last election lacks, or has less of, the attribute of having voted compared to a person who did vote.

With an **interval measurement** the intervals between the categories or values assigned to the observations do have meaning. The value of a particular observation is important not just in terms of whether it is larger or smaller than another value (as in ordinal measures) but also in terms of how much larger or smaller it is. For example, suppose we record the year in which certain events occurred. If we have three observations—1950, 1962, and 1977—we know that the event in 1950 occurred twelve years before the one in 1962 and twenty-seven years before the one in 1977. A one-unit change (the interval) all along this measurement is identical in meaning: the passage of one year's time.

Another characteristic of an interval level of measurement that distinguishes it from the next level of measurement (ratio) is that in the former case the zero point is assigned arbitrarily and does not represent the absence of the attribute being measured. For example, many time and temperature scales have arbitrary zero points. Thus, the year 0 A.D. does not indicate the beginning of time—if this were true, there would be no B.C. dates. Nor does 0°C indicate the absence of heat; rather, it indicates the temperature at which water freezes. For this reason, with interval-level measurements we cannot calculate ratios; that is, we cannot say that 60°F is twice as warm as 30°F because it does not represent twice as much warmth.

The final level of measurement is a **ratio measurement.** This type of measurement involves the full mathematical properties of numbers. That is, the values of the categories order the categories, tell something about the intervals between the categories, and state precisely the relative amounts of the variable that the categories represent. If, for example, a researcher is willing to claim that an observation with ten units of a variable possesses exactly twice as much of that attribute as an observation with five units of that variable, then a ratio-level measurement exists. The key to making this assumption is that a value of zero on the variable actually represents the absence of that variable. Because ratio measures have a true zero point, it makes sense to say that one measurement is [X] times another. It makes sense to say a sixty-year-old person is twice the age of a thirty-year-old person (60/30 = 2), whereas it does not make sense to say that 60°C is twice as warm as 30°C.[25] Political science researchers

DEBATING THE LEVEL OF MEASUREMENT

Suppose we ask individuals three questions designed to measure social trust, and we believe that individuals who answer all three questions a certain way have more social trust than persons who answer two of the questions a certain way, and these individuals have more social trust than individuals who answer one of the questions a certain way. We could assign a score of 3 to the first group, 2 to the second group, 1 to the third group, and 0 to those who did not answer any of the questions in a socially trusting manner. In this case, the higher the number, the more social trust an individual has.

What level of measurement is this variable? It might be considered to be ratio level if one interprets the variable as simply the number of questions answered indicating social trust. But does a person who has a score of 0 have no social trust? Does a person with a score of 3 have three times as much social trust as a person with a score of 1? Perhaps then, the variable is an interval-level measure if one is willing to assume that the difference in social trust between individuals with scores of 2 and 3 is the same as the difference between individuals with scores of 1 and 2. But what if the effect of answering more questions in the affirmative is not simply additive, in other words, that a person who has a score of 3 has a lot more social trust than someone with a score of 2 and that this difference is more than the difference between individuals with scores of 1 and 2? In this case, then, the measure would be ordinal level, not interval level.

have measured many concepts at the ratio level. People's ages, unemployment rates, percentage of the vote for a particular candidate, and crime rates are all measures that contain a zero point and possess the full mathematical properties of the numbers used. However, more political science research has probably relied on nominal- and ordinal-level measures than on interval- or ratio-level measures. This has restricted the types of hypotheses and analysis techniques that political scientists have been willing and able to use.

Identifying the level of measurement of variables is important, since it affects the data analysis techniques that can be used and the conclusions that can be drawn about the relationships between variables. However, the decision is not always a straightforward one, and uncertainty and disagreement often exist among researchers concerning these decisions. Few phenomena inherently require one particular level of measurement. Often a phenomenon can be measured with any level of measurement, depending on the particular technique designed by the researcher and the claims the researcher is willing to make about the resulting measure.

Working with Precision: Too Little or Too Much

Researchers usually try to devise as high a level of measurement for their concepts as possible (nominal being the lowest level of measurement and ratio the highest). With a higher level of measurement, more advanced data analysis techniques can be used and more precise statements can be made about the relationships between variables. Thus researchers measuring attitudes or concepts with multiple operational definitions often construct a scale or index from nominal-level measures that permits at least ordinal-level comparisons between observations. We discuss the construction of indexes and scales in greater detail below.

It is easy to transform ratio-level information, for example, age in number of years, into ordinal-level information (age groups), but if you start with the ordinal-level measure, age groups, you will not have each person's actual age. If you decide you want to use a person's actual age, you will have to go out and collect that data. Similarly, a researcher investigating the effect of campaign spending on election outcomes could use a ratio-level variable measuring how much each candidate spent on his or her campaign. This information could be used to construct a new variable indicating how much more one candidate spent than the other or simply whether or not a candidate spent more than his or her opponent. Candidate spending could also be grouped into ranges.

Nominal and ordinal variables with many categories or interval- and ratio-level measures using more decimal places are more precise than measures with fewer categories or decimal places, but sometimes the result may provide more information than can be used. Frequently researchers start out with ratio-level measures or with ordinal and nominal measures with quite a few categories but then collapse or combine the data to create groups or fewer categories so that they have enough cases in each category for statistical analysis or to make comparisons easier to follow. For example, one might want to present comparisons simply between Democrats and Republicans rather than presenting data broken down into categories of strong, moderate, and weak for each party.

It may seem contradictory now to point out that extremely precise measures also may create problems. For example, measures with many response possibilities take up space if they are questions on a written questionnaire or more time to explain if they are included in a telephone survey. Such questions may confuse or tire survey respondents. A more serious problem is that they may lead to measurement error. Think about the possible responses to a question asking respondents to use a 100-point scale (called a thermometer scale) to indicate their support for or opposition to a political candidate, assuming that 50 is considered the neutral position and 0 is least favorable or coldest and 100 most favorable. Some respondents may not use the whole scale (to them no candidate ever deserves more than an 80 or less than a 20), whereas others respondents may use the ends and the very middle of the scale and ignore the scores in between. We might predict that a person who gives a candidate a 100 is more likely to vote for that candidate than a person who gives the same candidate an 80, but in reality they like the candidate pretty much the same way and would be equally likely to vote for the candidate. Another problem with overly precise measurements is that they may be unreliable. If asked to rate candidates on more than one occasion, respondents could vary the number that they choose, even if they don't change their opinion.

One last example illustrates how concerns about measurement error affected the measurement and operationalization of a concept. Poe and Tate defined human rights abuses to involve instances of state terrorism: murders, disappearances, and imprisonment. They noted that there are two approaches to measuring repression of human rights: an events-based approach, which would involve counting the occurrences of these events, and thus would be a ratio-level measure; and a standards-based approach, in which countries would be ranked on an ordinal scale based on the extent to which these events occur:

1. Countries[are] under a secure rule of law, people are not imprisoned for their views, and torture is rare or exceptional. . . . Political murders are extremely rare.

2. There is a limited amount of imprisonment for nonviolent political activity. However, few persons are affected, torture and beating are exceptional. . . . Political murder is rare.

3. There is extensive political imprisonment, or a recent history of such imprisonment. Execution or other political murders may be common. Unlimited detention, with or without trial, for political views is accepted.

4. The practices of [level 3] are expanded to larger numbers. Murders, disappearances are a common part of life. . . . In spite of its generality, on this level terror affects primarily those who interest themselves in politics or ideas.

5. The terrors [of level 4] have been expanded to the whole population. . . . The leaders of these societies place no limits on the means or thoroughness with which they pursue personal or ideological goals.[26]

Poe and Tate rejected the events-based approach on two grounds: (1) accurate data are not available and (2) the count of one event, for example, arrests, may be affected by the occurrence of another event, executions.[27] They concluded that the events-based approach would not be a valid measure, even though superficially it would appear to be a more precise measure. They were able to check the reliability of their data using the alternative-form method because they had reports for many countries from two sources: Amnesty International and the U.S. State Department. Coders read the reports and assigned scores. In most cases, scores based on the two reports were highly correlated.[28]

Multi-item Measures

Many measures consist of a single item. For example, the measures of party identification, whether or not one party controls Congress, the percentage of the vote received by a candidate, how concerned about an issue a person is, the policy area of a judicial case, and age are all based on a single measure of each phenomenon in question. Often, however, researchers need to devise measures of more complicated phenomena that have more than one facet or dimension. For example, the concepts internationalism, political ideology, political knowledge, dispersion of political power, and the extent to which a person is politically active are complex phenomena or concepts that may be measured in many different ways. In this situation, researchers often develop a measurement strategy that allows them to capture numerous aspects of a complex phenomenon while representing the existence of that phenomenon in particular cases with a single representative value. Usually this involves the construction of a multi-item index or scale representing the several dimensions of the phenomenon. These multi-item measures are useful because they enhance the accuracy of a measure, simplify a researcher's data by re-

TABLE 4-3
Hypothetical Index for Measuring Freedom in Countries

	Country A	Country B	Country C	Country D	Country E
Does the country possess:					
Privately owned newspapers	1	0	0	0	1
Legal right to form political parties	1	1	0	0	0
Contested elections for significant public offices	1	1	0	0	0
Voting rights extended to most of the adult population	1	1	0	1	0
Limitations on government's ability to incarcerate citizens	1	0	0	0	1
Index Score	5	3	0	1	2

Note: Hypothetical data. The score is 1 if the answer is yes, 0 if no.

ducing them to a more manageable size, and increase the level of measurement of a phenomenon. In the remainder of this section we describe several common types of indexes and scales.

Indexes

A **summation index** is a method of accumulating scores on individual items to form a composite measure of a complex phenomenon. An index is constructed by assigning a range of possible scores for a certain number of items, determining the score for each item for each observation, and then combining the scores for each observation across all the items. The resulting summary score is the representative measurement of the phenomenon.

A researcher interested in measuring how much freedom exists in different countries, for example, might construct an index of political freedom by devising a list of items germane to the concept, determining where individual countries score on each item, and then adding these scores to get a summary measure. In Table 4-3 such a hypothetical index is used to measure the amount of freedom in countries A through E.

The index in Table 4-3 is a simple, additive one; that is, each item counts equally toward the calculation of the index score, and the total score is the summation of the individual item scores. However, indexes may be constructed with more complicated aggregation procedures and by counting some items as more important than others. In the preceding example a researcher might consider some indicators of freedom as more important than others and wish to have them contribute more to the calculation of the final index score. This could be done either by weighting (multiplying) some item scores by a number indicating their importance or by assigning a higher score than 1 for those attributes considered more important.

Indexes are often used with public opinion surveys to measure political attitudes. This is because attitudes are complex phenomena and we usually do not know enough about them to devise single-item measures. So we often ask several questions of people about a single attitude and aggregate the answers to represent the attitude. A researcher might measure attitudes toward abortion, for example, by asking respondents to choose one of five possible responses—strongly agree, agree, undecided, disagree, and strongly disagree—to the following three statements: (1) Abortions should be permitted in the first three months of pregnancy; (2) Abortions should be permitted if the woman's life is in danger; (3) Abortions should be permitted whenever a woman wants one.

An index of attitudes toward abortion could be computed by assigning numerical values to each response (such as 1 for strongly agree, 2 for agree, 3 for undecided, and so on) and then adding the values of a respondent's answers to these three questions. (The researcher would have to decide what to do when a respondent did not answer one or more of the questions.) The lowest possible score would be a 3, indicating the most extreme pro-abortion attitude, and the highest possible score would be a 15, indicating the most extreme anti-abortion attitude. Scores in between would indicate varying degrees of approval of abortion.

Finifter and Mickiewicz, the researchers who attempted to measure attitudes toward political change in the former Soviet Union, developed this type of index of attitudes. Using five of their questionnaire items pertaining to political change, they assigned a score of +1 to pro-reform answers and a –1 to status quo (opposite) answers and summed each individual's answers to the five questions. What resulted were index scores representing individual answers to all five questions that ranged in value from +5 to –5. This single index score of attitudes toward political change was then used in further analysis.

Another example of a multi-item index appears in a study of attitudes toward feminism in Europe.[29] To determine the extent and distribution of attitudinal support for feminism across European society, Lee Ann Banaszak and Eric Plutzer constructed a measure of pro-feminism attitudes. Respondents were asked six questions about various aspects of a feminist belief system (for example, achieving equality between women and men in their work and careers, fighting against people who would like to keep women in a subordinate role, achieving gender equality in responsibilities for child care) and were given scores ranging from 0 to 3 for the degree of agreement with each pro-feminist statement. Responses across the six items were then summed to yield a pro-feminism index score that varied from 0 to 18.

Indexes are typically fairly simple ways of producing single representative scores of complicated phenomena such as political attitudes. They are probably more accurate than most single-item measures, but they may also be flawed in important ways. Aggregating scores across several items assumes, for example, that each item is equally important to the summary measure of the concept and that the items used faithfully encompass the domain of the concept. Although individual item scores can be weighted to change their contribution to the summary measure, there is often little information upon which to base a weighting scheme.

Several standard indexes are often used in political science research. The FBI crime index and the consumer price index, for example, have been used by many researchers. Although simple summation indexes are generally more accurate than single-item measures of complicated phenomena would be, it is often unclear how valid they are or what level of measurement they represent. For example, is the index of pro-feminism attitudes an ordinal-level measure, or could it be an interval-level or even a ratio-level measure?

Scales

Although indexes are generally an improvement over single-item measures, their construction also contains an element of arbitrariness. Both the selection of particular items making up the index and the way in which the scores on individual items are aggregated are based on the researcher's judgment. Scales are also multi-item measures, but the selection and combination of items in them is accomplished more systematically than is usually the case for indexes. Over the years several different kinds of multi-item scales have been used frequently in political science research. We discuss two of them: Likert scales and Guttman scales.

A **Likert scale** score is calculated from the scores obtained on individual items. Each item generally asks a respondent to indicate a degree of agreement or disagreement with the item, as with the abortion questions discussed earlier. A Likert scale differs from an index, however, in that once the scores on each of the items are obtained, only some of the items are selected for inclusion in the calculation of the final score. Those items that allow a researcher to distinguish most readily those scoring high on an attribute from those scoring low will be retained, and a new scale score will be calculated based only on those items.

For example, consider the researcher interested in measuring the liberalism of a group of respondents. Since definitions of liberalism vary, the researcher cannot be sure how many aspects of liberalism need to be measured. With

Likert scaling the researcher would begin with a large group of questions thought to express various aspects of liberalism with which respondents would be asked to agree or disagree. A provisional Likert scale for liberalism, then, might look like this:

	Strongly Disagree (1)	Disagree (2)	Undecided (3)	Agree (4)	Strongly Agree (5)
The government should ensure that no one lives in poverty.	___	___	___	___	___
Military spending should be reduced.	___	___	___	___	___
It is more important to take care of people's needs than it is to balance the federal budget.	___	___	___	___	___
Social Security benefits should not be cut.	___	___	___	___	___
The government should spend money to improve housing and transportation in urban areas.	___	___	___	___	___
Wealthy people should pay taxes at a much higher rate than poor people.	___	___	___	___	___
Busing should be used to integrate public schools.	___	___	___	___	___
The rights of persons accused of a crime must be vigorously protected.	___	___	___	___	___

In practice, a set of questions like this would be scattered throughout a questionnaire so that respondents do not see them as related. Some of the questions might also be worded in the opposite way (that is, so an "agree" response is a conservative response) to ensure genuine answers.

The respondents' answers to these eight questions would be summed to produce a provisional score. The scores in this case can range from 8 to 40. Then the responses of the most liberal and the most conservative people to each question would be compared; any questions with similar answers from the disparate respondents would be eliminated—such questions would not distinguish liberals from conservatives. A new summary scale score for all the respondents would be calculated from the questions that remained.

Likert scales are improvements over multi-item indexes because the items that make up the multi-item measure are selected in part based on the respondents' behavior rather than on the researcher's judgment. Likert scales suffer two of the other defects of indexes, however. The researcher cannot be sure that all the dimensions of a concept have been measured, and the relative importance of each item is still determined arbitrarily.

The **Guttman scale** also uses a series of items to produce a scale score for respondents. Unlike the Likert scale, however, a Guttman scale presents respondents with a range of attitude choices that are increasingly difficult to agree with; that is, the items composing the scale range from those easy to agree with to those difficult to agree with. Respondents who agree with one of the "more difficult" attitude items will also generally agree with the "less difficult" ones. (Guttman scales have also been used to measure attributes other than attitudes. Their main application has been in the area of attitude research, however, so an example of that type is used here.)

Let us return to the researcher interested in measuring attitudes toward abortion. He or she might devise a series of items ranging from "easy to agree with" to "difficult to agree with." Such an approach might be represented by the following items:

Do you agree or disagree that abortions should be permitted:

1. When the life of the woman is in danger.

2. In the case of incest or rape.

3. When the fetus appears to be unhealthy.

4. When the father does not want to have a baby.

5. When the woman cannot afford to have a baby.

6. Whenever the woman wants one.

This array of items seems likely to result in responses consistent with Guttman scaling. A respondent agreeing with any one of the items is likely to also agree with those items numbered lower than that one. This would result in the "stepwise" pattern of responses characteristic of a Guttman scale.

Suppose six respondents answered this series of questions, as shown in Table 4-4. Generally speaking, the pattern of responses is as expected; those who agreed with the "most difficult" questions were also likely to agree with the "less difficult" ones. However, the responses of three people (2, 4, and 5) to the question about the father's preferences do not fit the pattern. Consequently,

TABLE 4-4
Guttman Scale of Attitudes toward Abortion

Respon-dent	Life of Woman	Incest or Rape	Un-healthy Fetus	Father	Afford	Any-time	No. of Agree Answers	Revised Scale Score
1	A	A	A	A	A	A	6	5
2	A	A	A	D	A	D	4	4
3	A	A	A	D	D	D	3	3
4	A	A	D	A	D	D	3	2
5	A	D	D	A	D	D	2	1
6	D	A	D	D	D	D	1	0

Note: Hypothetical data. A = Agree, D = Disagree.

the question about the father does not seem to fit the pattern and would be removed from the scale. Once that has been done, the stepwise pattern becomes clear.

With real data, it is unlikely that every respondent would give answers that fit the pattern perfectly. For example, in Table 4-4 respondent 6 gave an "agree" response to the question about incest or rape. This response is unexpected and does not fit the pattern. Therefore, we would be making an error if we assigned a scale score of "0" to respondent 6. There are statistical procedures to calculate how well the data fit the scale pattern. When the data fit the scale pattern well (number of errors is small), researchers assume that the scale is an appropriate measure and that the respondent's "error" may be "corrected" (in this case, either the "agree" in the case of incest or rape or the "disagree" in the case of the life of the woman). There are standard procedures to follow to determine how to correct the data to make it conform to the scale pattern. We emphasize, however, that this is done only if the changes are few.

Guttman scales differ from Likert scales in that, in the former case, generally only one set of responses will yield a particular scale score. That is, to get a score of 3 on the abortion scale, a particular pattern of responses (or something very close to it) is necessary. In the case of a Likert scale, however, many different patterns of responses can yield the same scale score. A Guttman scale is also much more difficult to achieve than a Likert scale, since the items must have been ordered and be perceived by the respondents as representing increasingly more difficult responses to the same attitude.

Both Likert and Guttman scales have shortcomings in their level of measurement. The level of measurement produced by Likert scales is, at best, ordinal (since we do not know the relative importance of each item and so we cannot be sure that a "5" answer on one item is the same as a "5" answer on another), and the level of measurement produced by Guttman scales is usually assumed to be ordinal.

The procedures described so far for constructing multi-item measures are fairly straightforward. There are other, more advanced statistical techniques for summarizing or combining individual items or variables. For example, it is possible that several variables are related to some underlying concept. **Factor analysis** is a statistical technique that may be used to uncover patterns across measures. It is especially useful when a researcher has a large number of measures and when there is uncertainty about how the measures are interrelated.

An example is the analysis by Daniel D. Dutcher, who conducted research on the attitudes of owners of streamside property toward the water quality improvement strategy of planting trees in a wide band (called riparian buffers) along the sides of streams. He asked landowners to rate the importance of twelve items thought to affect the willingness of landowners to create and maintain riparian buffers. He wanted to know whether the attitudes could be grouped into distinct dimensions that could be used as summary variables instead of using each of the twelve items separately. Using factor analysis, he found that the items factored into three dimensions. These dimensions and the items included in each dimension are listed in Table 4-5. The first dimension, which he labeled "maintaining property aesthetics," included items such as maintaining a view of the stream, neatness, and maintaining open space. A second dimension contained items related to concern over water quality. The third dimension related to protecting property against damage or loss.[30] Factor analysis is just one of many techniques developed to explore the dimensionality of measures and to construct multi-item scales. The readings listed at the end of this chapter include some resources for students who are especially interested in this aspect of variable measurement.

Through indexes and scales, researchers attempt to enhance both the accuracy and the precision of their measures. Although these multi-item measures have received most use in attitude research, they are often useful in

TABLE 4-5

Items Measuring Landowner Attitudes toward Riparian Buffers on Their Streamside Property Sorted into Dimensions Using Factor Analysis

Maintaining property aesthetics
- Maintaining my view of the stream
- Maintaining the look of a pastoral or meadow stream
- Maintaining open space
- Wanting the neighbors to think that I'm doing my part to keep up the traditional look of the neighborhood

Contributing to stream and bay quality
- Being confident that maintaining or creating a streamside forest on my property is necessary to protect the stream
- Contributing to the improvement of downstream areas, including the Chesapeake Bay
- Maintaining or improving stream-bank stability on my property

Protecting property against damage or loss
- Keeping vegetation from encroaching on fields or fences
- Minimizing the potential for flood damage to lands or buildings
- Discouraging pests (deer, woodchucks, snakes, insects, etc.)
- Initial costs, maintenance costs, or loss of income

Source: Adapted from Table 3.4 in Daniel D. Dutcher, "Landowner Perceptions of Protecting and Establishing Riparian Forests in Central Pennsylvania" (Ph.D. diss., Pennsylvania State University, May 2000), p. 64.

other endeavors as well. Both indexes and scales require researchers to make decisions regarding the selection of individual items and the way in which the scores on those items will be combined to produce more useful measures of political phenomena.

Conclusion

To a large extent, a research project is only as good as the measurements that are developed and used in it. Inaccurate measurements will interfere with the testing of scientific explanations for political phenomena and may lead to erroneous conclusions. Imprecise measurements will limit the extent of the comparisons that can be made between observations and the precision of the knowledge that results from empirical research.

Despite the importance of good measurement, political science researchers often find that their measurement schemes are of uncertain accuracy and precision. Abstract concepts are difficult to measure in a valid way, and the practical constraints of time and money often jeopardize the reliability and precision of measurements. The quality of a researcher's measurements makes an important contribution to the results of his or her empirical research and should not be lightly or routinely sacrificed.

Sometimes the accuracy of measurements may be enhanced through the use of multi-item measures. With indexes and scales, researchers select multiple indicators of a phenomenon, assign scores to each of these indicators, and combine those scores into a summary measure. Although these methods have been used most frequently in attitude research, they can also be used in other situations to improve the accuracy and precision of single-item measures.

Notes

1. Lane Kenworthy and Jonas Pontusson, "Rising Inequality and the Politics of Redistribution in Affluent Countries," *Perspectives on Politics* 3 (September 2005): 449–471.

2. See Walter Dean Burnham, "The Turnout Problem," in A. James Richley, ed., *Elections American Style* (Washington, D.C.: Brookings Institution, 1987). Also Michael P. McDonald and Samuel L. Popkin, "The Myth of the Vanishing Voter," *American Political Science Review* 95 (December 2001): 963–974.

3. Stephen C. Poe and C. Neal Tate, "Repression of Human Rights to Personal Integrity in the 1980s: A Global Analysis," *American Political Science Review* 88 (December 1994): 853–872.

4. David Bradley, Evelyne Huber, Stephanie Moller, Françoise Nielsen, and John D. Stephens, "Distribution and Redistribution in Postindustrial Democracies," *World Politics* 55 (January 2003): 193–228.

5. Jeffrey A. Segal and Albert D. Cover, "Ideological Values and the Votes of U.S. Supreme Court Justices," *American Political Science Review* 83 (June 1989): 557–565.

6. Valerie J. Hoekstra, *Public Reaction to Supreme Court Decisions* (New York: Cambridge University Press, 2003).

7. B. Dan Wood and Richard W. Waterman, "The Dynamics of Political Control of the Bureaucracy," *American Political Science Review* 85 (September 1991): 801–828.

8. David E. Lewis, *Presidents and the Politics of Agency Design: Political Insulation in the United States Government Bureaucracy, 1946–1997* (Stanford: Stanford University Press, 2003).

9. Stephen Ansolabehere, Shanto Iyengar, Adam Simon, and Nicholas Valentino, "Does Attack Advertising Demobilize the Electorate?" *American Political Science Review* 88 (December 1994): 829–838.

10. Question wording for the variable EQWLTH from GSS 1998 Codebook. Retrieved from www.icpsr.umich.edu/GSS/.

11. Wood and Waterman, "Dynamics of Political Control," 805.

12. Segal and Cover, "Ideological Values."

13. Ibid., 559.

14. For information on the LIS database, see www.lis.ceps.lu.

15. Kim Fridkin Kahn and Patrick J. Kenney, "Do Negative Campaigns Mobilize or Suppress Turnout? Clarifying the Relationship between Negativity and Participation," *American Political Science Review* 93 (December 1999): 877–890.

16. Martin P. Wattenberg and Craig Leonard Brians, "Negative Campaign Advertising: Demobilizer or Mobilizer?" *American Political Science Review* 93 (December 1999): 891–900.

17. Ansolabehere, Iyengar, Simon, and Valentino, "Does Attack Advertising Demobilize the Electorate?"

18. Edward G. Carmines and Richard A. Zeller, *Reliability and Validity Assessment,* Series on Quantitative Applications in the Social Sciences, No. 07–001, Sage University Papers (Beverly Hills, Calif.: Sage Publications, 1979).

19. Philip E. Converse, "The Nature of Belief Systems in Mass Publics," in David E. Apter, ed., *Ideology and Discontent* (New York: Free Press, 1964); D. M. Vaillancourt, "Stability of Children's Survey Responses," *Public Opinion Quarterly* 37 (Fall 1973): 373–387; J. Miller McPherson, Susan Welch, and Cal Clark, "The Stability and Reliability of Political Efficacy: Using Path Analysis to Test Alternative Models," *American Political Science Review* 71 (June 1977): 509–521; and Philip E. Converse and Gregory B. Markus, "The New CPS Election Study Panel," *American Political Science Review* 73 (March 1979): 32–49.

20. Raymond E. Wolfinger and Steven J. Rosenstone, *Who Votes?* (New Haven: Yale University Press, 1980), Appendix A.

21. Kenneth D. Bailey, *Methods of Social Research* (New York: Free Press, 1978), 58.

22. As discussed in Ross E. Burkhart and Michael S. Lewis-Beck, "Comparative Democracy: The Economic Development Thesis," *American Political Science Review* 88 (December 1994): Appendix A.

23. Ada W. Finifter and Ellen Mickiewicz, "Redefining the Political System of the USSR: Mass Support for Political Change," *American Political Science Review* 86 (December 1992): 857–874.

24. Ibid.

25. The distinction between an interval-level and a ratio-level measure is not always clear, and some political science texts do not distinguish between them. Interval-level measures in political science are rather rare; ratio-level measures (money spent, age, number of children, years living in the same location, for example) are more common.

26. Poe and Tate, "Repression of Human Rights," 867.

27. Ibid., 868.

28. Ibid., 855.

29. Lee Ann Banaszak and Eric Plutzer, "The Social Bases of Feminism in the European Community," *Public Opinion Quarterly* 57 (Spring 1993): 29–53.

30. Daniel D. Dutcher, "Landowner Perceptions of Protecting and Establishing Riparian Forests in Central Pennsylvania" (Ph.D. diss., Pennsylvania State University, May 2000).

Terms Introduced

ALTERNATIVE-FORM METHOD. A method of calculating reliability by repeating different but equivalent measures at two or more points in time.

CONSTRUCT VALIDITY. Validity demonstrated for a measure by showing that it is related to the measure of another concept.

CONTENT VALIDITY. Validity demonstrated by ensuring that the full domain of a concept is measured.

CORRELATION MATRIX. A table showing the relationships among discrete measures.

DICHOTOMOUS VARIABLE. A variable having only two categories that for certain analytical purposes can be treated as a quantitative variable.

FACE VALIDITY. Validity asserted by arguing that a measure corresponds closely to the concept it is designed to measure.

FACTOR ANALYSIS. A statistical technique useful in the construction of multi-item scales to measure abstract concepts.

GUTTMAN SCALE. A multi-item measure in which respondents are presented with increasingly difficult measures of approval for an attitude.

INTERITEM ASSOCIATION. A test of the extent to which the scores of several items, each thought to measure the same concept, are the same. Results are displayed in a correlation matrix.

INTERVAL MEASUREMENT. A measure for which a one-unit difference in scores is the same throughout the range of the measure.

LEVEL OF MEASUREMENT. The extent or degree to which the values of variables can be compared and mathematically manipulated.

LIKERT SCALE. A multi-item measure in which the items are selected based on their ability to discriminate between those scoring high and those scoring low on the measure.

MEASUREMENT. The process by which phenomena are observed systematically and represented by scores or numerals.

NOMINAL MEASUREMENT. A measure for which different scores represent different, but not ordered, categories.

OPERATIONAL DEFINITION. The rules by which a concept is measured and scores assigned.

ORDINAL MEASUREMENT. A measure for which the scores represent ordered categories that are not necessarily equidistant from each other.

RATIO MEASUREMENT. A measure for which the scores possess the full mathematical properties of the numbers assigned.

RELIABILITY. The extent to which a measure yields the same results on repeated trials.

SPLIT-HALVES METHOD. A method of calculating reliability by comparing the results of two equivalent measures made at the same time.

SUMMATION INDEX. A multi-item measure in which individual scores on a set of items are combined to form a summary measure.

TEST-RETEST METHOD. A method of calculating reliability by repeating the same measure at two or more points in time.

VALIDITY. The correspondence between a measure and the concept it is supposed to measure.

Suggested Readings

Carmines, Edward G., and Richard A. Zeller. *Reliability and Validity Assessment.* Series on Quantitative Applications in the Social Sciences, No. 07-001. Sage University Papers. Beverly Hills, Calif.: Sage Publications, 1979.

DeVellis, Robert F. *Scale Development.* Newbury Park, Calif.: Sage Publications, 1991.

Hatry, Harry P. *Performance Measurement: Getting Results.* Washington, D.C.: Urban Institute Press, 1999.

Kim, Jae-On, and Charles W. Mueller. *Introduction to Factor Analysis.* Beverly Hills, Calif.: Sage Publications, 1978.

Lodge, Milton. *Magnitude Scaling.* Beverly Hills, Calif.: Sage Publications, 1983.

Maranell, Gary M. *Scaling: A Sourcebook for Behavioral Scientists.* 4th ed. Hawthorne, N.Y.: Longman, 1983.

Mertler, Craig A., and Rachel A. Vannatta. *Advanced and Multivariate Statistical Methods: Practical Applications and Interpretation,* 3d ed. Glendale, Calif.: Pyrczak Publishing, 2005.

Rabinowitz, George. "Nonmetric Multidimensional Scaling and Individual Difference Scaling." In William D. Berry and Michael S. Lewis-Beck, eds. *New Tools for Social Scientists,* 77–107. Beverly Hills, Calif.: Sage Publications, 1986.

Robinson, John P., Jerrold G. Rusk, and Kendra B. Head. *Measures of Political Attitudes.* Ann Arbor, Mich.: Institute for Social Research, 1969.

CHAPTER 5

Research Design

CBS News described some political ads aired during the 2004 presidential campaign:

> *In one ad, pictures of Osama bin Laden and Mohammed Atta flash within seconds of an image of John Kerry, as a male voice asks, "Would you trust Kerry up against these fanatic killers?"*

> *In another, President Bush is accused of being in financial cahoots with the "corrupt" Saudi royal family and linked to alleged backers of terrorism.*

> *"Kerry had a secret meeting with the enemy during the Vietnam War," says a third commercial. President Bush put American soldiers in a "quagmire," another ad charges.*[1]

As we noted in Chapter 1, some political scientists argue that negative advertisements demobilize the electorate.[2] But the CBS network, like many other observers and scholars, believes this approach to campaigning works. They quote Kathleen Hall Jamieson, dean of the Annenberg School for Communication at the University of Pennsylvania: "People are more influenced by negative information than positive information; they are more likely to sway attitudes."[3]

Well, then, what are the effects of negative political advertisements? Do they just cause people to change their evaluations of candidates without affecting their propensity to vote? Do they encourage voting by stimulating interest in campaigns? Or do they simply annoy people to the extent that they want to have little or nothing to do with electoral politics? More specifically, do these types of ads *cause* people to become more or less interested in participating? The discussion in Chapter 1 showed that this is an ongoing and lively issue in political science. It is now time to think about how one might approach a problem of this sort. We require a plan or strategy for collecting and analyzing information in such a way that we can have confidence that our conclusions rest on solid evidence and not on faulty reasoning or mere opinion.

A **research design** is a plan that shows how a researcher intends to study an empirical question. It indicates what specific theory or propositions will be tested, what units of analysis (for example, people, nations, states, organizations) are appropriate for the tests, what measurements or observations (that is, data) are needed, how all this information will be collected, and which analytical and statistical procedures will be used to examine the data. All the parts of a research design should work to the same end: drawing sound conclusions that are supported by evidence.

A poor research design may produce insignificant and erroneous conclusions, no matter how original and brilliant the investigator's ideas and hypotheses happen to be. In this chapter we discuss various types of designs along with their advantages and disadvantages. As important, we show how a poor research strategy can result in uninformative or misleading results.

Many factors affect the choice of a design. One is the purpose of the investigation. Is it intended to be exploratory, descriptive, or explanatory? Another factor is the practical limitation on how researchers test their hypotheses. Some research designs may be unethical, others impossible to implement for lack of data or sufficient time and money. Researchers frequently must balance what is possible to accomplish against what would ideally be done to investigate a particular hypothesis. Many common research designs entail unfortunate but necessary compromises. As a result, the conclusions that may be drawn from them are more tentative and incomplete than anyone would like.

All research designs to test hypotheses are attempts to (1) establish a relationship between two or more variables, (2) demonstrate that the results are generally true in the real world, (3) reveal whether one phenomenon precedes another in time, and (4) eliminate as many alternative explanations for a phenomenon as possible. In this chapter we explain how various designs allow or do not allow researchers to accomplish these four objectives.

Causal Inferences and Controlled Experiments

Causal versus Spurious Relationships

Let us return to the question of the effects of campaign advertising on voting behavior. A tentative hypothesis is that negative ads, repeated over and over, bore, frustrate, or even anger potential voters and make them think that none of the candidates is worthy of their vote. We might expect that the more citizens are subjected to commercials and advertisements that vilify candidates, the more disinclined they will be to vote. Therefore, turnout will be lower in a campaign flooded with negative ads than in one in which the candidates stick to the issues. We might even be tempted to make the stronger claim that negative political advertising causes a decline in participation.

How could we support such assertions? Just after an election, it might be possible to interview a sample of citizens, ask them if they had heard or been

TABLE 5-1

Voting Intention by Ad Exposure

Y Voted?	X Yes, exposed	X No, not exposed
Yes		100%
No	100%	

Note: Hypothetical data.

aware of attack ads, and then determine whether or not they had voted. We might even find a relationship or connection between exposure and turnout. Let's say, for instance, that all those who report viewing negative commercials tell us that they did *not* vote, whereas all of those who were not aware of these ads did cast ballots. We might summarize the hypothetical results in a simple table. Let X stand for whether or not people saw the campaign ads and Y for whether or not they voted. What this table symbolizes is a relationship or association between X and Y.

This strategy is frequently called opinion or survey research and involves an investigator observing behavior indirectly by asking people questions about what they believe and how they act. Since it does not entail direct observation of their actions, we can only take the respondents' word about whether or not they voted or saw attack ads.

What can we make of these findings? (See Table 5-1.) Yes, there is a relationship. Note that 100 percent of the people who said they were exposed also said they did not vote, and vice versa for those who did not watch any ads. But does that mean that negative advertising causes a decline in turnout? After all, those who missed the ads might differ in other ways from those who saw them. Perhaps they have a higher level of education and that accounts for their higher turnout rate. Or maybe they have a generally strong sense of civic duty and would always vote no matter what the campaigns do or say.

At the same time, people with less education might watch a lot of television and *coincidentally* don't bother voting in any election. If conditions of these sorts hold, we may observe a connection between advertisement exposure and turnout, but it would not be a **causal relationship.** And outlawing negative campaigning would not necessarily have any effect on turnout because the one does not cause the other. In this case the association would be an example of a false or **spurious relationship.**

A spurious relationship arises because two things, such as viewing negative ads and voting, are both affected by some third factor and thus appear to be related. Once this additional factor has been identified and controlled, the original relationship weakens or disappears altogether. To take a trivial example, we might well find a positive relationship between the number of operations in hospitals and the number of patients who die in them. But this doesn't mean that operations cause deaths. Rather, it is probably the case that people with serious illnesses or injuries need operations *and* because of their conditions are prone to die.

Figure 5-1 illustrates causal and spurious relationships.[4]

Distinguishing real, causal relationships from spurious ones is an important part of any scientific research. To explain phenomena, we must know how and why two things are connected, not simply that they are associated. Thus one of the major goals in designing research is to come up with a way to make valid causal inferences. Ideally such a design does three things:

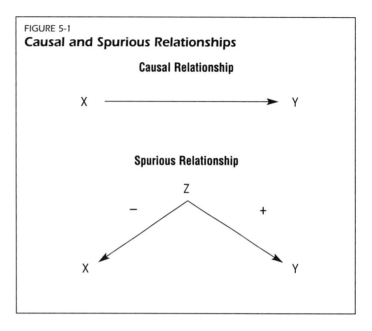

FIGURE 5-1
Causal and Spurious Relationships

1. Covariation. The research must demonstrate that the alleged cause (call it X) does in fact covary with the supposed effect, Y. Our simple study of advertising and voting does this because, as we saw in Table 5-1, viewing negative advertisements is connected to nonvoting, and not viewing the ads is associated with voting. Public opinion polls or surveys can relatively easily identify associations. But to make a causal inference, more is needed.

2. Time order. The research must show that the cause preceded the effect: X must come before Y in time. After all, can an effect appear before its cause? In our survey of citizens, we might reasonably assume that the television ads preceded the decision to vote or not. But however reasonable this assumption might be, we have not demonstrated it empirically. In other observational settings it may be difficult if not impossible to tell if X came before or after Y. Still, even if we can be confident of the time order, we have to demonstrate that a third condition holds.

3. Elimination of possible alternative causes, sometimes termed confounding factors. The research must be conducted in such a way that all possible joint causes of X and Y have been eliminated. To be sure that negative television advertising directly depresses turnout, we need to rule out the possibility that the two are connected by some third factor such as education or interest in politics.

Figure 5-2 shows the possibilities presented by the third requirement. The first diagram (Causal Relationship) shows a true causal connection between

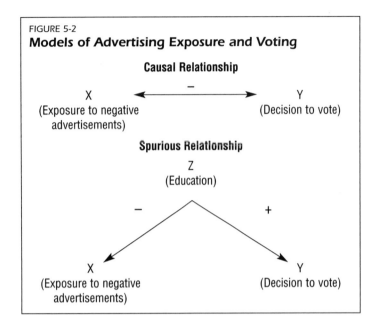

FIGURE 5-2
Models of Advertising Exposure and Voting

Causal Relationship

X — Y
(Exposure to negative   (Decision to vote)
advertisements)

Spurious Relationship

Z
(Education)

−   +

X   Y
(Exposure to negative   (Decision to vote)
advertisements)

X (ad exposure) and Y (voting). The arrow indicates causality: X causes Y. If this is the way the world really is, then attack advertisements have a direct link to nonvoting. The minus sign (−) means the greater the exposure, the less the inclination to vote. (It is called a negative relationship.) The arrowhead indicates the direction of causality, because in this example X causes Y, and not vice versa. In the second diagram (Spurious Relationship), by contrast, the X and Y are not directly related. There is no causal arrow between them. Yet an apparent association is produced by the action of a third factor, Z. The arrow with the negative sign means that people with higher levels of education do not see as many commercials (if any) as those with less schooling. At the same time, citizens with lots of education are likely to vote, whereas those with less education are not as apt to go the polls. (The arrow with the positive sign [+] means education has a positive causal effect on the decision to vote.) Hence, the presence of the third factor, Z (education), creates the impression of a causal relationship between X and Y, but this impression is misleading, because once we take into account the third factor—in language we use later, "once we control for Z"—the original relationship weakens or disappears.

Given the possibility of spuriousness, how do we make valid causal inferences? The answer leads to research design, because how we frame problems and plan their solutions greatly affects the confidence we can have in our results. Asking a group of people about what they have seen and heard in the media and relating their answers to their reported behavior is known in common parlance as *polling,* or opinion research. A more formal term is **survey research**: the direct or indirect collection of information from individuals by asking them questions, having them fill out forms, or other means. (We discuss survey research in Chapter 10.) This approach is perhaps the most common in the social sciences, and it is the one followed in the preceding hypothetical example. A difficulty with survey research, however, is that it is not the best way to make dependable causal inferences. For this purpose many social scientists think laboratory experiments lead to more valid conclusions.

Randomized Controlled Experiments

Experimentation allows the researcher to control exposure to an experimental variable (often called a test factor or an independent variable), the assignment of subjects to different groups, and the observation or measurement of responses and behavior. As we will see, experimental designs theoretically allow researchers to make causal inferences with greater confidence in their dependability than do nonexperimental approaches such as surveys. Although some political scientists do conduct experiments, as Stephen Ansolabehere, Shanto Iyengar, Adam Simon, and Nicholas Valentino's study of the effects of negative campaign advertising illustrates, most research in the field uses surveys or other methods.[5] The nature of the phenomena of greatest interest to political science, such as who votes in actual elections, requires the use of these methods. Nevertheless, an understanding of experimental design is important because it provides a standard for how to make and evaluate causal inferences and explanations.

As we noted earlier, making a valid causal claim involves showing three things: covariation, time order, and the absence of confounding factors. In theory an experiment can unambiguously accomplish all these objectives. How? Let's look at five basic characteristics of a **classical experimental design**:[6]

1. The experimenter establishes two groups: an **experimental group** (there can be more than one) that receives or is exposed to an experimental treatment, or **test stimulus or factor;** and a **control group,** so named because its subjects do not undergo the experimental manipulation. For example, Ansolabehere and his colleagues had some citizens (the experimental group) watch a negative political ad and others (the control group) watch a nonpolitical commercial. The investigators determined who watched the political ad and who watched the nonpolitical commercial. They did not rely on self-reports of viewership. This control over the two groups is directly analogous to a biologist exposing some laboratory animals to a chemical and exposing others to a placebo (an inactive ingredient or material).

2. Equally important, the researcher randomly assigns individuals to the groups. The subjects do not get to decide which group they join. The random assignment to groups is called **randomization,** and it means that membership is a matter of chance, not self-selection. Moreover, if we start with a pool of subjects, random assignment ensures that at the outset the experimental and control groups are virtually identical in all respects. They will, in other words, contain similar proportions, or

averages, of females and males; blondes, brunettes, and redheads; Republicans and Democrats; political activists and nonvoters; and on and on. On average the groups will not differ in any respect, because they have been created by random placement.* Randomization, as we will see, is what makes experiments such powerful tools for making causal inferences.

3. The researcher controls the administration or introduction of the experimental treatment (the test factor)—that is, the researcher can determine when, where, and under what circumstances the experimental group is exposed to the stimulus.

4. In an experiment, the researcher establishes and measures a dependent variable—the response of interest—both before and after the stimulus is given. The measurements are often called pre- and post-experimental measures, and they indicate whether or not there has been an **experimental effect.** An experimental effect, as the term suggests, reflects differences between the two groups' responses to the test factor.

5. The environment of the experiment—that is, the time, location, and other physical aspects—is under the experimenter's direction. Such control means that he or she can control or exclude **extraneous factors,** or influences, besides the independent variable that might affect the dependent variable. If, for instance, both groups are studied at the same time of day, any differences between the control and experimental subjects cannot be attributed to temporal factors.

To see how these characteristics tie in with the requirements of causal inferences, let us conduct a hypothetical randomized experiment to see if negative political advertising depresses the intention to vote. This case is purely hypothetical, but it resembles the research conducted by Ansolabehere and his associates, and more to the point it shows the inferential power of experiments. (The example also shows some of the weaknesses of this design.)

Our hypothesis states that exposure to negative television advertising will cause people to lose interest in politics and thus to be less inclined to vote. Stated this way, the test factor, or experimental variable, is seeing a negative ad ("yes" or "no"), and the response is the stated intention to vote ("likely" or "not likely"). Now, we recruit from somewhere a pool of subjects and randomly assign them to either an experimental (or treatment) group or a control group. It is crucial that we make the assignments randomly. We do not, for example, want to put mostly females in one group and males in the other because if we find a difference in propensity to vote, how will we be able to

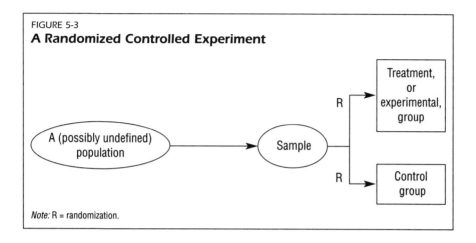

FIGURE 5-3
A Randomized Controlled Experiment

Note: R = randomization.

tell if it arose because of the advertisement or gender? We illustrate the procedure in Figure 5-3.

Note that we draw subjects from some population, perhaps by advertising in a newspaper or giving extra credit in an American government class. This pool of subjects does not constitute a random sample of any population. After all, the subjects volunteered to participate; we did not pick them randomly. But, and here is the key, once we have a pool of individuals we can *then* randomly assign them to the groups. Assume the first subject arrives at the test site. We could flip a coin and, depending on the result, assign that person to the experimental group or to the control section. When the next person comes we flip the coin again, and based on just that result we send the person to one or the other of the groups. If our pool consists of 100 potential subjects, our coin tossing should result in about 50 in each group.

Now suppose we administer a questionnaire to the members of both groups in which we ask about demographic characteristics (for example, age, sex, family income, years of education, place of birth) and about their political beliefs and opinions (party identification, attitude toward gun control, ideology, knowledge of politics). Of course, we would also ask about the main dependent variable, the intention to vote. If we compare the groups' averages on the variables, we should find that they are about the same. The experimental group may consist of 45 percent males, be on average 33.5 years old, and generally (75 percent, say) not care much for liberals. But the control group should also reflect these characteristics and tendencies. There may be only 40 percent males and the average age 35 years, but the differences reflect only chance (or, as we see in Chapter 7, sampling error). Of greatest importance, the proportions on the response variable, the intention to vote in the next election, should be approximately the same. Thus at the beginning of the experiment we have two nearly identical groups.

After the initial measurement of the variables (the **pre-test**), we start the experiment. So as not to reveal the exact purpose of our research—which if revealed might affect participants' responses to our experiment—we tell the informants that we are interested in television news. Those assigned to the experimental treatment go to Room 101, those in the control panel to Room 106. Both groups now watch identical fifteen-minute news broadcasts. So far both groups have been treated the same. Any differences between them are the result of happenstance.

RANDOMIZATION VERSUS RANDOM SAMPLE
Be sure to remember the difference between *randomization* and a *random sample*. The former, a process, means assigning individuals randomly to groups; it is something an investigator does. A random sample, by contrast, refers to how a small set of subjects is selected from a larger population.

Next, the first set of subjects sees a thirty-second negative ad that we have constructed to be as realistic as possible, while the others see a thirty-second commercial about toothpaste, also as true to life as we can make it. The different treatment constitutes the experimental manipulation (seeing versus not seeing the negative advertisement). After the commercials have aired we show both groups another fifteen-minute news clip. When the broadcast is over we administer (part of) the questionnaire again and measure the propensity to vote. This calculation gives us an indication of the size of the experimental variable's effect, if there is one. Hypothetical results from this experiment are shown in Table 5-2.

Both control and experimental subjects had about the same initial stated intention of voting (68 and 70 percent, respectively), as we would expect, because the participants had been randomized. But the **post-test** measurement shows quite a change for the experimental group—the percentage intending to vote has dropped from 70 to 20 percent, a decline of 50 percentage points. But the control group has changed hardly at all.

So we might conclude that the experimental factor did indeed cause a decline in intention to vote. How can we make this inference? Well, the research design satisfied all the conditions necessary for making such claims. In Table 5-2 we show that the two factors (or variables) covary: those who have seen a negative ad are much less likely to vote than those who did not (20 percent versus 66 percent). We have also established the time order, since we literally determined the timing of the experimental treatment and the subsequent post-test measurement. Finally, and most convincing of all, we have been able to rule out any possible al-

TABLE 5-2
Results of Hypothetical Media Experiment

Group	Before Experiment (Pre-test) Measure of Intention to Vote	After Experiment (Post-test) Measure of Intention to Vote
Experimental	70%	20%
Control	68%	66%

Note: Hypothetical data.

FIGURE 5-4

A Randomized Controlled Experiment: General Framework

		Pre-test		Post-test
R	Experimental Group	M_{exp1}	X	M_{exp2}
R	Control Group	$M_{control1}$		$M_{control2}$

X = Experimental manipulation

M = Measurements

Experimental effect = $(M_{exp2} - M_{exp1}) - (M_{control2} - M_{control1})$

R = Random assignment of subjects to groups

ternative explanation of the covariation, given that our randomization and experimental manipulation ensured that the groups differed *only* because one received the treatment and the other did not. Since that was the only difference, the gap between the post-test percentages of the two groups could be attributed only to viewing the commercial.

To simplify later explanations, we introduce a general framework for describing research designs, as shown in Figure 5-4. We let X stand for an experimental manipulation (for example, showing a negative campaign commercial) and let the letter M denote the average measurements on the response variable. In the present case M means the percentage of members in a group who said they intend to vote in the next election. The subscripts identify the group and time to which the measure applies. An experimental or treatment effect can be defined in a couple of ways. We are most interested in the change in the experimental group proportions from before and after the test factor was administered compared with those of the control group. Presumably the control group's responses did not change except perhaps by a slight amount due to random errors in collecting and recording the data. Hence, the difference between the post-test and pre-test measures for control subjects should be about zero. But in the experimental group we hypothesized that there will be a noticeable difference in measurements. In this case we would expect that the post-experiment percentage intending to vote will be smaller than the pre-experiment number. We might, then, say that the experimental effect is $E = M_{exp2} - M_{exp1}$, or $E = (M_{exp2} - M_{exp1}) - (M_{control2} - M_{control1})$, which is the same because in this particular case the second term is expected to be zero.

The purpose of an experiment is to isolate and measure the effects of the independent variable on a response. Researchers want to be able to separate the effect of the independent variables from the effects of other factors that might also influence the dependent variable. Control over the random assignment of subjects to experimental and control groups is the key feature of experiments because it helps them to exclude, rule out, or control for the effects of factors that might create a spurious relationship.

Randomization and the Assignment of Subjects

As we have stressed, the way researchers assign subjects to control and experimental groups is important. The best way is **assignment at random,** under the assumption that extraneous factors will affect all groups equally and thus be canceled out. Random assignment is an especially attractive choice when a researcher is not able to specify possible extraneous factors in advance, or when there are so many that it is not possible to assign subjects to experimental and control groups in a manner that ensures the equal distribution of these factors.

Even if a researcher does assign subjects at random, extraneous factors may not be totally randomly distributed and therefore can affect the outcome of the experiment. This might be true especially if the number of subjects is small. Prudent researchers do not assume that all significant factors are randomly distributed just because they have randomized the study's participants. So, in addition to random assignment, investigators use pre-tests to check to see if the control and experimental groups are, in fact, equivalent with regard to factors known to influence the outcome or suspected of doing so. An especially important measure in this regard is $M_{exp1} - M_{control1}$, which should be about zero.

If the researcher knows ahead of time that certain features are related to differences in the dependent variable, he or she can use **precision matching** to control for them. This requires creating matched pairs of subjects who are as similar as possible and assigning one to the experimental group and the other to the control group. Thus no one has to depend solely on randomization to eliminate or control for these factors. One problem with this method is that when many factors are to be controlled, it becomes difficult to match subjects on all relevant characteristics and a larger pool of prospective subjects is required. A second problem is that the researcher may not know ahead of time all extraneous factors. To guard against bias in the assignment of pairs, each member of the matched pairs should be randomly assigned to the control and experimental groups.

One of the biggest obstacles to experimentation in social science research is the inability of researchers to control assignment of subjects to experimental and control groups. This is especially true when public policies are involved. Even though the point of conducting an experiment is to test whether a treatment or program has a beneficial effect, it is often politically difficult to assign subjects to a control group. People assume that the experimental treatment must be beneficial—otherwise the treatment or program would not have been proposed as a response to a public policy problem. Social scientists generally lack sufficient authority or incentives to offer subjects in order to set up experimental and control groups.

Interpreting and Generalizing the Results of an Experiment

Most readers, we hope, have followed the logic of our arguments. But they must be flabbergasted at the unrealism of the hypothetical example we introduced and wonder how anyone could make a definitive statement about negative advertising based on these data, even if we had carried out this experiment on real people using real television commercials. Someone might say, "This test is invalid." It may be, but before jumping to that conclusion we need to consider carefully and closely the term *validity*.

Internal Validity

Statistical theory tells us that experiments conducted properly can lead to valid inferences about causality. In this context, however, *validity* has a particular meaning, namely, that the manipulation of the experimental or independent variable itself, and not some other variable, influenced the dependent variable. This kind of validity is known as **internal validity,** which means the research procedure demonstrated a true cause-and-effect relationship that was not created by spurious factors. Social scientists generally believe that the type of research design we have been discussing—a randomized controlled experiment—has strong internal validity. But it is not foolproof.

Several things can affect internal validity. As we have argued, the principal strength of experimental research is that the researcher has enough control over the environment to make sure that exposure to the experimental stimulus is the only difference between experimental and control groups and that at the outset all comparison groups have the same traits. Sometimes **history,** or events other than the experimental stimulus that occur between the pre-test and post-test measurements of the dependent variable, will affect the dependent variable. For example, suppose that after being selected and assigned to a room, the experimental subjects happen to hear a radio program that undercuts their faith in the electoral process. Such a possibility might arise if a long lag occurred between the first measurement of their attitudes and the start of the experiment. This situation is shown schematically in Figure 5-5. In this instance no one would be able to say if the radio program (Z in the figure) or the experimental treatment produced the effect on voting intentionality.

Another potential confounding influence is **maturation,** or a change in subjects over time, which might produce differences between experimental and control groups. To take a different example, subjects may become tired, confused, distracted, or bored during the course of an experiment. These

FIGURE 5-5
The Effects of "History" on an Experiment

		Pre-test			Post-test
R	Experimental Group	M_{exp1}	Z	X	M_{exp2}
R	Control Group	$M_{control1}$	Z		$M_{control2}$

    X = Experimental manipulation

    R = Randomization

    Z = Nonexperimental event

    M = Measurements

changes may affect their reaction to the test stimulus and introduce an unanticipated effect on post-treatment scores.

The standard experimental research design involves measurement or observation, which is sometimes called testing. However, **testing,** the process of measuring the dependent variable prior to the experimental stimulus, may itself affect the post-treatment scores of subjects. For example, simply asking individuals about politics on a pre-test may alert them to the purposes of the experiment. And that in turn may cause them to behave in unanticipated ways. Similarly, suppose a researcher wanted to see if watching a presidential debate makes viewers better informed than nonviewers. If the researcher measures the political awareness of the experimental and control groups prior to the debate, he or she runs the risk of sensitizing the subjects to certain topics or issues and contributing to a more attentive audience than would otherwise be the case. In this case, we would not know for sure whether any increase in awareness was due to the debate, the pre-test, or a combination of both. Fortunately, some research designs have been developed to separate these various effects.

Selection bias can also lead to problems. Such bias can creep into a study if subjects are picked (intentionally or not) according to some criteria and not randomly. For example, the experimental group might consist of volunteers who differ significantly from nonvolunteers. Sometimes a person might be picked for participation in an experiment because of an extreme measurement (very high or very low) of the dependent variable. Extreme scores may not be stable; when measured again, they may move back toward average scores. Thus changes in the dependent variable may be attributed erroneously to the experimental factor. This problem is called **statistical regression.**[7]

As we stressed, in assigning subjects to experimental and control groups, a researcher hopes that the two groups will be equivalent. If subjects selectively drop out of the study, experimental and control groups that were the same at the start may no longer be equivalent. Thus **experimental mortal-**

ity, or the differential loss of participants from comparison groups, may raise doubts about whether the changes and variation in the dependent variable are due to manipulation of the independent variable. A common selection problem occurs when subjects volunteer to participate in a program. Volunteers may differ significantly from nonvolunteers; they may be more compliant and eager to please, healthier, or more outgoing.

Sometimes, **instrument decay,** or change in the instrument used to measure the dependent variable, occurs during the course of an experiment so that the pre-test and post-test measures are not made in the same way. For example, a researcher may become tired and not take post-test measurements as carefully as pre-test ones. Or different persons with different biases may conduct the pre-test and post-test. Thus changes in the dependent variable may be due to measurement changes, not to the experimental stimulus.

Another possible problem comes from **demand characteristics,** or aspects of the research situation that cause participants to guess the purpose or rationale of the study and adjust their behavior or opinions accordingly. It has been found that people often want to "help" or contribute to an investigator's goals by acting in ways that will support the main hypotheses.[8] Perhaps something about our experiment on political advertising tips off subjects that we, the researchers, expect to find that negative ads depress turnout. The subjects (perhaps unconsciously) may adjust their feelings in order to prove the proposition and hence please us. In this case, the desire to satisfy the researchers' objectives, not the commercials themselves, affects the disposition to vote. This is not a minor issue. You may have heard about double-blind studies in medical research. The goal of this kind of design is to make it impossible for the patients and attendants to identify who is receiving a real experimental medicine and who is receiving a traditional medicine or a placebo.

External Validity

In short, a lot of things can go wrong even in a carefully planned experiment. Nevertheless, experimental research designs are better able to resist threats to internal validity than are other types of research designs. (In fact, they provide an ideal against which other research strategies may be compared.) Moreover, we discuss below some ways to mitigate these potential errors. Yet, even if we devised the most rigorous laboratory experiment possible to test for media effects on political behavior, some readers still might not be convinced that we have found a cause-and-effect relationship that applies to the real world. What they are concerned about without being aware of the term is **external validity,** the extent to which the results of an experiment can be generalized across populations, times, and settings.

One possible objection to experimental results is that the effects may not be found using a different population. Refer back to Figure 5-3, which showed that participants are selected from some population and then assigned to one of two groups. But the population from which they have been drawn may not reflect any meaningful broader population. Suppose, for instance, we conducted our advertising experiment on sophomores from a particular college. Results might be valid for second-year students attending that school but not for the public at large. Indeed, the conclusions might not apply to other classes at that or any other university. To take another example, findings from an experiment investigating the effects of live television coverage on legislators' behavior in state legislatures with fewer than one hundred members may not be generalizable to larger state legislative bodies or to Congress. In general, if a study population is not representative of a larger population, our ability to generalize about the larger population will be limited.

Another question is whether slightly different experimental treatments will result in similar findings or in findings that are fundamentally different. For example, a small increase in city spending for neighborhood improvements may not result in a more positive attitude of residents toward their neighborhood or the city. A slightly larger increase, however, may have an effect, perhaps because it has resulted in more noticeable improvements.

Threats to the external validity of an experiment also may result if the artificiality of the experimental setting or treatments makes it hard to generalize about findings in more natural settings. In our case, we showed the experimental group just one ad. But in the real world, voters are exposed to dozens and dozens of brief television advertisements at home and elsewhere. They may or may not pay attention to them, or the ads may be "filtered" by comments from family members or friends. None of these conditions was part of our study. Furthermore, as we noted, subjects may react to a stimulus differently when they know they are being studied than when they are in a natural setting.

Despite the difficulty of generalizing the results, experiments are still attractive to researchers because they provide control over the subjects and their exposure to various levels of the experimental or independent variable, and they do permit valid causal inferences, even if of limited generality.[9] Before considering alternative approaches, let us consider another study that used experimentation.

Shanto Iyengar, Mark D. Peters, and Donald R. Kinder attempted to determine the extent to which exposure to televised network news coverage of particular public policy issues can increase the public's awareness and concern about those issues.[10] To test the effect of the hypothesized independent variable (exposure to television news), the researchers used a randomized design. First, they recruited participants for their experiments (paying each participant $20) and had them fill out a questionnaire that included measures of the importance of various national problems (the pre-test). The participants were then randomly divided into experimental and control groups and, over a four-day period, exposed to videotape recordings of the preceding evening's network newscast.

Unknown to the participants, these newscasts had been edited to include (or exclude) stories from previous newscasts dealing with a particular public policy issue. In one session the experimental group saw newscasts that included stories about alleged weaknesses in U.S. defense capability, while the control group saw newscasts with no defense-related stories. In another session one experimental group again saw newscasts with stories about inadequacies in U.S. defense preparedness, while a second experimental group saw stories about environmental pollution and a third experimental group saw stories about inflation. The day after the last viewing session, participants completed a second questionnaire that again included measures of the importance of various policy issues. All groups except for one reported a significant increase in concern about an issue after they had been exposed to news stories about that issue. The exception involved concern about inflation. In that case, the level of concern about inflation was already so high (a score of 18.5 out of 20) that exposure to the newscasts about inflation had no appreciable effect.

The investigators were sensitive to the measurement **instrument reactivity** and the external validity issues discussed previously. They believed that they had achieved fairly natural viewing conditions (exposure to the news stories took place over several days, in small groups, in an informal setting, without any pressure to pay close attention), that the participants showed no signs of knowing what the experiment was about, and that the participants were fairly representative of a larger adult population. They believed that their experimental results demonstrated that network news coverage could significantly alter the public's sense of the importance of different political issues.

One might argue that the second part of this experiment was not a classical experiment in that there was no control group, only three experimental groups using a pre-test/post-test design. The pre-test/post-test design allows the researcher to compare changes in groups receiving different treatments. Any differences in change may be attributed to differences in treatments. Without a control group, however, change between a pre-test and a post-test does not absolutely establish the factors that caused the change. The researcher can never be sure what would have happened to subjects if no treatment had been given. Nevertheless, there may be good reasons for omitting a control group. Because the researchers had already conducted an experiment that included a control group and that demonstrated the impact of newscasts on issue concern, it was reasonable for them to omit a control group from their second experiment.

Other Versions of Experimental Designs

Now that we have discussed experimental research in general and some problems associated with it, we briefly describe some variations on this approach. Each variation represents a different attempt to retain control over the experimental situation while also dealing with threats to internal and external validity. Although you may not have an opportunity to use these designs, knowledge of them will help you to understand published research and to determine whether the research design used supports the author's conclusions.

Simple Post-test Design

The most basic experiment, the **simple post-test design,** involves two groups and two variables, one independent and one dependent, as before. And subjects are randomly assigned to one or the other of two groups. One group, the experimental group, is exposed to a treatment or stimulus, and the other, the control group, is not or is given a placebo. Then the dependent variable is measured for each group. Using the previous notation, this design may be diagrammed as in Figure 5-6.

Someone using this design can justifiably make causal inferences because he or she can make sure that the treatment occurred prior to measurement of the dependent variable. Furthermore, he or she knows that any difference between the two groups on the measure of the dependent variable may be attributed to the difference in the treatment—in other words, to the introduction of the independent variable—between the groups. Why? This design still requires random assignment of subjects to the experimental and control groups and therefore assumes that extraneous factors have been controlled (that is, were the same for both groups). It also assumes that prior to the application of the experimental stimulus, both groups were equivalent with respect to the dependent variable.

Let us illustrate with a basic example. Suppose we wanted to test the hypothesis that watching a national nominating convention on television makes

---

FIGURE 5-6
Simple Post-test Experiment

		Post-test
R Experimental Group	X	M_{exp}
R Control Group		$M_{control}$

X = Experimental manipulation

M = Measurements

R = Random assignment of subjects to groups

---

people better informed politically. Using this research design, we would randomly assign our subjects to a group that will watch a convention or to a group that will not and then measure how well informed the members of the two groups are after the convention is over. Any difference in the level of awareness between the two groups after the convention would be attributed to the effect of watching convention coverage.

The simple post-test experimental design assumes that the random assignment of subjects to the experimental and control groups creates two groups that are equivalent in all significant ways prior to the introduction of the experimental stimulus. If the assignment to the experimental and control groups is truly random, and the size of the two groups is large, this is ordinarily a safe assumption. However, if the assignment to groups is not truly random or the sample size is small, or both, then post-treatment differences between the two groups may be the result of pre-treatment differences and not the result of the independent variable. Because it is impossible with this design to tell how much of the post-treatment difference is simply a reflection of pre-treatment differences, an experimental research design using a pretest such as we described in the classical experimental design (shown in Figure 5-4) is considered to be a stronger design.

Repeated-Measurement Design

Naturally the pre-test comes before the experiment starts and the post-test comes afterward, but exactly how long before and how long afterward? Researchers seldom know for sure. Therefore, an experiment may contain several pre-treatment and post-treatment measures, especially when a researcher does not know how quickly the effect of the independent variable should be observed or when the most reliable pre-test measurement of the dependent variable should be taken. Figure 5-7 shows the layout for such an experiment.

An example of a repeated-measurement design would be an attempt to test the relationship between watching a presidential debate and support for the

FIGURE 5-7
Repeated-Measurement Design

		Pre-test				Post-test		
Experimental Group	R	M_1	M_2	$M_3 \ldots$	X	M_1	M_2	$M_3 \ldots$
Control Group	R	M_1	M_2	$M_3 \ldots$	X	M_1	M_2	$M_3 \ldots$

X = Experimental manipulation

R = Randomization

M = Measurements

candidates. Suppose we started out by conducting a classical experiment, randomly assigning some people to a group that watches a debate and others to a group that does not. On the pre- and post-tests we might obtain the following scores:

	Pre-debate Support for Candidate X	Treatment	Post-debate Support for Candidate X
Experimental Group	60	Yes	50
Control Group	55	No	50

These scores seem to indicate that the control group was slightly less supportive of Candidate X before the debate (that is, the random assignment did not work perfectly) and that the debate led to a decline in support for Candidate X of 5 percent $(60 - 50) - (55 - 50)$.

Suppose, however, that we had the following additional measures:

	Pre-test				Post-test		
	First	Second	Third	Treatment	First	Second	Third
Experimental Group	80	70	60	Yes	50	40	30
Control Group	65	60	55	No	50	45	0

It appears now that support for Candidate X eroded throughout the whole period for both the experimental and control groups and that the rate of decline was consistently more rapid for the experimental group (that is, the two groups were not equivalent prior to the debate). Viewed from this perspective, it seems that the debate had no effect on the experimental group, since the rate of decline both before and after the debate was the same. Hence the existence of multiple measures of the dependent variable, both before and after the introduction of the independent variable, would lead in this case to a more accurate conclusion regarding the effects of the independent variable.

Multigroup Design

To this point we have discussed mainly research involving one experimental and one control group, although in a previous example an experiment included three experimental groups rather than a single experimental and single control group. In a **multigroup design** more than one experimental or control group is created so that different levels of the experimental variable can be compared. This is useful if the independent variable can assume several values or if the researcher wants to see the possible effects of manipulat-

---

FIGURE 5-8
Multigroup Experimental Design

		Pre-test		Post-test
R	Experimental Group A	M_{exp1A}	X	M_{exp2A}
R	Experimental Group B	M_{exp1B}	X	M_{exp2B}
R	Experimental Group C	M_{exp1C}	X	M_{exp2C}
		.	.	.
		.	.	.
		.	.	.
R	Control Group	$M_{control1}$		$M_{control2}$

X = Experimental manipulation

R = Random assignment of subjects to groups

M = Measurements

---

ing the independent variable in several different ways. Multigroup designs may involve a post-test only or both a pre-test and a post-test. They may also include a time series component. In Figure 5-8 we show a diagram of the layout and analysis of this design. It uses the same notation as before, with R signifying randomization, M the various pre- and post-test measures, and so forth. Given the various experimental groups, one can make several comparisons among the levels of the independent variable.[11] (Read, for instance, M_{exp1A} as the "average measurement of members of experimental group A at the first time period, that is, pre-test.")

Here's an example. The proportion of respondents who return questionnaires in a mail survey is usually quite low. Investigators have attempted to increase response rates by including an incentive or a token of appreciation inside the survey. Since these items add to the cost of the survey, researchers want to know whether or not the incentives increase response rates and, if so, which ones are most effective and cost-efficient.

To test the effect of various incentives, we could use a multigroup post-test design. If we wanted to test the effects of five treatments, we could randomly assign subjects to six groups. One group would receive no reward (the control group), whereas the other groups would each receive a different reward—for example, 25¢, 50¢, $1.00, a pen, or a key ring. Response rates (the post-treatment measure of the dependent variable) for the groups could then be compared. In Table 5-3 we present a set of hypothetical results for such an experiment.

The response rates indicate that rewards increase response rates and that monetary incentives are more effective than are token gifts. Furthermore,

TABLE 5-3
Mail Survey Incentive Experiment

(Random Assignment)	Treatment	Response Rate (percent)
Experimental Group 1	25¢	45.0
Experimental Group 2	50¢	51.0
Experimental Group 3	$1.00	52.0
Experimental Group 4	pen	38.0
Experimental Group 5	key ring	37.0
Control Group	no reward	30.2

Note: Hypothetical data.

it seems that the dollar incentive is not cost effective, since it did not yield a sufficiently greater response rate than the 50¢ reward to warrant the additional expense. Other experiments of this type could be conducted to compare the effects of other aspects of mail questionnaires, such as the use of prepaid versus promised monetary rewards or the inclusion or exclusion of a pre-stamped return envelope.

Field Experiments

Laboratory experiments, whatever their power for making causal inferences, cannot be used to study a lot of (if not most of) the phenomena that interest political scientists. But some of the basic ideas of experimental design can be taken into the field.

Field experiments, or quasi-experiments, are experimental designs applied in a natural setting. As we noted, in a true experiment the investigator does two things: (1) randomly assigns participants to groups (for example, experimental and control), and (2) manipulates the experimental variable. In a field experiment, by contrast, there is no random assignment of participants to groups, but the investigator does try to manipulate one or more independent variables. The causal inferences are not as strong, but the design may be more practical.

As in any experimental research design, researchers attempt to control the selection of subjects and the manipulation of the independent variable. But in the field experiment the behaviors of interest are observed in a natural setting, increasing the likelihood that extraneous factors such as historical events will intrude and affect experimental results. Most important, because the subjects are not randomized, the groups do not necessarily start out the same in all relevant respects. Although it is possible to choose a natural setting that is isolated in some respects (and thereby approximates a controlled environment), in general the researcher can only hope that the environment remains unchanged during the course of his or her experiment.

Still, field experiments should not be considered totally inferior to laboratory experiments. The artificial environment of a laboratory or controlled setting may seriously affect the external validity of a study's conclusions. Something that can be demonstrated in a laboratory may have limited applicability in the real world. Therefore, a program or treatment that is effective in a controlled setting may be ineffective in a natural setting. Field experiments are more likely to produce results that reflect the real-world impact of a program or treatment than are researchers' controlled experimental manipulations.

An interesting example of a field experiment in political science was the New Jersey experiment in income maintenance funded by the Office of Economic Opportunity and conducted from 1967 to 1971.[12] This effort was the forerunner of other large-scale social experiments designed to test the effects of new social programs. The experiment is a good illustration of the difficulty of testing the effects of public policies on a large scale in a natural setting.

At the time of the experiment, dissatisfaction with the existing welfare system was high because of its cost and because it was thought to discourage the poor from lifting themselves out of poverty. The system was also blamed for discouraging marriage and for breaking up families. Families headed by able-bodied men generally were excluded from welfare programs, and welfare recipients' earned income was taxed at such a rate that many thought there was little incentive for recipients to work.

In 1965 a negative income tax was proposed that provided a minimum, nontaxable allowance to all families and that attempted to maintain work incentives by allowing the poor to keep a significant fraction of their earnings. For example, a family of four might be guaranteed an income of $5,000 and be allowed to keep 50 percent of all its earnings up to a break-even point, where it could choose to remain in the program or opt out. If the break-even income was $10,000, a family earning $10,000 could receive a $5,000 guaranteed minimum plus half of $10,000 ($5,000) for a total of $10,000, or it could keep all the $10,000 earned and choose not to receive any income from the government. Critics of the proposed program argued that a guaranteed minimum income would encourage people to reduce their work effort. Others expressed concern about how families would use their cash allowances. Numerous questions about the administration of the program were raised as well. Because of these uncertainties, researchers designed the New Jersey income-maintenance experiment to test the consequences of a guaranteed minimum income system with actual recipients in a natural setting.

The research design included two experimental factors. One was the income guarantee level, expressed as a percentage of the poverty line. The level is the

TABLE 5-4
Experimental Design of New Jersey Income-Maintenance Experiment

Guarantee (percentage of poverty line)	Tax Rate		
	30%	50%	70%
125		X	
100		X	X
75	X	X	X
50	X	X	

Note: X represents experimental conditions actually tested.

amount of money a family received if other income was zero. The other factor was the rate at which each dollar of earned income was taxed.

Table 5-4 shows the experimental conditions of the two independent variables of interest to the researchers. Policy analysts were originally interested in income guarantee levels of 50, 75, 100, and 125 percent of the poverty level and tax rates of 30, 50, and 70 percent. The 4 × 3 **factorial design** displayed in Table 5-4 would allow researchers to examine the effect of the variation in one factor while the other was held constant and to measure the effects of different combinations of the independent variables. For example, it would allow researchers to examine the effect of varying the tax rate from 30 to 50 to 70 percent while the guarantee remained set at 75 percent of the poverty line.

Certain theoretical combinations of the experimental conditions were not chosen for study because they were unrealistic policy options or because they increased the cost of the study. Therefore, actual income-maintenance results were investigated for only eight of the twelve possible experimental conditions, and families participating in the study were assigned to one of those eight conditions. In assigning families to "cells" representing experimental treatments, there was a trade-off between the number of families that could be included in the study and the number of families assigned to each cell, since some cells were more costly than others. The cells representing the most likely national policy options (the 100–50 and 75–50 plans) were assigned more families to make sure that enough families completed the experiment. Finally, for some of the less generous treatments, the researchers experienced difficulty in finding eligible families willing to participate in the experiment. Families placed in these cells were likely to receive at most a small payment because they were near the break-even point. This situation created resentment within the community because all the families participating in the program had hoped to gain benefits beyond the nominal payment they could expect each time they completed an income report. If the researchers had had complete control over their subjects, assignment problems would have been fewer. But in research involving human subjects, such control is understandably lacking.

Only families headed by able-bodied males were eligible for the experiment because of the great interest in the possible impact of the program on the work effort of poor families. Information about the work behavior of females with dependent children was not considered a good indicator of the work response of able-bodied males to public assistance. Very little was known about the work response of males because, as a group, able-bodied men and their families were generally not entitled to public assistance.

In the rest of this section we explore some of the issues and problems faced by the researchers during the course of this field experiment and discuss its outcome.

GENERALIZABILITY. To limit possible extraneous factors, families were originally chosen from a fairly homogeneous setting—New Jersey. Because a nationally dispersed sample was not chosen, however, the ability to generalize findings to a national program was limited. Generalizability was also affected by the three-year duration of the experiment. Families knew that the program was not permanent, and this may have affected their behavior.

INSTRUMENTATION DIFFICULTIES. The experiment also encountered problems with income measurement. Participants were asked to report their gross income, but families had trouble distinguishing between net and gross income. Families in the experimental groups learned how to fill out the reports correctly more quickly than did control group families because the experimental group families were asked to report income every month. Control group families (that is, other low-income families) were asked to report income only every three months. As a result, the accuracy of income data changed over time and differentially for experimental and control group families. This one-month/three-month difference arose because researchers were afraid that too much contact with control families would change their behavior (instrument reactivity) and make them less than true controls. This is an example of the trade-offs that researchers must make to avoid the numerous threats to the validity of experiments.

UNCONTROLLED ENVIRONMENT. In field experiments, unlike in laboratory experiments, researchers are not in complete control of subjects' environments. This point was illustrated dramatically during the New Jersey income-maintenance experiment. In the middle of the experiment, New Jersey adopted a public assistance program called Aid to Families with Dependent Children Unemployed Parent (AFDC-UP). Eligible families included those with dependent children and an unemployed parent, male or female. One reason that New Jersey had been chosen as an appropriate location for the income-maintenance experiment in the first place was precisely because it did not offer AFDC-UP. When it became available, however, AFDC-UP provided an attractive alternative to some of the experimental cell conditions, and many families dropped out of the experiment.

Another problem arose because there were not enough eligible families in the New Jersey communities that were chosen to provide sufficient ethnic diversity. As a consequence, an urban area in northeastern Pennsylvania was included. However, the families in that area faced conditions that differed from those of the New Jersey families, and the families varied on some important characteristics, such as home ownership. One purpose of the study was to examine whether ethnic groups responded differently to the income-maintenance program. Because whites were represented mostly from one site, it became difficult to separate ethnic differences from site-induced differences.

ETHICAL ISSUES.   Even though participation in the program was voluntary, the researchers were concerned about the effect of termination of the experiment on families that had been receiving payments. At the start of the experiment, families were given a card with the termination date of payments printed on it. Researchers debated tapering off payments and providing families with reminders as the end of the experiment approached. They decided to remind the families once toward the end, and research field offices remained open as referral agencies in case families needed help. But none requested help. Answers to a questionnaire three months after the last payment indicated that the experiment caused no serious adverse effects on the families that had participated.

MAJOR FINDINGS.   Among white male heads of families receiving negative income tax payments, there was only a 5 to 6 percent reduction in average hours worked. For black male heads of families the average hours worked increased, although not significantly. For Spanish-speaking male heads of families the average hours worked decreased but also not significantly. Researchers were unable to explain this unexpected finding, and therefore it may be unreliable. Black working wives did not change their behavior, whereas Spanish-speaking and white working wives reduced their work effort considerably. Experimental families made larger investments in housing and durable goods than control families. There was also an indication that experimental families experienced increased educational attainment.

Because of the many difficulties encountered, the income-maintenance experiment failed to provide accurate cost estimates for alternative negative income tax plans or clear findings on the work disincentive of various tax rates. Because of these shortcomings, the experiment was not able to provide conclusive evidence in favor of or against a negative income tax plan.

The New Jersey income-maintenance experiment illustrates the difficulty of studying a significant political phenomenon both experimentally and in a natural setting. The researchers who conducted this experiment developed plausible, significant, and testable hypotheses and used an imaginative research design to test those hypotheses. They identified the most interesting experimental treatments, attempted to assign their subjects to those treatments to accomplish pre-treatment equivalence, and conducted their experiment over a fairly lengthy period of time and in a natural setting to increase the external validity of their findings. Still, their efforts to isolate the effects of the independent variables in question were stymied by the real-world behavior of their subjects and by their inability to control completely both the experimental treatment and the environment in which it was operating. Researchers with fewer resources and even less control over both their subjects and the introduction of experimental treatments find it even more difficult to conduct meaningful experimental inquiries.

We have spent a considerable amount of time describing several experimental research designs to illustrate how experimental designs can help researchers draw appropriate conclusions about the effects of independent variables. Experimental designs are potentially useful because they allow researchers to isolate the effects of independent variables by controlling the assignment of subjects to experimental treatments, the introduction of the experimental stimulus itself, and the presence of extraneous influences. As a result, well-conducted experiments permit the evaluation of research hypotheses and the accumulation of causal knowledge.

Unfortunately, many of the sorts of hypotheses and behavioral phenomena of interest to political scientists do not lend themselves to the use of experimental research designs. Political scientists are limited by their inability to control completely the variables or the subjects of interest. Suppose, for example, that a researcher wanted to test the hypothesis that poverty causes people to commit robberies. Following the logic of experimental research, the researcher would have to randomly assign people to two groups, measure the number of robberies committed by members of the two groups prior to the experimental treatment, force the experimental group to become poor, and then at some later date measure again the number of robberies committed. Clearly, no researcher would be permitted to have this much control over a subject's life. Although the logic of experimental research designs is compelling, many researchers interested in explaining significant political phenomena have had to develop and employ other research designs.

Nonexperimental Designs

Because laboratory and field experiments are difficult to carry out, particularly when one wants to study aggregates like cities, counties, organizations, or countries, social scientists usually rely on nonexperimental approaches that are more practical. Although these methods are not as strong for making causal inferences, they allow the exploration of more realistic problems and even the study of nonindividual units of analysis such as events, groups, and aggregates (for example, states or countries). Whatever the case, a **nonexperimental design** is a strategy for collecting information and data that will be used to test hypotheses and, if possible, make causal inferences. Such a design is characterized by at least one of the following: presence of a single group, lack of control over the assignment of subjects to groups, lack of control over the application of the independent variable, or inability to measure the dependent variable before and after exposure to the independent variable occurs. Because of these factors, causal inferences made using nonexperimental designs are not as strong as those possible through the classical randomized controlled experiment. Think of them as alternative plans or strategies for collecting data in a nonlaboratory setting.

We overview some possibilities in Table 5-5. Look, for example, at the row labeled "Surveys." A survey or poll can include anywhere from 100 to 5,000 (or more!) individuals, but polling fewer than 100 people is possible and not necessarily unsound. Furthermore, many research projects combine elements of different designs as in a panel study with an intervention interpretation (see below).

Whatever the design, the purpose of research is to collect information or data in order to test hypotheses, look for relationships among variables, and where possible make causal inferences. To compensate for the inferential shortcomings of the nonexperimental designs (especially the lack of random assignment), it is frequently necessary to achieve a rough approximation of randomization by statistical means. For instance, as demonstrated in later chapters, especially Chapters 12 and 13, surveys can be used to gather quantitative and qualitative data, which can then be manipulated mathematically to control for the effects of one or more extraneous variables while seeing how the main independent variable influences the dependent variable. A few of these designs are described in more detail here and in subsequent chapters. In reviewing them, we compare their features to the characteristics of ideal experiments mentioned earlier.

Small-N Designs

CASE STUDIES AND COMPARATIVE ANALYSIS.   In a **small-N design** the researcher examines one or a few cases of a phenomenon in considerable detail, typically using several data collection methods, such as personal interviews, document analysis, and observation. When just one instance of a phenomenon is under investigation, the design is often called a *case study*; when two or more are involved the term *comparative* or *comparative case study* or analysis is frequently used. The units of analysis or the subjects of the study can be people

TABLE 5-5
Nonexperimental Research Designs

Design	Typical Number of Units or Cases (N)[a]	Examples of Units of Analysis	Purpose	Examples
Small-N Designs				
Single case (case study)	1	Event, nation, group, county, individual[b]	Usually one "thing" is studied in detail.	Study of French Revolution; study of U.S. House Ways and Means Committee
Comparative analysis	2 to 20	Events, nations, groups, counties, individuals	Two or more things are compared in relative detail.	Comparison of French and Russian revolutions; study of five Latin American countries; in-depth interviews with selected members of the British parliament
Focus group	10 to 20	Individuals	Often used in market research to probe reactions to stimuli such as commercials.	Test campaign ad's effectiveness
Cross-sectional Design				
Surveys (polls)	100 to 5,000	Individuals	A large number of people are measured on several variables to search for (possibly causal) relationships.	Study of U.S. public opinion and war in Iraq
Aggregate data analysis	20 to 500	Aggregates:[c] states, counties, cities, countries	Variables are often averages or percentages of geographical areas, but the goal is to search for (possibly causal) relationships.	Study of the death penalty and crime rates in different states; study of the relationship between union strength and welfare spending in developed countries
Longitudinal (Time-Series) Designs				
Trend analysis	20 to 300	Aggregates, individuals, cohorts[d]	Measurements on same variables at different time periods to examine changes in levels.	Study of changing levels of trust in government; level of unemployment and occurrence of civil strife in Europe, 1900–2000
Panel study	200 to 5,000	Individuals, households, cohorts	The *same* units are measured at different times to investigate relationships, changes in strength of relationships, and causality.	Study of changes in opinions toward prime minister of England
Intervention analysis	20 to 300	Individuals, cohorts, aggregates	Variables are measured at different times both before *and* after an intervention to see if it affected trends.	Study of number of traffic fatalities before and after enactment of seatbelt law in Texas

[a] These numbers are merely suggestive; some designs involve fewer or more cases.

[b] Biography, for example.

[c] Data are usually summations or averages of aggregations of individuals (often in geographical areas) such as median income in cities or counties.

[d] Individuals who experience the same event or experience or characteristics such as a birth cohort (people born in a specific year or period) or an event cohort (e.g., those who *first* voted in 1972).

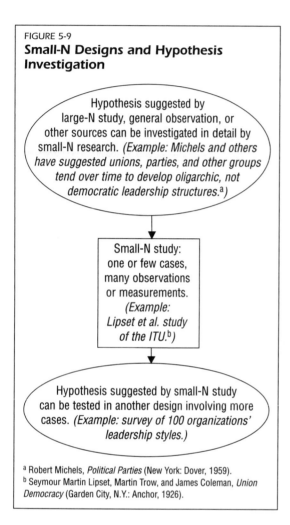

FIGURE 5-9
Small-N Designs and Hypothesis Investigation

Hypothesis suggested by large-N study, general observation, or other sources can be investigated in detail by small-N research. *(Example: Michels and others have suggested unions, parties, and other groups tend over time to develop oligarchic, not democratic leadership structures.[a])*

Small-N study: one or few cases, many observations or measurements. *(Example: Lipset et al. study of the ITU.[b])*

Hypothesis suggested by small-N study can be tested in another design involving more cases. *(Example: survey of 100 organizations' leadership styles.)*

[a] Robert Michels, *Political Parties* (New York: Dover, 1959).
[b] Seymour Martin Lipset, Martin Trow, and James Coleman, *Union Democracy* (Garden City, N.Y.: Anchor, 1926).

(for example, prime ministers), events (such as the outbreak of the Korean War), institutions (for example, the U.S. Senate), nations or alliances (for example, NATO), decisions (such as the decision to invade Iraq), or policies (for example, *Roe v. Wade* dealing with abortion). The point is that one or a few cases or instances are studied in depth. As the sociologist Theda Skocpol explains, these types of designs involve "too many variables and not enough cases,"[13] meaning that the investigator collects lots of data on one or a few units.

A small-N study may be used for exploratory, descriptive, or explanatory purposes. Exploratory case studies are sometimes conducted when little is known about a phenomenon. Researchers initially may observe only one or a few cases of that phenomenon, and careful observation of this small set of cases may suggest possible general explanations for the observed behavior or attributes. These explanations—in the form of hypotheses—can then be tested more systematically by observing more cases. (See Figure 5-9.) Carefully scrutinizing the origins of political dissent within one group at one location may suggest general explanations for dissent; or following a handful of incumbent representatives when they return to their districts may suggest hypotheses relating incumbent attributes, district settings, and incumbent-constituency relations.[14]

Descriptive studies may be used to find out and describe what happened in a single or select few situations with a view toward finding avenues for further research. Here, the emphasis is not on developing general explanations for what happened. Alternatively, in some situations a single case may provide a critical test of a theory.[15] Recall from Chapter 2 that verification and falsification are crucial activities in science; finding a single exception may cast doubt on a previously accepted proposition. Therefore, if you can find a well-documented instance in which a widely accepted or important proposition does not hold, you may make a significant contribution. Finally, according to Robert K. Yin, case studies are most appropriately used to answer "how" or "why" questions.[16] These questions direct our attention toward explaining events. The strongest case studies start out with clearly identified theories that are expected to explain the events.

For years some scholars considered the use of case studies a suspect or even inferior research strategy, partly because of the limited sample sizes. Moreover, it might be thought that they are useless in causal analysis. But social scientists now recognize this type of design as a "distinctive form of empirical inquiry" and an important design to use for developing and evaluating public policies as well as for developing explanations for and testing theories of political phenomena.[17] Figure 5-9 illustrates some of these ideas. Suppose, for instance, a theory suggests that organizations such as labor unions tend to develop an oligarchic or elite leadership structure. But as we explain in a moment, a detailed investigation of a particular union can show if this proposition holds and if not, why not. By the same token, research on a single revolution might suggest propositions that could be verified later.

Proponents argue that a small-N design has some distinct advantages over experimental and cross-sectional designs for testing hypotheses under certain conditions. For example, a case study may be useful in assessing whether a statistical correlation between independent and dependent variables, discovered using a cross-sectional design with survey data (see below), is really causal.[18] By choosing a case in which the appropriate values of the independent and dependent variables are present, researchers can try to determine the timing of the introduction of the independent variable and how the independent variable caused the dependent variable. That is, they can learn whether a link exists between the variables and, therefore, can more likely offer an explanation for the statistical association. Benjamin Page and Robert Shapiro concluded their study of the statistical relationship between public opinion and public policy with numerous case studies.[19]

These studies differ from experimental designs in that the researcher is not able to assign subjects or cases to experimental and control groups nor can he or she manipulate the independent variable. Furthermore, the researcher does not control the context or environment as in a laboratory experiment. Yet the careful selection of a case or cases can lead to the approximation of a quasi-experimental situation. For example, a historian or political scientist may choose cases with different values of an independent variable but with the same values for important control variables. Cases with similar environments can be chosen. Lack of complete control over the environment or context of a phenomenon can be seen as useful. If it can be shown that a theory works and is applicable in a real situation, then the theory may more readily be accepted. This may be especially important, for example, in testing theories underlying public policies and public programs.

A good example of a case study is *Union Democracy* by Seymour Martin Lipset, Martin Trow, and James Coleman.[20] Sociologists and political scientists had long observed that voluntary organizations conform to Robert Michels's "iron law of oligarchy."[21] Lipset and his colleagues, however, observed that the

International Typographical Union (ITU) did not conform to the normal oligarchical pattern in which one group "controls the administration, usually retains power indefinitely, rarely faces organized opposition, and when faced with such opposition often resorts to undemocratic procedures to eliminate it."[22] The ITU had an institutionalized two-party system that regularly presented candidates for chief union posts elected in biennial elections. In *Union Democracy*, the authors attempted to explore this anomaly and in doing so helped explain the workings of democratic processes in general.

This kind of research may involve more than one case, in which instances they are often called *comparative case* studies. A comparative or multiple case study is more likely to have explanatory power than a single case study because it allows for replication; that is, it enables a researcher to test a single theory more than once. For some cases, similar results will be predicted; for others, different results will be predicted.[23] Multiple cases should not be thought of as a sample. Cases are not chosen using a statistical procedure to form a representative sample from which the frequency of a particular phenomenon will be calculated and inferences about a larger population drawn. Rather, cases are chosen for the presence or absence of factors that a political theory has indicated to be important.

As an example of the logic and layout of a comparative study and its potential utility in causal analysis, suppose a political scientist wanted to know why socialism never emerged as a major political force in the United States, especially in contrast to European nations such as Great Britain, a situation that has intrigued innumerable scholars for the past hundred years.[24] The most commonly cited "causes" of the failure of socialism to take root in America include "its relatively high levels of social equalitarianism, [enormous] economic productivity, and social mobility (particularly into elite strata), alongside the strength of religion, the weakness of the central state, the earlier timing of electoral democracy, ethnic and racial diversity, and . . . the absence of fixed social classes."[25] Imagine the researcher trying to sort out these possibilities by comparing France and the United States.

One strategy is the application of Mill's *method of difference* as introduced by the English philosopher John Stuart Mill: "If an instance in which the phenomenon under investigation occurs, and an instance in which it does not occur, have every circumstance in common save one, that one occurring only in the former: the circumstance in which alone the two instances differ, is the effect, or cause, or a necessary part of the cause, of the phenomenon."[26] For example, our investigator would first identify a country in which the condition (socialism) is present and one in which it is not. He or she would then look for similarities and differences in these antecedents. As the hypothetical data in Table 5-6 suggest, it might be that in the nineteenth century the United States and France shared

TABLE 5-6
Mill's Method of Difference

Case (Country)	Condition or Effect (Socialist Movements)	Antecedent 1 (Industrialized)	Antecedent 2 (Urbanized)	Antecedent 3 (Common Language)	Antecedent 4 (Historically Strong and Fixed Social Classes)
United States	yes	yes	yes	yes	no
France	no	yes	yes	yes	yes

similar experiences such as extensive industrialization and urbanization and that in both countries citizens spoke common languages (French and English) but they differed in that the French had a fixed and rigid social class system (including a landed aristocracy), whereas the United States did not. Since the two countries have parallel backgrounds except for their class structures, we may infer that this difference, rather than the other factors, explains why socialism has not had much of an influence on American politics.

Needless to say, this comparison is woefully inadequate and simplistic. In fact, no real analysis would take exactly this form. Instead, the table is a reconstruction (see Chapter 2) of the logic of comparison using this method. (Mill, by the way, introduced several other comparative methods, but knowing them is not essential for understanding the gist of comparative analysis.[27]) If you were to attempt research of this sort, you would have to consider many more factors and make difficult decisions about when an antecedent is or is not present. But the method of difference and similar designs underlies a great deal of political research.[28]

Despite the important contribution that case studies can make to our understanding of political phenomena, there are some concerns about the knowledge they generate.[29] One potential problem is the lack of rigor used in presenting evidence and the possibility for bias in using it. Typically, researchers sift through enormous quantities of detailed information about their cases. But how does one know all the important possible antecedents have been identified? Has something significant been omitted? Or the researcher may be the only one to record certain behavior or phenomena. Still, the potential for bias of this sort is not limited to case studies.

Another frequently raised criticism of case studies is the problem of generalization. One response to this criticism is to use multiple case studies. In fact, as Yin points out, the same criticism can be leveled against a single experiment—scientific knowledge is usually based on multiple experiments rather than on a single experiment.[30] Yet people do not say that performing a single experiment is not worthwhile. Furthermore, Yin states: "Case studies, like experiments, are generalizable to theoretical propositions and not to populations or universes. In this sense, the case study, like the experiment, does not represent a 'sample,' and the investigator's goal is to expand and generalize theories (analytic generalization) and not to enumerate frequencies (statistical generalization)."[31]

A third potential drawback of case studies is that they may require long and arduous efforts to describe and report the results owing to the need to present adequate documentation. (Think about the complexity of untangling the differences between French and American societies.) This criticism may stem from confusing the case study with particular methods of data collection, such as participant observation (discussed in Chapter 8), which often requires a long period of data collection.[32] But case studies should not be ruled out as an appropriate research design due to this historic association.

Finally, in spite of the enthusiasm for case studies, considerable debate remains about just how strong causal inferences can be in these designs. Consider Table 5-6 again.[33] If our hypothetical study had discovered that France and the United States had the same values on *all* the independent variables, we would conclude that none of the hypotheses in this instance holds. And, as we stressed in Chapter 2, that might be an important conclusion given that falsification of propositions is one of the goals of science. But instead the conclusion seems to be that having a deep-seated social class system is at least a necessary condition for the emergence of socialism. Yet the result is hardly definitive. If, for example, the "real" cause of the dependent variable is not identified and explicitly included in the analysis, this design cannot detect it. Or it is plausible that the nonexplanatory variables (for example, industrialization, common language) interact or have a simultaneous joint effect on the dependent variable. In a design of the sort illustrated in Table 5-6 it is impossible to know.[34] Furthermore, this argument assumes causation is deterministic: once X appears, Y *always* follows. Yet, many, if not most scholars regard probabilistic causation—if X appears, then Y *probably* follows—as a more realistic description of the way the world works. Is it possible that deep social-economic cleavages do not always produce socialist movements? The method of difference provides no foolproof answer.[35]

Still, in many circumstances the case study design can be an informative and appropriate research design. The design permits a deeper understanding of causal processes, the explication of general explanatory theory, and the development of hypotheses regarding difficult-to-observe phenomena. Much of our understanding of politics and political processes comes from case studies of individual presidents, senators, representatives, mayors, judges, statutes, campaigns, treaties, policy initiatives, and wars. The case study design should be viewed as complementary to, rather than inconsistent with, other experimental and nonexperimental designs.

FOCUS GROUPS. A focus group consists of a small number of individuals (about twenty, say) who meet in a single location and discuss with a leader a topic or research stimulus such as a proposed campaign brochure. A focus group can superficially resemble an experiment, but usually no effort is made to assign participants randomly to treatment and control groups or to systematically introduce an experimental variation. The deliberations may or may not be (surreptitiously) recorded or observed by others on the research team. This approach lends itself nicely to market research but for the reasons just mentioned is seldom used to make causal inferences.

Focus groups have become somewhat controversial in politics because, critics assert, the results often help candidates, groups, and parties take safe or noncontroversial positions on issues. Some seemingly daring policy proposals, for example, have been thoroughly researched in focus groups, which eased the minds of political consultants that their candidates stood very little risk in adopting them. Yet, these small group discussions can be used to create hypotheses that can then be tested in larger surveys. Suppose you wanted to conduct a poll on your campus about physician-assisted suicide. You might begin with a focus group discussion to see generally what students thought about the issue. The verbal reports might then assist you in developing specific items for a questionnaire.

Cross-sectional Designs:
Surveys and Aggregate Data Analysis

Perhaps the most common nonexperimental research design is cross-sectional analysis. In a **cross-sectional design,** measurements of the independent and dependent variables are taken at approximately the same time,[36] and the researcher does not control or manipulate the independent variable, the assignment of subjects to treatment or control groups, or the conditions under which the independent variable is experienced. If the units of analysis are individuals, the study is often called a *survey* or poll; if the subjects are geographical entities such as states or nations or other groupings of units, the term *aggregate analysis* is frequently applied. In either situation the units are simply measured or observed and the data recorded. The measurements are taken at roughly the same time (in an ideal world they would be taken simultaneously). In surveys the respondents themselves report their exposure to various factors. In aggregate analysis the investigator only observes which units have what values on the variables. There is no physical assignment of subjects to treatments. The measurements are used to construct, with the help of statistical methods, post-treatment quasi-experimental and quasi-control groups. Here we use the term *quasi* to indicate that the researcher does not control the assignment to the experimental or control groups. Rather the researcher sorts subjects into naturally occurring groups based on the subjects' values on an independent variable, and the measurements of the dependent variable are used to assess the differences between these groups. Data analysis, rather than physical manipulation of variables, is the basis for making causal inferences.

Although this approach makes it far more difficult to measure the causal effects that can be attributed to the presence or introduction of independent variables (treatments), it allows observation of phenomena in more natural, realistic settings; increases the size and representativeness of the populations studied; and allows the testing of hypotheses that do not lend themselves easily to experimental treatment. In short, cross-sectional designs improve external validity at the expense of internal validity.

The example presented at the beginning of the chapter illustrates a particularly simple cross-sectional study. Recall that we tried to assess the effects of negative campaigning on the likelihood of voting by interviewing (that is, surveying) a sample of citizens and then categorizing respondents according to their answers to questions such as "Did you happen to see or read any campaign ads? How many? How many seemed to attack the opponent?" We could then sort respondents by their self-reported level of exposure to negativity. Because there is no random assignment (only self-reports), we do not control who is in each group by forcing people to view differing levels of negative advertising. The groups are simply observed. Figure 5-10 shows the layout of this study. If the groups differed by their rate of voting, we would have identified a relationship but would not demonstrate cause and effect.

Suppose, for instance, we find that M_1 (the percentage of those who viewed six or more negative ads who also reported voting) is less than M_2, which in turn is less than M_3. Because of our research design and our inability to ensure that those with less and those with more exposure were alike in every other way, we could not necessarily conclude that campaign tone determines the propensity to vote. And this is true no matter how large our sample. With a survey design, then, we typically have to use data analysis techniques to control for potential confounders that may affect both the independent and dependent variables. If we wanted to control for these factors, we would have to in-

---

FIGURE 5-10
Logic of Survey Design

Questionnaire: "Did you see any negative commercials? How many?"	Assignment based on responses $\longrightarrow$	NR_1: More than 6 ads	M_1
		NR_2: 1 to 6 ads	M_2
		NR_3: None	M_3

NR_i = nonrandomized group based on questionnaire responses; that is, quasi-treatment and control groups

M_i = measurement on dependent variable: percentage of group who report voting

FIGURE 5-11
Design with Control Variable

Step 1: Independent Variable: "Did you see any negative commercials? How many?"	Step 2: Control Variable: "Did you graduate from high school?"	Assignment based on responses to independent *and* control variables →	NR_1: High school + More than 6 ads	M_1
			NR_2: No High school + More than 6 ads	M_2
			NR_3: High school + 1 to 6 ads	M_3
			NR_4: No High school + 1 to 6 ads	M_4
			NR_5: High school + No ads	M_5
			NR_6: No High school + No ads	M_6

NR_i = nonrandomized group based on questionnaire responses; that is, quasi-treatment and control group

M_i = measurement on dependent variable: percentage of group who report voting

clude appropriate questions in the survey and then use statistics to hold them constant. Suppose we thought that education independently affected the propensities to watch a lot of television and to not vote. We would include in the survey a question about the level of the respondents' schooling, as indicated in Figure 5-11. Here we have formed six, nonrandomized groups and can compare their participation rates, M, as before. Presumably if education is creating the (spurious) relationship between voting and viewing habits, the M values for those with the same level of education would all be about the same except for sampling and measurement error. If, however, advertising does affect motivations even after controlling for education, the average measurements in the groups would vary. Under the hypothesis being considered, M_1 and M_2 would be less than, say, M_3 and M_4.

It might be possible to analyze this same problem with cross-sectional aggregate data, but we would confront the same difficulties. For example, we might treat "campaigns for the House of Representatives" as the unit of analysis. Then for each campaign in our survey we would try to determine (1) the proportion of radio, print, and television ads that attack the opposition—presumably we would have an objective standard for judging advertisements as negative, positive, or neutral—and (2) the percentage of the electorate, in each district, that cast a ballot. We might find that turnout was substantially lower when campaigns used a lot of negative advertising (say, more than 50 percent), compared with those races involving less negativity. Again, we would identify an association—the more negative the campaign, the lower the participation. But, and this is the essential point, we would be unable to infer causality. It is possible, after all, that the "negative" House races *also* or coincidentally took place in regions where turnout is generally low because of historical and economic factors, whereas the "clean" elections occurred where levels of participation are generally higher. Of course, we could control for past turnout, but that would not eliminate the possibility that some other, unmeasured variable is creating a spurious relationship between campaign tone and voting.

In essence, the limitations of the cross-sectional design—that is, lack of control over exposure to the independent variable and inability to form pure experimental and control groups—force us to rely on data analysis techniques to isolate the impact of the independent variables of interest. This process requires researchers to make their comparison groups equivalent by holding relevant extraneous factors constant and then observing the relationship between independent and dependent variables, a procedure described more fully in Chapters 12 and 13). Yet holding these factors constant is problematic, since it is difficult to be sure that all relevant variables have been explicitly identified and measured. Even if a causal variable is not recognized and brought into the analysis, its effects are still operative.

A more detailed and real example of these ideas is Edward Tufte's use of cross-sectional aggregate data in a study of whether compulsory automobile safety inspection programs help reduce traffic fatalities.[37] Tufte's hypothesis was that "states with inspection programs have fewer automobile deaths than states without inspection programs." He measured the relevant variables at the same time, even though he used the average of auto fatality rates in three years as the dependent variable. The average number of traffic fatalities per 100,000 people for states with mandatory inspections was 26.1. For states without inspections, it was 31.9. Hence, he observed, in our notation, that $M_{\text{quasi-experimental}}$ was less than $M_{\text{quasi-control}}$.

Given Tufte's data, would we be safe in concluding that inspection programs caused the lower death rate? Possibly, but there are some problems with this conclusion. First, because the study lacks a pre-test of the dependent variable, we don't know whether a difference has always existed in the auto death rates of these two groups of states. Even before adopting inspection programs, some states may have had very low death rates for reasons that have nothing to do with car inspections. (We discuss alternative approaches to this static design in the next section.) Second, because Tufte did not control the assignment of the states to the two groups, he could not be sure that all relevant extraneous factors were distributed at random. Tufte did statistically control for some relevant differences among states, yet states with inspection programs may still have differed systematically from states without programs in some other way that was related to traffic fatalities. Hence we cannot be certain that any portion of the difference in fatality rates between the two groups of states can be attributed to the effect of an automobile inspection program.

That said, the difference in average death rates between the two groups of states may in fact understate the benefits of inspection programs, especially if many of the inspection programs were weak or poorly implemented. This possibility is plausible, since the treatment given in the states with inspection was not controlled by Tufte and could not be carefully observed. Clearly the lack of a pre-test and of control over the assignment of cases to the quasi-experimental and quasi-control groups creates difficulties for researchers who use the cross-sectional design.

Let's wrap up this section by recalling from Chapter 3 the inferential problems raised by aggregate data analysis. To wit, conclusions based on aggregates may or may not apply to individuals. For example, political scientists Donald Matthews and James Prothro used data from the 1950s on counties in the American south and found that the level of voting registration of African Americans was negatively correlated with median years of schooling among whites.[38] That is, the higher the level of education in a county, the lower the proportion of blacks who were registered. The data, however, pertain to counties. And, as the authors knew full well, it can be misleading, even fallacious, to extrapolate to the behavior of individuals from information gathered on aggregations. In Chapter 3 we called this type of inference an ecological fallacy. The point is that aggregate data analysis can be extremely helpful for the study of many social and political topics. But you must always remember the units of analysis to which conclusions apply. If you have measured, say, counties, your conclusion will, strictly speaking, apply to them and not to individuals or other kinds of units.

Large Longitudinal (Time Series) Designs

Longitudinal or time series designs are characterized by the availability of measures of variables at different points in time. As with the other nonexperimental designs, the researcher does not control the introduction of the independent variable(s) and must rely on data collected by others to measure the dependent variable. On the other hand, they have two distinct advantages: (1) Change in the level of variables or conditions can be measured and modeled, and (2) it is sometimes easier to decide time order or which comes first, X or Y. For political scientists interested in the dynamics of political phenomena, these are important considerations. Being static, cross-sectional studies do not lend themselves as well to these purposes.

Additional benefits of longitudinal studies include the fact that they can in principle estimate three kinds of effects: age, period (history), and cohort. Age effects can be considered a direct measure of (chronological) time, and its effects can be assessed like other variables. As in cross-sectional work, an investigator may be interested in the effect of age on political predispositions or ideology. (It is commonly asserted that as people age they become more politically conservative. Presumably something in the aging process—experiences, changes in hormones—affects people's perceptions and attitudes.) But in longitudinal analysis a **period effect** may be thought of as an indicator of history during a time period, and this "history," not chronological age, is what matters. (During the late 1960s, for example, events such as Watergate and the Vietnam War adversely affected many citizens' trust in government, whether they were young or old. When that era passed, the effects on newer generations dissipated. So people who lived through those stormy times might have much different beliefs and opinions than younger people.)

Another way of interpreting time series effects is to consider cohorts.[39] A **cohort** is defined as a group of people who all experience a significant event at roughly the same time. A birth cohort, for instance, consists of those born in a given year or period; an event cohort includes those who shared a common experience such as their first entry into the labor force at a particular time. It is often hypothesized that one cohort will, because of its shared background, behave differently from individuals in a different cohort. To take one example, people born in the years immediately after World War II (the baby boomers) may have different political attitudes and affiliations than those who were born in the 1980s. Note that cohort and period effects are inescapably related because "cohort (year of birth) = period (duration of an event) – age (years since birth)."[40] There are, in short, a number of ways of understanding longitudinal research. The choice depends on the analyst's interests.

TREND ANALYSIS. As the term implies the analysis of a trend starts with measurements on a dependent variable of interest taken at different time periods (usually 20 or more) and attempts to determine whether and why the level of the variable is changing. A simple example shows changes in party identification—the feeling of being "close" to one of the parties—over the last 30 years. Figure 5-12 indicates that the percentages of citizens identifying with the Democratic and Republican Parties have declined significantly. If you are a party activist or journalist, this information is important. Still, it doesn't explain *why* partisanship seems to be slipping.

For that purpose an investigator needs to introduce additional variables and measure them over time. This type of analysis takes (roughly speaking) this form:

$$Y_1 = f(Y_{t-1} + Y_{t-2} \ldots + X's_t + X's_{t-1} + X's_{t-2} \ldots),$$

where the Y's and X's are measures of the dependent (Y) and independent (X) variables at the current (latest) time (t) and at previous times (t – 1, t – 2, . . .) and "*f*" means "is a function of . . ." or "is produced by . . ."[41] When data are measured at many different points, as illustrated in Figure 5-12, statistical procedures called time series analysis are often used. Although these techniques are somewhat advanced, they build on the ideas presented here and in later chapters.

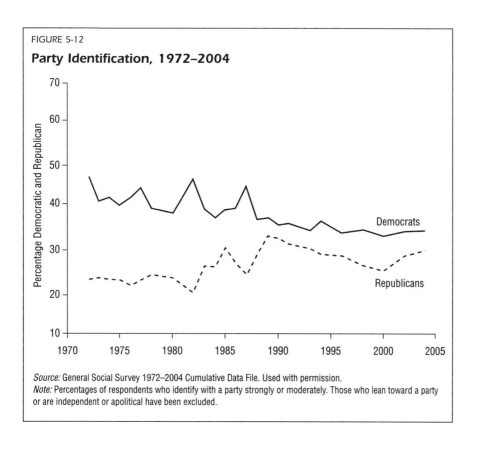

FIGURE 5-12

Party Identification, 1972–2004

Democrats

Republicans

Source: General Social Survey 1972–2004 Cumulative Data File. Used with permission.
Note: Percentages of respondents who identify with a party strongly or moderately. Those who lean toward a party or are independent or apolitical have been excluded.

An example of this approach is Christopher Gelpi, Peter D. Feaver, and Jason Reifler's investigation of Americans' support for overseas military operations, described in Chapter 1. In particular, the authors noted that in April 2003 "76 percent of the public approved of President Bush's handling of the war; by the time of the U.S. election in November 2004, that number had dropped to 47 percent; a year later, it had dropped below 35 percent."[42] But instead of merely documenting this change, the authors attempted to explain it by examining trends in the numbers of American casualties and perceptions of progress in the war effort. Refer to Figure 1-2, on page 23. It graphs President Bush's approval ratings and military deaths in various phases of the war. Notice that the trend in the dependent variable (Bush's approval ratings) has been plotted along with trends in causality rates. Such a technique might have uncovered covariation among the variables. The data, however, did not back up the initial hypothesis that support would decline as casualties increased. This finding led to a more nuanced analysis in which the investigators examined the effects of changes in media coverage, perceptions of success, and events on the ground.

161

Note also that, although the previous examples pertain to changing proportions in samples of individuals, trends in aggregate variables (for example, crime or poverty rates in urban areas) can also be investigated.

Panel Studies

Suppose a public opinion analyst wants to learn if and how changes in a dependent variable, such as preferences for a particular candidate, are affected by changes in one or more independent variables, such as increasing attention to a campaign. A **panel study** is a cross-sectional design that introduces a time element. A researcher taking this approach measures the variables of interest on the same units of analysis at several different times. This strategy may be used to observe (cumulative or net) changes over time and to provide an analogue to a pre-test of some phenomenon prior to natural exposure to the experimental stimulus. A panel study is similar to a cross-sectional study, however, in that the subjects are measured at the same times in each "wave," and the researcher has no control over which subjects are exposed to the experimental stimulus.

Let us return to one of the hypothetical examples of a classical pre-test/post-test experiment described earlier in this chapter. In that example, we were interested in finding out whether or not exposure to a candidate's televised campaign commercials increased voters' ability to identify the important issues in a campaign. If we used the pre-test/post-test experimental design to test this hypothesis, we would measure pre-exposure issue awareness, randomly assign people to an experimental or a control group, expose only the experimental group to the commercials, and then measure post-exposure issue awareness again.

When using a panel research design, a researcher would proceed in a slightly different way. First, pre-exposure issue awareness would be measured for a group of subjects (presumably before any commercials have been broadcast). The researcher would wait for time to pass, the campaign to begin, and the commercials to be broadcast. Then the researcher would interview the same respondents again and measure both the amount of exposure to commercials and the post-exposure issue awareness for everyone. Finally, the researcher would use the measure of commercial exposure to construct quasi-experimental and quasi-control groups and compare the change in the amount of issue awareness for the two groups.

The major difference between the panel study and the classical experiment is that in the former the researcher waits for exposure to the experimental stimulus to occur naturally and then uses the amount of exposure reported by the respondents to create naturally occurring quasi-experimental and quasi-control groups. Hence, the researcher observes rather than controls exposure to the experimental stimulus.

Because the panel study has a pre-test and a quasi-control group, the researcher can claim greater confidence in his or her conclusions than is possible with the cross-sectional design. However, the lack of control over who is exposed to the independent variable, and under what conditions, creates the problem of nonequivalent experimental and control groups. In our example, those who are naturally exposed to more commercials may be more interested in politics and hence more likely to develop issue awareness than those who are not, for reasons that have nothing to do with exposure to commercials.

Panel studies are particularly useful in studies of change in individuals over time. One difficulty with panel studies, however, is that individuals may die, move away, or decide to drop out of the study—what researchers refer to as **panel mortality.** If these persons differ from those who remain in the study, study findings may become biased and unrepresentative.

Panel studies have often been used in election campaigns to investigate the changes in voter beliefs, attitudes, and behavior that may be attributed to aspects of a campaign. A panel study of opinion change during the 1980 presidential campaign, for example, relied on surveys conducted with the same national sample of voting-age citizens in January/February, June, and September of 1980.[43] Larry M. Bartels was interested in the effect of media exposure during a presidential campaign on "each of 37 distinct perceptions and opinions regarding the presidential candidates, their character traits, their issue positions, the respondents' own issue preferences, and (in the case of incumbent Jimmy Carter) various aspects of job performance."[44] Since Bartels had available both January/February measures of the dependent variables, which were presumably unaffected by campaign news exposure, and later measures of the same variables after four and seven months of campaign coverage, change in voter perceptions and opinions could be analyzed. Measures of exposure to television network news and daily newspapers during the campaign allowed the creation of quasi-experimental and quasi-control comparison groups, and measures of other attributes, such as party identification, permitted statistical control of other significant political factors. As a result, the researcher was able to demonstrate significant media effects during a campaign with considerable confidence, even without experimental control over the introduction of the media exposure in question.

INTERVENTION ANALYSIS.   In one version of a nonexperimental time series design, called **intervention analysis** or interrupted time series analysis, measurements of a dependent variable are taken both before and after the introduction of an independent variable. Here we speak figuratively: the occurrence of the independent variable is *observed,* not literally introduced or administered. We could observe, for instance, the annual poverty rate both before and after the ascension of a leftist party to see if regime change makes

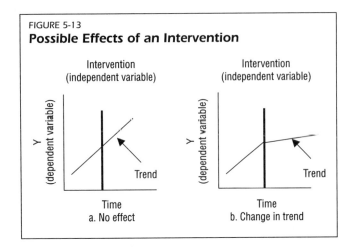

FIGURE 5-13
Possible Effects of an Intervention

Intervention
(independent variable)

Intervention
(independent variable)

Y
(dependent variable)

Y
(dependent variable)

Trend

Trend

Time
a. No effect

Time
b. Change in trend

any difference on living standards. The pre-measurements allow a researcher to establish trends in the dependent variable that are presumably unaffected by the independent variable so that appropriate conclusions can be drawn about post-treatment measures. Refer to Figure 5-13. Panel a shows an increase in a dependent variable over time. (Suppose it is the poverty rate in metropolitan areas.) At a specific moment or period, an intervention takes place (perhaps the enactment of a job training program). But the trend line remains undisturbed: Y grows at the same rate before and after the "appearance" of the independent variable. In this case the intervention did not interrupt or alter the trend. (We would conclude, for example, that the program did not affect the increase in poverty.) Now consider panel b. It shows an increase in Y until the intervention occurs, at which point the growth in the trend begins to abate. In this instance the introduction of the factor appears to have caused the trend to flatten (for example, the introduction of the job training program slowed the growth in poverty).

Intervention analysis works best when the independent variable (that is, the intervention) occurs at a particular moment or during a fairly brief period, affects a dependent variable that is routinely measured, or is known about in advance so that appropriate pre-test measurements can be made. This design would work well if, as in the previous example, we wished to evaluate the impact of a new program or policy initiative. To take another case we might try to evaluate the impact of sobriety checks on alcohol-related traffic accidents in states by examining the number of such accidents in the years before and after the introduction of the checks. If we observed a decline in accidents, we might conclude that sobriety checks had been effective. But whether the checks caused the decrease would remain unclear because other unmeasured things (the age distribution of the population, perhaps) may also be changing during the time period under study. If so, we cannot know if the checkpoints, the other factor(s), or both are affecting the fatality rate. The problem, of course, lies in the fact that there is no control group with which to compare the unit or units of analysis that experienced the independent variable. The results of a time series can often be improved if the researcher can identify quasi-experimental and quasi-control groups and produce a time series of measurements of the dependent variable for each. In this way the researcher can have more confidence that the observed shift in the dependent variable is the result of the introduction or presence of the independent variable.

For example, some states may adopt health insurance programs for children whereas others do not. Researchers can compare children's health trends in both types of states using regularly collected indicators of children's health to determine whether or not the health of children in the states with insurance programs has improved relative to the health of those in states without the programs. Even though the researcher controls neither the assignment of states to the groups with or without the program nor the content or implementation of the programs, this situation is often referred to as a "natural" experiment because of the presence of before and after measurements for both quasi-experimental and quasi-control groups.

As another example, suppose that we are interested in whether or not an aggressive media campaign organized by an interest group has an effect on popular support for a public policy initiative, such as mandatory, comprehensive health care coverage. We might first obtain a series of public opinion polls measuring popular support for mandatory coverage. Using a measure of overall media exposure, we could then separate the respondents into two groups: those most likely to be exposed to the media campaign and those least likely to be exposed to the media campaign. By continuing the time series of popular support for health care during and after the media campaign and comparing the entire series for the quasi-experimental and quasi-control groups, we could assess the influence of the media campaign on popular support for the comprehensive health care initiative.

In Figure 5-14 we see the hypothetical results of such a time-series-with-quasi-control-group design. Before the introduction of the independent variable, the less exposed (quasi-control) group was more supportive of mandatory health care coverage, with a positive trend among the less exposed group and a negative trend among the more exposed group. (This could be because the less exposed group was more Democratic and less affluent than the more exposed group and because both groups were already responding to Washington and interest group rhetoric before the campaign began.) During the media campaign the downward trend in support in the quasi-experimental group was reversed, and by the end of the campaign the most exposed group was just as supportive of mandatory health care coverage as the less exposed group. After the media campaign concluded, the level of support among the most exposed group began to decline again, while the level of support in the less exposed group remained fairly constant. This is strong evidence that the media campaign had an effect, albeit one that diminished with time.

As with trend analysis, these types of problems are usually tackled with mathematical procedures, not graphs alone. Yet the pictures illustrate the logic underlying these more advanced methods.

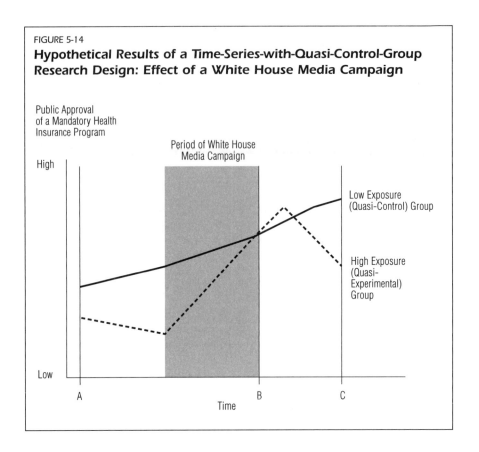

FIGURE 5-14

Hypothetical Results of a Time-Series-with-Quasi-Control-Group Research Design: Effect of a White House Media Campaign

Alternative Research Strategies

To study a phenomenon such as the effects of the media on voters or the behavior of federal justices, political scientists most commonly use one of the experimental or nonexperimental designs described above. This propensity stems from their commitment to verify hypotheses empirically and strive for valid causal inferences. But these approaches hardly exhaust the list of possibilities. We now briefly describe two alternatives that flow from the quest for scientific knowledge but that rely on different tactics.

Formal Modeling

Anyone who has ever seen or perhaps built a model airplane knows full well that these replicas do not fly passengers or carry cargo or drop bombs. They are simply representations of reality. Nevertheless, they can be quite useful and not just for entertainment. At the very least they suggest what a real airplane looks like, and many can even be used for scientific purposes. Aeronautical engineers, for example, use models to see how certain wing shapes affect a plane's stability or what a sudden downdraft will do to the wing's structural integrity. So, even though model planes may be totally unrealistic in one sense, they can still be useful, even essential, devices for learning about flight. Models, it turns out, are also quite useful in political science.

A **formal model** (frequently termed an *analytic model,* or just a *model*) is a simplified and abstract representation of reality that can be expressed verbally, mathematically, or in some other symbolic system, and that purports to show how variables or parts of a system tie together.[†] This definition may seem a bit vague, and so we describe the components of a model and then provide an example.[45]

The main parts of a model are (1) a set of "primitives," or undefined terms or words whose meaning is taken for granted, (2) a collection of assumptions, that is, statements or assertions whose validity is again taken for granted but are explicitly stated, (3) a body of rules or logic for linking the parts of the model together and making deductions, and (4) various derived propositions that are true by virtue of the rules used to deduce them. They may not be supported empirically, as we will see. A modeler has to start some place, so the most elemental parts of the system have to be taken as given. It is also necessary to make some assumptions. As an example, many political scientists assume—they do not prove with data—that people are "rational" in a sense to be discussed below. These primitives and assumptions are akin to the axioms used to construct geometries. What can make a model powerful is the application of rules, such as the rules of algebra or symbolic logic, that allow one to move from base terms and assumptions to validly derived statements. As noted, the conclusions in a model are true if the rules have been applied correctly, period. These statements may or may not hold in the real world. But most modelers believe that they will be approximately empirically valid and that, if the model is a good one, the understanding they provide more than compensates for inaccuracies in the predictions.

For an example of formal modeling let us return to the question of why people do or do not vote. This time we approach the question from a theoretical perspective. All models by definition simplify reality, and the model that follows is an especially unsophisticated or bare-bones version presented only for illustrative purposes.[46] Consider a single citizen. We assume that this person has desires or wants, which in modeling are often called "utilities." This individual may desire lower taxes, an end to federal gun control, and less military spending.[†] Denote the sum of the utilities as U. See? The model is already becoming abstract, and we are taking the meaning of "utility" more or less for granted. In any event, we can introduce some additional notation to clarify the amount of utility this person would get from two parties or candidates. For instance, let U_A be the value that Party (or Candidate) A brings to the voter on these three issues *if* it comes to power. Similarly let U_B stand for the utility the other party (or candidate) brings if it takes office. Now the model *assumes* that voters are rational, which means that they try to maximize or get the most value or utility from their actions. This conception or definition of rationality is usually termed subjective or psychological because what is getting maximized are internal mental states such as expected happiness and contentment, states that are hard to measure objectively. It is as though people mentally calculate the difference between the utilities from A and B and use the result to guide their decisions: vote for A if U_A is greater than U_B, vote for B if U_B is higher, and abstain if U_A equals U_B. (Why? See below for the answer.) No one does exactly this, of course, but people may behave as though they did.[47] More formally, we might symbolize the comparison of utilities, which we can call a party differential, as follows:

$$U_{AVB} = U_A - U_B$$

THE MEANING OF *VALIDITY*

It is necessary to separate standards of evaluation for formal models and experiments. With experiments we try to conduct research in such a fashion that we can make valid causal inferences based on empirical observation. The type of validity involved in models is different. In this case we say deductions are valid if and only if they have been properly deduced from the premises. These deductions may or may not be empirically valid. Just as the claim that the sum of the angles in a right triangle is 180 degrees follows from the axioms in Euclidean geometry and not the state of nature, derivations of a modeling process depend on starting premises and logic and not empirical reality. As we will see, we may want to test a model's predictions. If we find that they do not hold, it does not necessarily mean the model is incorrect, just that it does not describe the real world.

Note this important point: if U_{AVB} equals zero, the person sees no difference between the parties or candidates (the utilities from each are identical), has no incentive to vote, and thus abstains.[48]

To this point we have introduced a few primitive terms (for example, *utility*) and one key assumption—namely, voters are rational, meaning that they vote for the alternative that brings them the most pleasure, or in the modeler's language, they maximize utility. From these elements we use the rules of algebra to derive predictions of how a person will behave. (It is just simple algebra that if U_A equals U_B, their difference is zero. And, by assumption, if an action brings zero utility, it is not taken.) True, the conclusion appears rather trivial—people vote for their most preferred party—but we can expand the model to reach a startling conclusion.

The model implicitly assumes that people have preferences and can easily act on them. But that presumption may be too simplistic because it does not take into account so-called information and transaction costs. That is, potential voters have to take time to find out where the parties stand on the relevant issues. That task may seem relatively trivial, but it's not. Politics has to compete with a lot of matters of importance to voters, including job and family matters, desire for relaxation and entertainment, and health issues. To become informed about electoral politics requires at the least a fair amount of time spent reading or listening to information about the campaign. There may even be monetary costs, as in the expense of newspapers and magazines. In addition, parties and candidates frequently obscure their stands on issues or try to distort those of their opponents. And to make matters worse, potential voters have to inconvenience themselves to get registered, find the polling place, and take the time to actually vote.[49] A lot of these costs may not seem like much, but research tells us that they probably affect political behavior. So let us factor in a term C for the cost of becoming informed and going to the poll. Admittedly, this is an abstract term, since people do not literally summarize their expenses in a single number. Still, it does represent symbolically what people must *feel* when allocating their time and energy to certain tasks.

Now we can make the prediction that if C is greater than or equal to the absolute value of U_{AVB}, our hypothetical citizen will feel that the cost of voting outweighs any utility derived from one candidate's bringing more utility than another. In symbols, if

$$|U_{AvB}| \leq C,$$

then the costs of voting exceed the benefits that one or the other party (or candidate) brings and again we predict abstention. The rationality assumption (that is, that people maximize utility) explains why: if the costs of acting outweigh the benefits, action is not rational.

And there is another possible difficulty that we can model. Everyone must know that in any sizable election the probability that any one vote will be decisive is minuscule. For example, how likely is it that any one person's vote will determine the outcome of a contest for state representative that brings thousands of people to the polls? Elections are just not that close. (How many elections are decided by a margin of 5,001 to 5,000?) So anyone must reason that it is not going to make a huge difference in the outcome if he or she stays home on election day. That is to say, the chances of reaping the benefits from a favorite party's taking power are not affected by anything any particular voter does. And our citizen's participating or not will in all likelihood not affect a favorite party's chances of winning or losing. There are just too many votes being cast for a single vote to be decisive.

To see what taking this small probability into account does to the model of turnout, let the likelihood that a vote is decisive be P. This number will be exceedingly small in any realistic election, say one in ten thousand. And if we discount a person's utility derived from one party over another, the result will also be exceedingly small, almost approaching zero, as a matter of fact. The discounting (or multiplication) of utility by P (that is, $P|U_{AvB}|$) is called the expected or subjectively expected utility of A versus B.

Example: If the expected benefit of A over B is 100 units of utility, and the probability of casting the deciding vote and thus actually bringing about that amount of utility is one in 10,000, then the result is 0.01. Now this should be compared with the cost of voting. After all, doesn't everyone compare expected gains with the costs of obtaining them? Any cost of becoming informed and voting over that amount, say 1 unit of utility, will mean the rational voter has no incentive to participate. And since we assume all voters are rational, and that all make similar calculations, we deduce that *no one* will ever vote!

Once again this result may seem far-fetched. But it is a conclusion of many models of voting and has even earned the title "the paradox of voting."[50] This paradox has been so troubling that it has occupied dozens and dozens of social scientists who have tried to figure out why this logically deduced (from the primitives and assumptions of the model) flies in the face of the reality that many, many people do, in fact, vote.

Perhaps we can rescue our model by introducing another kind of utility that goes above and beyond that obtained by seeing one party or another elected. This additional value might come from the pleasure one receives just by participating in politics and knowing that widespread apathy could undermine democracy.[51] Let us call this new factor *E*, which means the extra utility or value brought by being active in civic life. It enters the potential voter's "calculation" as an additional or extra utility. Table 5-7 summarizes our model.

We can interpret the table a couple of ways. The deductions follow logically from the premises. They are true despite what happens in the real world. If you are troubled by this situation, you might maintain that the model has a kind of internal validity, but its external validity is low because the results do not generalize to any meaningful population. This is a reasonable argument, but it brings us to a discussion of the value of formal models in political science.[52]

Many social scientists think that modeling as a research method has many advantages. Models, they believe, lead to clear and precise thinking. As Morris P. Fiorina puts it, modeling requires that we put "all the cards on the table."[53] For a model to be useful, definitions have to be unambiguous and assumptions made explicit. If these conditions hold and the rules are known and accepted, other researchers can verify the deductions. We may not like the way a term has been defined or an assumption introduced. But if we at least know what the model says, we can suggest alternatives and find and correct errors. Verbal theories or histories, by contrast, often contain hidden or vaguely defined terms and assumptions, and because of the ambiguities, people often talk past one another. Models also tend to be compact. They do away with all but the essential aspects of a problem. True, they may oversimplify, but simplification may be necessary when studying complex political phenomena. Besides, almost any kind of research involves such narrowing of problems to manageable proportions and the reliance on simplifying postulates, a point sometimes forgotten by those critical of this technique.

TABLE 5-7

Simple Formal Model of Turnout Based on Rationality for Citizen X

Primitive terms
- U_{AvB}: utility or value Citizen X gets from voting for Party (Candidate) A versus Party (Candidate) B.
- P: Citizen X's estimated probability that his or her vote will be decisive.
- C: the cost to Citizen X of becoming informed, registered, and voting.
- E: benefits to Citizen X of democracy and participating in elections.

Assumption
- Rationality: citizens act to maximize utility.

Predictions
- Decide to vote or not:

  1. $P|U_{AvB}| + E > C$ Vote (see 3–4 below)
  2. $P|U_{AvB}| + E \leq C$ Do not vote, but go to 6–7 below

- Decide how to vote (if 1 above holds):

  3. $U_{AvB} > 0$ Vote for A
  4. $U_{AvB} \leq 0$ Vote for B
  5. $U_{AvB} = 0$ Do not vote, but go to 6–7 below

- If parties do not differ ($U_{AvB} = 0$):

  6. $E > C$ Toss coin in voting booth (because $|U_{AvB}| = 0$)
  7. $E \leq C$ Do not vote

Finally, political scientists apply formal modeling techniques in a surprisingly wide variety of areas, from international relations (for example, coalition formation, the outbreak of war, arms races) to the behavior of organizations and groups (legislative and judicial decision making, roll call voting, and committee assignments in Congress) to individual behavior (candidate preferences, candidate campaign tactics). Indeed, the procedures now occupy a prominent place in many professional journals and graduate school curricula.

We conclude this section by noting that despite the growing popularity of modeling in the social sciences, it has numerous critics who are especially upset at the use of assumptions to make a model "work." The rationality premise, for instance, causes concern because it is defined in ways that many find odd. Is a chain smoker who buys cigarettes at the lowest possible cost rational? Formal modelers would say "yes, if he or she acts to maximize (subjective or psychological) utility." But probably no doctor would agree.[54]

Simulation

A research design closely connected to modeling is simulation. For our purposes we define a **simulation** as a representation of a system by a device in order to study the system's behavior over time. The units of analysis are not discrete individuals like those in a sample survey. Rather, the fundamental interest lies in a process or structure, such as a large organization, a legislative body, or a party system, which has several or many components. To the extent that individuals enter into a simulation, their behavior as a collectivity (for example, a crowd, a committee, a coalition) is the main interest. The device for investigating the phenomenon can be a computer program, a board game, role playing, or some other method that allows the investigator to see how the system's components interact and change. At least for investigative purposes the system or process is usually thought of as "closed," or not subject to external forces.

A major consideration in simulation studies is time. They emphasize the dynamic interaction of the internal elements. If one part changes, what happens to the others? Are there feedback loops or paths of reciprocal causation? Does the system evolve to an equilibrium state, or does it spin out of control? Consequently, simulations are normally "run." From a starting point or set of "givens" the pieces are made to interact in order to see two things: how each affects and is affected by the others, and what happens to the overall state of the system at different time intervals, or iterations.

Creating simulations requires knowing as much about the subject matter as possible. They are, therefore, not very helpful for exploratory work but are useful for discovering emergent properties that cannot be found in static models or easily calculated by conventional means. If the investigator has a good idea of a system's constituents and how they interrelate but cannot easily predict what happens when they start functioning together over a period of time, a series of runs may provide insights not available by computation or deduction. As a result, they work best or are most informative when a researcher has a solid, extensive body of knowledge with which to work.

You may be familiar with simulations, although not by that name. Some games, like Monopoly, are simulations. Role-playing exercises such as mock parliaments or model courtrooms are simulations that teach participants legislative or judicial behavior. But simulations are also widely used for academic and applied purposes to study complex phenomena that cannot be investigated with laboratory or survey tools. Many simulations appear in the press, as in models of the economy constructed by economists in an attempt to determine future levels of employment or inflation. In making these models they set certain parameters and then let one or more factors vary. What will happen, they might ask, if people suddenly decide to retire early? What will happen to productivity? To health care costs? To Social Security trust funds? As another example, the Congressional Budget Office, a nonpartisan research arm of Congress, routinely tries to predict the likely effects of alternative policy options (a tax cut, perhaps) on the state of government spending or on the economy as a whole. The natural sciences, as you may know, frequently resort to simulations to investigate complex phenomena such as the effect of accumulating greenhouse gases on global climate.

To appreciate the possibilities, let's take a simplified example based on research conducted by Richard J. Stoll.[55] For generations, international relations theorists have wondered if and how in a world of self-interested, independent nations (a "Hobbesian world," it is sometimes called), a balance of power could be "automatically" achieved and maintained, especially in the absence of an external authority like a world government. For this study we assume the world consists of several dozen nations of varying power, all of which are locked in a struggle for survival. We want to know what will happen in this kind of an anarchic system if states start attacking one another. Will they eventually reach equilibrium or a balance of power? Will they mutually annihilate each other? Will one country come to dominate the others? These are interesting questions, because, after all, everyone lives in a world of sovereign states having different amounts of power and interests, and the potential for deadly conflict is always present.

The mechanism we use to assemble and "operate" the simulation depends on our purposes, the needed calculations, and similar considerations. For this example, we will use a computer program. This approach allows us to try different configurations and initial values, insert random errors, and run hundreds or even thousands of trials to see what on average happens. We cannot, of course, make an actual simulation here, but in Figure 5-15 we sketch a flowchart of the sequential steps our hypothetical program would go through on each iteration. A flowchart does not literally describe the inner workings of the machine, but it does suggest how a simulation operates.

We start by "creating" a set of countries and assigning them initial values. In this instance we might, as Stoll did, let the computer set up a table with, say, five rows and ten columns to make fifty internal countries. The program would also give each nation, which is just a storage location in the computer, values for different variables, such as its beginning power. (At the outset, the power scores might be allocated randomly.) We would also write computer programming code to reflect the rules of the game. Only contiguous states, for example, can instigate attacks on one another. These program statements represent what we think we know about an anarchic international system. If we have been woefully unrealistic, critics can tell us and adjustments can be made. The model described here takes type of regime for granted, but it might be desirable to write a rule to prevent democracies from attacking one another.[56]

Then the simulation begins with the first round or iteration. An aggressor is selected at random, its power in relation to its neighbor is calculated, alliances are formed, and power is recalculated. Based on the rules a decision is made to instigate war. The damage (cost) to the participants is calculated next, and new power scores allotted according to the rules. Some countries may be eliminated in the process. Now the program makes a decision: is more than one nation left? If yes, the simulation goes to the next round, during which an initiator is selected and new calculations are made. We might let the maximum number of iterations be any number, such as 500 or 1,000. Should countries remain after this maximum number of trials, we would conclude that the system has been preserved. If, however, only one state remains, the game ends with total domination by one superpower. The process can be run again and again to see what happens on average. Stoll based his study on 270 replications.

It is probably apparent that a simulation is something like a "thought experiment" in which a researcher wonders "what would happen if A does ____

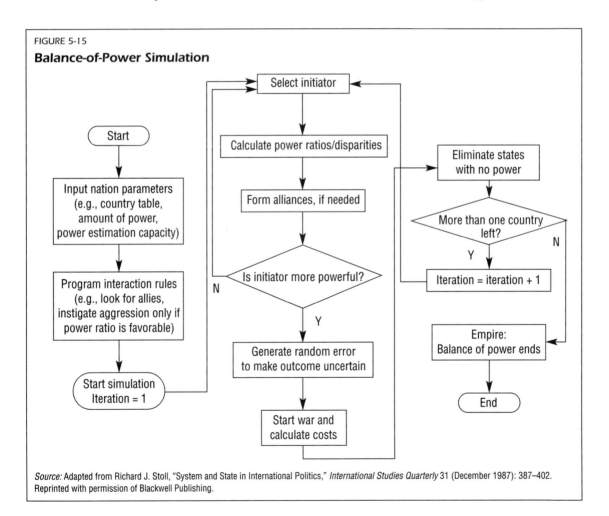

FIGURE 5-15
Balance-of-Power Simulation

Source: Adapted from Richard J. Stoll, "System and State in International Politics," *International Studies Quarterly* 31 (December 1987): 387–402. Reprinted with permission of Blackwell Publishing.

and then B responds with ____." Such an undertaking may seem sterile or fruitless, but that is only because we have provided just a bare sketch of what is and can be done with simulations. As we noted, they have become an integral part of the social sciences.[57] A great deal of what we think we know about the world comes partly from simulation studies.

Conclusion

In this chapter we discussed why choosing a research design is an important step in the research process. A design enables the researcher to achieve his or her research objectives and can lead to valid, informative conclusions. We presented two basic types of research designs—experimental and nonexperimental—along with a couple of alternative approaches. We discussed their advantages and disadvantages. Experimental designs—which allow the researcher to have control over the independent variable, the units of analysis, and their environment—are often preferred over nonexperimental designs because the former enable the researcher to establish sounder causal explanations. Therefore, experimental designs are generally stronger in internal validity than are nonexperimental ones. However, it may not always be possible or appropriate to use an experimental design. Thus nonexperimental observation may also be used to test hypotheses in a meaningful fashion and often in a way that increases the external validity of the results. In these instances, causal assertions rest on weaker grounds and frequently have to be approximated by statistical means (see Chapters 12 and 13). Yet the basic objectives of research designs, whether experimental or nonexperimental, are the same.

A single research design may not be able to eliminate all threats to internal and external validity. Researchers often use several designs together so that the weaknesses of one can be overcome by the strengths of another. Also, findings based on research with a weak design are likely to be accepted more readily if they corroborate findings from previous research that used different designs.

Notes

*. If you have trouble following this idea, imagine that you have a large can of marbles, most of which are red but a few of which are blue. Now, draw randomly from the can a single marble and put it in a box. Then draw another marble—again randomly—and put this one in a second box. Repeat this process nineteen more times. At the end you should have two boxes of twenty marbles each. If you have selected them randomly, there should be approximately the same proportion of red and blue marbles in *each* box. If you started with a can holding 90 percent red marbles and 10 percent blue, for example, each of the two boxes should hold about eighteen red marbles and two blue ones. These may not be the exact numbers, but the boxes would differ only slightly—one might have three blue marbles and the other just one—but these differences will be due solely to chance.

†. The word *model* used alone can be misleading because social scientists employ it for many different purposes. But in this context we give it a particular meaning, namely, the activities described in the discussion that follows.

‡. In this version of rationality, the wishes do not have to be logically consistent. The desire for lower taxes may be inconsistent with less military spending. But political scientists generally take preferences as given. This inconsistency is a sore point with some critics of formal models.

1. CBS News, "Going Negative," September 27, 2004. Retrieved January 17, 2004, from www .cbsnews.com/stories/2004/09/27/politics/main645881.shtml.

2. See, for example, Stephen Ansolabehere, Shanto Iyengar, Adam Simon, and Nicholas Valentino, "Does Attack Advertising Demobilize the Electorate?" *American Political Science Review* 88 (December 1994): 829–838.

3. CBS News, "Going Negative." Also see Paul Freedman and Ken Goldstein, "Measuring Media Exposure and the Negative Effects of Campaign Ads," *American Journal of Political Science* 43 (October 1999): 1189–1208.

4. See Chapter 13 for a more thorough discussion of spurious relationships.

5. Ansolabehere, Iyengar, Simon, and Valentino, "Does Attack Advertising Demobilize the Electorate?"

6. See Donald T. Campbell and Julian C. Stanley, *Experimental and Quasi-Experimental Designs for Research* (Chicago: Rand-McNally, 1966), 5–6; and Paul E. Spector, *Research Designs* (Beverly Hills, Calif.: Sage Publications, 1981), 24–27. Four components of an ideal experiment are identified by Kenneth D. Bailey in *Methods of Social Research* (New York: Free Press, 1978), 191.

7. A good reference is Donald T. Campbell and David A. Kenny, *A Primer on Regression Artifacts* (New York: Guilford Press, 1999), especially chap. 3.

8. Martin T. Orne, "On the Social Psychology of the Psychological Experiment: With Particular Reference to Demand Characteristics and Their Implications," *American Psychologist* 17 (November 1962): 776–783.

9. This discussion is based on Bailey, *Methods of Social Research,* 204–206; and Jarol B. Mannheim and Richard C. Rich, *Empirical Political Science* (Englewood Cliffs, N.J.: Prentice-Hall, 1981), 76–77.

10. Shanto Iyengar, Mark D. Peters, and Donald R. Kinder, "Experimental Demonstrations of the 'Not-So-Minimal' Consequences of Television News Programs," *American Political Science Review* 76 (December 1982): 848–858.

11. A major subfield in applied statistics is the design and analysis of this and more complicated types of experiments. For some introductions, see Steven R. Brown and Lawrence E. Melamed, *Experimental Design and Analysis* (Thousand Oaks, Calif.: Sage Publications, 1998); and Larry Toothacker, *Multiple Comparisons* (Thousand Oaks, Calif.: Sage Publications, 1990). One of the best books is the classic by Donald T. Campbell and Julian C. Stanley, *Experimental and Quasi-Experimental Designs for Research* (Chicago: Rand-McNally, 1966).

12. This discussion is based on Joseph A. Pechman and P. Michael Timpare, eds., *Work Incentives and Income Guarantees: The New Jersey Negative Income Tax Experiment* (Washington, D.C.: Brookings Institution, 1975), esp. chaps. 2 and 3.

13. Theda Skocpol, *States and Social Revolutions* (New York: Cambridge University Press, 1979), 36.

14. See Richard F. Fenno Jr., *Home Style: House Members in Their Districts* (Boston: Little, Brown, 1978).

15. Robert K. Yin, *Case Study Research: Design and Methods,* rev. ed. (Beverly Hills, Calif.: Sage Publications, 1989), 47.

16. Ibid., 17–19.

17. Ibid., 21.

18. Alexander L. George, "Case Studies and Theory Development: The Method of Structured, Focused Comparison," in Paul Gordon Lauren, ed., *Diplomacy: New Approaches in History, Theory and Policy* (New York: Free Press, 1979), 46; and Skocpol, *States and Social Revolutions,* chap. 1.

19. Benjamin Page and Robert Shapiro, "Effects of Public Opinion on Policy," *American Political Science Review* 77 (March 1983): 186.

20. Seymour Martin Lipset, Martin Trow, and James Coleman, *Union Democracy* (Garden City, N.Y.: Anchor, 1962).

21. Robert Michels, *Political Parties* (New York: Dover, 1959).

22. Lipset, Trow, and Coleman, *Union Democracy,* 1.

23. Yin, *Case Study Research,* 53.

24. See, among countless other sources, Werner Sombart, *Why There Is No Socialism in the United States* (White Plains, N.Y.: International Arts & Sciences Press, 1976; first published in German in 1906); and Seymour Martin Lipset, *The First New Nation: The United States in Historical and Comparative Perspective* (New York: Basic Books, 1963).

25. Seymour Martin Lipset and Gary Marks, *It Didn't Happen Here: Why Socialism Failed in the United States* (New York: Norton, 2000): 16.

26. John Stuart Mill, *Ratiocinative and Inductive: Being a Connected View of the Principles of Evidence and the Methods of Scientific Investigation* (New York: Harper & Brothers Publishers, 1858) 225. (Available on Google Book Search, http://books.google.com/, accessed February 11, 2007.)

27. See Merrilee H. Salmon, *Logic and Critical Thinking,* 2d ed. (San Diego: Harcourt Brace Jovanovich, 1989), 109–115.

28. Skocpol's *States and Social Revolutions* is an excellent example of seminal research that explicitly uses Mill's method.

29. Yin, *Case Study Research,* 21–22.

30. Ibid., 21.

31. Ibid., 23.

32. Ibid.

33. The literature discussing the pros and cons of small-N research, especially its applicability to causal inference includes, among many others, Gary King, Robert Keohane, and Sidney Verba, *Designing Social Inquiry: Scientific Inference in Qualitative Research* (Princeton: Princeton University Press, 1991); James Mahoney, "Strategies of Causal Analysis in Small-N Analysis," *Sociological Methods and Research* 28 (May 2000): 387–424; and Stanley Lieberson, "Small N's and Big Conclusions: An Examination of the Reasoning in Comparative Studies Based on a Small Number of Cases" *Social Forces* 70 (December 1991): 307–320.

34. Lieberson, "Small N's and Big Conclusions," 312–313.

35. For an extended discussion, see King, Keohane, and Verba, *Designing Social Inquiry.* For a more optimistic picture of the possibilities of causal inference in small-N research, see Douglas Dion, "Evidence and Inference in Comparative Case Study," *Comparative Politics* 30 (January 1998): 127–145.

36. Although the measurements may be taken over a period of days or even weeks, cross-sectional analysis treats them as though they were obtained simultaneously.

37. Edward R. Tufte, *Data Analysis for Politics and Policy* (Englewood Cliffs, N.J.: Prentice-Hall, 1974), 5–17.

38. Donald Matthews and James Prothro, "Socio-Economic Factors and Negro Voter Registration in the South," *American Political Science Review* 57 (March 1963): 36–38.

39. The relationships among age, period, and cohort can be expressed as follows: Cohort (e.g., year of birth) = period (e.g., duration of an event) – age (e.g., years since birth). See Scott Menard, *Longitudinal Research* (Newbury Park, Calif.: Sage University Paper Series on Quantitative Applications in the Social Sciences, 1991), 7.

40. Ibid.

41. In practice, relationships of this sort are thought of as probabilistic, not deterministic, so a random error term would be added.

42. Christopher Gelpi, Peter D. Feaver, and Jason Reifler, *International Security* 30 (Winter 2005/06): 8–9.

43. Larry M. Bartels, "Messages Received: The Political Impact of Media Exposure," *American Political Science Review* 87 (June 1993): 267–285.

44. Ibid., 269.

45. This discussion is based on Morris P. Fiorina, "Formal Models in Political Science," *American Journal of Political Science* 19 (February 1975): 133–159. Also see Michael Laver, *Private Desires, Political Action: An Invitation to the Politics of Rational Choice* (Beverly Hills, Calif.: Sage Publications, 1997).

46. A classic work that builds a more thorough model of voter turnout (and many other political phenomena) is Anthony Downs, *An Economic Theory of Democracy* (New York: Harper and Row, 1957). Also see William H. Riker and Peter C. Ordeshook, "A Theory of the Calculus of Voting," *American Political Science Review* 62 (March 1968): 25–42.

47. Isn't this just a formalization of statements such as "Well, on the whole I just prefer Bush over his opponent?"

48. This deduction comports with the commonly heard complaint from nonvoters: "There ain't a dime's worth of difference between _____ and _____."

49. Not everyone in society can bear these costs equally. College graduates may find them less onerous than high school dropouts, which explains in part why the less educated vote less regularly than do those with more education.

50. John A. Ferejohn and Morris P. Fiorina, "The Paradox of Not Voting: A Decision Theoretic Analysis," *American Political Science Review* 68 (June 1974): 525–536.

51. In *An Economic Theory of Democracy,* Downs added exactly this sort of ad hoc term into his model. He called it the "long-run participation value."

52. Donald P. Green and Ian Shapiro provide a major critique of formal models in *Pathologies of Rational Choice Theory: A Critique of Applications in Political Science* (New Haven: Yale University Press, 1994). Their analysis particularly faults many models for leading to trivial conclusions and empirically invalid predictions. These points have been rebutted by several social scientists in Jeffrey Friedman, ed., *The Rational Choice Controversy* (New Haven: Yale University Press, 1996).

53. Fiorina, "Formal Models in Political Science," 137.

54. Green and Shapiro, *Pathologies of Rational Choice Theory,* and Jeffrey Friedman, *The Rational Choice Controversy,* explore this issue in greater detail.

55. Richard J. Stoll, "System and State in International Politics: A Computer Simulation of Balancing in an Anarchic World," *International Studies Quarterly* 31 (December 1987): 387–402.

56. Although we are presenting this simulation for illustrative purposes and do not care much about its realism, the last sentence reminds us of an important point. Science is a cumulative process that builds and rebuilds on the work of others. If we have stated our model's components clearly, then others can see where it might be weak, and we or they might make additions and corrections. In this instance, international relations theorists have noted that democracies generally do not go to war with one another, and these theorists have devoted much time and effort trying to figure out why not. Whatever the answer, we might want to explicitly include this consideration in our simulated international system.

57. For discussions and examples of social science simulations, see Robert Axelrod, "Advancing the Art of Simulation," in Rosario Conte, Rainer Hegelsmann, and Pietro Terna, eds., *Simulating Social Phenomena* (New York: Springer, 1997): 21–40; and Nigel Gilbert, "Simulation: An Emergent Perspective," Centre for Research in Social Simulation, n.d. (based on lectures given in 1995 and 1996), www.soc.surrey.ac.uk/research/cress/resources/emergent.html.

Terms Introduced

ASSIGNMENT AT RANDOM. Random assignment of subjects to experimental and control groups.

CAUSAL RELATIONSHIP. A connection between two entities that occurs because one produces, or brings about, the other with complete or great regularity.

CLASSICAL EXPERIMENTAL DESIGN. An experiment with the random assignment of subjects to experimental and control groups with a pre-test and post-test for both groups.

COHORT. A group of people who all experience a significant event in roughly the same time frame.

CONTROL GROUP. A group of subjects who do not receive the experimental treatment or test stimulus.

CROSS-SECTIONAL DESIGN. A research design in which measurements of independent and dependent variables are taken at the same time; naturally occurring differences in the independent variable are used to create quasi-experimental and quasi-control groups; extraneous factors are controlled for by statistical means.

DEMAND CHARACTERISTICS. Aspects of the research situation that cause participants to guess the purpose or rationale of the study and adjust their behavior or opinions accordingly.

EXPERIMENTAL EFFECT. Effect, usually measured numerically, of the independent variable on the dependent variable.

EXPERIMENTAL GROUP. A group of subjects who receive the experimental treatment or test stimulus.

EXPERIMENTAL MORTALITY. A differential loss of subjects from experimental and control groups that affects the equivalency of groups; threat to internal validity.

EXPERIMENTATION. A research design in which the researcher controls exposure to the test factor or independent variable, the assignment of subjects to groups, and the measurement of responses.

EXTERNAL VALIDITY. The ability to generalize from one set of research findings to other situations.

EXTRANEOUS FACTORS. Factors besides the independent variable that may cause change in the dependent variable.

FACTORIAL DESIGN. Experimental design used to measure the effect of two or more independent variables singly and in combination.

FIELD EXPERIMENTS. Experimental designs applied in a natural setting.

FORMAL MODEL. A simplified and abstract representation of reality that can be expressed verbally, mathematically, or in some other symbolic system, and that purports to show how variables or parts of a system are interconnected.

HISTORY. A change in the dependent variable due to changes in the environment over time; threat to internal validity.

INSTRUMENT DECAY. A change in the measurement device used to measure the dependent variable, producing change in measurements; threat to internal validity.

INSTRUMENT REACTIVITY. Reaction of subjects to a pre-test.

INTERNAL VALIDITY. The ability to show that manipulation or variation of the independent variable actually causes the dependent variable to change.

INTERVENTION ANALYSIS. Also known as interrupted time series design. A nonexperimental design in which the impact of a naturally occurring event (intervention) on a dependent variable is measured over time.

MATURATION. A change in subjects over time that affects the dependent variable; threat to internal validity.

MULTIGROUP DESIGN. Experimental design with more than one control and experimental group.

NONEXPERIMENTAL DESIGN. A research design characterized by at least one of the following: presence of a single group, lack of researcher control over the assignment of subjects to control and experimental groups, lack of researcher control over application of the independent variable, or inability of the researcher to measure the dependent variable before and after exposure to the independent variable occurs.

PANEL MORTALITY. Loss of participants from a panel study.

PANEL STUDY. A cross-sectional study in which measurements of variables are taken on the same units of analysis at multiple points in time.

PERIOD EFFECT. An indicator or measure of history effects on a dependent variable during a specified time period.

POST-TEST. Measurement of the dependent variable after manipulation of the independent variable.

PRECISION MATCHING. Matching of pairs of subjects with one of the pair assigned to the experimental group and the other to the control group.

PRE-TEST. Measurement of the dependent variable prior to the administration of the experimental treatment or manipulation of the independent variable.

RANDOMIZATION. The random assignment of subjects to experimental and control groups.

REPEATED MEASUREMENT DESIGN. An experimental design in which the dependent variable is measured at multiple times after the treatment is administered.

RESEARCH DESIGN. A plan specifying how the researcher intends to fulfill the goals of the study; a logical plan for testing hypotheses.

SELECTION BIAS. Bias in the assignment of subjects to experimental and control groups; threat to internal validity.

SIMPLE POST-TEST DESIGN. Weak type of experimental design with control and experimental groups but no pre-test.

SIMULATION. A simple representation of a system by a device in order to study its behavior.

SMALL-N DESIGN. Type of experimental design in which one or a few cases of a phenomenon are examined in considerable detail, typically using several data collection methods, such as personal interviews, document analysis, and observation.

SPURIOUS RELATIONSHIP. A relationship between two variables caused entirely by the impact of a third variable.

STATISTICAL REGRESSION. Change in the dependent variable due to the temporary nature of extreme values; threat to internal validity.

SURVEY RESEARCH. The direct or indirect solicitation of information from individuals by asking them questions, having them fill out forms, or using other means.

TEST STIMULUS OR FACTOR. The independent variable introduced and controlled by an investigator in order to assess its effects on a response or dependent variable.

TESTING. Effect of a pre-test on the dependent variable; threat to internal validity.

Suggested Readings

Campbell, Donald T., and Julian C. Stanley. *Experimental and Quasi-Experimental Designs for Research.* Chicago: Rand-McNally, 1966.

Cook, Thomas D., Donald T. Campbell, and Thomas H. Cook. *Quasi-Experimentation: Design and Analysis Issues.* New York: Houghton Mifflin, 1979.

Creswell, John W. *Research Design: Qualitative and Quantitative Approaches.* Thousand Oaks, Calif.: Sage Publications, 1994.

Downs, Anthony. *An Economic Theory of Democracy.* New York: Harper and Row, 1957.

Hakim, Catherine. *Research Design: Strategies and Choices in the Design of Social Research.* London: Allen and Unwin, 1987.

Hanley, Ryan Patrick. "What Is a Case Study and What Is It Good For?" *American Political Science Review* 98 (May 2004): 327–354.

Laver, Michael. *Private Desires, Political Action: An Invitation to the Politics of Rational Choice.* Thousand Oaks, Calif.: Sage Publications, 1997.

Menard, Scott. *Longitudinal Research.* Newbury Park, Calif.: Sage University Paper Series on Quantitative Applications in the Social Sciences, 1991.

Sambanis, Nicholas. "Using Case Studies to Expand Economic Models of Civil War," *Perspectives on Politics* 2 (June 2004): 259–279.

Sekhon, Jasjeet S. "Quality Meets Quantity: Case Studies, Conditional Probability, and Counterfactuals" *Perspective on Politics* 2 (June 2004): 281–293.

Spector, Paul E. *Research Designs.* Beverly Hills, Calif.: Sage Publications, 1981.

Vendaele, Walter. *Applied Time Series and Box-Jenkins Models.* New York: Academic Press, 1983.

Yin, Robert K. *Case Study Research: Design and Methods,* rev. ed. Beverly Hills, Calif.: Sage Publications, 1989.

CHAPTER 6
Conducting a Literature Review

So far we have discussed the nature of empirical political science and the initial stages of a typical study, including selecting a research strategy or design and formulating testable propositions. We noted earlier in the book that science is a cumulative activity with new discoveries usually building upon previous work. This previous work, or the literature, is defined as a collection of work with a similar research question or method. We also noted that one of the distinguishing characteristics of empiricism is the requirement that one's results be subjected to independent scrutiny and verification. We should stress, in addition, that it is practically impossible for students, even advanced graduates, to develop a completely new idea to investigate. For all these reasons, most researchers find it essential to spend time reading what others have written about a subject. We refer to this activity as conducting a **literature review,** or a systematic examination and interpretation of the literature for the purpose of informing further work on a topic.

A literature review can have many different purposes. At the bare minimum, it can be used to establish that the proposed study does not totally duplicate someone else's work. But a literature review can be used for several other, often complementary purposes. It may be used as a systematic survey of the literature to learn about what others have discovered, or it can help a researcher identify important research questions that have not been addressed fully by others. It may also be helpful in identifying the data or methods that others have used to answer specific questions, or the research strategies that have worked well or failed. Alternatively, a literature review can help narrow or focus a research topic, or efficiently guide the investigation to a fruitful conclusion. Reviewing the literature is also a critical component of motivating and developing a specific research question. In this chapter we discuss in some detail the reasons for, and methods of writing, a literature review. We also demonstrate techniques for searching for information using electronic databases and the Internet.

For many students, simply finding a research topic can be a time-consuming and frustrating experience. So, although we concentrate in this chapter on how to conduct a literature review and use it to improve your research project once a topic has been chosen (if only in a general sense), we begin with some suggestions to help you identify potentially interesting questions and problems to investigate.

Selecting a Research Topic

Potential research topics about politics come from many sources. These sources may be classified as personal, nonscholarly, or scholarly. Personal sources include your own life experiences and political activities and those of your family and friends, as well as class readings, lectures, and discussions.

You can also look to nonscholarly sources for research topics, including print, broadcast, and Internet sources. Becoming aware of current or recent issues in public affairs will help you develop interesting research topics. You can start by reading a daily newspaper or issues of popular magazines that deal with government policies and politics. The CD in the workbook and the Web site accompanying this book (http://psrm.cqpress.com) offer many possibilities and lists of other Web sites. The best print sources include national newspapers and magazines featuring in-depth political coverage. First, consider reading major urban daily newspapers like the *New York Times* and the *Washington Post.* Daily newspapers provide the most up-to-date printed political news and discussions. In addition, look at weekly magazines like *Atlantic, Harper's,* the *Economist,* the *American Prospect, National Review,* the *New Republic,* the *New Yorker,* and the *Weekly Standard.* Most of these weeklies have a decidedly partisan leaning (either conservative or liberal; Republican or Democratic), but—and this is a key point—they contain serious discussions of domestic and foreign government and politics and are wonderful sources of ideas and claims to investigate.[1] Each of these sources also features online material, much of which is free. An underappreciated source of potential research topics within these printed sources are the editorial and letters-to-the-editor pages. Although these pieces express opinions, the writers often support them with what they claim are empirical facts. Consider an editorial asserting that the "death penalty deters crime" and is therefore justified. It may be tricky, but you might test this claim by comparing the incidence of homicide in states with and without capital punishment.

Broadcast news sources can also inspire topical research projects. The best radio and television programs for this purpose are those that include long segments dedicated to political news, discussion, and debate. The best radio programs with a civic or political focus feature a variety of topics, like National Public Radio's *Morning Edition* or political talk shows like the *Rush Limbaugh Show.* The best political television shows tend to be programs featuring long interviews with political actors such as NBC's *Meet the Press,* the

Charlie Rose Show on PBS, or investigative journalism programs like CBS's *60 Minutes.*

Internet sources can include the print and broadcast sources discussed above, found through the publications' and broadcasts' sites on the Web. In addition to offering the same content that is printed or broadcast, many print and broadcast sources feature exclusive Internet material such as the *Washington Post*'s *White House Watch,* a daily blog focusing on the presidency. Other Internet sources include government, university, or organization Web sites; Web sites created by individuals; and political blogs. Students with broad interest in international affairs might access Web pages maintained by the U.S. State Department, the United Nations, or nongovernmental organizations such as the International Red Cross. These sites might include reports or publications, calendars of events, or other useful materials. All universities maintain Web sites, and departments with an interest in politics often include links to political Internet sources, current research, syllabi, and data archives. Web sites maintained by individuals can be useful, especially if the individual has a good reputation for providing reliable information, such as a government official or a university professor. An increasingly relevant source for political information on the Internet is the political blog. Blogs like *Daily Kos* or *InstaPundit* have become fixtures in the national political debate, raising topics or uncovering evidence that the traditional news media have not. Blogs, much like talk radio or magazines, often feature political discussion and debate from a particular ideological or partisan perspective.

Although each of the sources in this discussion—print, broadcast, and Internet—has value in identifying topics for study or investigation, the sources must be used carefully. Documents available from a Web site maintained by the U.S. State Department can generally be used as a primary source for inclusion in an analysis. News articles in the *Washington Post* can typically be trusted as accurate accounts of political events for use in a research project. Sources that rely on opinion rather than fact or an unbiased account of events, such as talk radio, political blogs, or sources with an ideological or partisan bias—like some political magazines—must be used with much more care. Although the *Rush Limbaugh Show* is a great source for topical issues and testable claims, it relies on the opinions of the host and callers for content, and it generally should not be used as a source of unbiased information.

Although personal and nonscholarly sources are good places to find potential research topics, you can satisfy the scientific requirement of relevance to the discipline only by surveying the

CHECK POPULAR BELIEFS

Consider scouring the popular press or mass media—the sorts of newspapers and magazines read by average citizens—for statements that are widely believed and repeated but that you suspect might be misleading or downright wrong. If an argument pertains to an important issue, and if you can discredit it with empirical evidence, you might well make a potential, even if modest, contribution to the public's knowledge of current events.

scholarly literature. The scholarly literature includes books and articles written by political scientists and other academics or political practitioners. Such literature establishes which topics and questions are important to political scientists. You can differentiate scholarly works from nonscholarly ones by looking for a few characteristics. Most important, professional articles and books published in political science or other disciplines will often go through a peer-review process. The most common peer-review standard is that a journal or book editor will send an article or book manuscript submitted for publication to one or more scholars with expertise in the topical area of the article. The review is performed in a blind fashion, in which the reviewers are not told the author's name to ensure that the review is fair. Otherwise, reviewers may be inclined to reject a piece authored by a particular author based on a personal grievance or accept a piece written by a friend. The blind process ensures that reviewers will assess only the quality of the work. The editor will rely on the peer reviewers' comments to suggest revisions of the work and assess whether or not the work makes a sufficient contribution to the literature to deserve publication. The peer-review process helps assure that the work published in scholarly journals and books is of the best possible quality and of the most value to the discipline. It also assures the reader that, although there still may be mistakes or invalid or unreliable claims, the article or book has been vetted by one or more experts on the topic.

Alternatively, some scholarly journals and books are reviewed only by the editorial staff. Although this method provides a check on the quality of the work, it is usually not as rigorous as a blind peer-review. The type of review a journal or book publisher uses will typically be explained in the journal or on the journal's or publisher's Web site.

In addition to a peer-review process, some other indicators can differentiate scholarly from nonscholarly work. Scholarly articles and books are usually written by academics, journalists, political actors, or other political practitioners, so looking for a description of the authors is the place to start. Scholarly books are published by both university presses and commercial presses for a professional audience rather than a general audience.[2] As such, the work will include complex analyses and be written with the assumption that the reader is familiar with the literature and method. Scholarly work will also cite other scholarly sources, which can be easily verified by scanning the works cited, footnotes, or endnotes. If you are still unsure about whether or not a particular work is scholarly, consult with a reference librarian or your instructor. In addition, here is a short list of some of the major publications, many (if not all) of which are available online:

American Journal of Political Science. Primarily articles on American government and politics.

American Political Science Review. The official journal of the American Political Science Association.

British Journal of Political Science. Although emphasizing a comparative perspective, this publication contains important research on American political institutions and behavior.

Comparative Politics. Begin here when looking for scholarly studies of all aspects of cross-national politics and government.

International Organization. Contains important articles on international relations. One of the leading journals in the field.

Journal of Conflict Resolution. A widely cited journal with articles on, among other topics, international relations, war and peace, and individual attitudes and behavior. Authors use a variety of methods and research designs.

Journal of Politics. Primarily articles on American government and politics.

Legislative Studies Quarterly. Articles about legislative organization and functioning and electoral behavior.

Political Analysis. For students with a serious interest in methods and statistics. Articles frequently contain important substantive results.

Political Research Quarterly (formerly *Western Political Quarterly*). Broad coverage of political science and public administration.

Polity. Articles on American politics, comparative politics, international relations, and political philosophy.

Social Science Quarterly. Articles on a wide range of topics in the social sciences.

In addition, research is frequently presented at professional conferences before it is published. If you want to be informed up to the minute or if a research topic is quite new, it may be worthwhile to investigate papers given at professional conferences.

The *Index to Social Sciences and Humanities Proceedings* indexes published proceedings. However, the proceedings of the annual meetings of the American Political Science Association (APSA) and regional political science associations are not published. Summer issues of APSA's *PS* contain the preliminary program for the forthcoming annual meeting. The program lists authors and titles of papers. The *International Studies Newsletter* publishes preliminary programs for International Studies Association meetings. Copies of programs for other political science and related conferences (frequently announced in

PS) may be obtained from the sponsoring organization. Abstracts for some fields—for example, *Sociological Abstracts*—include papers presented at conferences. After promising papers presented at professional conferences have been located, copies of the papers may usually be obtained through online archives like APSA's *Proceedings*[3] or by writing to the authors directly. Although these unpublished articles represent the most up-to-date work in the discipline, care should be taken in relying on them as they have not yet been vetted by a peer-review process and may be less reliable than published work.

To guide you further in finding topics and searching for appropriate sources, this book's companion Web site lists additional professional journals as well as indexes and bibliographies, data banks, guides to political resources, and the like.[4] A reference librarian will undoubtedly be able to provide additional information and guidance on particular library sources available.

Still another source of ideas for research papers is a textbook used in substantive courses, such as comparative politics or international relations. These works can be particularly valuable for pointing out controversies within a field. For example, as the discussion of judicial behavior in the first chapter illustrated, political scientists argue about what underlies judges' decisions, political ideology, or adherence to legal precedent and principles. You might do a case study of a particular justice to see which side this person's rulings seem to support.

Why Conduct a Literature Review?

Most research topics are initially much too broad and general to be manageable. It would be virtually impossible to write something meaningful on "international terrorism" or even "the causes of terrorism in the Middle East" without first knowing a great deal about the subject. Good research, therefore, involves reviewing previous work on the topic to motivate and sharpen a research question. Among the many reasons for doing so are (1) to see what has and has not been investigated; (2) to develop general explanations for observed variations in a behavior or phenomenon; (3) to identify potential relationships between concepts and to identify researchable hypotheses; (4) to learn how others have defined and measured key concepts; (5) to identify data sources that other researchers have used; (6) to develop alternative research designs; and (7) to discover how a research project is related to the work of others. Let us examine some of these reasons more closely.

Often someone new to empirical research will start out by expressing only a general interest in a topic, such as terrorism or the effects of campaign advertising or public opinion and international relations, but the specific research question has yet to be formulated (for example, "What kinds of people become terrorists?" or "Do negative televised campaign advertisements sway voters?" or "Does the public support isolationism or internationalism?"). A review of previous research can help sharpen a topic by identifying research questions that others have asked.

After reading the published work in an area, you may decide that previous reports do not adequately answer the question. Thus a research project may be designed to answer an old question in a new way. An investigation may replicate a study to confirm or challenge a hypothesis or expand our understanding of a concept. Replication is one of the cornerstones of scientific work. By testing the same hypothesis in different ways or confirming the results from previous research using the same data and methods, we increase our confidence that the results are correct. Replication can therefore help build consensus or identify topics that require further work.

At other times, research may begin with a hypothesis or with a desire to explain a relationship that has already been observed. Here a literature review may reveal reports of similar observations made by others and may also help you develop general explanations for the relationship by identifying theories that explain the phenomenon of interest. The research will be more valuable if you can provide a general explanation of the observed or hypothesized relationship rather than simply a report of the empirical verification of a relationship.

In addition to seeking theories that support the plausibility and increase the significance of a hypothesis, you should be alert for competing or alternative hypotheses. You may start with a hypothesis specifying a simple relationship between two variables. Since it is uncommon for one political phenomenon to be related to or caused by just one other factor or variable, it is important to look for other possible causes or correlates of the dependent variable. Data collection should include measurement of these other relevant variables so that in subsequent data analysis you may rule out competing explanations or at least indicate more clearly the nature of the relationship between the variables in the original hypothesis.

Collecting Sources for a Literature Review

After selecting a research topic using the sources described in the previous section, you must begin collecting sources for use in writing a literature review. Although personal and nonscholarly sources can be quite helpful in selecting a research topic, and a literature review can encompass virtually anything published on your topic, we strongly encourage you to become familiar with the scholarly literature. Relying on scholarly sources rather than nonscholarly ones will improve the quality of a literature review, but, as a practical concern, many instructors may not accept or give much credit for citations from nonscholarly sources unless their content constitutes part of your topic. After all, a literature review is supposed to establish what is known about a topic with that knowledge having been attained and communicated according to professional or scientific principles.

Students commonly ask, "How many sources must I find to write my literature review?" The answer, unfortunately, is not easy. The decision of how many books and articles to include in a literature review depends on the purpose and scope of the project as well as available resources. If your project is focused largely on reporting the work of others, you will probably need to include more sources than if your project is focused mostly on your own analysis. Furthermore, a more complex topic, or a topic with a larger literature, may require a more in-depth literature review than will a more straightforward topic or one with a smaller literature. Finally, consider how much time and effort you are willing to dedicate to collecting sources. Although we cannot provide an easy answer to the question of how many sources are necessary, we can explain how available time and effort could be best directed and used most efficiently.

PYRAMID CITATIONS
Each time you find what appears to be a useful source, look at its list of notes and references. One article, for example, may cite two more potentially useful papers. Each of these in turn may point to two or more additional ones, and so on. By starting with a small list, you can quickly assemble a huge list of sources. Moreover, you increase your chances of covering all the relevant literature.

Identifying the Relevant Scholarly Literature

It would be impossible for anyone to identify, let alone read and or write about, every book or article with relevance to any particular research project. With that caveat in mind, you can think of the first step in collecting sources—identifying the relevant literature—as limiting the search to only those books and articles with the most direct relevance to the research topic of interest. You can begin to narrow the field of potential sources in many ways. The first step is to search comprehensive **electronic databases**, such as *Web of Science* or *Google Scholar,* or databases that include links to full text articles, such as *JSTOR*. These databases allow you to quickly locate a large number of articles that investigate similar topics.

Web of Science is a particularly useful starting point for building a literature review because you can search the Social Sciences Citation Index database of social science journal articles generally, using a keyword search; you can search for articles written by a particular social scientist; or, perhaps most important for starting a literature review, you can search for all of the articles in the database that cite an article of interest, and for articles that subsequently cited those articles. Two quick examples highlight the value of these searches. First, suppose you are interested in understanding judicial behavior. By typing the phrase *judicial behavior* into the "quick search" field, and searching only the Social Sciences Citation Index, we found 363 articles written between 1956 and 2007. This is far too many articles to read, but after reading through article abstracts in this larger topic, you might narrow the search to a particular kind of judicial behavior. For example, say that, after reading a few abstracts and articles, you found you were intrigued by judicial decision making on the Supreme Court in particular. By entering the phrase *Supreme Court decision making* into the "search within these results" search field, we narrowed the search to only seventy-two articles. Suppose next, that, after additional scanning of abstracts and articles, you determined that what you were really interested in was why justices dissent from majority decisions. Following the same procedure as above yields four articles to begin your literature review.

A second way to use the Web of Science is to begin with a single article instead of searching for topics. Suppose that at the beginning of your search you instead decided to search for articles related to an article you have already found—from your course syllabus, for example. Imagine that while reading Jeffrey A. Segal and Albert D. Cover's "Ideological Values and the Votes of U.S. Supreme Court Justices" (a brief discussion of the article is found in Chapter 1), you found that the topic interested you and you thought you might like to find out what else was written on the topic. Because you discussed the article in class, and your professor told you of the importance of the article in establishing our current understanding of judicial decision making, you decided that it was an important article to include in your literature review. With this single article, you could use the Web of Science to quickly find other work in the literature investigating similar research questions. For example, by using an advanced search, and selecting the Social Sciences Citation Index and the default of all years, you could find Segal and Cover's article by searching for the author's names and part of the article's title, as shown in Figure 6-1. Clicking on the underlined "1" will reveal the results from the search. Click on "view full text" or "AE Get Article" buttons to find the full text of the article if it is available through the library's database subscriptions, or click on the article title to find a wealth of information about the article. Figure 6-2 shows the full citation, which will be needed for the works cited or notes, the number of articles the article cited, and the number of articles that subsequently cited Segal and Cover's article. Segal and Cover cited thirty-four references in their article; by clicking on "Cited References: 34," you will find all thirty-four references with electronic links to those references included in the Web of Science database. This feature makes it easy to review the base of knowledge that was in place before Segal and Cover's article. As of this writing, Web of Science has identified 156 articles in its database that have cited Segal and Cover's article. By clicking on "Times Cited: 156," you can find a link to each of these articles. This is particularly important because once you find an essential reference like Segal and Cover's arti-

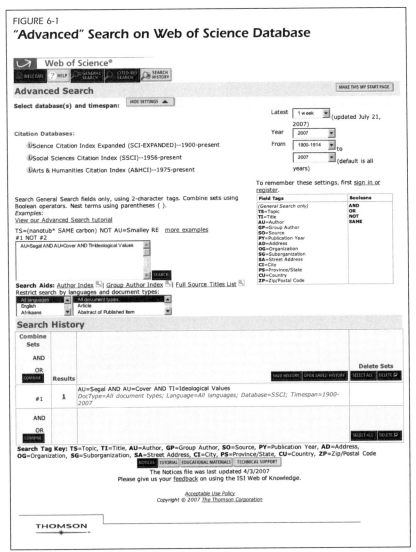

FIGURE 6-1

"Advanced" Search on Web of Science Database

cle—an article that most work in this literature includes as a reference—you can easily identify a large number of articles for your own literature review. In this example, we located 190 references to relevant books and articles in a matter of minutes.

The larger lesson from this example is that once you find a relevant article, you can sharpen the direction of your search for relevant literature by examining the literature review and works cited in that article. Since the article is directly relevant to the research topic of interest, the sources used in the article

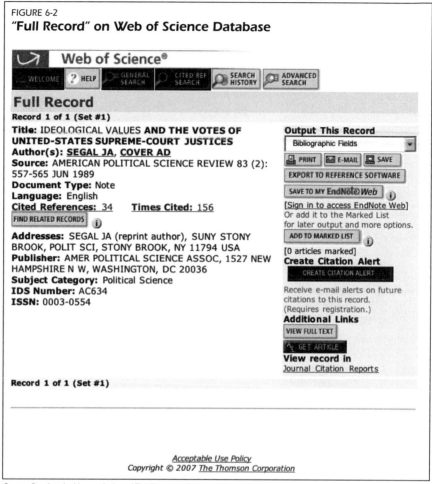

FIGURE 6-2
"Full Record" on Web of Science Database

Web of Science®

WELCOME | ? HELP | GENERAL SEARCH | CITED REF SEARCH | SEARCH HISTORY | ADVANCED SEARCH

Full Record

Record 1 of 1 (Set #1)

Title: IDEOLOGICAL VALUES **AND THE VOTES OF UNITED-STATES SUPREME-COURT JUSTICES**
Author(s): SEGAL JA, COVER AD
Source: AMERICAN POLITICAL SCIENCE REVIEW 83 (2): 557-565 JUN 1989
Document Type: Note
Language: English
Cited References: 34   **Times Cited:** 156
FIND RELATED RECORDS

Addresses: SEGAL JA (reprint author), SUNY STONY BROOK, POLIT SCI, STONY BROOK, NY 11794 USA
Publisher: AMER POLITICAL SCIENCE ASSOC, 1527 NEW HAMPSHIRE N W, WASHINGTON, DC 20036
Subject Category: Political Science
IDS Number: AC634
ISSN: 0003-0554

Output This Record

Bibliographic Fields

PRINT | E-MAIL | SAVE
EXPORT TO REFERENCE SOFTWARE
SAVE TO MY EndNote Web

[Sign in to access EndNote Web]
Or add it to the Marked List for later output and more options.
ADD TO MARKED LIST
[0 articles marked]
Create Citation Alert
CREATE CITATION ALERT
Receive e-mail alerts on future citations to this record. (Requires registration.)
Additional Links
VIEW FULL TEXT
GET ARTICLE
View record in
Journal Citation Reports

Record 1 of 1 (Set #1)

Source: Reprinted with permission of The Thomson Corporation.

will likely be related as well. By building a list of sources in this fashion, you can save a great deal of time and effort as well as collect sources with a greater certainty that important work will not be overlooked. Remember, however, that even though both of the above example strategies will help you find relevant articles quickly, articles without much relevance may also come up in a search. Two articles that share a common search term do not necessarily have much related content. Furthermore, one article's citing another does not necessarily mean that the two articles investigate the same topic or even mention the other article in the text. Therefore, you should search for relevant literature using multiple search parameters and tools.

You could also search for articles on judicial behavior using a database like JSTOR, a comprehensive electronic archive of academic journals and publications. Although not every campus has access to it, and it does not include full text articles for many important sources, JSTOR is widely available and searching it illustrates guidelines for searching other databases.

Figure 6-3 shows the preparation of a JSTOR search for articles containing the phrase *judicial behavior.* Note the limitations on the search. Because a lot of work deals with judicial behavior, we hunted for only *articles* with the phrase in their abstracts. We also confined the search to the last decade (1998–2007). Finally, although we limited our search to political science journals, many useful articles may be found in journals specializing in related disciplines such as economics or psychology.

The results of the search appear in Figure 6-4. For the particular criteria used, the database returned twenty-five articles. As previously mentioned, JSTOR provides the full text of articles in the database, but it also provides links to the abstract (which we discuss in greater detail in the next section) and the full citations. An advantage of finding articles with a full text Web site like JSTOR is that you can save electronic copies of articles for review without incurring printing expenses. And similar to Web of Science, JSTOR includes a feature that allows you to locate all of the articles in the JSTOR database that cite an article, or all of the references in the Google Scholar database. One limitation to JSTOR and similar databases is that the archive's collection generally lags three to five years behind the current date, although others have shorter or longer lags. You should not rely on these sites alone when searching for research materials because the most current work will not be available.

FINDING A TERM ON A PAGE
Most browsers (for example, Internet Explorer, or IE) have a "hot key" combination that allows you to search for a particular word or phrase on a displayed web page. (IE's is "Ctrl-F.") Take advantage of this shortcut when viewing a massive document that has small text or lots of content.

Many other institutes and organizations maintain Web sites that are open to the public. Once you access these sites, you will find that most of them are really lists of lists. That is, they do not contain many documents themselves but point you to places that do. For example, if you want information about campaigns, you would need to start at one of the sites devoted to political science or the social sciences. Once there, you would find that many of these sites offer connections to more specific sites, for example, sites concerned with parties, campaign headquarters, polling firms, survey research, and the like. You will find the information you need at these other sites.

Identifying Useful Popular Sources

Although most students likely use the Internet on a regular basis, many of them may not have used it to search for nonscholarly political sources. One of the benefits of the revolution in global communications is that it places an

FIGURE 6-3

"Advanced" Search on JSTOR

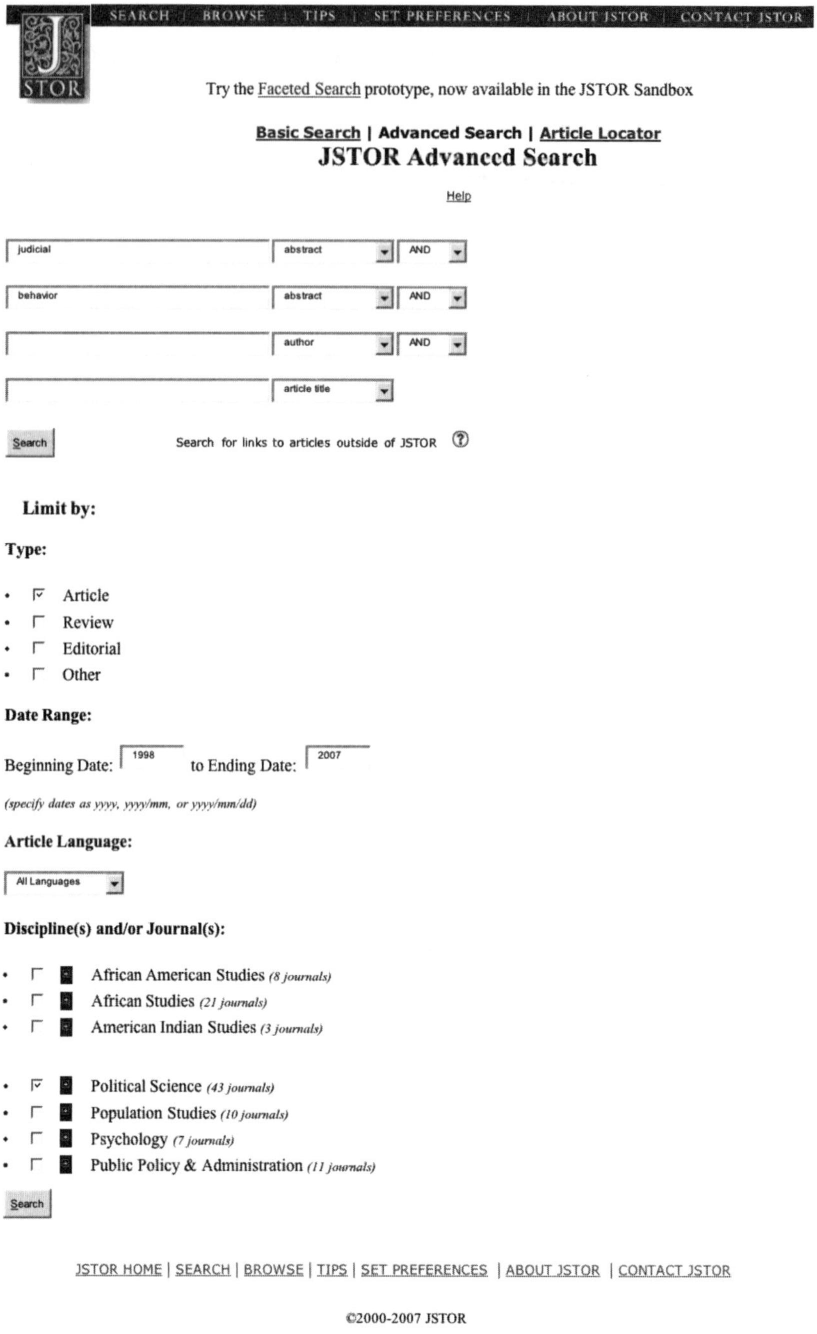

SEARCH | BROWSE | TIPS | SET PREFERENCES | ABOUT JSTOR | CONTACT JSTOR

Try the Faceted Search prototype, now available in the JSTOR Sandbox

Basic Search | Advanced Search | Article Locator
JSTOR Advanced Search

Help

judicial	abstract ▾	AND ▾
behavior	abstract ▾	AND ▾
	author ▾	AND ▾
	article title ▾	

Search   Search for links to articles outside of JSTOR ⑦

Limit by:

Type:

- ☑ Article
- ☐ Review
- ☐ Editorial
- ☐ Other

Date Range:

Beginning Date: [1998]   to Ending Date: [2007]

(specify dates as yyyy, yyyy/mm, or yyyy/mm/dd)

Article Language:

[All Languages ▾]

Discipline(s) and/or Journal(s):

- ☐ ■ African American Studies *(8 journals)*
- ☐ ■ African Studies *(21 journals)*
- ☐ ■ American Indian Studies *(3 journals)*

- ☑ ■ Political Science *(43 journals)*
- ☐ ■ Population Studies *(10 journals)*
- ☐ ■ Psychology *(7 journals)*
- ☐ ■ Public Policy & Administration *(11 journals)*

Search

JSTOR HOME | SEARCH | BROWSE | TIPS | SET PREFERENCES | ABOUT JSTOR | CONTACT JSTOR

©2000-2007 JSTOR

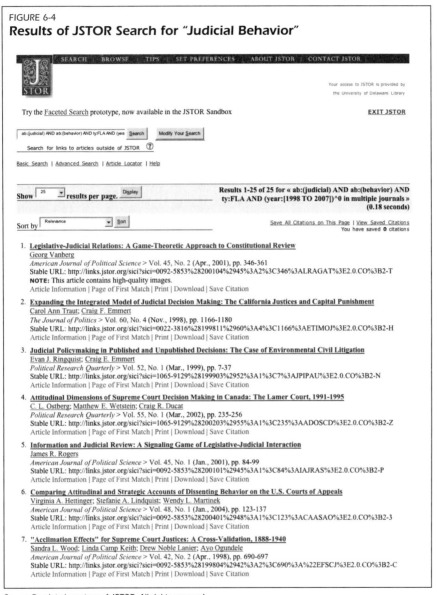

FIGURE 6-4

Results of JSTOR Search for "Judicial Behavior"

almost limitless supply of information literally at your fingertips. Scouring the Internet also allows you to find many kinds of documents and data that a traditional library search will not turn up or that are simply not available on many campuses. To see the advantages—and also the problems—with using the Internet to conduct a literature search, let's suppose you are interested in "public opinion and foreign policy." As stressed in Chapter 4, your research will go more smoothly and doubtlessly be more informative if you concentrate on a specific set of questions or hypotheses. Let's start with a straightforward question: what kinds of people backed and opposed the 2003 war in Iraq? Democrats? Liberals? Northerners? Laborers? College graduates? Our goal, in a nutshell, is to find the correlates, or factors, that distinguish support and opposition to the conflict.

It is tempting to think that you need only to access a **search engine,** a computer program that systematically visits and searches Web pages, and type in a few **search terms,** or keywords. However powerful the facilities may be, the search process is not always simple. Most students are familiar with popular search engines such as *Google* or *Yahoo.* These search engines may be a good place to start if you are trying to see what sources are available on a topic and you are not looking for a specific reference. These search engines, however, can be quite indiscriminate in what they return and leave the user with pages of unsuitable or redundant findings. As an example, we used Google and Yahoo to search for the phrases and words *public opinion, Iraq,* and *war.* On this particular occasion (July 2007), the Google search produced more than 18 million sources and the Yahoo search found more than 26 million. That is, of course, way too many to read. Search programs often order the results by the frequency of appearance of search words in the title and in the text near the top of the page or by the regularity with which the page is visited. But these may not be the best criteria for your purposes. Use of the Internet clearly has drawbacks *unless* careful planning and thought have preceded the search. Like many of life's activities, the more time spent searching for literature review materials, the easier it becomes. Nevertheless, following a few practical guidelines will expedite the process.

Most search engines and databases enable you to narrow a search to meet your specific needs. Usually you want to see only the documents that contain all the words—or even specific phrases, such as *international terrorism—* on a list. Advanced search features allow you to

Searching the Internet

Here are some quick tricks for surfing the Internet:

- When first visiting a site, particularly one with search features, click the "help" button, which usually provides specific instructions for searching that site.

- If possible, pyramid your search by going first to a political science page and then from there looking for more specific sites.

- If you have a clear topic in mind, start with a specific Internet site, such as those sponsored by research organizations or universities. Doing so will reduce the number of false hits.

- Open a simple word-processing program such as Notepad or WordPad. Highlight and copy selected text from a web page to facilitate collecting information. Be sure to properly document the source of this material. This technique is especially helpful for copying complicated and long Internet addresses (URLs).

- On a complicated page with lots of text and images, use your browser's Find option to locate the word or phrase of interest.

- Take advantage of advanced search options. If possible, limit your search to specified time periods, to certain types of articles, to particular authors or subjects, and to data formats.

- Check this book's Web site (http://psrm.cqpress .com) for links to specific topics.

use connectors and modifiers to specify exactly what words or phrases should be included in the document and which ones should be excluded. If you enter the desired words without adding modifiers, in all likelihood the search engine will look for pages that contain any of the listed words but not necessarily all of them.

A more focused tool for finding popular political sources is a service such as *LexisNexis*. LexisNexis offers a subscription service that allows you to search a variety of sources, from state laws or legal documents to newspapers to agency reports. Although some research libraries have access to all of the databases LexisNexis offers, many do not. Most, however, have access to the newspaper database. Searching LexisNexis for news accounts of events provides a fast and easy way to find reliable information. Like the search tools discussed previously, LexisNexis offers a general keyword search, or searches can be more limited or focused by searching within news categories, regions, or even for specific newspapers (see Figure 6-5). For example, students of political science might be most interested in news accounts of political activity and events taking place in Washington, D.C. By limiting the search to *U.S. News* and specifying that the search includes only *District of Columbia News Sources,* you will be able to click on "Source List." The source list includes a collection of nine Washington, D.C., news sources: the wire services *Associated Press State and Local Wire* and *States News Service*; the *Washington Post*, its Internet content, and the *Washington Times*; two newspapers focusing on Congress (*The Hill* and *Roll Call*); the weekly political news magazine *Insight on the News*; and the *Washington, D.C. Employment Law Letter.* So, unlike general search engines such as Google and Yahoo, LexisNexis allows you to easily restrict the search to the sources that will be most useful in finding accounts of political events from news organizations.

Although the Internet allows for a wide search of material, not all information found on the Internet is reliable. Virtually anyone or any group, no matter what their credentials, can create a Web site. The only way to know for sure that the information you are looking at is dependable is to be familiar with the site's sponsor. In general, sites presented by individuals, even those with impressive-looking titles and qualifications, may not have the credibility or scholarly standing that your literature review requires. In contrast, you can usually have confidence in sources cited in professional publications or by established authors or reputable organizations. Note, too, that even sources in the form of opinion can be dependable. Many associations that hold strong political or ideological positions nevertheless offer useful information that is worth citing. If in doubt about the reliability of any sources, check with your instructor or adviser. He or she should be able to help you assess whether or not accessed information is usable.

FIGURE 6-5
A LexisNexis Academic Search

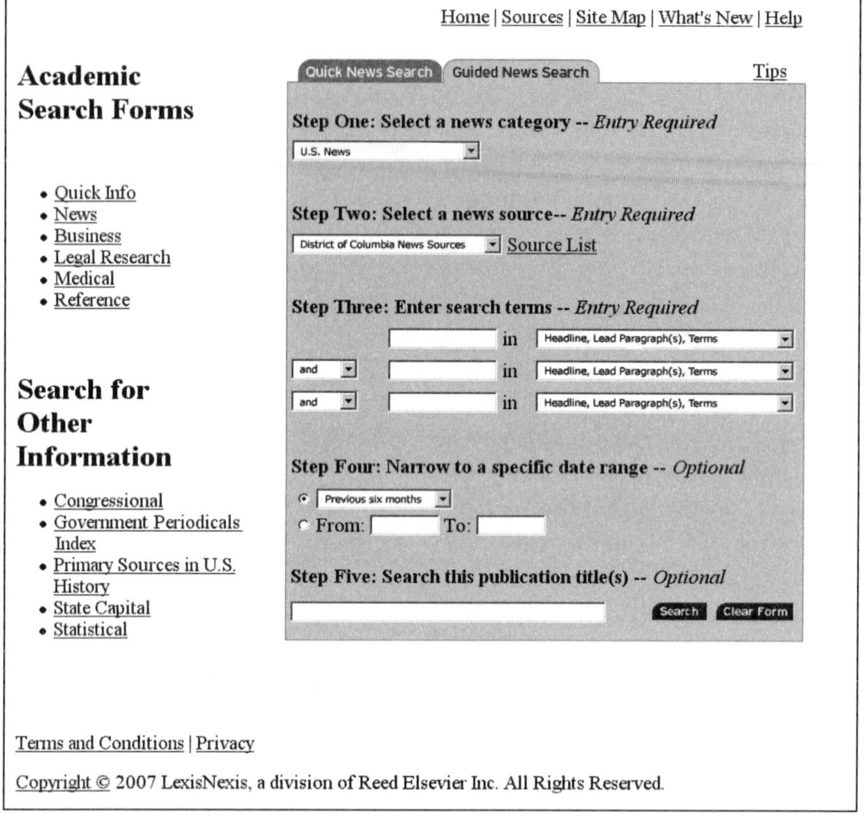

Home | Sources | Site Map | What's New | Help

**Academic
Search Forms**

Quick News Search | Guided News Search   Tips

Step One: Select a news category -- *Entry Required*
U.S. News

- Quick Info
- News
- Business
- Legal Research
- Medical
- Reference

Step Two: Select a news source-- *Entry Required*
District of Columbia News Sources Source List

Step Three: Enter search terms -- *Entry Required*

	in	Headline, Lead Paragraph(s), Terms
and	in	Headline, Lead Paragraph(s), Terms
and	in	Headline, Lead Paragraph(s), Terms

**Search for
Other
Information**

Step Four: Narrow to a specific date range -- *Optional*
Previous six months
From: To:

- Congressional
- Government Periodicals Index
- Primary Sources in U.S. History
- State Capital
- Statistical

Step Five: Search this publication title(s) -- *Optional*

Search   Clear Form

Terms and Conditions | Privacy

Copyright © 2007 LexisNexis, a division of Reed Elsevier Inc. All Rights Reserved.

Source: Used with permission of LexisNexis.

Internet sources must be cited properly, partly because so much variation exists in the quality of these sources but also, and even more important, because academic standards dictate that proper citations be provided for any work consulted. In this way authors are fully credited for their data and ideas, and readers can check the accuracy of the information and the quality of a literature review.

At a minimum, the citation should include the author or creator of the page and the title of the article as well as the complete Internet address at which the article was found. Following is a generic form for citing a Web page in a bibliography or note:

Author [last name, first name, or full organization name]. (Date of publication, if available) Title [Online]. Available: Full Web address (Date page was accessed).

For example,

Stroupe, Kenneth S. Jr., and Larry J. Sabato. (2004) Politics: The Missing Link of Responsible Civic Education. www.centerforpolitics.org/ downloads/civicengagement-stroupe-final.pdf (14 August 2007)

indicates that your information is from a report by Kenneth S. Stroupe Jr. and Larry J. Sabato, and is available on a Web page administered by the Center for Politics that you accessed online at www.centerforpolitics.org/downloads/civic engagement-stroupe-final.pdf on August 14, 2007.

Citation style will depend on the standards set by your institution or instructor, but include at least enough detail to let a reader retrieve the page and verify information.

Reading the Literature

Once you have identified references for possible inclusion in a literature review, the next step is to figure out how the references fit together in a way that (1) explains the base of knowledge, or what we know about a topic from previous work,

ORGANIZING REFERENCES
The first time you conduct a comprehensive literature search, the number of citations you discover may overwhelm you. Managing them systematically is often quite a challenge. It may help to put each relevant citation on a separate three-by-five-inch index card. If the citation proves to be useful, then complete bibliographic information can be entered later on the card in the form you will be using for your bibliography. These cards can be sorted according to various needs. This method preserves the fruits of a literature search in a form that will be useful to you, and it saves the step of writing the citation information onto a list and then transferring it to a card.

with respect to the research question and (2) establishes how the current project is going to build on that knowledge. The best way to understand the base of knowledge is to read the work that answers the central research questions and understand how each contributed to a comprehensive understanding of the important research questions. To read an entire literature would take far too much time, so it is wise to rely on shortcuts whenever available.

First, by following the suggestions in the preceding section on collecting references, care should be taken in selecting references. Once references are identified and collected, you can rely on the abstract on the first page of most articles and the preface at the beginning of most books to serve as a short description of the organization and conclusions contained therein. A good abstract will include a great deal of important information about the contents of an article, including the research question, the theory and hypotheses, the data and methods used to test the hypotheses, and the results and conclusions. Most article abstracts are only 200–300 words long, so they offer an easy way to assess quickly whether an article is worth reading further. A good preface will include the same kind of information, but a book's length makes this summary much more cursory or general. A preface will also include more

199

attention to organization of the chapters. Reading book reviews in scholarly journals is another way to learn quickly the value of a book to a given project. For most books, you can find a review that will relay the book's theoretical importance or help place the book in the context of the literature and help you understand how it fits in the existing literature and what it adds to the base of knowledge—in addition to assessing the quality of the research.

Use of an abstract, a preface, or a book review will help narrow a list of references. This smaller list can then be culled for those references that are essential in motivating the current research project and those that add depth, range, or a unique perspective to the literature review. In addition, the first few pages of political science articles contain most of the description of the key research design components that this book has described thus far. These pages will explain in greater detail the research question, theory and hypotheses, data and methods, and will include a literature review. The conclusion or discussion of findings will summarize the results and explain how they add to the base of knowledge. Students with limited time for reading articles should read the first few pages and the conclusion and then, if more information is needed, proceed with the rest of the article. Finally, although many political science articles include complex methods and tables, the text describing the results usually includes a more jargon-free description of the results that does not require an advanced understanding of statistics. The same time-saving tips can be applied to books by concentrating on the book's introduction and conclusion as well as selected chapters identified in the table of contents.

Nonscholarly references like magazine or newspaper articles, or Web site content, generally are much shorter than references from the scholarly literature and require fewer shortcuts. These sources can typically be read quickly and in most cases do not provide an abstract.

Writing a Literature Review

After you have identified the relevant literature, and started reading the literature, it is time to begin crafting the literature review. In this section we explain how you can integrate a collection of related materials into an effective literature review. Essential to this process is limiting the discussion of materials to the most relevant previous work, focusing the literature review on concepts and ideas rather than around individual books, articles, or authors. This is important because organizing the literature review in this way will make it easier to establish the base of knowledge and demonstrate how the current research project can extend or add to that knowledge—with a new perspective, new data, or a different method—by resolving conflicting results in the literature or by replicating, and thereby validating, previous research.

When thinking about a literature review as motivating a research project in one of these ways, you will see that the literature review is an integral part of a research project that requires a great deal of attention to establish the direction of the project.

The key to organizing and writing an effective literature review is to focus on concepts, ideas, and methods shared across the literature. Many students are used to writing about multiple references with a focus on the individual references, discussing each collected reference in turn. For example, imagine that you have collected ten articles for a literature review. You might decide that the easiest way to organize a review that incorporates all ten articles would be to take the first article, perhaps selected because it was the most relevant, and summarize the most important parts of the article: the research question, theory, hypotheses, data, methods, and results. After summarizing the first article, you then move on to the second article and write a similar summary in the next paragraph, then the third, and so on, until all ten articles have been summarized. We call this approach to a literature review the boxcar method because such a review links the independent discussions of each article much like a series of boxcars on a train. Although this may be the easiest method for including multiple references in a literature review, it is an ineffective method because it does not explain how the ten articles fit together to establish the base of knowledge to which the current project will add nor does it establish how the current project will add to that knowledge. By tacking together independent discussions of articles, you will find it difficult to discuss common themes across references, conflicting results or conclusions, or questions left unanswered in the literature.

A more effective way to write a literature review is to focus on the concepts, ideas, and methods in the relevant literature. For example, imagine that you have the same ten articles from the previous example, but instead of discussing each independently, you begin by identifying the common themes across all ten articles. The first step might be to group the articles according to their research questions. It is likely that all ten articles address a similar broad topic but do not share the exact same research questions. You can begin to establish the base of knowledge by identifying, for example, three common research questions between the ten articles (four articles answering question one, three articles answering question two, and three articles answering question three). These three research questions represent the three areas of study the previous literature has undertaken in building our understanding of the broader topical area. Beginning the literature review with a discussion of these three research questions, and citing the articles that use each, will be an effective start at defining the base of knowledge.

Next, you might regroup the articles based on the data or different research designs used. Perhaps three of the articles used experiments and seven of the articles used case studies. Researchers commonly discuss in their literature reviews the different research designs used in the literature because, as explained in Chapter 10, different approaches have advantages and disadvantages and will be better or worse for making certain kinds of conclusions (consider issues with external and internal validity, for example).

In addition to differences in research design, each of the three experiments and seven case studies also likely used different data. Some of the case studies may have relied on personal interviews; others may have used participant-observation methods. Likewise, some of the experiments may have collected data from college students, and others may have collected data from the general population. Much like differences in research designs, differences in the data might lead to different conclusions.

As a final example, you might sort the ten references by the results or conclusions. It is unlikely that all ten articles came to the same conclusion. In fact, results of at least one of the ten articles likely contradict the results of the others. Identifying commonalities and contradictions in the literature review allows a researcher to identify ideas that have been established through replication as accepted widely in the literature and areas of disagreement that are ripe for further clarification and explanation. Conflicting results can provide a wonderful motivating factor for new research and establish for the reader the importance and relevance of the current research project.

Compared with the boxcar method, the latter example describes a much more sophisticated literature review because it integrates previous research along conceptual and methodological lines and provides a more effective organization for the researcher to explain the base of knowledge and how the current project fits into that literature. As we noted earlier, the boxcar method may be attractive because it seems easier, but the integrated literature review will better inform the current research project and the reader—and practically speaking, will earn a better grade for students on research projects.

Anatomy of a Literature Review

To further demonstrate how you might write a highly effective literature review, we include in Figure 6-6 a literature review from an article discussed in Chapter 1: "Does Attack Advertising Demobilize the Electorate?" by Stephen Ansolabehere, Shanto Iyengar, Adam Simon, and Nicholas Valentino. In this section we dissect this literature review to highlight the value of integrating references in a literature review by focusing on concepts and ideas rather than articles or books. This literature review begins with the first paragraph in the article and continues on to page two of the article. As we will see, the authors do an excellent job of explaining previous work on the effect of campaign advertising on voters, explaining the received wisdom from this work, identifying the shortcomings of previous work, and explaining how the article will correct those shortcomings.

Note first that this is a scholarly article from a highly respected political science journal. The article is written following the style and citation guidelines for the *American Political Science Review* (*APSR*). The *APSR* and many other journals use parenthetical notation to identify for the reader, at a glance, the names of the cited authors, the year of the cited publication, and a page number if relevant. The interested reader will find that the names and dates match a full citation in the works cited at the end of the literature review. Other journals may use a different citation style, such as endnotes or footnotes, but in all cases the author must provide citations acknowledging others' work and a full citation within the article. You should do the same, or your literature review will fail to give credit where credit is due and leave you open to charges of plagiarism.

In the first paragraph, the authors begin by identifying the conventional wisdom that "It is generally taken for granted that political campaigns boost citizens' involvement—their interest in the election, awareness of and information about current issues, and sense that individual opinions matter."[5] This sentence succinctly captures the essence of the received wisdom about the relationship between campaigns and voters and is followed by citations of those responsible for laying the early groundwork in developing this understanding. The second and third sentences extend the discussion of the conventional wisdom and cite two more recent studies that tested these ideas and found similar results.

The second paragraph explains that the authors question this conventional wisdom and cites various changes to the nature of campaigns since the 1940s—primarily the role of television. As in the first paragraph, after introducing a new idea in the literature review, the authors included parenthetical notes citing the work responsible for the idea. In this section the authors cite four references for the role television has played and one reference that documents the increasing importance of paid political advertising to campaign operatives. The third paragraph discusses similar themes and cites work that examines the value of rhetorical skill and the ability to withstand media scrutiny during an election. Finally, the third paragraph explains that campaigns have become "increasingly hostile and ugly" and cites two references to establish the point. As you can see, the first three paragraphs of this literature review are organized around concepts and ideas that are essential to understanding the base of knowledge about the relationship between campaign advertising and voters.

An important part of the fourth and fifth paragraphs is that they are the transition between establishing that the nature of campaigns has changed since early work on the topic and establishing that some work has attempted to mea-

FIGURE 6-6
A Well-constructed Literature Review

American Political Science Review   Vol.88, No. 4   December 1994

DOES ATTACK ADVERTISING DEMOBILIZE THE ELECTORATE?

STEPHEN ANSOLABEHERE *Massachusetts Institute of Technology*
SHANTO IYENGAR, ADAM SIMON, and
NICHOLAS VALENTINO *University of California, Los Angeles*

We address the effects of negative campaign advertising on turnout. Using a unique experimental design in which advertising tone is manipulated within the identical audiovisual context, we find that exposure to negative advertisements dropped intentions to vote by 5%. We then replicate this result through an aggregate-level analysis of turnout and campaign tone in the 1992 Senate elections. Finally, we show that the demobilizing effects of negative campaigns are accompanied by a weakened sense of political efficacy. Voters who watch negative advertisements become more cynical about the responsiveness of public officials and the electoral process.

It is generally taken for granted that political campaigns boost citizens' involvement—their interest in the election, awareness of and information about current issues, and sense that individual opinions matter. Since Lazarsfeld's pioneering work (Berelson, Lazarsfeld, and McPhee 1954; Lazarsfeld, Berelson, and Gaudet 1948), it has been thought that campaign activity in connection with recurring elections enables parties and candidates to mobilize their likely constituents and "recharge" their partisan sentiments. Voter turnout is thus considered to increase directly with "the level of political stimulation to which the electorate is subjected" (Campbell et al. 1966, 42; Patterson and Caldeira 1983).

The argument that campaigns are inherently "stimulating" experiences can be questioned on a variety of grounds. American campaigns have changed dramatically since the 1940s and 1950s (see Ansolabehere et al. 1993). It is generally accepted that television has undermined the traditional importance of party organizations, because it permits "direct" communication between candidates and the voters (see Bartels 1988; Polsby 1983; Wattenberg 1984, 1991). All forms of broadcasting, from network newscasts to talk show programs, have become potent tools in the hands of campaign operatives, consultants, and fund-raisers. In particular, paid political advertisements have become an essential form of campaign communication. In 1990, for example, candidates spent more on televised advertising than any other form of campaign communication (Ansolabehere and Gerber 1993).

We are now beginning to realize that the advent of television has also radically changed the nature and tone of campaign discourse. Today more than ever, the entire electoral process rewards candidates whose skills are rhetorical, rather than substantive (Jamieson 1992) and whose private lives and electoral viability, rather than party ties, policy positions, and governmental experience, can withstand media scrutiny (see Brady and Johnston 1987; Lichter, Amundson, and Noyes 1988; Sabato 1991). Campaigns have also turned increasingly hostile and ugly. More often than not, candidates criticize, discredit, or belittle their opponents rather than promoting their own ideas

and programs. In the 1988 and 1990 campaigns, a survey of campaign advertising carried out by the *National Journal* found that attack advertisements had become the norm rather than the exception (Hagstrom and Guskind 1988, 1992).

Given the considerable changes in electoral strategy and the emergence of negative advertising as a staple of contemporary campaigns, it is certainly time to question whether campaigns are bound to stimulate citizen involvement in the electoral process. To be sure, there has been no shortage of hand wringing and outrage over the depths to which candidates have sunk, the viciousness and stridency of their rhetoric, and the lack of any systematic accountability for the accuracy of the claims made by the candidates (see Bode 1992; Dionne 1991; Rosen and Taylor 1992). However, as noted by a recent Congressional Research Service survey, there is little evidence concerning the effects of attack advertising on voters and the electoral process (see Neale 1991).

A handful of studies have considered the relationship between campaign advertising and political participation, with inconsistent results. Garramone and her colleagues (1990) found that exposure to negative advertisements did not depress measures of political participation. This study, however, utilized student participants and the candidates featured in the advertisements were fictitious. In addition, participants watched the advertisements in a classroom setting. In contrast to this study, an experiment reported by Basil, Schooler, and Reeves (1991) found that negative advertisements reduced positive attitudes toward both candidates in the race, thereby indirectly reducing political involvement. This study, however, was not conducted during an ongoing campaign and utilized a tiny sample, and the participants could not vote for the target candidates. Finally, Thorson, Christ, and Caywood (1991) reported no differences in voting intention between college students exposed to positive and negative advertisements.

We assert that campaigns can be either mobilizing or demobilizing events, *depending upon the nature of the messages they generate.* Using an experimental design that manipulates advertising tone while holding all other features of the advertisements constant, we demonstrate that exposure to attack advertising in and of itself significantly decreases voter engagement and participation. We then reproduce this result by demonstrating that turnout in the 1992 Senate campaigns was significantly reduced in states where the tone of the campaign was relatively negative. Finally, we address three possible explanations for the demobilizing effects of negative campaigns.

EXPERIMENTAL DESIGN

There is a vast literature, both correlational and experimental, concerning the effects of televised advertisements (though not specifically negative advertisements) on public opinion (for a detailed review, see Kosterman 1991). This literature, however, is plagued by significant methodological shortcomings. The limitations of the opinion survey as a basis for identifying the effects of mass communications have been well documented (see Bartels 1993; Hovland 1959). Most importantly, surveys cannot reliably assess exposure to campaign advertising. Nor is most of the existing experimental work fully valid. The typical experimental study, by relying on fictitious candidates as the "target" stimuli, becomes divorced from the real world of campaigns. Previous experimental studies thus shed little evidence on the interplay between voters' existing information and preferences and their reception of campaign advertisements. When experimental work has focused on real candidates and their advertisements, it is difficult to capture the effects of particular characteristics of advertising because the manipulation confounds several such characteristics (Ansolabehere and Iyengar 1991; Garramone 1985; Pfau and Kenski 1989). That is, a Clinton spot and Bush spot differ in any number of features (the accompanying visuals, background sound, the voice of the announcer, etc.) in addition to the content of the message. Thus there are many possible explanations for differences in voters' reactions to these spots.

To overcome the limitations of previous research, we developed a rigorous but realistic experimental design for assessing the effects of advertising tone or valence[1] on public opinion and voting. Our studies all took place during ongoing political campaigns (the 1990 California gubernatorial race, the 1992 California Senate races, and the 1993 Los Angeles mayoral race) and featured "real" candidates who were in fact advertising heavily on television and "real" voters (rather than college sophomores) who on election day would have to choose between the candidates whose advertisements they watched. Our experimental manipulations were professionally produced and could not (unless the viewer were a political consultant) be distinguished from the flurry of advertisements confronting the typical voter. In addition, our manipulation was unobtrusive; we embedded the experimental advertisement into a 15-minute local newscast.

The most-distinctive feature of our design is its ability to capture the casual effects of a particular feature of campaign advertisement—in this case, advertising tone or valence. The advertisements that we produced were identical in all respects but tone and the candidate sponsoring the advertisement. In the 1992 California Senate primaries, for example, viewers watched a 30-second advertisement that either promoted or attacked on the general trait of "integrity." The visuals featured a panoramic view of the Capitol Building, the camera then zooming in to a closeup of an unoccupied desk inside a Senate office. In the "positive" treatments (using the example of candidate Dianne Feinstein), the text read by the announcer was as follows:

> For over 200 years the United States Senate has shaped the future of America and the world. Today, California needs honesty, compassion, and a voice for all the people in the U.S. Senate. As mayor of San Francisco, Dianne Feinstein *proposed* new government ethics rules. She *rejected* large campaign contributions from special interests. And Dianne Feinstein *supported* tougher penalties on savings-and-loan crooks.
>   California *needs* Dianne Feinstein in the U.S. Senate.

In the "negative" version of this Feinstein spot, the text was modified as follows:

> For over 200 years the United States Senate has shaped the future of America and the world. Today, California needs honesty, compassion, and a voice for all the people in the U.S. Senate. As state controller, Gray Davis *opposed* new government ethics rules. He *accepted* large campaign contributions from special interests. And Gray Davis *opposed* tougher penalties on savings-and-loan crooks.
>   California *can't afford a politician* like Gray Davis in the U.S. Senate.

By holding the visual elements constant and by using the same announcer, we were able to limit differences between the conditions to differences in tone.[2] With appropriate modifications to the wording, the identical pair of advertisements was also shown on behalf of Feinstein's primary opponent, Controller Gray Davis, and for the various candidates contesting the other Senate primaries.

In short, our experimental manipulation enabled us to establish a much tighter degree of control over the tone of campaign advertising than had been possible in previous research. Since the advertisements watched by viewers were identical in all respects and because we randomly assigned participants to experimental conditions, any differences between conditions may be attributed only to the tone of the political advertisement (see Rubin 1974).

The Campaign Context

Our experiments spanned a variety of campaigns, including the 1990 California gubernatorial election, both of the state's 1992 U.S. Senate races, and the 1993 mayoral election in Los Angeles. In the case of the senatorial campaigns, we examined three of the four primaries and both general election campaigns.

sure this new relationship. The authors cite "Neale 1991" when claiming that "there is little evidence concerning the effects of attack advertising on voters and the electoral process."[6] They also cite three studies that examined the same research question as Ansolabehere et al., "Garramone and her colleagues (1990)," "Basil, Schooler, and Reeves (1991)," and "Thorson, Christ, and Caywood (1991)." According to the authors, the previous work was inconclusive because it found conflicting results. Garramone et al. found that negative advertising did not depress turnout; Basil, Schooler, and Reeves found that negative advertisements indirectly reduced political participation; and Thorson, Christ, and Caywood reported that negative advertisements had no effect on the intention of voting. With each citation, the authors also identified some of the problems in each research design that might lead to suspect results. Given these conflicting results, the authors propose in the sixth paragraph that they will attempt to provide clarity by improving upon previous work by correcting research design flaws.

The first new paragraph on the second page, under the "Experimental Design" heading, provides further detail about the flaws of previous work using two different approaches: survey research and experimental research. The authors first point the interested reader to another reference that has documented the literature on television advertising and public opinion, "Kosterman 1991." They then turn their attention to survey research and identify the main drawback of this approach: a lack of measurement of direct exposure to advertising, as documented by two cited references. Next, the authors discuss the flaws of previous experimental work, primarily issues of external validity, and point to three cited references. The following paragraph begins the description of this article's research design.

With this example, you can see that there is a logical order to the literature review: establish conventional wisdom; establish that the nature of politics has changed—while the conventional understanding has not; and identify flaws in previous research that can be corrected. Discussing the literature in this manner makes a convincing case to the reader that this research project will be an important addition to the literature because it will improve our understanding of a topic that until now has been misunderstood. Also by organizing the literature review in this way, the authors have found a clear motivation for designing their research project as they have. Throughout the literature review, the authors integrated twenty-nine references by focusing on the concepts, ideas, and methods that were shared across the literature. Finally, the authors established that this is an important area of study (as others have an interest in writing in this area) and that our understanding is not complete (as there is disagreement through conflicting results and conclusions). Although different literature reviews will vary in the organizational style they use, we recommend that students working on their own literature reviews try to follow this topical style of integrating references; it will make even a brief discussion, like the two pages in the Ansolabehere et al. article, very powerful.

Conclusion

No matter what the original purpose of your literature review, it should be thorough. In your research report you should discuss the sources that provide explanations for the phenomena you are studying and that support the plausibility of your hypotheses. You should also discuss how your research relates to other research and use the existing literature to document the significance of your research. You can look to the example in the previous section or to an example of a literature review contained in the research report in Chapter 14. Another way to learn about the process is to read a few articles in any of the main political science journals that we listed earlier in this chapter and take some time to review the literature reviews carefully, looking for effective styles that would suit your own project.

Notes

1. The *Political Science Research Methods* CD contains several text documents that illustrate this point and allow the reader to extract empirical and testable claims from verbal arguments.
2. Larry P. Goodson, Bradford Dillman, and Anil Hira, "Ranking the Presses: Political Scientists' Evaluations of Publisher Quality," *PS: Political Science and Politics* 32 (June 1999): 257–262.
3. See www.apsanet.org/conferencepapers/.
4. A source of much of the commentary in this chapter on journals, indexes, bibliographies, and abstracts is Bill Katz and Linda Sternberg Katz, *Magazines for Libraries,* 6th ed. (New York: Bowker, 1989).
5. Stephen Ansolabehere, Shanto Iyengar, Adam Simon, and Nicholas Valentino, "Does Attack Advertising Demobilize the Electorate?" *American Political Science Review* 88 (December 1994): 829.
6. Ibid.

Terms Introduced

ELECTRONIC DATABASE. A collection of information (of any type) stored on an electromagnetic medium that can be accessed and examined by certain computer programs.

LITERATURE REVIEW. A systematic examination and interpretation of the literature for the purpose of informing further work on a topic.

SEARCH ENGINE. A computer program that visits Web pages on the Internet and looks for those containing particular directories or words.

SEARCH TERM. A word or phrase entered into a computer program (a search engine) that looks through Web pages on the Internet for those that contain the word or phrase.

Suggested Readings

Cooper, Harris M. *Integrating Research: A Guide for Literature Reviews,* 2d ed. Newbury Park, Calif.: Sage Publications, 1989.

Lester, James D. "Citing Cyberspace: How to Search the Web." Retrieved November 5, 1999, from http://longman.awl.com/englishpages.

Pan, M. Ling. *Preparing Literature Reviews: Qualitative and Quantitative Approaches,* 2d ed. Glendale, Calif.: Pyrczak Publishing, 2004.

Sarah Byrd Askew Library. "Guide for Citing Electronic Information." Retrieved November 16, 2000, from www.wpunj.edu/library.

Schmidt, Diane E. *Writing in Political Science: A Practical Guide,* 3d ed. New York: Longman, 2005.

CHAPTER 7
Sampling

Here's a conundrum. Anyone even faintly interested in American politics surely knows how important public opinion polls have become in the political process. The results can determine if a potential candidate will run for office, if a new program will receive support in a legislative body, if and how long an unpopular policy will be pursued. What is equally true is that this technique underlies a vast amount of social and political research. Indeed, much of what social scientists know (or think they know) about behavior and attitudes comes from polling. On the other hand, nearly every poll is based on a sample, and the sample sizes in most polls are tiny, especially compared to the populations from which they are drawn. A typical newspaper survey, for example, might contain the responses of at most 700 or 800 Americans, not the nearly 300 million citizens living in the country. So the question naturally arises: how can anyone make claims about what people in general think, given the small numbers of interviewees? Isn't it necessary to talk to everyone or at least to a substantial portion of the population? Skepticism about the knowledge derived from polls abounds. Probably every candidate who has ever trailed in the polls has expressed doubt about their accuracy and claimed in one way or another, "The only poll that matters comes on election day when all the ballots have been cast."

Our task in this chapter, then, is to address this issue. There are really two general questions. First, exactly what are samples and how are they collected? Second, what kind of information do they supply? Do they really provide precise measures of opinions, or do they just offer rough approximations? That is, how much confidence can we place in statements about a population given observations derived from a very few of its members? We begin answering these questions with a description of sampling techniques and reserve for later in the chapter a discussion of inferences based on samples.

The Basics of Sampling

The essence of the empirical or positivist methodology is the verification of propositions with observations and data. In Chapter 1, for example, we looked at Gelpi, Feaver, and Reifler's study of public opinion and the war in Iraq, in which they investigated a widespread perception: as casualties in a military conflict increase, popular support for it declines. The authors found some merit to this claim but identified other factors, especially beliefs about the chances of success, that are stronger predictors of public support. Where and how, one might wonder, do the observations for their research come from? In this instance the source would be, theoretically, every citizen's answer to a series of questions about President George W. Bush, the conduct of the war, the prospects for success, and the war's connection to terrorism. Of course, putting the question to every citizen would be impractical. So most researchers collect information on a much smaller set of individuals.[1] As we just noted, however, that strategy raises another issue: if an investigation of public opinion rests on 100 or even 1,000 observations, can it really say anything about the millions of Americans who comprise the general public? Can it, in other words, lead to reliable and valid conclusions? To answer this question we need to know a little about sampling.

The fundamentals are quite simple, at least in theory (see Figure 7-1). Suppose we want to assess Americans' level of support for the war in Iraq. At the outset we need to clarify what we mean by *Americans.* More formally we need to define or specify an appropriate population. In the figure the population is defined to be all adult (aged eighteen and older) citizens not residing in institutional settings (for example, prisons, hospitals) in the United States in 2006. A **population** is any well-defined set of units of analysis. It does not necessarily refer to people. A population might be all the adults living in a geographical area, such as a country or state, or working in an organization. But it could equally well be a set of counties, corporations, government agencies, events, magazine articles, or years. What is important is that the population be carefully and fully defined and that it be relevant to the research question.[2] The polygon in Figure 7-1 represents the population of adult American citizens. Since there are millions and millions of citizens the diagram only symbolizes this huge number. In this hypothetical analysis our claims refer to these people, not to Germans or Mexicans or children or any other group.

Since it is impossible to interview everyone, a more practical approach is to select just a "few" members of the population for further investigation. This is where sampling comes in. A **sample** is any subset of units collected in some manner from a population. (In the figure, the sample consists of just five out of millions of people.) The sample size and *how* its members are chosen determines the quality (that is, the accuracy and reliability) of inferences about the

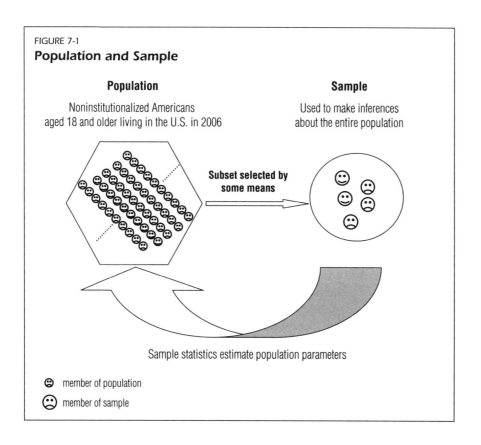

FIGURE 7-1

Population and Sample

Population

Noninstitutionalized Americans
aged 18 and older living in the U.S. in 2006

Sample

Used to make inferences
about the entire population

**Subset selected by
some means**

Sample statistics estimate population parameters

☻ member of population

☹ member of sample

whole population. The important thing to clarify is the method of selection and the number of observations to be drawn.

Once a sample has been gathered, its features or characteristics of interest can be examined and measured. The attributes of most interest in empirical research are numerical or quantitative indicators such as percentages or averages. These measures, or **sample statistics** as they are known, are used to approximate the corresponding population values, or parameters. That's the idea behind the arrow: we use sample statistics to estimate population characteristics (parameters). It may be intuitively obvious that the sample statistics will not exactly equal the corresponding values, but, as we demonstrate in this chapter, if we follow suitable procedures, they will be reasonably close.

Population or Sample?

A researcher's decision whether to collect data for a population or for a sample is usually made on practical grounds. If time, money, and other costs were not considerations, it would almost always be better to collect data for a population, because we would then be sure that the observed cases accurately reflected the population characteristics of interest. However, in many if not most instances it is simply not possible or feasible to study an entire population. Imagine, for instance, the difficulty of attempting to interview every adult in even a small city. Since research is costly and time consuming, researchers must weigh the advantages and disadvantages of using a population or a sample. The advantages of taking a sample are often savings in time and money. The disadvantage is that information based on a sample is usually less accurate or more subject to error than is information collected from a population.

The point is more subtle than it first appears. In the late 1990s Congress and President Bill Clinton debated the merits of taking a sample instead of interviewing the entire population when conducting the 2000 Census. Many members of Congress (mainly Republicans) argued that the Constitution requires a complete enumeration of the people. But Clinton and the Census Bureau argued that trying to tally everyone leads to so many errors that many groups are undercounted, in particular, undocumented aliens and inhabitants of inner cities and rural areas. It would be more accurate, they maintained, to draw careful samples of target populations and conduct quality interviews and measurements. (We should note that politics per se formed the context of this dispute; after all, innumerable government grants as well as seats in the House of Representatives are awarded according to population size.)

Some studies simply do not lend themselves to sampling. For example, case studies, which are often quite useful and lead to scientific understanding, involve the detailed examination of just one or a few units.[3] Usually these are selected by a kind of judgmental process described below.

Consider, as an example, a political scientist who wants to test some hypotheses regarding the content of televised political campaign commercials. The project requires an examination of the content of numerous commercials, which is the unit of analysis. From the standpoint of accuracy, it would be preferable to have data on the total population of televised commercials (in other words, to have available for measurement every campaign commercial that ever aired). But undertaking this type of analysis is simply impossible because no such data bank exists anywhere, nor does anyone even know how many such commercials have been televised across the country since they first appeared. Consequently, the researcher will have to rely on a sample of readily available commercials to test the hypotheses—a decision that is practical, necessary, and less costly, but perhaps subject to error.[4]

For reasons of necessity and convenience, political science researchers often collect data on a sample of observations. In fact, public opinion and voting behavior researchers almost always rely on samples. This means, however, that they must know how to select good samples. They also must appreciate the implications of relying on samples for testing hypotheses if they want to claim that their findings for the sample accurately reflect what they would find if they were to test the hypothesis on the whole population.

Fundamental Concepts

As noted in the previous section, a sample is simply a subset of a larger population, just as a sample of blood is a subset of all the blood in your body at one moment in time. If the sample is selected properly, the information it yields may be used to make inferences about the whole population. Since sampling is always used in public opinion surveys, it is often thought of in connection with that activity. Sampling arises whenever a researcher takes measurements on a subset of a population, however defined, covered by the hypothesis being investigated. Whatever empirical findings emerge from a sample from a specified population will apply to that and only that population. It would be a mistake, for instance, to sample campaign speeches from the last four presidential elections and then generalize to *all* American presidential rhetoric. By contrast, a sample drawn from a population of campaign speeches given by all presidents could be used to generalize to that population of speeches.

Before proceeding further, we should note that what usually matters most is that the data are obtained according to well-established rules. To understand why, we need to review some terms commonly used in discussions of sampling.[5]

Social scientists are interested mainly in certain characteristics of populations, such as averages, differences between groups, and relationships among variables. If any one of these traits can be quantified as a number, we call it a **population parameter.** Population parameters are typically denoted by capital English or Greek letters. Often θ is used to refer to a population parameter in general. A proportion, such as the proportion of Americans who support the war in Iraq at a particular time, typically is designated P or π (the Greek letter pi). The purpose of sampling is to collect data that provide an accurate inference about a population parameter. An **estimator,** then, is a sample statistic based on sample observations that estimates the numerical value of a population characteristic, or parameter. A specific estimator of a population characteristic or attribute calculated from sample data is called a **sample statistic**. Like population parameters, these are typically denoted by symbols or letters. Frequently, we use a hat ($\wedge$) over a character to denote a sample statistic; in some situations, a lowercase letter is used, for example, a lowercase p for a sample proportion. Sometimes, though, another symbol is used. The population mean (average) is almost always symbolized by μ (the lowercase Greek mu). But in this case nearly everyone lets $\bar{Y}$, not $\hat{\mu}$, stand for the sample mean.

An **element** (frequently called a unit of analysis) is a single occurrence, realization, or instance of the objects or entities being studied. Elements in political science research are often individuals, but they also can be states, cities, agencies, countries, campaign advertisements, political speeches, wars, social or professional organizations, crimes, or legislatures, just to name a few.

As noted previously, a population is a collection of elements defined according to a researcher's theoretical interest. Sometimes this is referred to as the theoretical population. It may, for example, consist of all campaign speeches given by major candidates for president in the last four presidential elections. Or it may be all international armed conflicts that have occurred in the past two hundred years. The key is to be clear and specific. You may refer to presidential campaign speeches as the focus of your research, but at some point you should make clear which speeches in what time periods constitute the population.

DON'T BE INTIMIDATED BY SYMBOLS AND FORMULAS
Empirical social scientists frequently use symbols or letters as a shorthand way to describe terms. These devices allow for greater precision in expression. But there is no need to panic when you come across them. Authors are usually very clear in describing exactly what a symbol means.

In this book we generally use capital English and lowercase Greek letters to designate population parameters (for example, the mean or proportion) and the same symbols with hats over them to denote corresponding sample statistics. One important exception is that in this book $\bar{Y}$ always stands for the sample mean. Also P refers to the population proportion and p refers to the sample proportion.

For reasons that we discuss shortly, a population may be stratified—that is, subdivided or broken up into groups of similar elements—before a sample is drawn. Each **stratum** is a subgroup of a population that shares one or more characteristics. For example, we might divide the population of campaign speeches in the last four presidential elections into four strata, each stratum containing speeches from one of the four elections. In a study of students' attitudes, particularly at a university, the student body may be stratified by academic class, major, and grade point average (GPA). The chosen strata are usually characteristics or attributes thought to be related to the dependent variables under study.

The particular population from which a sample is *actually* drawn is called a **sampling frame,** and it must be specified clearly. Technically speaking, all elements that are part of the population of interest to the research question should be part of the sampling frame. If they are not, any data collected may not be representative of the population. Often, however, sampling frames are incomplete, as the following example illustrates.

Suppose a researcher evaluates community opinion about snow removal by interviewing every fifth adult entering a local supermarket. The sampling frame would consist of all adults entering the supermarket while the researcher was standing outside. This sampling frame could hardly be construed as including all adult members of the community unless all adult members of the community made a trip to the supermarket *when* the researcher was there. Furthermore, use of such a sampling frame would probably introduce bias into the results. Perhaps many of the people who stayed at home rather than going to the supermarket considered the trip too hazardous because of poor snow removal. The closer the sampling frame is to the population of interest or theoretical population, the better.

Sometimes lists of elements exist that constitute the sampling frame. For example, a university may have a list of all students, or the Conference of Mayors may have a list of current mayors of cities with 50,000 residents or more. The existence of a list may be enticing to a researcher, since it removes the need to create one from scratch. But lists may represent an inappropriate sampling frame if they are out of date, incorrect, or do not really correspond to the population of interest. A common example would be if a researcher used a telephone directory as the sampling frame for interviewing sample households within the service area. Households with unlisted numbers would be missed, some numbers would belong to commercial establishments or no longer be working, and recently assigned numbers would not be included. Consequently, the telephone book could constitute an inaccurate or inappropriate sampling frame for the population in that area. Researchers should carefully check their sampling frames for potential omissions or erroneously included elements. Consumers of research should also carefully examine sampling frames to see that they match the populations researchers claim to be studying.

An example of a poll that relied on an incomplete sampling frame is the infamous *Literary Digest* poll of 1936. Despite being based on a huge sample, it predicted that the winner of the presidential election would be Alf Landon, not Franklin D. Roosevelt. This poll relied on a sample drawn from telephone directories and automobile registration lists compiled by the investigators. At that time telephone and automobile ownership were not as widespread as they are today. Thus the sampling frame overrepresented wealthy individuals.[6] The problem was compounded by the fact that in the midst of the Great Depression an unprecedented number of poor people voted, and they voted overwhelmingly for Roosevelt, the eventual landslide winner.

A newer problem with the use of telephone directories is that so many people have unlisted numbers that reliance on a printed list will quite possibly lead to a biased sample. To deal with this problem, a procedure known as **random digit dialing** (RDD) has been developed. In effect, numbers are dialed randomly.[7] In this way, all telephone owners can potentially be contacted, whether they have listed numbers or not.[8] Keep in mind, however, that not all households have telephones. The existence of millions of cell phones has further complicated the situation. Hence, even a sampling frame consisting of randomly generated telephone numbers is still an incomplete listing of households. It is estimated that 90 to 98 percent of all households have telephones.[9] Therefore, if the survey population is *all* households in the United States, a telephone sample will not be entirely representative of that population.

In many instances a list of the complete population may not exist, or it may not be feasible to create one. It may be possible, however, to make a list of groups. Then the researcher could sample this list of groups and enumerate the elements only in those groups that are selected. In this case, the initial sampling frame would consist of a list of groups, not elements.

For example, suppose you wanted to collect data on the attitudes and behavior of civic and social service volunteers in a large metropolitan area. Rather than initially developing a list of all such volunteers—a laborious and time-consuming task—you could develop a list of all organizations known to use volunteers. Next, a subset of these organizations could be selected, and then a list of volunteers could be obtained for only this subset.

A **sampling unit** is an entity listed in a sampling frame. In simple cases the sampling unit is the same as an element. In more complicated sampling designs it may be a collection of elements. In the previous example, organizations are the sampling units.

Types of Samples

Researchers make a basic distinction among types of samples according to how the data are collected. We mentioned earlier that political scientists often select a sample, collect information about elements in the sample, and then use those data to talk about the population from which the sample was drawn. In other words, they make inferences about the whole population from what they know about a smaller group. If a sampling frame is incomplete or inappropriate, **sample bias** will occur. In such cases the sample will be unrepresentative of the population of interest, and inaccurate conclusions about the population may be drawn. Sample bias may also be caused by a biased selection of elements, even if the sampling frame is a complete and accurate list of the elements in the population.

Suppose that in the survey of opinion on snow removal mentioned earlier every adult in the community did enter the supermarket while the researcher was there. And suppose that instead of selecting every fifth adult who entered, the researcher avoided individuals who appeared to be in a hurry or in poor humor (perhaps because of snowy roads). In this case the researcher's sampling frame was fine, but the sample itself would probably be biased and not representative of public opinion in that community. Because of the concern over sample bias, it is important to distinguish between two basic types of samples: probability and nonprobability samples. A **probability sample** is simply a sample for which each element in the total population has a known probability of being included in the sample. This knowledge allows a researcher to calculate how accurately the sample reflects the population from which it is drawn. By contrast, a **nonprobability sample** is one in which each element in the population has an unknown probability of being selected. Not knowing the probabilities of inclusion rules out the use of statistical theory to make inferences. Whenever possible, probability samples are preferred to nonprobability samples.

In the next several sections we consider different types of probability samples: simple random samples, systematic samples, stratified samples (both proportionate and disproportionate), cluster samples, and telephone samples. We then examine nonprobability samples and their uses.

Simple Random Samples

In a **simple random sample** each element and combination of elements has an *equal* chance of being selected. A list of all the elements in the population must be available, and a method of selecting those elements must be used that ensures that each element has an equal chance of being selected.[10] We review two common ways of selecting a simple random sample so that you can see how elements are given an equal chance of selection.

Note first that despite the seeming simplicity of the idea of "random," it can be quite difficult in practice to draw a truly simple random sample. Try writing down one hundred (much less a thousand) random integers. If you are like most people, the chances are that subtle patterns will creep into the list. You may subconsciously, for example, have a slight predilection for sevens, in which case your list will contain too many of them and too few of other numbers. This is not just an academic issue but a practical problem that confronts researchers in all fields.

Here's an instructive case that illustrates the difficulty. As the war in Vietnam wore on and opposition to it grew, the U.S. government tried to make the draft fairer so that members of all segments of society, not just the poor and people of color, would be liable for duty. Obviously, no one could reach into the population and pick men at random. Another method was needed. The Selective Service, the agency in charge of the draft, came up with a random lottery. The basic idea seemed simple enough: the likelihood that a young man would be drafted into the army was to be determined randomly by writing every day of the year on separate slips of paper, placing the slips in separate capsules, and putting all the capsules in a barrel. After turning the drum, days would be drawn randomly. On December 1, 1969, in full view of a national television audience, the dates were drawn one after another and given a number. The first date, September 14, was assigned the number one, and eligible men born on that day would be the first to be drafted, if they were not otherwise exempted. Another date was drawn, assigned the number two, and those with that birthday were second in line. The process was repeated for all the days of the year. The Selective Service estimated that anyone with a number of 200 or higher would probably not be called. So, if a person's birthday was drawn early on, there was a good chance that he would have to serve. Others would be more fortunate. Randomness supposedly ensured the system's fairness.

But observers quickly noticed that people with low numbers (and hence likely to be drafted) tended to be born in the latter months of the year. In fact, there was what we call in Chapter 13 a substantial negative correlation between day of birth (1–366) and draft number. In the minds of many, the selection process was clearly not random. If you were born in, say, January or February you had somewhat *less* chance of being called up than someone born in October, November, or December, not the same chance! What happened? The process may have been too mechanical. The capsules, which were placed in the drum sequentially so the latter days of the year were on top, may have been insufficiently mixed. This would increase the likelihood that the last days put in were most likely to be the first taken out.[11]

Problems of this sort plague research. So, after describing simple random samples, we explain a few alternative methods. One way of selecting elements at random from a list is by assigning a number to each element in the sample frame and then using a random numbers table, which is simply a list of random numbers, to select a sample of numbers. A computer can also create random numbers for this purpose. However it is done, those units having the chosen numbers associated with them are included in the sample.

Suppose, for instance, we have a population of 1,507 elements and wish to draw from it a sample of 150. We first number each member of the population, 1, 2, 3, and so on, up to 1,507. Then we can start at a random place in a random numbers table and look across and down the columns of numbers to identify our selections. Today, computers are typically used to create random numbers (see Table 7-1).[12] Each time a number between 0001 and 1,507 appears, the element in the population with that number is selected. If a number appears more than once, that number is ignored after the first time, and we simply go on to another number. (This is called sampling without replacement.) For example, if we combine the adjacent cells of the first two columns in Table 7-1 (a table of random integers), we would have the following, random numbers: 4633, 2339, 9816, 2038, and 0869. Because only 0869 falls between 0001 and 1,507, it (or more precisely, the element to which it is assigned) would be included in the sample. Doing the same for the next two columns, we would include elements 0554 and 1240. As long as we do not deliberately look for a certain number, we may start anywhere in the table and use any system to move through it. As we suggested earlier, it would not be

TABLE 7-1
Fifty Random Numbers

46	33	35	65	86	18	16	15	43	77
23	39	49	87	40	97	45	85	63	23
98	16	97	48	06	86	93	11	07	24
20	38	05	54	41	28	32	55	29	93
08	69	12	40	80	32	45	85	33	35

Note: The fifty pseudo-random integers lie between 0 and 99 and were computer generated.

acceptable to generate four-digit numbers in one's head, however, since the numbers would likely be biased in some way. Of course, for a real project we would automate the entire process by having a computer select the 150 random numbers that meet our criterion.

The second way of drawing a random sample is the by-the-lot method. In this procedure all the elements in the population are "tossed" in a hat (or some analogous procedure is used), and elements are drawn out randomly until the desired sample size has been reached. This procedure requires that the elements "in the hat" be continuously and thoroughly mixed so that each remaining element has an equal chance of being selected. This procedure can be quite cumbersome when the population size is large. It does, however, eliminate the necessity of assigning a number to each element in the sampling frame.

Whichever method of selection is used, simple random sampling requires a list of the members of the population. Whenever an accurate and complete list of the target population is available and is of manageable size, a simple random sample can usually be drawn. For example, a random sample of members of Congress could be drawn from a list of all 100 senators and 435 representatives. A simple random sample of countries could be chosen from a list of all the countries in the world, or a random sample of American cities with more than 50,000 people could be selected from a list of all such cities in the United States. The problem, as we will see, is that obtaining such a list is not always easy or even possible.

Systematic Samples

Assigning numbers to all elements in a list and then using random numbers to select elements may be a cumbersome procedure. Fortunately, a **systematic sample,** in which elements are selected from a list at predetermined intervals, provides an alternative method that is sometimes easier to apply. It too requires a list of the target population. But the elements are chosen from the list systematically rather than randomly. That is, every jth element on the list is selected, where j is the number that will result in the desired number of elements being selected. This number is called the **sampling interval,** or the "space" or number of elements between elements that are drawn.

Suppose we wanted to draw a sample of 100 names from a list of all 5,000 students attending a college. If we were going to use systematic sampling, we would first calculate the sampling interval by dividing the number of elements in the list by the desired sample size. In this case, we would divide 5,000 by the desired sample size of 100 to get a sampling interval of 50 ($j = 50$). Next we would systematically go through the list and select every fiftieth student, thereby selecting 100 names. To determine where to begin on the list, we would need to make a **random start**—that is, we would select a number at random. In our example, we would choose a number between 1 and 50 at random using a table of random digits or some other process. Thus, if we chose the number 31 at random, then students 31, 81, 131, and so on would be included in our sample.

Systematic sampling is useful when dealing with a long list of population elements. It is often used in product testing. Suppose you have been given the job of ensuring that a firm's tuna fish cans are sealed properly before they are delivered to grocery stores. And assume that your resources permit you to test only a sample of tuna fish cans rather than the entire population of tuna fish cans. It would be much easier to systematically select every 300th tuna fish can as it rolls off the assembly line than to collect all the cans in one place and randomly select some of them for testing.

Despite its advantages, systematic sampling may result in a biased sample in at least two situations.[13] One occurs if elements on the list have been ranked according to a characteristic. In that situation the position of the random start will affect the average value of the characteristic for the sample. For example, if students were ranked from the lowest to the highest GPA, a systematic sample with students 1, 51, and 101 would have a lower GPA than a sample with students 50, 100, and 150. Each sample would yield a GPA that presented a biased picture of the student population.

The second situation leading to bias occurs if the list contains a pattern that corresponds to the sampling interval. Suppose you were conducting a study of the attitudes of children from large families and you were working with a list of the children by age in each family. If the families included in the list all had six children and your sampling interval was six (or any multiple of six), then systematic sampling would result in a sample of children who were all in the same position among their siblings. If attitudes varied with birth order, then your findings would be biased.

A survey of soldiers conducted during World War II offers a good example of a case in which a pattern in the list used as the sampling frame interfered with the selection of an unbiased systematic sample.[14] The list of soldiers was arranged by squad, with each squad roster arranged by rank. The sampling interval and squad size were both ten. Consequently, the sample consisted of all persons who held the same rank, in this case squad sergeant. Clearly, sergeants might not be representative of all soldiers serving in World War II.

Stratified Samples

A **stratified sample** is a probability sample in which elements sharing one or more characteristics are grouped, and elements are selected from each group in proportion to the group's representation in the total population. Stratified samples take advantage of the principle that the more homogeneous the population, the easier it is to select a representative sample from it. Also, if a population is relatively homogeneous, the size of the sample needed to produce a given degree of accuracy will be smaller than for a heterogeneous population. In stratified sampling, sampling units are divided into strata with each unit appearing in only one stratum. Then a simple random sample or systematic sample is taken from each stratum.

A stratified sample can be either proportionate or disproportionate. In proportionate sampling, a researcher uses a stratified sample in which each stratum is represented in proportion to its size in the population—what researchers call a **proportionate sample.** For example, let's assume we have a total population of 500 colored balls: 50 each of red, yellow, orange, and green and 100 each of blue, black, and white. We wish to draw a sample of 100 balls. To ensure a sample with each color represented in proportion to its presence in the population, we first stratify the balls according to color. To determine the number of balls to sample from each stratum, we calculate the **sampling fraction,** which is the size of the desired sample divided by the size of the population. In this example the sampling fraction is 100/500, or one-fifth of the balls. Therefore, we must sample one-fifth of all the balls in each stratum.

Since there are 50 red balls, we want 10 red balls, or one-fifth. We could select these 10 red balls at random or select every fifth ball with a random start between 1 and 5. If we followed this procedure for each color, we would end up with a sample of 10 each of red, yellow, orange, and green balls and 20 each of blue, black, and white balls. Note that if we select a simple random sample of 100 balls, there is a finite chance (albeit slight) that all 100 balls will be blue or black or white. Stratified sampling guarantees that this cannot happen, which is why stratified sampling results in a more representative sample.

Systematic sampling of an entire stratified list, rather than sampling from each stratum, will yield a sample in which each stratum's representation is roughly proportional to its representation in the population. Some deviation from proportional representation will occur, depending on the sampling interval, the random start, and the number of sampling units in a stratum.

In selecting characteristics on which to stratify a list, you should choose characteristics that are expected to be related to or affect the dependent variables in your study. If you are attempting to measure the average income of households in a city, for example, you might stratify the list of households by education, sex, or race of household head. Because income may vary by education, sex, or race, you would want to make sure that the sample is representative with respect to these factors. Otherwise the sample estimate of average household income might be biased.

If you were selecting a sample of members of Congress to interview, you might want to divide the list of members into strata consisting of the two major parties, or the length of congressional service, or both. This would ensure that your sample accurately reflected the distribution of party and seniority in Congress. Or if you were selecting a sample of television news stories from NBC, CBS, Fox, CNN, and ABC to analyze, you might want to divide the population of news stories into four strata based on the network of origin to ensure that your sample contained an equal number of stories from each of the networks.

Some lists may be inherently stratified. Telephone directories are stratified to a degree by ethnic groups, because certain last names are associated with particular ethnic groups. Lists of Social Security numbers arranged consecutively are stratified by geographical area, because numbers are assigned based on the applicant's place of residence.

In the examples of stratified sampling we have considered so far, we assured ourselves of a more representative sample in which each stratum was represented in proportion to its size in the population. There may be occasions, however, when we wish to take a **disproportionate sample.** In such cases we would use a stratified sample in which elements sharing a characteristic are underrepresented or overrepresented in our sample.[15]

For example, suppose we are conducting a survey of 200 students at a college in which there are 500 liberal arts majors, 100 engineering majors, and 200 business majors, for a total of 800 students. If we sampled from each major (the strata) in proportion to its size, we would have 125 liberal arts majors, 25 engineering majors, and 50 business majors. If we wished to analyze the student population as a whole, this would be an acceptable sample. But if we wished to investigate some questions by looking at students in each major separately, we would find that 25 engineering students was too small a sample with which to draw inferences about the population of engineering students.

To get around this problem we could sample disproportionately—for example, we could include 100 liberal arts majors, 50 engineering majors, and 50 business majors in our study. Then we would have enough engineering students to draw inferences about the population of engineering majors. The problem now becomes evaluating the student population as a whole, since our sample is biased due to an undersampling of liberal arts majors and an oversampling of engineering majors. Suppose engineering students have high GPAs. Our sample estimate of the student body's GPA would be biased upward because we have oversampled engineering students. Therefore, when we wish to analyze the total sample, not just a major, we need some method of adjusting our sample so that each major is represented in proportion to its real representation in the total student population.[16]

Table 7-2 shows the proportion of the population of each major and the mean GPA for each group in a hypothetical sample of college students. To calculate an unbiased estimate of the overall mean GPA for the college, we could use a **weighting factor,** a mathematical factor used to make a disproportionate sample representative. In this example, we would multiply the mean GPA

TABLE 7-2
Stratified Sample of Student Majors

	Liberal Arts	Engineering	Business	Total
Number of students	500	100	200	800
Proportion or weight	.625	.125	.25	1.00
Size of sample	100	50	50	200
Sample mean grade-point average	2.5	3.3	2.7	2.65

Note: Hypothetical data.

for each major by the proportion of the population of each major (that is, the weighting factor).[17] Thus the mean GPA would be .625(2.5) + .125(3.3) + .25(2.7) = 2.65.

Disproportionate stratified samples allow a researcher to represent more accurately the elements in each stratum and ensure that the overall sample is an accurate representation of important strata within the target population. This is done by weighting the data from each stratum when the sample is used to estimate characteristics of the target population. Of course, to accomplish disproportionate stratified sampling, the proportion of each stratum in the target population must be known.

Cluster Samples

Thus far we have considered examples in which a list of elements in the sampling frame exists. There are, however, situations in which a sample is needed but no list of elements exists and to create one would be prohibitively expensive. A **cluster sample** is a probability sample in which the sampling frame initially consists of clusters of elements. Cluster sampling is used to address the problem of having no list of the elements in the target population. Since only some of the elements are to be selected in a sample, it is unnecessary to be able to list all elements at the outset.

In cluster sampling, groups or clusters of elements are identified and listed as sampling units. Next a sample is drawn from this list of sampling units. Then, for the sampled units only, elements are identified and sampled. For example, suppose we wanted to take an opinion poll of 1,000 persons in a city. Since there is no complete list of city residents, we might begin by obtaining a map of the city and identifying and listing all blocks. This list of blocks becomes the sampling frame from which a small number of blocks are sampled at random or systematically. (The individual blocks are sometimes called the primary sampling units.) Next we would go to the selected blocks and list all the dwelling units in those blocks. Then a sample of dwelling units would be drawn from each block. Finally, the households in the sampled dwellings would be contacted, and someone in each household would be interviewed for the opinion poll. Suppose there are 500 blocks, and from these 500 blocks 25 are chosen at random. On these 25 blocks, a total of 4,000 dwelling units or households are identified. One-quarter of these households will be contacted because a sample of 1,000 individuals is desired. These 1,000 households could be selected with a random sample or a systematic sample.

Note that even though we did not know the number of households ahead of time, each household has an equal chance of being selected. The probability that any given household will be selected is equal to the probability of one's block being selected times the probability of one's household being selected, or $25/500 \times 1000/4000 = 1/80$. Thus cluster sampling conforms to the requirements of a probability sample.

Our example involved only two samples or levels (the city block and the household). Some cluster samples involve many levels or stages and thus many samples. For example, in a national opinion poll the researcher might list and sample states, list and sample counties within states, list and sample municipalities within counties, list and sample census tracts within municipalities, list and sample blocks within census tracts, and finally list and sample households—a total of six stages.

Cluster sampling allows researchers to get around the problem of acquiring a list of elements in the target population. Cluster sampling also reduces fieldwork costs for public opinion surveys, because it produces respondents who are close together. For example, in a national opinion poll, respondents will not come from every state. This reduces travel and administrative costs.

A drawback to cluster sampling is greater imprecision. Errors that arise by virtue of taking samples instead of enumerating an entire population occur at each stage of the cluster sample. For example, a sample of states will not be totally representative of all states, a sample of counties will not be totally representative of all counties, and so on. The random errors at each level must be added together to arrive at the total sampling error for a cluster sample.

In cluster sampling, the researcher must decide how many elements to select from each cluster. In the previous example, the researcher could have selected two individuals from each of the 500 blocks (requiring no selection of blocks), or 1,000 individuals from one of the blocks (making the selection of the particular block terribly important), or some other combination in between (40 individuals from 25 blocks, 25 individuals from 40 blocks, and so on). But how does the researcher decide how many units to sample at each stage?

We know that samples are more accurate when drawn from homogeneous populations. Generally, elements within a group are more similar than are elements from two different groups. Thus households on the same block are more likely to resemble each other than households on different blocks. Sample size can be smaller for homogeneous populations than for heterogeneous populations and still be as accurate. (If a population is totally homogeneous, a sample of one element will be accurate.) Therefore, sampling error could be reduced by selecting many blocks but interviewing only a few households from each block. Following this reasoning to the extreme, we could select all 500 blocks and sample two households from each block. This approach, however, would be very expensive, since every household in the city would have to be identified and listed, which defeats the purpose of a cluster sample. The desire to maximize the accuracy of a sample must be balanced by the need to reduce the time and cost of creating a sampling frame— a major advantage of cluster sampling. Sometimes the stratification of clusters can reduce sampling error by creating more homogeneous sampling units. States can be grouped by region, census tracts by average income, and so forth, before the selection of sample elements occurs.

Systematic, stratified (both proportionate and disproportionate), and cluster samples are acceptable and often more practical alternatives to the simple random sample. In each case the probability of a particular element's being selected is known; consequently, the accuracy of the sample can be determined. The type of sample chosen depends on the resources a researcher has available and the availability of an accurate and comprehensive list of the elements in a well-defined target population.

Nonprobability Samples

A nonprobability sample is a sample for which each element in the total population has an unknown probability of being selected. Probability samples are usually preferable to nonprobability samples because they represent a large population fairly accurately, and it is possible to calculate how close an estimated characteristic is to the population value. In some situations, however, probability sampling may be too expensive to justify (in exploratory research, for example), or the target population may be too ill-defined to permit probability sampling (this was the case with the television commercials example discussed earlier). Researchers may feel that they can learn more by studying carefully selected and perhaps unusual cases than by studying representative ones. A brief description follows of some of the types of nonprobability samples.

With a **purposive sample** a researcher exercises considerable discretion over what observations to study, because the goal is typically to study a diverse and usually limited number of observations rather than to analyze a sample representative of a larger target population. Richard F. Fenno Jr.'s *Home Style,* which describes the behavior of eighteen incumbent representatives, is an example of research based on a purposive sample.[18] Likewise, a study of journalists that concentrated on prominent journalists in Washington, D.C., or New York City would be a purposive rather than a representative sample of all journalists.

In a **convenience sample,** elements are included because they are convenient or easy for a researcher to select. A public opinion sample in which interviewers haphazardly select whomever they wish is an example of a convenience sample. A sample of campaign commercials that consists of those advertisements that a researcher is able to acquire or a study of the personalities of politicians who have sought psychoanalysis is also a convenience sample, as is any public opinion survey consisting of those who volunteer their opinions. Convenience samples are most appropriate when the research is exploratory or when a target population is impossible to define or locate. But like other nonprobability samples, convenience samples provide estimates of the attributes of target populations that are of unknown accuracy.

A **quota sample** is a sample in which elements are sampled in proportion to their representation in the population. In this respect, quota sampling is similar to proportionate stratified sampling. The difference between quota sampling and stratified sampling is that the elements in the quota sample are not chosen in a probabilistic manner. Instead, they are chosen in a purposive or convenient fashion until the appropriate number of each type of element (quota) has been found. Because of the lack of probability sampling of elements, quota samples are usually biased estimates of the target population. Even more important, it is impossible to calculate the accuracy of a quota sample.

A researcher who decided to conduct a public opinion survey of 550 women and 450 men and who instructed his interviewers to select whomever they pleased until these quotas were reached would be drawing a quota sample. A famous example of an error-ridden quota sample is the 1948 Gallup Poll that predicted that Thomas Dewey would defeat Harry Truman for president.[19]

In a **snowball sample,** respondents are used to identify other persons who might qualify for inclusion in the sample.[20] These people are then interviewed and asked to supply appropriate names for further interviewing. This process is continued until enough persons are interviewed to satisfy the researcher's needs. Snowball sampling is particularly useful in studying a relatively select, rare, or difficult-to-locate population such as draft evaders, political protesters, drug users, or even home gardeners who use sewage sludge on their gardens—a group estimated to constitute only 3 to 4 percent of households.[21]

We have discussed the various types of samples that political science researchers use in their data collection. Samples allow researchers to save time, money, and other costs. However, this benefit is a mixed blessing, for by avoiding these costs researchers must rely on information that is less accurate than if they had collected data on the entire target population. Now we consider the type of information that a sample provides and the implications of using this information to make inferences about a target population.

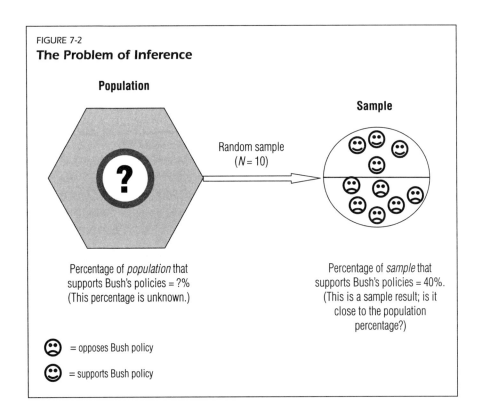

FIGURE 7-2
The Problem of Inference

Population

Sample

Random sample
(*N* = 10)

Percentage of *population* that
supports Bush's policies = ?%
(This percentage is unknown.)

Percentage of *sample* that
supports Bush's policies = 40%.
(This is a sample result; is it
close to the population
percentage?)

= opposes Bush policy

= supports Bush policy

Samples and Statistical Inference: A Gentle Introduction

We have spent a lot of time in earlier parts of this book talking about the use of statistics in research. For example, statistical techniques sometimes let us compensate for the lack of physical control or manipulation of variables in experiments. Equally important, statistics underlies the process of making inferences about populations from samples. Here we describe some of the basic ideas in nonmathematical terms.

Let's return to a previous example. We want to measure support for President Bush's handling of the war in Iraq. Figure 7-2 illustrates our problem. On the one hand, at any given time a presumably *unknown* proportion of Americans back the president's policies, but we have little or no idea what that percentage is. (In an earlier section we called this percentage a *parameter*.) If we had absolutely no knowledge of American politics, we might think it could be anywhere from 0 to 100 percent. Usually, however, an investigator has a rough idea of the public's attitudes, but even so it could run from, say, 25 to 75 percent. That still leaves a lot of room for doubt, so we need to do better than this ballpark guess.

Suppose we were to draw (at random) a sample of ten adult Americans and count the number who are supportive. (Look at the right side of Figure 7-2.) Here we see that four out of ten, or 40 percent, of the respondents are supportive. This number, the sample statistic or estimator, provides an estimate of the population proportion. Not having any other information, we might take it as our best approximation of public opinion on the matter. But just how good is it? Can we really say anything about the attitudes of millions of Americans based on a sample of just ten people? Before making a judgment let's examine sampling and inference in a bit more detail.

Samples provide only estimates or approximations of population attributes. Occasionally these "guesses" may be right on the money. Most of the time, however, they will differ from the "true" or population parameter. When we report a sample statistic, we always assume there will be a margin of error, or a difference between the reported and actual values. For example, a finding that 40 percent of a random sample backs the president's goals and policies does not mean that exactly 40 percent of the public are sympathetic. It means merely that *approximately* 40 percent are. In other words, researchers sacrifice some precision whenever they rely on samples instead of enumerating and measuring the entire population. How much precision is lost (that is, how accurate our estimate is) depends on how the sample has been drawn and its size.

Where does the loss of precision or accuracy come from? The answer is chance or luck of the draw. If you flip a coin ten times, you probably won't get exactly five heads, even if the coin is fair or the probability of heads is one-half. Randomness just seems to be a feature of nature, at least at the scales we observe it. Similarly, a random sample of ten (or even much larger) is not likely to produce precisely the value of a corresponding population parameter. But, if we follow proper procedures and certain assumptions have been met (for example, the sample is a simple random sample from an infinite population), a sample statistic approximates the numerical value of a population parameter. If a population percentage really is 40, it is unlikely (*but not impossible*) for a sample result to be, say, 5 or 10 or 99 percent or some other "extreme" value. More likely the sample estimate will be something like 30 or 35 or 45 or 50 percent. The difficulty is figuring out how far off the estimate is likely to be in any individual case. Here is where statistics helps.

The major goal of **statistical inference** is to make supportable conjectures about the unknown characteristics of a population based on sample statistics. The study of statistics partly involves defining much more precisely what *supportable* means. To make this clear we introduce in a nontechnical way three concepts:

- Expected values
- Standard errors
- Sampling distributions

Although these terms may appear at first sight to have technical meanings, they can be given commonsense interpretations.

Expected Values

Let's change to a simpler example. A candidate for the state senate wants to know how many independents live in her district, which has grown rapidly in the past ten years. Although the Bureau of Elections reports that 25 percent of registered voters declined to name a party, she believes that the records are badly out of date. She asks you to conduct a poll to estimate the proportion of citizens, aged eighteen and older, who registered as independents rather than Democrats or Republicans.

Suppose you interview ten *randomly* chosen adults living in the district and discover that two of them registered as independents.[22] Based on this finding, you could report that 20 percent of voters are registered independents. Intuitively, however, you know that this estimate may be off by quite a bit, because you interviewed only ten people. The true proportion may be very different.

Now suppose for the moment that the Bureau of Elections' records are still accurate: one-fourth, or 25 percent, of the population are registered as independents, or, in more formal terms, $P = .25$, where P stands for the value of the population parameter. Of course, no one knows the population value because at the time of your poll it is unobserved, but we will pretend that we do in order to illustrate the ideas of sampling and inference. Your first estimate, .20, then, is a little bit below the true value. This difference is called the **sampling error,** which is the discrepancy between an observed and a true value that arises purely by chance or happenstance.

What you need is some way to measure the amount of error or uncertainty in the estimate so that you can tell your client what the margin of error is. That is, you want to be able to say, "Yes, my estimate is probably not equal to the real value, but chances are that it is close." What exactly do words like *chances are* and *close* mean?

To answer those questions, imagine taking another, totally independent sample of ten adults from the same district and calculating the proportion of independents. (We will assume that not much time has passed since the first sample, so the probability of being an independent is still 25 percent.) This time the estimate turns out to be .30. As with your first sample, such a result is possible if $P = .25$ because your data come from a small sample, and samples are unlikely to reproduce exactly the characteristics of the populations from which they are drawn.

Repeating the procedure once more you find that the next estimated proportion of independents is .40. This estimate, while quite high, is still possible. And after you take a fourth independent sample, you find that the estimated proportion, .20, is again wide of the mark. So far two of your estimates

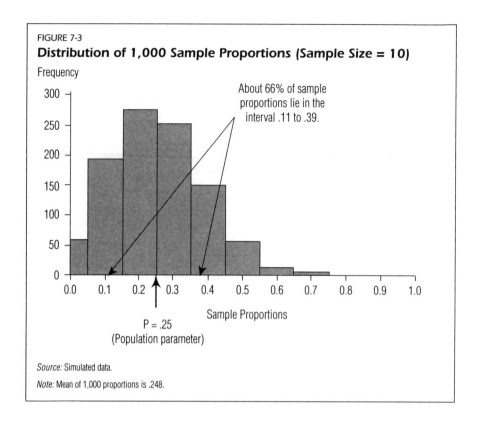

FIGURE 7-3

Distribution of 1,000 Sample Proportions (Sample Size = 10)

Frequency

About 66% of sample proportions lie in the interval .11 to .39.

P = .25
(Population parameter)

Sample Proportions

Source: Simulated data.

Note: Mean of 1,000 proportions is .248.

have been too large, two too low, and none exactly on target. But notice that the average of the estimates, $(.20 + .30 + .40 + .20)/4 = .275$, is not too far from the real value of .25.

What would happen, you might wonder, if you repeated the process indefinitely? That is, what would happen if you took an infinite number of independent samples of $N = 10$ and calculated the proportion of independents in each one?[23] (Throughout this discussion we use N to denote the size of a sample.) After a while you would have an extended list of sample proportions or percentages. What would their distribution look like? Figure 7-3 gives an idea. In brief, we programmed a computer to take 1,000 samples (each of size 10) from a hypothetical population in which $P = .25$. This technique permits us to investigate the behavior of a huge number of sample outcomes.

The separate sample proportions are spread around the true value ($P = .25$) in a bell-curve-shaped distribution, that is, a curve with a single peak and more or less symmetric or equal tails. A few of the estimates are quite low, even close to zero, while a few more of them are way above .25. (The frequencies can be determined by looking at the y-axis, the vertical line.) Yet the vast majority are in the range .05 to .45, and the center of the distribution (the average of the 1,000 sample proportions) is near .25, the population value. Indeed, the average of the 1,000 proportions in this particular data set is .248, which lies very close to the true value! This is no coincidence, as we will see.

This illustration highlights an important point about samples and the statistics calculated from them. If statistics are calculated for each of many, many independently and randomly chosen samples, their average or mean will equal the corresponding true, or population, quantity, no matter what the sample size. Statisticians refer to this mean as the **expected value** (E) of the estimator. This idea can be stated more succinctly. If θ represents a population parameter or characteristic such as a proportion or mean, then $\hat{\theta}$ stands for a sample estimator of that characteristic. We can then write

$$E(\hat{\theta}) = \theta,$$

which reads, "The expected (or long run) value of the estimator equals the corresponding value for the population from which the sample has been drawn."

In the case of a sample proportion based on a simple random sample, we have

$$E(p) = P,$$

where p is the estimated proportion, and the equation reads, "The expected (or long run, or average) value of sample proportions equals the population proportion."

In plain words, although any particular estimate result may not equal the parameter value of the population from which the data come,[24] if the sampling procedure were to be repeated an infinite number of times, and a sample estimate calculated each time, then the average, or mean, of these results would equal the true value. This fact gives us confidence in the sampling method, not in any particular sample statistic. Since Figure 7-3 includes only 1,000 estimates, not an infinite number, it only illustrates what can be demonstrated mathematically for many types of sample statistics such as a proportion.

Measuring the Variability of the Estimates: Standard Errors

Besides the expected value, statistical theory also tells us that in situations like this one the sample proportions will fall above and below the true value in a predictable manner as suggested by Figure 7-3. That is, there is variation or variability in the outcomes. As we just observed, most of the sample proportions fall between .05 and .45 (or 5 and 45 percent). A few will be much larger or smaller, but they will be the exceptions. Consequently we can use a graph like that shown in Figure 7-3 to determine approximately the likelihood of getting a particular sample result *if* the true value of the population from which the samples have been drawn is .25. For example, what are the chances of getting a sample proportion of .29 if the population proportion is .25? The answer: very likely. Why? Because statistics tells us that most sample results will be close to the true value. But suppose a sample proportion turns out to be .75. If the true number is .25, is this a likely result? Look at the figure. It suggests that a value that far from .25 occurs only rarely. So the answer might be "A sample proportion of .75 is possible but not very probable." (You can use the areas in the rectangles to "guesstimate" the chances.)

The fact that statistics behave in this manner helps us make inferences. To anticipate the material in later chapters let us continue to hypothesize that the true proportion is .25. Now assume that a sample of 10 produces a proportion of .19. Given that such a result is reasonably possible—look at Figure 7-3 once again—we might conclude that this hypothesis cannot be rejected. However, if the sample result turned out to be .9, we would be justified in concluding that the hypothesis does not hold water and should be rejected. Why? Because .9 is an unlikely result given that $P = .25$.

Of course, we could be making a mistake. It is possible that the true proportion is .25 even though our sample estimate is way above that number. If we did reject the hypothesis (that is, $P = .25$), we would be wrong. Yet the chances of making this kind of error are relatively small. That's what people mean when they say they have **confidence** in an estimate. (Confidence does not equal certainty, just as in legal trials judgments are based on the standard of reasonable doubt, not absolute, infallible knowledge.)

The mathematical term for the variation around the expected value is the standard error of the estimator, or **standard error** for short. Loosely speaking, the standard error provides a numerical indication of the variation in our sample estimates. (Like all statistical indicators, it has its own symbol, $\sigma_{\hat{\theta}}$.) We do not show the calculations but simply assert that the standard error in this example is .14.[25] As of now this number has no obvious meaning, but, as shown later, it can be used to make probability statements such as "roughly two-thirds of the sample proportions lie in the interval between .11 and .39," which means that approximately two out of three sample proportions will fall in that interval. So now you can tell your client, the senator, that based on just the first (and presumably the only) sample you have taken, the true proportion of independents in the district is probably somewhere between 11 and 39 percent with the best bet being 25 percent. When she asks what you mean by *probably,* you can tell her, "I am about 66 percent sure." (You might be able to recognize this point by looking at the frequencies represented by the bars in the graph.)

Not surprisingly, your first estimate may not be too helpful to the campaign, which must decide how to target its limited resources. After all, if the percentage of independents in the district is as low as 11 percent, the senator might

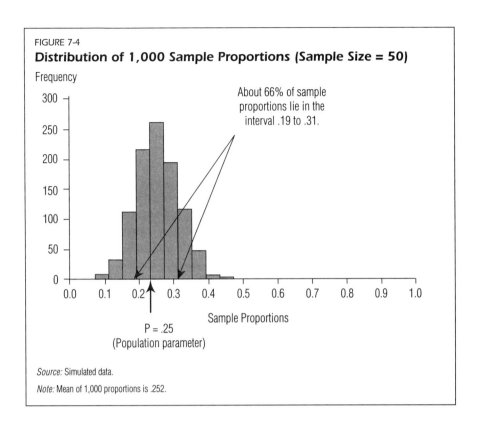

FIGURE 7-4

Distribution of 1,000 Sample Proportions (Sample Size = 50)

About 66% of sample proportions lie in the interval .19 to .31.

P = .25
(Population parameter)

Sample Proportions

Source: Simulated data.

Note: Mean of 1,000 proportions is .252.

follow one strategy, but if it is 39 percent or more, she could do something else. As a result, the senator would like you to narrow the range of uncertainty. What can you do? The answer may be obvious: take a larger sample.

Imagine that you increase the sample size to 50 registered voters (N = 50) from the population in which P is still .25, and then note the estimated proportion. This time you might find that 15 respondents out of 50, or 30 percent, are independents. Because of our omniscience we know this estimate is a bit too large. But, as before, let us repeat the process. If you drew 1,000 independent random samples, each containing 50 observations, and plotted the distribution of the estimated proportions, you would get a graph similar to the one in Figure 7-4.

For the hypothetical data shown in this figure, the mean of the 1,000 sample proportions is .252, a value quite near the true number.[26] The figure illustrates once again what can be shown mathematically, namely, that the distribution of p's is approximately bell shaped with the expected or long-run value of sample estimates equal to the true proportion of the population from which the samples have been collected. Also notice that the distribution is not as spread out as the one depicted in Figure 7-3. In our statistical language the standard error is

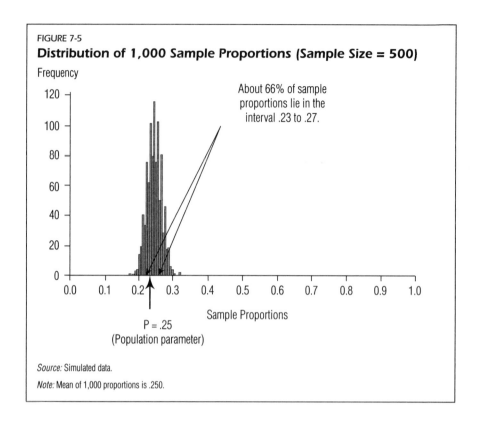

FIGURE 7-5

Distribution of 1,000 Sample Proportions (Sample Size = 500)

Frequency

About 66% of sample proportions lie in the interval .23 to .27.

P = .25
(Population parameter)

Sample Proportions

Source: Simulated data.

Note: Mean of 1,000 proportions is .250.

smaller, .06 versus .14 previously. Hence, about two-thirds of the sample proportions fall in the interval .19 to .31, which is about half the width of the one based on ten cases; very few fall in the tails of the distribution. So increasing the sample size gives us perhaps more confidence that .252 is near the true value.

To cement the point, let us repeat the simulation using a much larger sample, N = 500. The result appears in Figure 7-5. It too shows that the average of the sample proportions is close to the true value and that the variability of the estimates, the sampling error, has been greatly reduced.

This finding illustrates the generalization that the sample size affects the magnitude of sampling variation: the larger the sample, the smaller the standard error. That statement, in turn, implies that as sample sizes increase, the range of sample estimators decreases. (This fact may be consistent with your intuition that large samples should be more accurate than small ones.) But keep in mind the expected value of sample estimators does not depend on the sample size. Instead, it is the confidence placed in them that does.

Table 7-3 summarizes the results for samples of size 10, 50, and 500. For instance, the row labeled "10" contains the results of taking 1,000 independent samples (each of N = 10) from a population in which P = .25. The average of the

TABLE 7-3
Properties of Samples of Different Sizes

Sample Size	Average (Mean)[a]	Standard Error[b]	66% Confidence Interval	Minimum Proportion	Maximum Proportion	Range of Proportions
10	.248	.14	.11–.39	0	.70	.70
50	.252	.06	.19–.31	.1	.48	.38
500	.250	.02	.23–.27	.19	.32	.13

[a]Each mean is the average of 1,000 sample proportions taken from a population in which the true probability is P = .25
[b]This term measures the variation or variability of the sample proportions. It indicates the magnitude of sampling error.

1,000 sample proportions is .248, the standard error is .14, the interval containing about 66% of the sample proportions is .11 to .39, and the lowest and highest proportions are 0 and .70. Similarly, the next row contains the results of 1,000 samples of $N = 50$. For these sets of simulated data the average of the sample proportions is always close to the true value no matter how large the sample, again illustrating the argument about expected values. But, and herein lies the crux of the argument, the measures of variability of the proportions decrease considerably as the sample sizes get larger. The numbers may seem small to you, but notice that the variability of the sample results based on 10 cases (.14) is more than twice as large as the corresponding number for the samples of size 50 (.06).

What does all this mean in plain English? Small sample sizes are not invalid or worthless. The expected values of many of their sample statistics will equal the population parameters. But estimates based on small samples may be imprecise in the sense that any specific estimate may be much too high or low to be useful. Whether or not this is the case depends on the study's purposes.

Sampling Distributions

Let's tie things together by introducing the last concept on our list, sampling distributions. Figures 7-3, 7-4, and 7-5 may be graphical representations of sampling distributions, but they offer an inkling of what the phrase *sampling distribution* means.

A **sampling distribution** of a sample statistic is a theoretical expression that describes the mean, variation, and shape of the distribution of an infinite number of occurrences of the statistic when calculated on samples of size N drawn independently and randomly from a population. Think of it as a statistical tool for calculating the probability that sample statistics fall within certain distances of the population parameter. The sample information cannot, of course, tell us exactly where within the range of values the population parameter lies. But it allows us to make an educated guess.

A general method for making such an inference is quite simple. Let $\hat{\theta}$ be a sample statistic that estimates a population parameter, θ. (In the previous example when we knew the population parameter was a proportion, we called these p and P, respectively.) Since θ is unknown, we want to surround an estimate of it with a range or interval of values that with some known probability includes it. (For example: Is the population parameter "probably" between 20 and 30?) This range can be found by adding and subtracting some multiple of the standard error.

For some sample statistics obtained from random samples, we can say that the interval $\hat{\theta} \pm 1.96\sigma_{\hat{\theta}}$ has a 95 percent probability of containing the population value. (To be clear about the point, the previous statement means that if we drew 100 independent samples from a population having a parameter θ, we believe that about 95 out of the 100 estimated intervals would include this value. The parameter does not "bounce" around; it is a constant. Rather, the intervals vary from sample to sample.) Obviously you need to know how to calculate the standard error and where numbers like "1.96" come from. We explain this in Chapter 11. For now only the basic idea is important. By adding and subtracting some multiple of the standard error to the estimator we can obtain a confidence interval for the statistic and interpret this range to mean "there is such and such a probability that the calculated interval includes the population value."

This reasoning underlies poll results reported in the media. You may occasionally see reports of sample-based information that imply that the sample results are precise estimates of the target population. During the 2004 presidential election campaign, for example, you may have seen newspaper headlines declaring, "Bush Leads Kerry, 52% to 48%." Such reports can be misleading. No probability sample can produce exact estimates of the voting intentions of the population. Although the results may have been 52 percent for Bush, we now know that these estimates are subject to sampling error. True, the calculation of these errors and related confidence intervals flows directly from knowledge of the sampling distribution of statistics and mathematical theory, which in many cases is well understood. But there is always some doubt about an estimate.

How Large a Sample?

Ideally, sample estimates of the target population are equal to the purposes of the research. Of course, the margin of error should be small. However, the only way to eliminate it entirely is to collect data from the entire target population (in other words, not to rely on a sample at all). Since doing so is usually impractical, sampling error is the price researchers pay for reducing the costs involved in measuring some attribute of the target population.

Note too that sampling error occurs not just in public opinion polls but whenever a researcher relies on a sample. For example, any measurements of attributes in samples of members of Congress, convention delegates, census tracts, or nation-states are estimates of the target population of members, convention delegates, census tracts, or nation-states, and therefore are subject to sampling error.

As we learned earlier, the key to controlling sampling error is the sample size. Generally, the larger the sample, the smaller the sampling error, as measured by the standard error. Given that the sample size figures so prominently in sampling distributions, you might think that by increasing N you could reduce uncertainty to near zero. However, the relationship between sample size and sampling error is exponential rather than linear. For example, to cut sampling error in half, the sample size must be quadrupled. This means that researchers must balance the costs of increasing sample size with the size of the sampling error they are willing to tolerate.

Table 7-4 shows the relationship between sample size and the margin of error for Gallup Poll–type samples.[27] In public opinion research, increasing sample size may be too costly. Survey analysts usually draw samples of 1,500 to 2,000 people (regardless of the size of the target population). This yields a margin of error (about ±3 percent) at a cost that is within reach for at least some survey organizations. But reducing the sampling error appreciably in this kind of research would mean incurring costs that are prohibitive for most researchers.

TABLE 7-4

The Relationship between Sample Size and Sampling Error

Sample Size	Confidence Interval (percent)
4,000	± 2
1,500	± 3
1,000	± 4
600	± 5
400	± 6
200	± 8
100	± 11

Source: Charles W. Roll Jr. and Albert H. Cantril, *Polls: Their Use and Misuse in Politics* (New York: Basic Books, 1972), 72. Copyright © 1972 by Basic Books, Inc. Reprinted by permission of Basic Books, a member of Perseus Books, L.L.C.

Note: This table is based on a 95 percent confidence level and is derived from experience with Gallup Poll samples.

The question of how large a sample should be thus depends not so much on bias or no bias (after all, the expected value of a statistic based on even a very small sample is unbiased, as we saw above) as on how narrow an interval a researcher needs for a given level of confidence. For exploratory projects in which a rough approximation is adequate, a sample need not be huge. But when researchers attempt to make fine distinctions they must collect more data.

We can illustrate this point with still another example. Assume that we want to estimate a population mean, and suppose further that we want to be 99 percent certain about our estimate. (Notice that we have established a specific level of confidence—99 percent certainty.) To achieve this level of confidence, how wide off the mark can our estimate be and still be useful? Once we answer this question, we can choose an appropriate sample size. For example, if we want to say with 99 percent certainty that the interval $25,500 to $28,500 contains the true mean, then we would need a sample of a certain size (perhaps 200). But if we want to be 99 percent certain that the mean lies between $26,500 and $26,600—a mere $100 difference—then we will need a much larger sample.[28]

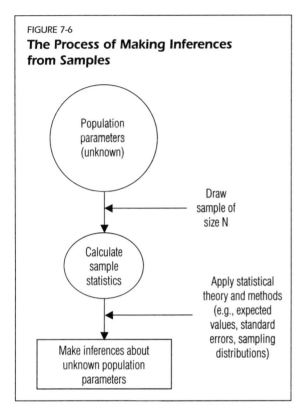

FIGURE 7-6

The Process of Making Inferences from Samples

Population parameters (unknown)

Draw sample of size N

Calculate sample statistics

Apply statistical theory and methods (e.g., expected values, standard errors, sampling distributions)

Make inferences about unknown population parameters

Decisions about sample sizes involve trade-offs. As a sample gets larger, the margin of error in an estimate decreases. But the cost of the study goes up. It may be that the senate candidate wants an estimated proportion to be within 1 or 2 percent of the true value; but does she have sufficient funds to collect a large enough survey? If not, she might have to settle for a wider confidence interval.[29]

Conclusion

In this chapter we discussed what it means to select a sample out of a target population, the various types of samples that political scientists use, and the kinds of information they yield. Figure 7-6 provides an intuitive summary of sampling in the research process.

The following guidelines may help researchers who are deciding whether or not to rely on a sample as well as students who are evaluating research based on sample data:

1. If cost is not a major consideration, and the validity of the measures will not suffer, it is generally better to collect data for the complete target population than for just a sample of that population.

2. If cost or validity considerations dictate that a sample be drawn, a probability sample is usually preferable to a nonprobability sample. The accuracy of sample estimates can be determined only for probability samples. If the desire to represent a target population accurately is not a major concern or is impossible to achieve, then a nonprobability sample may be used.

3. Probability samples yield estimates of the target population. All samples are subject to sampling error. No sample, no matter how well drawn, can provide an exact measurement of an attribute of, or relationship within, the target population.

4. Fortunately, statistical theory gives us methods for making systematic inferences about unknown parameters and for objectively measuring the probabilities of making inferential errors. This information allows the researcher and the scientific community to judge the tenability of many empirical claims.

Notes

1. It would not only be impractical but perhaps inaccurate as well. Social scientists generally believe that for *some* purposes, trying to contact and study every person in a population may lead to more errors than if a smaller sample were investigated.

2. A related concern is the size of the population. For most purposes we always assume it is infinite. In fact, no population of real "things" has an infinite number of members, but we nevertheless treat populations as if they were infinite.

3. For a good discussion, see Gary King, Robert O. Keohane, and Sidney Verba, *Designing Social Inquiry* (Princeton: Princeton University Press, 1994).

4. Richard A. Joslyn, *Mass Media and Elections* (Reading, Mass.: Addison-Wesley, 1984).

5. This discussion of terms used in sampling is drawn primarily from Earl R. Babbie, *Survey Research Methods* (Belmont, Calif.: Wadsworth, 1973), 79–81.

6. Ibid., 74–75.

7. Random digit dialing has become increasingly sophisticated. For more details on how it works and for a list of references, see Johnny Blair and others, "Sample Design for Household Telephone Surveys," Survey Research Center University of Maryland, available online at www.bsos.umd.edu/src/sampbib.html.

8. For various methods of random digit dialing, see E. L. Landon Jr. and S. K. Banks, "Relative Efficiency and Bias of Plus-One Telephone Sampling," *Journal of Marketing Research* 14 (August 1977): 294–299; K. M. Cummings, "Random Digit Dialing: A Sampling Technique for Telephone Surveys," *Public Opinion Quarterly* 43 (Summer 1979): 233–244; R. M. Groves and R. L. Kahn, *Surveys by Telephone* (New York: Academic Press, 1979); and J. Waksberg, "Sampling Methods for Random Digit Dialing," *Journal of the American Statistical Association* 73 (March 1978): 40–46.

9. James H. Frey, *Survey Research by Telephone* (Beverly Hills, Calif.: Sage, 1983), 22.

10. When used to describe a type of sample, *random* does not mean haphazard or casual; rather, it means that every element has a known probability of being selected. Strictly speaking, to ensure an equal chance of selection, *replacement* is required, that is, putting each selected element back on the list before the next element is selected. In *simple* random sampling, however, elements are selected without replacement. This means that on each successive draw, the probability of an element's being selected increases because fewer and fewer elements remain. But for each draw the probability of being selected is equal among the remaining elements. If the sample size is less than one-fifth the size of the population, the slight deviation from strict random sampling caused by sampling without replacement is acceptable. See Hubert M. Blalock Jr., *Social Statistics,* 2d ed. (New York: McGraw-Hill, 1972), 513–514.

11. This episode has been studied extensively. See, for example, Stephen E. Fienberg, "Randomization and Social Affairs: The 1970 Draft Lottery," *Science* 171 (January 22, 1971): 255–261; and Stephen E. Fienberg, "Randomization for the Selective Service Draft Lotteries," in Frank Mosteller and others, eds., *Statistics by Example: Finding Models* (Reading, Mass.: Addison-Wesley, 1973): 1–13.

12. These numbers are sometimes called "pseudo-random" because computers use an algorithm to generate a string of digits, and if you knew that algorithm and its starting place, you could recreate the list exactly. But once produced, the numbers pass all sorts of mathematical tests of randomness. For a more advanced discussion, see Micah Altman, Jeff Gill, and Michael P. McDonald, *Numerical Issues in Statistical Computing for the Social Sciences* (New York: Wiley-Interscience, 2003), esp. chaps. 2 and 3.

13. Blalock, *Social Statistics,* 515.

14. Babbie, *Survey Research Methods,* 93.

15. There are two reasons for using disproportionate sampling in addition to obtaining enough cases for statistical analysis of subgroups: the high cost of sampling some strata and differences in the heterogeneity of some strata that result in differences in sampling error. A researcher might want to minimize sampling when it is costly or increase sampling from heterogeneous strata while decreasing it from homogeneous strata. See Blalock, *Social Statistics,* 518–519.

16. Ibid., 521–522.

17. We could have obtained the same results by multiplying the GPA of each student by the weighting factor associated with the student's major and then calculating the mean GPA for the whole sample.

18. Richard F. Fenno Jr., *Home Style: House Members in Their Districts* (Boston: Little, Brown, 1978).

19. Babbie, *Survey Research Methods,* 75.

20. Snowball sampling is generally considered to be a nonprobability sampling technique, although strategies have been developed to achieve a probability sample with this method. See Kenneth D. Bailey, *Methods of Social Research* (New York: Free Press, 1978), 83.

21. J. W. Bergsten and S. A. Pierson, "Telephone Screening for Rare Characteristics Using Multiplicity Counting Rules," *1982 Proceedings of the American Statistical Association Section on Survey Research Methods* (Alexandria, Va.: American Statistical Association, 1982), 145–150.

22. The following remarks assume that we have a simple random sample, meaning (just as a reminder) that each member of the sample has been selected randomly and independently of all the others. We assume the same throughout the discussion in this section.

23. This procedure, called sampling with replacement, is premised on the assumption that, at least theoretically, people will sooner or later be interviewed twice or more. We ignore this nuance, because it does not affect the validity of the conclusions in this case.

24. Indeed, in all likelihood it will not exactly equal the population value.

25. It is easy enough to calculate from the formula:

$$\sigma_p = \sqrt{P(1-P)/N}.$$

In the present case the standard error is

$$\sigma_p = \sqrt{P(1-P)/N} = \sqrt{(.25)(.75)/10} = .14.$$

26. Note, too, that it is close to the value obtained from the 1,000 samples, where $N = 10$. So the average of the Ps based on samples of 10 is not much different from the average based on samples of 50.

27. The decision about appropriate sample size depends on many factors, including the type of sample, attributes being measured, heterogeneity of the population, and complexity of the data analysis plan. A more complete discussion of these factors may be found in Royce Singleton Jr., Bruce C. Straits, Margaret M. Straits, and Ronald J. McAllister, *Approaches to Social Research* (New York: Oxford University Press, 1988), 158–163; and Edwin Mansfield, *Basic Statistics with Applications* (New York: Norton, 1986), 287–294.

28. Sample size is not the only factor that affects statistical inferences. For a somewhat advanced discussion, see Daniel D. Boos and Jacqueline M. Hughes-Oliver, "How Large Does *n* Have To Be for *Z* and *t* Intervals?" *American Statistician* 54 (May 2000): 121–128.

29. Sampling error also depends on the type of sample drawn. For a given sample size, a simple random sample provides a more accurate estimate of the target (that is, a smaller margin of error) than does a cluster sample. Sampling error is also smaller for an attribute that is shared by almost all elements in the sample than for one that is distributed across only half of the sample elements. Finally, sampling error is reduced if the sample represents a significant proportion of the target population—that is, if the sampling fraction is greater than one-fourth of the target population. Because this is unusual, however, the effect of the sampling fraction on sampling error is generally minuscule.

Terms Introduced

CLUSTER SAMPLE. A probability sample that is used when no list of elements exists. The sampling frame initially consists of clusters of elements.

CONFIDENCE. The degree of belief that an estimated range of values—more specifically, a high or low value—includes or covers the population parameter.

CONVENIENCE SAMPLE. A nonprobability sample in which the selection of elements is determined by the researcher's convenience.

DISPROPORTIONATE SAMPLE. A stratified sample in which elements sharing a characteristic are underrepresented or overrepresented in the sample.

ELEMENT. A particular case or entity about which information is collected; the unit of analysis.

ESTIMATOR. A statistic based on sample observations that is used to estimate the numerical value of an unknown population parameter.

EXPECTED VALUE. The mean or average value of a sample statistic based on repeated samples from a population.

NONPROBABILITY SAMPLE. A sample for which each element in the total population has an unknown probability of being selected.

POPULATION. All the cases or observations covered by a hypothesis; all the units of analysis to which a hypothesis applies.

POPULATION PARAMETER. The incidence of a characteristic or an attribute in a population (not a sample).

PROBABILITY SAMPLE. A sample for which each element in the total population has a known probability of being selected.

PROPORTIONATE SAMPLE. A probability sample that draws elements from a stratified population at a rate proportional to the size of the samples.

PURPOSIVE SAMPLE. A nonprobability sample in which a researcher uses discretion in selecting elements for observation.

QUOTA SAMPLE. A nonprobability sample in which elements are sampled in proportion to their representation in the population.

RANDOM DIGIT DIALING. A procedure used to improve the representativeness of telephone samples by giving both listed and unlisted numbers a chance of selection.

RANDOM START. Selection of a number at random to determine where to start selecting elements in a systematic sample.

SAMPLE. A subset of observations or cases drawn from a specified population.

SAMPLE BIAS. The bias that occurs whenever some elements of a population are systematically excluded from a sample. It is usually due to an incomplete sampling frame or a nonprobability method of selecting elements.

SAMPLE STATISTIC. The estimator of a population characteristic or attribute that is calculated from sample data.

SAMPLING DISTRIBUTION. A theoretical (nonobserved) distribution of sample statistics calculated on samples of size N that, if known, permits the calculation of confidence intervals and the test of statistical hypotheses.

SAMPLING ERROR. The difference between a sample estimate and a corresponding population parameter that arises because only a portion of a population is observed.

SAMPLING FRACTION. The proportion of the population included in a sample.

SAMPLING FRAME. The population from which a sample is drawn. Ideally it is the same as the total population of interest to a study.

SAMPLING INTERVAL. The number of elements in a sampling frame divided by the desired sample size.

SAMPLING UNIT. The entity listed in a sampling frame. It may be the same as an element, or it may be a group or cluster of elements.

SIMPLE RANDOM SAMPLE. A probability sample in which each element has an equal chance of being selected.

SNOWBALL SAMPLE. A sample in which respondents are asked to identify additional members of a population.

STANDARD ERROR. The standard deviation or measure of variability or dispersion of a sampling distribution.

STATISTICAL INFERENCE. The mathematical theory and techniques for making conjectures about the unknown characteristics (parameters) of populations based on samples.

STRATIFIED SAMPLE. A probability sample in which elements sharing one or more characteristics are grouped, and elements are selected from each group in proportion to the group's representation in the total population.

STRATUM. A subgroup of a population that shares one or more characteristics.

SYSTEMATIC SAMPLE. A probability sample in which elements are selected from a list at predetermined intervals.

WEIGHTING FACTOR. A mathematical factor used to make a disproportionate sample representative.

Suggested Readings

Govindarajulu, Zakkula. *Elements of Sampling Theory and Methods.* Upper Saddle River, N.J.: Prentice-Hall, 1999.

Levy, Paul S., and Stanley Lemeshow. *Sampling of Populations.* New York: Wiley, 1999.

Lohr, Sharon L. *Sampling.* Pacific Grove, Calif.: Brooks/Cole, 1998.

Rea, Louis M., and Richard A. Parker. *Designing and Conducting Survey Research.* San Francisco: Jossey-Bass, 1997.

Rosnow, Ralph L., and Robert Rosenthal. *Beginning Behavioral Research.* Upper Saddle River, N.J.: Prentice-Hall, 1998.

CHAPTER 8

Making Empirical Observations:

Direct and Indirect Observation

Types of Data and Collection Techniques

Political scientists tend to use three broad types of empirical observations, or data collection methods, depending on the phenomena they are interested in studying. Interview data, discussed in Chapter 10, are derived from written or verbal questioning of some group of respondents. This type of data collection may involve interviewing a representative cross-section of the national adult population or a select group of political actors, such as committee chairs in Congress. It may involve face-to-face interviews or interviews conducted over the phone or through the mail. It may involve highly structured interviews in which a questionnaire is followed closely or a less structured, open-ended discussion. Regardless of the particular type of interview setting, however, the essentials of the data collection method are the same: the data come from responses to the verbal or written cues of the researcher and the respondent knows these responses are being recorded.

In addition to interview data, political scientists use documents (newspapers, photographs, video clips, letters, and diaries) as well as statistical data that exist in various archival records. In this type of data collection, known as document analysis (the subject of Chapter 9), researchers rely heavily on the record-keeping activities of government agencies, private institutions, interest groups, media organizations, and even private citizens. Although some of these data may be based on interviews, they are reported in summary, or aggregate, form. We refer to these sources of data collectively as the written record. For example, unemployment statistics are derived from the Census Bureau's *Current Population Survey,* a household survey conducted each month. What sets document analysis apart from other data collection methods is that the researcher is usually not the original collector of the data and the original reason for the collection of the data may not have been to further a scientific research project. The record keepers may be unaware of how the data they collect will ultimately be used, and the phenomena they record are not generally personal beliefs and attitudes of individuals, which are more typically collected through interviews, although diaries and letters could provide such information.

Finally, data may be collected through observation, discussed further in this chapter. In this type of data collection the researcher collects data on political behavior by observing either the behavior itself (direct observation) or some physical trace of the behavior (indirect observation). Unlike interviewing, this method of data collection does not rely on people's verbal responses to verbal stimuli presented by the researcher, but on first-hand examination of activities, behavior, events, or the like. Furthermore, those whose behavior is being directly or indirectly observed may be unaware that they are being observed.

Qualitative versus Quantitative Uses of Data

There are two general approaches to data analysis: quantitative and qualitative analysis. Both approaches, while different, seek to explain trends with data. We define quantitative approaches as those that involve numeric manipulation. In this book, we focus primarily on quantitative analysis through the use of statistical analysis. Qualitative analysis is often defined as those approaches that are not included within quantitative approaches—or approaches that do not use numeric manipulation. Another way to think about qualitative analysis is that it relies on using quotations, comments, or anecdotes to provide evidence and support for arguments. Although we do not focus on qualitative analysis at length in this text, it is an important approach in the study of political science and social science more generally, and most if not all of the topics we cover at length with respect to performing high-quality quantitative research, like validity and reliability, are equally applicable to performing high-quality qualitative research. Furthermore, it is difficult to make a clear distinction between quantitative and qualitative approaches because the two approaches are commonly used together within a single study. For example, many books in political science may rely on a series of case studies (a qualitative method) to provide contextual richness to an analysis and a statistical analysis (a quantitative method) to provide a more generalizable conclusion.

For example, data from the written record may be used with both quantitative and qualitative approaches. Historical institutionalism uses a qualitative analysis of the written record whereas a content analysis is a quantitative approach to using the written record. In the former method, researchers usually rely on nominal or categorical explanations for behavior and outcomes with attention to the historical path that institutions followed to reach a particular behavior or outcome. In the latter method, researchers may use a wide range of variables coded from written documents in a statistical analysis. One might, for example, code the content of political speeches for the number of references to specific policy issues, length, or applause lines. Likewise, one can use data from interviews and observation in both qualitative and quantitative approaches. Qualitative approaches might rely on using statements drawn from interviews or conversations as evidence for conclusion whereas quantitative approaches would instead rely on a statistical analysis of variables created by coding responses during interviews or conversations.

Choosing among Data Collection Methods

A political scientist's choice of a data collection method depends on many factors. One important consideration is the validity of the measurements that a particular method will permit. For example, a researcher who wants to measure the crime rate of different cities may feel that the crime rates reported by local police departments to the FBI are not sufficiently accurate to support a research project. The researcher may be concerned that some departments overreport and some underreport various criminal acts or that some victims of crimes may fail to report the crimes to the police, hence rendering that method of collecting data and measuring the crime rate unacceptable. Therefore, the researcher may decide that a more accurate indication of the crime rate can be attained by interviewing a sample of citizens in different cities and asking them how much crime they have experienced themselves.

Also reflecting a concern over the validity of measurements, Susan J. Carroll and Debra J. Liebowitz note that scholars of women and politics have criticized the use of survey research to study the political participation of women.[1] One problem is that existing conceptions of what is considered political, and hence what is used in survey questions, may not fully capture the range of women's political activity. They suggest that researchers look at the issue inductively, that is, study women's activities and determine in what ways their activities are political. For this approach, observation, in-depth interviews, and focus groups, rather than structured questionnaires, are more appropriate data collection methods.

A political scientist is also influenced by the reactivity of a data collection method—the effect of the data collection itself on the phenomena being measured. When people know their behavior is being observed and know or can guess the purpose of the observation, they may alter their behavior. As a result, the observed behavior may be an unnatural reaction to the process of being observed. People may be reluctant, for example, to admit to an interviewer that they are anti-Semitic or have failed to vote in an election. Thus many researchers prefer unobtrusive or nonreactive measures of political behavior, because they believe that the resulting data are more natural or real.

The population covered by a data collection method is another important consideration for a researcher. The population determines whose behavior is observed. One type of data may be available for only a few people, whereas another type may permit more numerous, interesting, and worthwhile comparisons. A researcher studying the behavior of political consultants, for example, may decide that relying on the published memoirs of a handful of consultants will not adequately cover the population of consultants (not to mention the validity problems of the data) and that it would be better to seek out a broad cross-section of consultants and interview them. Or a researcher interested in political corruption may decide that interviewing a broad cross-section of politicians charged with various corrupt practices is not feasible and that data (of a different kind) could be obtained for a more diverse set of corrupt acts from accounts published in the mass media.

Additionally, cost and availability are crucial elements in the choice of a data collection technique. Some types of data collection are simply more expensive than others, and some types of observations are made more readily than others. Large-scale interviewing, for example, is very expensive and time consuming, and the types of questions that can be asked and behaviors that can be observed are limited. Data from archival records are usually much less expensive, since the record-keeping entity has borne most of the cost of collecting and publishing the data. With the increased use of computers, many organizations are systematically collecting data of interest to researchers. The disadvantage, however, is that because the data are not under the researcher's control, they must be made available by the record-keeping organization, which can refuse a researcher's request or take a long time filling it.

Although the costs of data generated through interviews or the written record may be high, the cost of first-hand observation through the expenditure of time (if the researcher does it) or money (if the researcher pays others to do it) will generally be even higher. Students will often find suitable data generated through interviews or the written record for free in publicly available data archives (see Chapters 9 and 10), but students wishing to use data generated through direct or indirect observation must usually rely on their own ability to make the observations. Most data observation is usually recorded in the form of personal notes, recordings, and transcripts made during observation. These data are less likely to be made publicly available because notes, in particular, are highly individualized and intended to help the person taking the notes remember observations. Hence, they would be less useful to others even if they were made publicly available. Therefore, observation is generally an example of **primary data,** that is, data recorded and used by the researcher making the observations, whereas data from interviews or the written record can be primary data or **secondary data**—data used by a researcher who did not personally collect the data. The high cost of direct and indirect observation means that most students will not have the resources to make their own observations for use in a research paper, except in the most limited fashion. For example, you might be able to use observations made during your internship with a political campaign in an analysis of election strategy, but you are not likely to have the time to make first-hand observations across multiple campaigns. In most cases, it will be more cost-effective to rely on other, more readily available sources of data for research projects.

In addition to these factors, researchers must consider the ethical implications of their proposed research. In most cases, the research topics you are likely to propose will not raise serious ethical concerns, nor will your choice of method of data collection hinge on the risk it may pose to human subjects. Nevertheless, you should be aware of the ethical issues and risks to others that can result from social science research, and you should be aware of the review process that researchers are required to follow when proposing research involving human subjects.

In accordance with federal regulations, universities and other research organizations require faculty and students to submit research proposals involving human subjects for review to an **institutional review board** (often called a human subject review board). There may be some variation in practice concerning unfunded research, but the proper course of action is to contact your institution's research office for information regarding the review policy on human subjects. There are three levels of review: some research may be exempt, some may require only expedited review, and some research will be subject to full board review. Even if your research project seems to fit one of the categories of research exempt from review, you must request and be granted an exemption.[2]

Three ethical principles—respect for persons, beneficence, and justice—form the foundation for assessing the ethical dimensions of research involving human subjects. These principles were identified in the *Belmont Report,* a report of the National Commission for the Protection of Human Subjects of Biomedical and Behavioral Research.[3] The principle concerning respect for persons asserts that individuals should be treated as autonomous agents and that persons with diminished capacity are entitled to protection. Beneficence refers to protecting people from harm as well as making efforts to secure their well-being. The principle of justice requires researchers to consider the distribution of the benefits and burdens of research.

The principle of respect for persons requires that subjects be given the opportunity to choose what shall or shall not happen to them. **Informed consent** means that subjects are to be given information about the research, including the research procedure, its purposes, risks, and anticipated benefits; alternative procedures (where therapy is involved); how subjects are selected; and the person responsible for the research. In addition, the subject is to be given a statement offering him or her the opportunity to ask questions and to withdraw from the research at any time. This information and statement should be conveyed in a manner that is comprehensible to the subject, and the consent of the subject must be voluntary.

An assessment of risks and benefits relates directly to the beneficence principle by helping to determine whether risks to subjects are justified and by providing information useful to subjects for their informed consent. The justice principle is often associated with the selection of subjects insofar as some populations may be more likely to be targeted for study.

In this chapter and Chapters 9 and 10, the relative advantages and disadvantages of each of the major data collection methods are examined with respect to the factors of validity, reactivity, population coverage, cost, and availability. We also point out the ethical issues raised by some applications of these data collection methods.

Observation

Although observation is more generally a research tool of anthropologists, psychologists, and sociologists, observation has been used by political scientists to study political campaigning, community politics, leadership and executive decision making, program implementation, judicial proceedings, the U.S. Congress, and state legislatures. In fact, any student who has had an internship, kept a daily log or diary, and written a paper based on his or her experiences has used this method of data collection.

Every day we "collect data" using observational techniques. We observe some attribute or characteristic of people and infer some behavioral trait from that observation. For example, we watch the car in front of us swerve between the traffic lanes and conclude that the driver has been drinking. We observe the mannerisms, voice pitch, and facial expressions of a student making a presentation in one of our classes and decide that the person is exceptionally nervous. Or we decide that most of the citizens attending a public hearing are opposed to a proposed project by listening to their comments to each other before the start of the hearing. The observational techniques used by political scientists are only extensions of this method of data collection. They resemble everyday observations but are usually more self-conscious and systematic.

Observations may be classified in at least four different ways: (1) direct or indirect, (2) participant or nonparticipant, (3) overt or covert, and (4) structured or unstructured. The most basic distinction is whether an observation is direct or indirect.[4] For example, a direct method of observing college students' favorite studying spots in classrooms and office buildings would involve walking around the buildings and noticing where students are. An indirect method of observing the same behavior would be to arrive on campus early in the morning before the custodial staff and measure the amount of food wrappers, soda cans, and other debris at various locations.

In **participant observation,** the investigator is a regular participant in the activities or group being observed. For example, someone who studies political campaigns by becoming actively involved in them is a participant observer. A researcher does not, however, have to become a full-fledged member of the group to be a participant observer. Some mutually acceptable role or identity must be worked out. For example, Ruth Horowitz did not become a gang member when she studied Chicano youth in a Chicago neighborhood.[5] She hung around with gang members but as a nonmember. She did not participate in fights and was able to decline when asked to conceal weapons for gang members. A nonparticipant observer does not participate in group activities or become a member of the group or community. For example, an investigator interested in hearings held by public departments of transportation or city council meetings could observe those proceedings without becoming a participant.

A third way to characterize observation is by noting whether it is overt or covert. In **overt observation,** those being observed are aware of the investigator's presence and intentions. In **covert observation,** the investigator's presence is hidden or undisclosed and his or her intentions are disguised. For example, observation was used in a study to measure what percentage of people washed their hands after using the restroom. Research involving covert observation of public behavior of private individuals is not likely to raise ethical issues as long as individuals are not or cannot be identified and disclosure of individuals' behavior would not place them at risk. However, elected or appointed public officials are not shielded by these limitations. Ethical standards and their application or enforcement have changed, and it is likely that many earlier examples of participant observation research, especially those involving covert observation, would not receive approval from human subject review boards today. For example, social scientists Mary Henle and Marian B. Hubble once hid under beds in students' rooms to study student conversations.[6]

In **structured observation,** the investigator looks for and systematically records the incidence of specific behaviors. In **unstructured observation,** all behavior is considered relevant, at least at first, and recorded. Only later, upon reflection, will the investigator distinguish between important and trivial behavior.

Direct Observation

The vast majority of observation studies conducted by political scientists involve **direct observation,** in which the researcher observes actual behavior, with the observation more likely to occur in a natural setting than in a laboratory. Again, observation may be structured or unstructured. The term **field study** is typically used to refer to open-ended and wide-ranging, rather than structured, observation in a natural setting. As a student, you are not likely to conduct your own observation research in a laboratory.

Observation in a laboratory setting gives a researcher the advantage of having control over the environment of the observed. Thus the researcher may be able to use a more rigorous experimental design than is possible in a natural, uncontrolled setting. Also, observation may be easier and more convenient to record and preserve, since one-way windows, videotape machines, and other observational aids are more readily available in a laboratory.

A disadvantage of laboratory observation is that subjects usually know they are being observed and therefore may alter their behavior, raising questions about the validity of the data collected. The use of aids that allow the observer to be physically removed from the setting and laboratories that are designed to be as inviting and as natural as possible may lead subjects to behave more naturally and less self-consciously.

An example of an attempt to create a natural-looking laboratory setting may be found in Stanley Milgram and R. Lance Shotland's book *Television and Antisocial Behavior.*[7] These researchers were interested in the effect of television programming on adult behavior, specifically in the ability of television drama to stimulate antisocial acts such as theft. They devised four versions of a program called "Medical Center," each with a different plot, and showed different versions to four different audiences. Some of the versions showed a character stealing money, and those versions differed in whether the person was punished for his theft or not. The participants in the study were then asked to go to a particular office at a particular time to pick up a free transistor radio, their payment for participating in the research study. When they arrived in the office (the laboratory), they encountered a sign that said the radios were all gone. The researchers were interested in how people would react and specifically in whether they would imitate any of the behaviors in the versions of "Medical Center" that they had seen (such as the theft of money from see-through plastic collection dishes). Their behavior was observed covertly via a one-way mirror. Once the subjects left the office, they were directed to another location where they were, in fact, given the promised radio. This experiment, reported in 1973, raises some serious ethical issues about deceiving research subjects.

Direct observation in natural settings has its own advantages. One advantage of observing people in a natural setting rather than in the artificiality of a laboratory setting is that generally people behave as they would ordinarily. Furthermore, the investigator is able to observe people for longer periods of time than would be possible in a laboratory. In fact, one of the striking features of field studies is the considerable amount of time an investigator may spend in the field. It is not uncommon for investigators to live in the community they are observing for a year or more. William F. Whyte's classic study of life in an Italian slum, *Street Corner Society,* was based on three years of observation (1937–1940), and Marc Ross's study of political participation in Nairobi, Kenya, took more than a year of field observation.[8] To study the behavior of U.S. representatives in their districts, Richard Fenno traveled intermittently for almost seven years, making thirty-six separate visits and spending 110 working days in eighteen congressional districts.[9] Ruth Horowitz spent three years researching youth in an inner-city Chicano community in Chicago.[10]

Sometimes researchers have no choice but to observe political phenomena as they occur in their natural setting. Written records of events may not exist, or the records may not cover the behavior of interest to the researcher. Relying on personal accounts of participants may be unsatisfactory because of participants' distorted views of events, incomplete memories, or failure to observe what is of interest to the researcher. Joan E. McLean suggests that researchers interested in studying the decision-making styles of women running for public office need to spend time with campaigns in order to gather information as decisions are being made, rather than rely on post-election questionnaires or debriefing sessions.[11]

Open-ended, flexible observation is appropriate if the research purpose is one of description and exploration. For example, Fenno's research purpose was to study "representatives' perceptions of their constituencies while they are actually in their constituencies."[12]

As Fenno explains, his visits with representatives in their districts

> *were totally open-ended and exploratory. I tried to observe and inquire into anything and everything these members did. I worried about whatever they worried about. Rather than assume that I already knew what was interesting, I remained prepared to find interesting questions emerging in the course of the experience. The same with data. The research method was largely one of soaking and poking or just hanging around.*[13]

In these kinds of field studies, researchers do not start out with particular hypotheses that they want to test. They often do not know enough about what they plan to observe to establish lists and specific categories of behaviors to look for and record systematically. The purpose of the research is to discover what these might be.

Some political scientists have used observation as a preliminary research method.[14] For example, James A. Robinson's work in Congress provided first-hand information for his studies of the House Rules Committee and of the role of Congress in making foreign policy.[15] Ralph K. Huitt's service on

Lyndon B. Johnson's Senate majority leader staff gave Huitt inside access to information for his study of Democratic Party leadership in the Senate.[16] And David W. Minar served as a school board member and used his experience to develop questionnaires in his comparative study of several school districts in the Chicago area.[17] As mentioned earlier, Carroll and Liebowitz suggest observing women's activities in order to identify behaviors with political effect that have not previously been included in measures of political activity. Subsequent surveys could then include questions that ask about such behaviors.

You may look upon an internship, volunteer work, or participation in a community or political organization as opportunities to conduct your own research using direct observation. More than likely yours will be a case study in which you are able to compare the real world with theories and general expectations suggested in course readings and lectures. If you are fortunate, your case may turn out to be a critical or deviant case study.

Most field studies involve participant observation. An investigator cannot be like the proverbial fly on the wall, observing a group of people for long periods of time. Usually he or she must assume a role or identity within the group that is being studied and participate in the activities of the group. As noted earlier, many political scientists who have studied Congress have worked as staff members on committees and in congressional offices. In addition to interviewing influential Latinos in Boston, Carol Hardy-Fanta joined the community group Familias Latinas de Boston while conducting her research on Latina women and politics. As she points out, this strategy complemented her research interviews:

> *Joining the community group Familias Latinas de Boston allowed me to gain an in-depth understanding of one community group over an extended period. Participating in formal, organized political activities such as manning the phone bank at the campaign office of a Latino candidate and attending political banquets, public forums, and conferences and workshops provided another means of observing how gender and culture interacted to stimulate—or suppress—political participation. I also joined protest marches and rallies and tracked down voter registration information in Spanish for a group at Mujeres Unidas en Acción. In addition, I learned much from informal interactions: at groups on domestic violence, during lunch at Latino community centers, and during spontaneous conversations with Latinos from many countries and diverse backgrounds. As I talked to people in community settings and observed how they interacted politically, the political roles of Latina women and the gender differences in how politics is defined emerged. Thus, multiple observations were available to check what I was hearing in the interviews about how to stimulate Latino political participation, and how Latina women and Latino men act politically.[18]*

Acceptance by the group is necessary if the investigator is going to benefit from the naturalness of the research setting. Negotiating an appropriate role for oneself within a group may be a challenging and evolving process. As Chicago gang researcher Ruth Horowitz points out, a researcher may not wish, or be able, to assume a role as a "member" of the observed group. Personal attributes (gender, age, ethnicity) of the researcher or ethical considerations (gang violence) may prevent this.[19] The role the researcher is able to establish also depends on the setting and the members of the group.

> *I was able to negotiate multiple identities and relationships that were atypical of those generally found in the research setting, but that nonetheless allowed me to become sufficiently close to the setting members to do the research. By becoming aware of the nature, content, and consequences of these identities, I was able to use the appropriate identity to successfully collect different kinds of data and at the same time avoid some difficult situations that full participation as a member might have engendered.[20]*

Investigators using participant observation often depend on members of the group they are observing to serve as **informants,** persons who are willing to be interviewed about the activities and behavior of themselves and of the group to which they belong. An informant also helps the researcher interpret behavior. A close relationship between the researcher and informant may help the researcher gain access to other group members, not only because an informant may familiarize the researcher with community members and norms, but also because the informant, through close association with the researcher, will be able to pass on information about the researcher to the community.[21] Some participant observation studies have one key informant; others have several. For example, Whyte relied on the leader of a street corner gang whom he called "Doc" as his key informant, while Fenno's eighteen representatives all could be considered informants.

Although a valuable asset to researchers, informants may present problems. A researcher should not rely too much on one or a few informants, since they may give a biased view of a community. And if the informant is associated with one faction in a multifaction community or is a marginal member of the community (and thus more willing to associate with the researcher), the researcher's affiliation with the informant may inhibit rather than enhance access to the community.[22]

Participant observation offers the advantages of a natural setting, the opportunity to observe people for lengthy periods of time so that interaction and changes in behavior may be studied, and a degree of accuracy or completeness impossible with documents or recall data such as that obtained in surveys. Observing a city council or school board meeting or a public hearing on the licensing of a locally undesirable land use will allow you to know and understand what happened at these events far better than reading official minutes or transcripts. However, there are some noteworthy limitations to the method as well.

The main problem with direct, participant observation as a method of empirical research for political scientists is that many significant instances of political behavior are not accessible for observation. The privacy of the voter in the voting booth is legally protected, U.S. Supreme Court conferences are not open to anyone but the justices themselves, political consultants and bureaucrats do not usually wish to have political scientists privy to their discussions and decisions, and most White House conversations and deliberations are carefully guarded. Occasionally, physical traces of these private behaviors become public—such as the Watergate tapes of Richard Nixon's conversations with his aides—and disclosures are made about some aspects of government decision making, such as congressional committee hearings and Supreme Court oral arguments, but typically access is the major barrier to directly observing consequential political behavior.

Another disadvantage of participant observation is lack of control over the environment. A researcher may be unable to isolate individual factors and observe their effect on behavior. Participant observation is also limited by the small number of cases that are usually involved. For example, Fenno observed only eighteen members or would-be members of Congress—too few for any sort of statistical analysis. He chose "analytical depth" over "analytical range"; in-depth observation of eighteen cases was the limit that Fenno thought he could manage intellectually, professionally, financially, and physically.[23] Whyte observed one street corner gang in depth, although he did observe others less closely. Because of the small numbers of cases, the representativeness of the results of participant observation has been questioned. But, as we tried to stress in our discussion of research designs (Chapter 5), the number of cases deemed appropriate for a research topic depends on the purpose of the research. Understanding how people function in a particular community may be the knowledge that is desired, not whether the particular community is representative of some larger number of communities.

Unstructured participant observation also has been criticized as invalid and biased. A researcher may selectively perceive behaviors, noting some, ignoring others. The interpretation of behaviors may reflect the personality and culture of the observer rather than the meaning attributed to them by the observed themselves. Moreover, the presence of the observer may alter the behavior of the observed no matter how skillfully the observer attempts to become accepted as a nonthreatening part of the community.

Fieldworkers attempt to minimize these possible threats to data validity by immersing themselves in the culture they are observing and by taking copious notes on everything going on around them no matter how seemingly trivial. Events without apparent meaning at the time of observation may become important and revealing upon later reflection. Of course, copious note taking leads to what is known as a "high dross rate"; much of what is recorded is not relevant to the research problem or question as it is finally formulated. It may be painful for the investigator to discard so much of the material that was carefully recorded, but it is standard practice with this method.

Another way to obtain more valid data is to allow the observed to read and comment on what the investigator has written and point out events and behavior that may have been misinterpreted. This check on observations may be of limited or no value if the person being observed cannot read or if the written material is aimed at persons well versed in the researcher's discipline and therefore is over the head of the observed.

Researchers' observations may be compromised if the researchers begin to overidentify with their subjects or informants. "Going native," as this phenomenon is known, may lead them to paint a more complimentary picture of the observed than is warranted. Researchers combat this problem by returning to their own culture to analyze their data and by asking colleagues or others to comment on their findings.

A demanding yet essential aspect of field study is note taking. Notes can be divided into three types: mental notes, jotted notes, and field notes. Mental note taking involves orienting one's consciousness to the task of remembering things one has observed, such as "who and how many were there, the physical character of the place, who said what to whom, who moved about in what way, and a general characterization of an order of events."[24] Because mental notes may fade rapidly, researchers use jotted notes to preserve them. Jotted notes consist of short phrases and key words that will activate a researcher's memory later when the full field notes are written down. Researchers may be able to use tape recorders if they have the permission of those being observed.

Taped conversations do not constitute "full" field notes, which should include a running description of conversations and events. For this aspect of field notes, John Lofland advises that researchers should be factual and concrete, avoid making inferences, and use participants' descriptive and interpretative terms. Full field notes should include material previously forgotten and subsequently recalled. Lofland suggests that researchers distinguish between verbal material that is exact recall, paraphrased or close recall, and reasonable recall.[25]

Field notes should also include a researcher's analytic ideas and inferences, personal impressions and feelings, and notes for further information.[26] Because events and emotional states in a researcher's life may affect observation, they should be recorded. Notes for further information provide guidance for future observation, either to fill in gaps in observations, call attention to things that may happen, or test out emerging analytic themes.

Full field notes should be legible and should be reviewed periodically, since the passage of time may present past observations in a new light to the researcher or reveal a pattern worthy of attention in a series of disjointed events. Creating and reviewing field notes is an important part of the observational method. Consequently a fieldworker should expect to spend as much time on field notes as he or she spends on observation in the field. Fortunately, computerized text analysis programs exist to help analyze field notes and interviews.

Indirect Observation

Indirect observation, the observation of physical traces of behavior, is essentially detective work.[27] Inferences based on physical traces can be drawn about people and their behavior. An unobtrusive research method, indirect observation is nonreactive: subjects do not change their behavior because they do not know they are being studied.

Physical Trace Measures

Researchers use two methods of measurement when undertaking indirect observation. An **erosion measure** is created by selective wear on some material. For example, campus planners at one university observed paths worn in grassy areas and then rerouted paved walkways to correspond to the most heavily trafficked routes. Other examples of natural erosion measures include wear on library books, wear and tear on selected articles within volumes, and depletion of items in stores, such as sales of newspapers.

The second measurement of indirect observation is the **accretion measure,** which is created by the deposition and accumulation of materials. Archaeologists and geologists commonly use accretion measures in their research by measuring, mapping, and analyzing accretion of materials. Other professions find them useful as well. Eugene Webb and his colleagues reported a study in which mechanics in an automotive service department recorded radio dial settings to estimate radio station popularity.[28] This information was then used to select radio stations to carry the dealer's advertising. The popularity of television programs could be measured by recording the drop in water level while commercials are aired, since viewers tend to use the toilet only during commercials when watching very popular shows. Or the reverse could be explored to test the popular wisdom that commercials shown during the Superbowl are more popular than the game itself. Similarly, declines in telephone usage could indicate television program popularity. The presence of fingerprints and nose prints on glass display cases may indicate interest as well as reveal information about the size and age of those attracted to the display. The effectiveness of various antilitter policies and conservation programs could also be measured using physical trace evidence, and the amount and content of graffiti may represent an interesting measurement of the beliefs, attitudes, and mood of a population.

One of the best known examples of the use of accretion measures is W. L. Rathje's study of people's garbage.[29] He studied people's behavior based on what they discarded in their trash cans. One project involved investigating whether poor people wasted more food than those better off: they did not.

Indirect observation typically raises fewer ethical issues than direct observation because the measures of individual behavior are taken after the individuals have left the scene, thus ensuring anonymity in most cases. However, Rathje's studies of garbage raised ethical concerns because some discarded items (such as letters and bills) identified the source of the garbage. Although a court ruled in Rathje's favor by declaring that when people discard their garbage they have no further legal interest in it, one might consider sorting through a person's garbage to be an invasion of privacy. In a study in which data on households were collected, consent forms were obtained, codes were used to link household information to garbage data, and then the codes were destroyed. Rathje's assistants in another garbage study were instructed not to examine any written material closely.

It is also possible that garbage may contain evidence of criminal wrongdoing. Twice during Rathje's research, body parts were discovered, although not in the bags collected as part of the study. Rathje took the position that evidence of victimless crimes should be ignored; evidence of serious crimes should be reported. Of course, the publicity surrounding Rathje's garbage study may have deterred disposal of such evidence. This raises the problem of reactivity: to what extent might people change their garbage-disposing habits if they know there is a small chance that what they throw away will be examined?

This example also illustrates the possibility that indirect observation of physical traces of behavior may border on direct observation of subjects if the observation of physical traces quickly follows their creation. In some situations, extra measures may have to be taken to preserve the anonymity of subjects.

Another good example of the use of accretion measures is Kurt Lang and Gladys Engel Lang's study of the MacArthur Day parade in Chicago in 1951.[30] Gen. Douglas MacArthur and President Harry S. Truman were locked in an important political struggle at the time, and the Langs wanted to find out how much interest there was in the parade. They used data on mass-transit passenger fares, hotel reservations, retail store and street vendor sales, parking lot usage, and the volume of tickertape on the streets to measure the size of the crowd attracted by MacArthur's appearance.

Validity Problems with Indirect Observation

Although physical trace measures generally are not subject to reactivity to the degree that participant observation and survey research are, threats to the validity of these measures do exist. Also, erosion and accretion measures may be biased. For example, certain traces are more likely to survive because the materials are more durable. Thus physical traces may provide a selective, rather than complete, picture of the past. Differential wear patterns may be due not to variation in use but to differences in material. Researchers studying garbage must be careful not to infer that garbage reflects all that is used or consumed. Someone who owns a garbage disposal, for example, generally discards less garbage than someone who doesn't.

Researchers should exercise caution in linking changes in physical traces to particular causes. Other factors may account for variation in the measures. Webb and his colleagues suggested that several physical trace measures be used simultaneously or that alternative data collection methods be used to supplement physical trace measures.[31] For example, physical trace measures of the use of recreational facilities, such as which trash cans in a park fill up the fastest, could be supplemented with questionnaires to park visitors on facility usage.

Caution should also be used in making inferences about the behavior that caused the physical traces. For example, wear around a particular museum exhibit could indicate either the number of people viewing the exhibit or the amount of time people spent near the exhibit shuffling their feet. Direct observation could determine the answer; but in cases where the physical trace measures occurred in the past, this solution is not possible.

Examples of the use of indirect observation in political science research are not numerous. Nevertheless, this method has been used profitably, and you may be able to think of cases where it would be appropriate. For example, you could assess the popularity of candidates by determining the number of yard signs appearing in a community. Or you could estimate the number of visitors and level of office activity of elected representatives by noting carpet wear in office entryways. Although this would not be as precise as counting visitors, it would allow you to avoid having to post observers or question office staff.

Indirect observation, when used ingeniously, can be a low-cost research method free from many of the ethical issues that surround direct observation. Let us now turn to a consideration of some of the ethical issues that develop in the course of fieldwork and in simple, nonexperimental laboratory observations.

Ethical Issues in Observation

Ethical dilemmas arise primarily when there is a potential for harm to the observed. The potential for serious harm to subjects in most observational studies is quite low. Observation generally does not entail investigation of highly sensitive, personal, or illegal behavior, because people are reluctant to be observed in those circumstances and would not give their informed consent. Nor do fieldwork and simple laboratory observation typically involve experimental manipulations of subjects and exposure to risky experimental treatments. Nonetheless, harm or risks to the observed may result from observation. They include (1) negative repercussions from associating with the researcher because of the researcher's sponsors, nationality, or outsider status; (2) invasion of privacy; (3) stress during the research interaction; and (4) disclosure of behavior or information to the researcher resulting in harm to the observed during or after the study. Each of these possibilities is considered here in turn.

In some fieldwork situations, contact with outsiders may be viewed as undesirable behavior by an informant's peers. Cooperation with a researcher may violate community norms. For example, a researcher who studies a group known to shun contact with outsiders exposes informants to the risk of being censured by their group.

Social scientists from the United States have encountered difficulty in conducting research in countries that have hostile relations with the United States.[32] Informants and researchers may be accused of being spies, and informants may be exposed to harm for appearing to sympathize with "the enemy." Harm may result even if hostile relations develop after the research has been conducted. Military, CIA, or other government sponsorship of research may particularly endanger the observed.

A second source of harm to the observed results from the invasion of privacy that observation may entail. Even though a researcher may have permission to observe, the role of observer may not always be remembered by the observed. In fact, as a researcher gains rapport, there is a greater chance that informants may view the researcher as a friend and reveal to him or her something that could prove to be damaging. A researcher does not always warn, "Remember, you're being observed!" Furthermore, if a researcher is being treated as a friend, such a warning may damage rapport. Researchers must consider how they will use the information gathered from subjects. They must judge whether use in a publication will constitute a betrayal of confidence.[33] Even when a subject being interviewed does not consider the research to be an invasion of privacy, there may be stress involved if the topic of conversation is emotionally painful for the subject.

Much of the harm to subjects in fieldwork occurs as a result of publication. They may be upset at the way they are portrayed, subjected to unwanted publicity, or depicted in a way that embarrasses the larger group to which they belong. Carelessness in publication may result in the violation of promises of confidentiality and anonymity. And value-laden terminology may offend those being described.[34]

Carole Johnson has prepared the following guidelines for the "ethical proofreading" of manuscripts prior to publication. To diminish the potential for harm to the observed, researchers should keep in mind these nine points:

1. *Assume that both the identities of the location studied and the identities of individuals will be discovered. What would the consequences of this discovery be to the community? To the individuals? What would the consequences be both within the community and outside the community? Do you believe that the importance of what you have revealed in your publication is great enough to warrant these consequences? Could you, yourself, live with these consequences should they occur?*

2. *Look at the words used in your manuscript. Are they judgmental or descriptive? How accurate are the descriptions of the phenomena observed? A judgment, for example, would be to say that a community is backward. A description might be to say that 10% of the adult population can neither read nor write. The latter is preferable both scientifically and ethically. . . .*

3. *Where appropriate in describing private or unflattering characteristics, consider generalizing first and then giving specifics. . . . This tends to make research participants feel less singled out. It also adds to the educational value of the writing.*

4. *Published data may affect the community studied and similar communities in a general way even though the identities of the community and individuals may remain unknown. In James West's book* Plainville, U.S.A., *for example, people were described as backward. Some people were said to live like animals. Some men were said to be as dirty as animals. West also related that many people from Plainville left the community to seek employment in the cities. What if such descriptive information about rural communities affected individuals' opportunities for employment due to the creation of negative stereotypes about people from rural areas? Therefore, ask yourself how your information might be used in a positive way? In a negative way? And again ask if the revelations are worth the possible consequences.*

5. *Will your research site be usable again or have you destroyed this site for other researchers? Have you destroyed other similar sites? Is such destruction worth the information obtained and disseminated?*

6. *What was your perspective toward subjects? What were your biases? How did your perspective and biases, both positive and negative, affect the way you viewed your subjects and wrote about them?. . .*

7. *In what ways can research participants be educated about the role of fieldworkers and the nature of objective reporting of fieldwork? It*

may be advisable to caution your subjects at various stages of the research that it is not easy to read about oneself as one is described by another.

8. *When conducting research within a larger project, know the expectations of other project members concerning what each member will be permitted to publish both in the short and long run, i.e., are there any limitations? If not, what limits ought ethically to be imposed? Who will have the final say about publication? Who will own the data? Who will have access to the data and on what terms? What will happen to the data after publication? Most important, see that agreements are set forth in writing in a legally enforceable contract.*

9. *Have several people do "ethical proofreading" of your manuscript. One or two of those people might be your subjects. They should read it for accuracy and should provide any general feedback they are inclined to offer. One or two of your colleagues should also read the manuscript. Preferably those colleagues should not be ones who are particularly supportive or sympathetic to your research but colleagues who can be constructively critical.[35]*

Ethical proofreading of manuscripts will protect informants from some of the worst examples of researcher carelessness or insensitivity, but it does not protect the observed from the harm that might arise during observation. Protecting the observed against harm and assessing the potential for harm to the observed prior to starting observation may be difficult. The risk to subjects posed by observation cannot be precisely estimated, nor may concrete measures to avoid all harm be easily specified and enforced. It is up to the researcher to behave ethically. An appropriate ethical framework for judging fieldwork should be "constructed on respect for the autonomy of individuals and groups based on the fundamental principle that persons always be treated as ends in themselves, never merely as means" to a researcher's own personal or professional goals.[36]

Conclusion

Observation is an important research method for political scientists. Observational studies may be direct or indirect. Indirect observation is less common but has the advantage of being a nonreactive research method. Direct observation of people by social scientists has produced numerous studies that have enhanced knowledge and understanding of human beings and their behavior. Fieldwork—direct observation by a participant observer in a natural setting—is the best known variety of direct observation, although direct observation may take place in a laboratory setting. Observation tends to produce data that are qualitative rather than quantitative. Because the researcher is the measuring device, this method is subject to particular questions about researcher bias and data validity. Since there is an evolving relationship between the observer and the observed, participant observation is a demanding and often unpredictable research endeavor. Part of the demanding nature of fieldwork stems from the difficult ethical dilemmas it raises.

As a student you may find yourself in the position of an observer, but it is more likely that you will be a consumer and evaluator of observational research. In this position you should base your evaluation on many considerations: Does it appear that the researcher influenced the behavior of the observed or was biased in his or her observation? How many informants were used? A few or only one? Does it appear likely that the observed could have withheld significant behavior of interest to the researcher? Are generalizations from the study limited because observation was made in a laboratory setting or because of the small number of cases observed? Were any ethical issues raised by the research? Could they have been avoided? What would you have done in a similar situation? These questions should help you evaluate the validity and ethics of observational research.

Notes

1. Susan J. Carroll and Debra J. Liebowitz, "Introduction: New Challenges, New Questions, New Directions," in Susan J. Carroll, ed., *Women and American Politics: New Questions, New Directions* (Oxford: Oxford University Press, 2003), 1–29.

2. Exemption categories are as follows: "1. Research conducted in established or commonly accepted educational settings, involving normal educational practices, such as (a) research on regular and special education instructional strategies or (b) research on the effectiveness of or the comparison among instructional techniques, curricula, or classroom management methods. 2. Research involving the use of educational tests (cognitive, diagnostic, aptitude, achievement), survey procedures, interview procedures, or observation of public behavior, unless (a) information obtained is recorded in such a manner that human subjects can be identified, directly or through identifiers linked to the subjects, AND (b) any disclosure of the human subjects' responses outside the research could reasonably place the subjects at risk of criminal or civil liability or be damaging to the subjects' financial standing, employability, or reputation. 3. Research involving the use of education tests, survey procedures, interview procedures, or observation of public behavior that is not exempt under category 2, if (a) the human subjects are elected or appointed public officials or candidates for public office or (b) federal statute(s) requires without exception that the confidentiality of the personally identifiable information will be maintained throughout the research and thereafter. 4. Research involving the collection or study of existing data, documents, records, pathological specimens, or diagnostic specimens, if these sources are publicly available or if the information is recorded by the investigator in such a manner that subjects cannot be identified directly or through identifiers linked to the subjects. 5. Research and demonstration projects that are conducted by or subject to the approval of department or agency heads and that are designed to study, evaluate, or otherwise examine (a) public benefit or service programs, (b) procedures for obtaining benefits or services under those programs, (c) possible changes in or alternatives to those programs or procedures, or (d) possible changes in methods or levels of payment for benefits or services under those programs. 6. Taste and food quality evaluation and consumer acceptance studies, (a) if wholesome foods without additives are consumed or (b) if a food is consumed that contains a food ingredient at or below the level and for a use found to be safe, or agricultural chemical or environmental contaminant at or below the level found to be safe, by the Food and Drug Administration or approved by the Environmental Protection Agency or the Food Safety and Inspection Service of the U.S. Department of Agriculture." From Title 45, *Code of Federal Regulations*, part 46.101(b), 6/18/91.

These exemptions do not apply to research involving prisoners, fetuses, pregnant women, or human in vitro fertilization. Exemption 2 does not apply to children except for research involving observations of public behavior when the investigator does not participate in the activities being observed.

3. National Commission for the Protection of Human Subjects of Biomedical and Behavioral Research (April 18, 1979), "The Belmont Report: Ethical Principles and Guidelines for the Protection of Human Subjects of Research." Retrieved November 15, 2000, from http://helix.nih.gov:8001/ohsr/mpa/belmont.php3.

4. Eugene J. Webb and others, *Nonreactive Measures in the Social Sciences,* 2d ed. (Boston: Houghton Mifflin, 1981).

5. Ruth Horowitz, "Remaining an Outsider: Membership as a Threat to Research Rapport," *Urban Life* 14 (January 1986): 409–430.

6. Mary Henle and Marian B. Hubble, "Egocentricity in Adult Conversation," *Journal of Social Psychology* 9 (May 1938): 227–234.

7. Stanley Milgram and R. Lance Shotland, *Television and Antisocial Behavior: Field Experiments* (New York: Academic Press, 1973).

8. William F. Whyte, *Street Corner Society: The Social Structure of an Italian Slum,* 3d ed. (Chicago: University of Chicago Press, 1981); and Marc H. Ross, *Grass Roots in an African City: Political Behavior in Nairobi* (Cambridge: MIT Press, 1975).

9. Richard F. Fenno Jr., *Home Style: House Members in Their Districts* (Boston: Little, Brown, 1978).

10. Ruth Horowitz, *Honor and the American Dream* (New Brunswick, N.J.: Rutgers University Press, 1983).

11. Joan E. McLean, "Campaign Strategy," in Susan J. Carroll, ed., *Women and American Politics: New Questions, New Directions* (Oxford: Oxford University Press, 2003), 53–71.

12. Fenno, *Home Style,* xiii.

13. Ibid., xiv.

14. Jennie-Keith Ross and Marc Howard Ross, "Participant Observation in Political Research," *Political Methodology* 1 (Winter 1974): 65–66.

15. James A. Robinson, *The House Rules Committee* (Indianapolis: Bobbs-Merrill, 1963); and James A. Robinson, *Congress and Foreign Policy-making* (Homewood, Ill.: Dorsey, 1962). Also, extensive first-hand observations of Congress are reported in many of the articles in Raymond E. Wolfinger, ed., *Readings on Congress* (Englewood Cliffs, N.J.: Prentice-Hall, 1971).

16. Ralph K. Huitt, "Democratic Party Leadership in the Senate," *American Political Science Review* 55 (June 1961): 333–344.

17. David W. Minar, "The Community Basis of Conflict in School System Politics," in Scott Greer and others, eds., *The New Urbanization* (New York: St. Martin's, 1968), 246–263.

18. Carol Hardy-Fanta, *Latina Politics, Latino Politics: Gender, Culture, and Political Participation in Boston* (Philadelphia: Temple University Press, 1993), xiv.

19. Horowitz, "Remaining an Outsider," 412.

20. Ibid., 413.

21. Ross and Ross, "Participant Observation," 70.

22. Ibid.

23. Fenno, *Home Style,* 255.

24. John Lofland, *Analyzing Social Settings: A Guide to Qualitative Observation and Analysis* (Belmont, Calif.: Wadsworth, 1971), 102–103.

25. Ibid., 105.

26. Ibid., 106–107.

27. Webb and others, *Nonreactive Measures,* 4.

28. Ibid., 10–11.

29. See discussion of Rathje's work in ibid., 15–17.

30. Kurt Lang and Gladys Engel Lang, *Politics and Television* (Chicago: Quadrangle, 1968).

31. See Webb and others, *Nonreactive Measures,* 27–32.

32. See Myron Glazer, *The Research Adventure: Promise and Problems of Field Work* (New York: Random House, 1973), 25–48, 97–124.

33. See Fenno, *Home Style,* 272.

34. For a discussion and examples of value-laden terminology in published reports of participant observers, see ibid.

35. Carole Garr Johnson, "Risks in the Publication of Fieldwork," in J. E. Sieber, ed., *The Ethics of Social Research: Fieldwork, Regulation and Publication* (New York: Springer-Verlag, 1982), 87–88. This passage is reprinted with permission from the publisher.

36. Joan Cassell, "Harms, Benefits, Wrongs and Rights in Fieldwork," in ibid., 14.

Terms Introduced

ACCRETION MEASURES. Measures of phenomena through indirect observation of the accumulation of materials.

COVERT OBSERVATION. Observation in which the observer's presence or purpose is kept secret from those being observed.

DIRECT OBSERVATION. Actual observation of behavior.

EROSION MEASURES. Measures of phenomena through indirect observation of selective wear of some material.

FIELD STUDY. Observation in a natural setting.

INDIRECT OBSERVATION. Observation of physical traces of behavior.

INFORMANT. Person who helps a researcher employing participant observation methods interpret the activities and behavior of the informant and the group to which the informant belongs.

INFORMED CONSENT. Procedures that inform potential research subjects about the proposed research in which they are being asked to participate. The principle that researchers must obtain the freely given consent of human subjects before they participate in a research project.

INSTITUTIONAL REVIEW BOARD. Panel to which researchers must submit descriptions of proposed research involving human subjects for the purpose of ethics review.

OVERT OBSERVATION. Observation in which those being observed are informed of the observer's presence and purpose.

PARTICIPANT OBSERVATION. Observation in which the observer becomes a regular participant in the activities of those being observed.

PRIMARY DATA. Data recorded and used by the researcher who is making the observations.

SECONDARY DATA. Data used by a researcher that was not personally collected by that researcher.

STRUCTURED OBSERVATION. Systematic observation and recording of the incidence of specific behaviors.

UNSTRUCTURED OBSERVATION. Observation in which all behavior and activities are recorded.

Suggested Readings

Fenno, Richard F., Jr. *Home Style: House Members in Their Districts.* Boston: Little, Brown, 1978. See the Introduction and Appendix, "Notes on Method: Participant Observation."

Glazer, Myron. *The Research Adventure: Promise and Problems of Field Work.* New York: Random House, 1972.

Horowitz, Ruth. "Remaining an Outsider: Membership as a Threat to Research Rapport." *Urban Life* 14 (January 1986): 409–430.

Ross, Jennie-Keith, and Marc Howard Ross. "Participant Observation in Political Research." *Political Methodology* 1 (Winter 1974): 63–88.

Shaffir, William B., Robert A. Stebbins, and Allan Turowitz, eds. *Fieldwork Experience: Qualitative Approaches to Social Research.* New York: St. Martin's, 1980.

Shrader-Frechette, Kristin. *Ethics of Scientific Research.* Lanham, Md.: Rowman and Littlefield, 1994.

Sieber, J. E. *Planning Ethically Responsible Research: A Guide for Students and Internal Review Boards.* Newbury Park, Calif.: Sage, 1992.

——, ed. *The Ethics of Social Research: Fieldwork, Regulation and Publication.* New York: Springer-Verlag, 1982.

Whyte, William F. *Street Corner Society: The Social Structure of an Italian Slum,* 3d ed. Chicago: University of Chicago Press, 1981. See Appendix A, "On the Evolution of *Street Corner Society.*"

CHAPTER 9
Document Analysis:
Using the Written Record

Political scientists have three main methods of collecting the data they need to test hypotheses: interviewing, document analysis, and observation. Of these, interviewing and document analysis are the most frequently used. In Chapter 8 we discussed observation techniques; here we describe how empirical observations can be made using the **written record,** which is composed of documents, reports, statistics, manuscripts, and other written, oral, or visual materials.

Political scientists turn to the written record when the political phenomena that interest them cannot be measured through personal interviews, with questionnaires, or by direct observation. For example, interviewing and observation are of limited utility to researchers interested in large-scale collective behavior (such as civil unrest and the budget allocations of national governments), or in phenomena that are distant in time (Supreme Court decisions during the Civil War) or space (defense spending by different countries).

The political phenomena that have been observed through written records are many and varied—for example, judicial decisions concerning the free exercise of religion, voter turnout rates in gubernatorial elections, the change over time in Soviet military expenditures, and the incidence of political corruption in the People's Republic of China.[1] Of the examples of political science research described in Chapter 1 and referred to throughout this book, Steven C. Poe and C. Neal Tate's investigation of governments' violation of human rights, Jeff Yates and Andrew Whitford's investigation into Supreme Court justices' decisions in cases involving presidential powers, Jeffrey A. Segal and Albert D. Cover's investigation of the ideology of Supreme Court justices, and Kim Fridkin Kahn and Patrick J. Kenney's study of national elections all depended on written records for the measurement of important political concepts.[2] Not all portions of the written record are equally useful to political scientists. Hence we discuss the major components of the written record of interest to political scientists and how researchers use those components to measure significant political phenomena.

Generally speaking, use of the written record raises fewer ethical issues than either observation or interviewing. Research involving the collection or study of existing data, documents, or records often does not pose risks to individuals, because the unit of analysis for the data is not the individual. Also, issues of risk are not likely to arise where records are for individuals, as long as individuals cannot be identified directly or though identifiers linked to them, or where the records are publicly available, as in the case of the papers of public figures such as presidents and members of Congress. However, allowing researchers access to their private papers may pose some risk to private individuals. Thus access to private papers may be subject to conditions designed to protect the individuals involved.

Types of Written Records

Some written records are ongoing and cover an extensive period of time; others are more episodic. Some are produced by public organizations at taxpayers' expense; others are produced by business concerns or by private citizens. Some are carefully preserved and indexed; other records are written and forgotten. In this section we discuss two types of written records: the episodic record and the running record.

The Episodic Record

Records that are not part of an ongoing, systematic record-keeping program but are produced and preserved in a more casual, personal, and accidental manner are called **episodic records.** Good examples are personal diaries, memoirs, manuscripts, correspondence, and autobiographies; biographical sketches and other biographical materials; the temporary records of organizations; and media of temporary existence, such as brochures, posters, and pamphlets. The episodic record is of particular importance to political historians, since much of their subject matter can be studied only through these data.

The papers and memoirs of past presidents and members of Congress could also be classified as part of the episodic record, even though considerable resources and organizational effort are invested in their preservation, insofar as the content and methods of organization of these documents vary and the papers are not available all in the same location.

To use written records, researchers must first gain access to the materials and then code and analyze them. Gaining access to the episodic record is sometimes particularly difficult.[3] Locating suitable materials can easily be the most time-consuming aspect of the whole data collection exercise.

Researchers generally use episodic records to illustrate phenomena rather than as a basis for the generation of a large sample and numerical measures for statistical analysis. Consequently, quotations and other excerpts from research materials are often used as evidence for a thesis or hypothesis. Over the years, social scientists have conducted some exceptionally interesting and imaginative studies of political phenomena based on the episodic record. We describe three particular studies that used the episodic record to illuminate an important political phenomenon.

DEVIANCE IN THE MASSACHUSETTS BAY COLONY.   In the 1960s the sociologist Kai T. Erikson studied deviance in the Puritans' Massachusetts Bay Colony during the seventeenth century.[4] He was interested in the process by which communities decide what constitutes deviant behavior. In particular, he wished to test the idea that communities alter their definitions of deviance over time and use deviant behavior to reaffirm and establish the boundaries of acceptable behavior. Contrary to the conventional view that deviant behavior is uniformly harmful, Erikson believed that the identification of and reaction to deviant behavior serve a useful social purpose for a community.

Obviously no one is still alive who could be interviewed about the Puritan form of justice in the colony. Consequently Erikson had to search existing historical documents for evidence relating to his thesis. He found two main collections germane to his inquiry: *The Records of the Governor and Company of the Massachusetts Bay in New England* and *The Records and Files of the Quarterly Courts of Essex County, Massachusetts, 1636–1682.*[5] With these documents Erikson was able to weave together a fascinating tale of crime and punishment, Puritan style, during the mid-1600s.

Erikson's primary concern was with the identification of acts judged deviant in the Massachusetts Bay Colony. From the records of the Essex County courts, he was able to collect information on all 1,954 convictions reached between 1651 and 1680. These data allowed Erikson to investigate the frequency of criminal behavior and to calculate a crude crime rate for the Bay Colony during this period.

Erikson's analysis of the historical records was not altogether straightforward. For example, he discovered that the Puritans were extremely casual about how they spelled people's names. One man named Francis Usselton made many appearances before the Essex County Court, and his name was spelled at least fourteen different ways in the court's records. This did not present insurmountable difficulties in his case because his name was so distinctive. However, Erikson had a more difficult time deciding whether Edwin and Edward Batter were the same man and whether "the George Hampton who stole a chicken in 1649" was the same man as "the George Hampden who was found drunk in 1651."[6]

A second problem with the Puritans' record keeping was that they often passed the same name from generation to generation. Hence it was sometimes unclear whether two crimes twenty years apart were committed by the same person or by a father and a son. Between 1656 and 1681, for example, John Brown was convicted of seven offenses. However, since John Brown's father and grandfather were also named John Brown, it was unclear who committed which crimes.

Despite these difficulties, Erikson's research is a testimonial to the ability of historical records to address important contemporary issues. Without the foresight of those who preserved and printed these records, an important aspect of life in Puritan New England would have been measurably more difficult to piece together.

ECONOMICS AND THE U.S. CONSTITUTION. In 1913 the historian Charles Beard published a book about the U.S. Constitution in which he made imaginative use of the episodic record.[7] Beard's thesis was that economic interests prompted the movement to frame the Constitution. He reasoned that if he could show that the framers and pro-Constitution groups were familiar with the economic benefits that would ensue upon ratification of the Constitution, then he would be able to argue that economic considerations were central to the Constitution debate. If, in addition, he could show that the framers themselves benefited economically from the system of government established by the Constitution, the case would be that much stronger. This thesis, which has stimulated a good deal of controversy, was tested by Beard with a variety of data from the episodic record.

The first body of evidence presented by Beard measured the property holdings of those present at the 1787 Constitutional Convention. These measures, which Beard admits are distressingly incomplete, are derived largely from six different types of sources: biographical materials, such as James Herring's multivolume *National Portrait Gallery* and the *National Encyclopedia of Biography;* census materials, in particular the 1790 census of heads of families, which showed the number of slaves owned by some of the framers; U.S. Treasury records, including ledger books containing lists of securities; records of individual state loan offices; records concerning the histories of certain businesses, such as the *History of the Bank of North America* and the *History of the Insurance Company of North America*; and collections of personal papers stored in the Library of Congress.

From these written records Beard was able to discover the occupations, land holdings, number of slaves, securities, and mercantile interests of the framers. This allowed him to establish a plausible case that the framers were not economically disinterested when they met in Philadelphia to "revise" the Articles of Confederation.

Beard coupled his inventory of the framers' personal wealth with a second body of evidence concerning their political views. His objective was to demonstrate that the framers realized and discussed the economic implications of the Constitution and the new system of government. By using the existing minutes of the debate at the convention, the personal correspondence and writings of some of the framers, and the *Federalist Papers,* by James Madison, Alexander Hamilton, and John Jay—which were written to persuade people to vote for the Constitution—Beard was able to demonstrate that the framers were concerned about, and cognizant of, the economic implications of the Constitution they wrote.

A third body of evidence allowed Beard to analyze the distribution of the vote for and against the Constitution. Where the data permitted, Beard measured the geographical distribution of the popular vote in favor of ratification and compared this with information about the economic interests of different geographical areas in each of the states. He also attempted to measure the personal wealth of those present at the state ratification conventions and then related those measures to the vote on the Constitution. These data were gleaned from the financial records of the individual states, the U.S. Treasury Department, and historical accounts of the ratification process in the states.

Through this painstaking and time-consuming reading of the historical record, Beard constructed a persuasive (although not necessarily proven) case for his conclusion that "the movement for the Constitution of the United States was originated and carried through principally by four groups of personal interests which had been adversely affected under the Articles of Confederation: money, public securities, manufactures, and trade and shipping."[8]

PRESIDENTIAL PERSONALITY. A third example of the use of the episodic record may be found in James David Barber's *The Presidential Character.* Because of the importance of the presidency in the American political system and the extent to which that institution is shaped by its sole occupant, Barber was interested in understanding the personalities of the individuals who had occupied the office during the twentieth century. Although he undoubtedly would have preferred to observe directly the behavior of the fourteen presidents who held office between 1908 and 1984 (when he conducted his study), he was forced instead to rely on the available written materials about them.

For Barber, discerning a president's personality means understanding his style, world view, and character. Style is "the President's habitual way of performing his three political roles: rhetoric, personal relations, and homework." A president's world view is measured by his "primary, politically relevant beliefs, particularly his conceptions of social causality, human nature, and the central moral conflicts of the time." And character "is the way the President orients himself toward life." Barber believes that a president's style, character, and world view "fit together in a dynamic package understandable in psychological terms" and that this personality "is an important shaper of his

Presidential behavior on nontrivial matters." But how is one to measure the style, character, and world view of presidents who are dead or who will not permit a political psychologist access to their thoughts and deeds? This is an especially troublesome question when one believes, as Barber does, that "the best way to predict a President's character, world view, and style is to see how they were put together in the first place . . . in his early life, culminating in his first independent political success."[9]

Barber's solution to this problem was to use available materials on the twentieth-century presidents he studied, including biographies, memoirs, diaries, speeches, and, for Richard Nixon, tape recordings of presidential conversations. Barber did not use all the available biographical materials. For example, he "steered clear of obvious puff jobs put out in campaigns and of the quickie exposés composed to destroy reputations."[10] He quoted frequently from the biographical materials as he built his case that a particular president was one of four basic personality types. Had these materials been unavailable or of questionable accuracy (a possibility that Barber glosses over in a single paragraph), measuring presidential personalities would have been a good deal more difficult, if not impossible.

Barber's analysis of the presidential personality was exclusively qualitative; the book contained not one table or graph. He used the biographical material to categorize each president as one of four personality types and to show that the presidents with similar personalities exhibited similar behavioral patterns when in office. In brief, Barber used two dimensions—activity-passivity (how much energy does the man invest in his presidency?) and positive-negative affect (how does he feel about what he does?)—to define the four types of presidential personality. (See Table 9-1.)

Barber's research is a provocative and imaginative example of the use of the episodic record—in this case, biographical material—as evidence for a series of generalizations about presidential personality. Although Barber did not empirically test his hypotheses in the ways that we have been discussing in this book, he did accumulate a body of evidence in support of his assertions and presented his evidence in such a way that the reader can evaluate how persuasive it is.[11]

The Running Record

Unlike the episodic record, the **running record** is more likely to be produced by organizations than by private citizens; it is carefully stored and easily accessed; and it is available for long periods of time. The portion of the running record that is concerned with political phenomena is extensive and growing. The data collection and reporting efforts of the U.S. government alone are impressive, and if you add to that the written records collected and preserved by state and local governments, interest groups, publishing

TABLE 9-1

Presidential Personality Types

Positive-Negative Affect	Activity-Passivity	
	Active	Passive
Positive	Franklin D. Roosevelt Harry S. Truman John F. Kennedy Gerald Ford Jimmy Carter	William Howard Taft Warren Harding Ronald Reagan
Negative	Woodrow Wilson Herbert Hoover Lyndon Johnson Richard Nixon	Calvin Coolidge Dwight Eisenhower

Source: Based on data from James David Barber, *The Presidential Character*, 3d ed. (Englewood Cliffs, N.J.: Prentice-Hall, 1985). Courtesy of James David Barber, James B. Duke Professor of Science, Emeritus, Duke University, Durham, N.C.

houses, research institutes, and commercial concerns, the quantity of politically relevant written records increases quickly. Reports of the U.S. government, for example, now cover everything from electoral votes to electrical rates, taxes to taxi cabs, and, in summary form, fill one thousand pages in the *Statistical Abstract of the United States,* published annually by the U.S. Bureau of the Census. What makes the running record especially attractive as a resource is that many data sets are now housed online. The *Statistical Abstract,* for example, can be found at www.census.gov/compendia/statab.

There are far too many sources of the running record to list them here, but a quick look at some popular sources will help you understand the array of sources that are available. If you are interested in elections and campaigns you can visit the *Federal Election Commission* at www.fec.gov and find financial records filed by candidates, interest groups, and political parties, or you can visit privately operated Web sites, like www.opensecrets.org, that offer processed reports in an easy to read and use format. Or you might visit the Web sites administered by the secretaries of state to find state-level election returns or summaries of election law changes over time, or the *America Votes* series to find election results for national and some state and local elections. Alternatively, if you are interested in the lawmaking process, Congress makes the text and legislative histories of bills, committee reports, hearings, congressional votes, and the *Congressional Record* available in print or online at www.thomas.gov with a useful search engine to find needed documents. Or you can search for similar material through nongovernmental sources like the Inter-university Consortium for Political and Social Research archive or in print in the *CQ Almanac* or in CQ Press's *Politics in America*. Finally, you can find a wealth of information related to foreign affairs in *World Resources,* published by the World Resources Institute in collaboration with the United Nations Environment Programme and the United Nations Development Programme, or in the Central Intelligence Agency's *World Fact Book* at https://www.cia.gov/library/publications/the-world-factbook/index.html. As you can imagine, the references listed here represent only a small fraction of the available records. Each reference has its own advantages and disadvantages, and you should take care to understand exactly what is and what is not included in each reference before using it.

THE POLICY AGENDAS PROJECT. In this section we provide a detailed example of how you can use sources of the running record in your own research projects by focusing on one such reference, the Policy Agendas Project. Available at www.policyagendas.org and maintained by Bryan Jones and John Wilkerson at the University of Washington and Frank Baumgartner at Penn-

| TABLE 9-2 |
Policy Agendas Project Policy Topics
Agriculture
Banking & Commerce
Civil Rights/Liberties
Defense
Education
Energy
Environment
Foreign Trade
Government Operations
Health
Housing & Community Development
International Affairs & Aid
Labor, Employment, & Immigration
Law, Crime, & Family
Macroeconomics
Public Lands
Science & Technology
Social Welfare
Transportation

sylvania State University, the Policy Agendas Project Web site offers many different sources of data linked together by public policy topics. The project seeks to provide users with an easy one-source way to track long-term policy changes at the national level of government across many different arenas. At the heart of this project is a comprehensive list of public policy topics (Table 9-2) and subtopics that includes a numeric code for virtually every public policy issue on the national agenda. As demonstrated in Figure 9-1, each topic is divided into dozens of subtopics to better organize the broad policy areas. Finally, each topic and subtopic is assigned a unique identification number that is used in each of the data sets available on the Web site.

The Web site is updated continually with new and more recent data. As of this writing, it included ten distinct data sets with links to several others. The currently available data sets are as follows: Budget, Congressional Quarterly Almanac, Congressional Hearings, Executive Orders, New York Times Index, Gallup's Most Important Problem, Public Laws, State of the Union Speeches, Supreme Court Cases, and Congressional Roll Call Votes. A partner site, *The Congressional Bills Project* (available at www.congressionalbills.org) is maintained by E. Scott Adler and John Wilkerson at the University of Washington and includes data on every congressional bill from 1947 to 2000. This Web

FIGURE 9-1
Sample Codes for the Policy Agendas Project

4. Agriculture

400: General (includes combinations of multiple subtopics)

Examples: DOA, USDA and FDA appropriations, general farm bills, farm legislation issues, economic conditions in agriculture, impact of budget reductions on agriculture, importance of agriculture to the U.S. economy, national farmland protection policies, agriculture and rural development appropriations, family farmers, state of American agriculture, farm program administration, long range agricultural policies, amend the Agriculture and Food Act, National Agricultural Bargaining Board.

401: Agricultural Trade

Examples: FDA inspection of imports, agriculture export promotion efforts, agricultural trade promotion programs, tobacco import trends, agricultural export credit guarantee programs, impact of imported meats on domestic industries, country of origin produce labeling, USDA agricultural export initiatives, value added agricultural products in U.S. trade, establish coffee export quotas, effects of Mexican produce importation, international wheat agreements, livestock and poultry exports, amend Agricultural Trade Development and Assistance Act of 1954, reemphasize trade development, promote foreign trade in grapes and plums, prohibit unfair trade practices affecting producers of agricultural products, extend Agricultural Trade Development, enact the Agriculture Trade Act of 1978, establish agricultural aid and trade missions to assist foreign countries to participate in US agricultural aid and trade programs, Food, Agriculture, Conservation and Trade Act Amendments.

See also: 1800 general foreign trade; 1502 agricultural commodities trading.

402: Government Subsidies to Farmers and Ranchers, Agricultural Disaster Insurance

Examples: agricultural price support programs, USDA crop loss assistance, farm credit system financial viability, federal agriculture credit programs, agricultural disaster relief programs, subsidies for dairy producers, farm loan and credit issues, reforming federal crop insurance programs, credit assistance for family operated farms, federal milk supply and pricing policies, renegotiation of farm debts, USDA direct subsidy payments to producers, establishing farm program payment yields, peanut programs, wheat programs, evaluation of the supply and demand for various agricultural commodities, beef prices, cotton acreage allotments, shortages of agricultural storage facilities, agricultural subterminal storage facilities, financial problems of farm banks, Agricultural Adjustment Act, farm vehicle issues, Wool Act, Sugar Act, feed grain programs, cropland adjustment programs.

See also: 1404 farm real estate financing.

403: Food Inspection and Safety (including seafood)

Examples: FDA monitoring of animal drug residues, consumer seafood safety, budget requests for food safety programs, food labeling requirements, grain inspection services, regulation of health and nutrition claims in food advertising and labeling, sanitary requirements for food transportation, regulation of pesticide residues on fruit, food irradiation control act, regulation of artificial food coloring, federal control over the contamination of food supplies, meat grading standards, meat processing and handling requirements, improvement of railroad food storage facilities, shortage of grain storage facilities, food packaging standards, food buyer protection, regulation of food additives, federal seed act, definition and standards of dry milk solids.

See also: 401 inspection of food imports.

404: Agricultural Marketing, Research, and Promotion

Examples: soybean promotion, research, and consumer information act, USDA commodity promotion programs, cotton research and promotion, wheat marketing problems, livestock marketing, new peanut marketing system, establishing a national commission on food marketing, fruit and vegetable marketing, industrial uses for agricultural products, meat promotion program, national turkey marketing act, federal marketing quotas for wheat.

405: Animal and Crop Disease and Pest Control

Examples: USDA regulation of plant and animal mailing to prevent the spread of diseases, control of animal and plant pests, pork industry swine disease eradication program, virus protection for sheep, grasshopper and cricket control programs on farmland, USDA response to the outbreak of citrus disease in Florida, eradication of livestock diseases, brucellosis outbreak in cattle, USDA integrated pest management program, toxic contamination of livestock, fire ant eradication program, proposed citrus blackfly quarantine, predator control problems, biological controls for insects and diseases on agricultural crops, eradication of farm animal foot and mouth diseases.

See also: 704 for pollution effects of pesticides; 403 for pesticide residues on foods.

498: Agricultural Research and Development

Examples: condition of federally funded agricultural research facilities, USDA nutrition research activities, USDA agricultural research programs, regulation of research in agricultural biotechnology programs, organic farming research, potential uses of genetic engineering in agriculture, agricultural research services, research on aquaculture.

499: Other

Examples: methodologies used in a nationwide food consumption survey, agricultural weather information services, federal agricultural census, designate a national grain board, home gardening, redefinition of the term "farm", farm cooperative issues.

site uses the same policy codes used on the Policy Agendas Project Web site for easy combination of data between the two sites.

Each of these data sets is a useful source of data in its own right, but the policy codes linking these data make this Web site especially important. In the paragraphs that follow we briefly discuss the contents of the data sets. Each data set is available in Microsoft Excel or text format along with a codebook. You can open the files using Excel or a program that will be able to read the text format data and transform them into a spreadsheet or table. The codebooks are important because they explain what the data include, how the data are formatted, and any special instructions that must be followed in using the data. If you wish to use the public policy codes, you will also need the public policy codebook, which provides the identification numbers for each policy topic and subtopic.

The Budget data set contains data on the budget of the U.S. government from financial years 1947 through 2006. These data allow you to track government spending across time and to compare spending in different policy areas.

The Congressional Quarterly Almanac data set includes data on each article appearing in the main chapters of the *CQ Almanac* from 1948 through 2003. These data include information about the policy topics, congressional committees and members, bills, and laws mentioned in each article. This data set is linked through identification numbers with the public laws data set maintained by the Policy Agendas Project. Since the *CQ Almanac* follows congressional action quite closely, the data set provides an accessible and easy way to track congressional action in different policy areas.

The Congressional Hearings data set includes data on congressional hearings from 1946 to 2005. The data set includes information about the committee or subcommittee holding the hearing and the policy topics discussed. These data will be useful to students who are, for example, interested in determining how active congressional committees are in responding to policy failures with hearings to determine what went wrong and how they might fix problems.

The Executive Orders data set includes information about every executive order issued between 1945 and 2003. This file includes data on the policy content of executive orders as well as characteristics such as the party of the president and whether the order came during a period of unified or divided government.

The New York Times Index data set includes a systematic random sample of *New York Times* articles from 1946 to 2003 coded for policy content. The data can be used to measure media attention to a policy issue over time or at specific points in time or to determine which issues were most important in a given time period.

The Gallup's Most Important Problem data set includes aggregated responses to the Gallup opinion poll's "most important problem" question from 1947 to 2004. The data allow students to measure the policy issue that was most important to the public during a specific time frame and include a relative ranking of each issue mentioned by respondents. Used in combination with other data available through the Policy Agendas Project, these data might be useful in determining how Congress or the president react to public opinion on the most important issues on the national agenda.

The Public Laws data set contains information about the laws passed between 1948 and 2004. Each law is coded for policy topic and includes an identification variable that links the law to the Congressional Quarterly Almanac data set. The Web site includes a subset of this data set, called the Most Important Law data set, which identifies the 576 most important laws passed between 1948 and 2004, based on the number of lines CQ dedicated to coverage of the laws. These data offer the opportunity to link policy demands with governmental outputs based on policy topic.

The State of the Union Speeches data set contains information on each State of the Union address delivered from 1946 to 2005. The speeches are coded on the statement level for policy content and other variables. These data offer a way to determine the issues the president views as important, based on the amount of attention given to the issues within a State of the Union speech.

The Supreme Court data set includes data on every Supreme Court case from 1953 until 1998. The data set includes measures of important characteristics such as the decision to accept or reject a case and a summary of the decision if the case was accepted for review. Used in combination with the Public Laws, Executive Orders, or Congressional Hearing data sets, these data could be used to explore inter-branch relations on public policy issues.

The Congressional Roll Call Votes data set codes every congressional roll-call vote from 1946 to 2000. Combining these data with the Gallup's Most Important Question data set would allow students to estimate the effect of public opinion on congressional voting.

As indicated earlier, combining these data sets will allow you to answer a variety of research questions about the development and passage of public policy. The Policy Agendas Project Web site offers a single source for many kinds of policy-related data, and the unique policy identification numbers allow for a seamless combination of the data.

The Running Record and Episodic Record Compared

There are three primary advantages to using the running record rather than the episodic record. The first is cost, in both time and money. Since the costs of collecting, tabulating, storing, and reporting the data in the running record are generally borne by the record keepers themselves, political scientists are usually able to use these data inexpensively. Researchers can often use the data stored in the running record by photocopying a few pages of a reference book, purchasing a government report or data file, or downloading data into a spreadsheet. In fact, the continued expansion of the data collection and record-keeping activities of the national government has been a financial boon to social scientists of all types.

A second, related advantage is the accessibility of the running record. Instead of searching packing crates, deteriorated ledgers, and musty storerooms, as users of the episodic record often do, users of the running record more often handle reference books, government publications, and computer printouts. Many political science research projects have been completed with only the data stored in the reference books and government documents of a decent research library.

A third advantage of the running record is that by definition it covers a more extensive period of time than does the episodic record. This permits the type of longitudinal analysis and before-and-after research designs discussed in Chapter 5. Although the episodic record helps explain the origins of and reasons for a particular event, episode, or period, the running record allows the measurement of political phenomena over time.

The running record presents problems, however. One is that a researcher is at the mercy of the data collection practices and procedures of the record-keeping organizations themselves. Researchers are rarely in a position to influence record-keeping practices; they must rely instead on what organizations such as the U.S. Census Bureau, Federal Election Commission, and the Policy Agendas Project decide to do. A trade-off often exists between ease of access and researcher influence over the measurements that are made. Some organizations—some state and local governments, for example—do not maintain records as consistently as researchers may like. One colleague found tracing the fate of proposed constitutional amendments to the Delaware State Constitution to be a difficult task. Delaware is the only state in which voters do not ratify constitutional amendments. Instead, the state legislature must pass an amendment in two consecutive legislative sessions in between which a legislative election has occurred. Thus constitutional amendments are treated like bills and tracking them depends on the archival practices of the state legislature. Even when clear records are kept, such as election returns for mayoral contests, researchers may face a substantial task in collecting the data from individual cities, because the returns from only the largest cities are reported in various statistical compilations.

Another related disadvantage of the running record is that some organizations are not willing to share their raw data with researchers. The processed data that they do release may reflect calculations, categorizations, and aggregations that are inaccurate or uninformative. Access to public information is not *always* easy. More problems may be encountered when trying to obtain public information that shares some of the characteristics of the episodic record, for example, such as information on the effect of specific public programs and agency activities. Emily Van Dunk, a senior researcher at the Public Policy Forum, a nonpartisan, nonprofit research organization that conducts research on issues of importance to Wisconsin residents, notes that obtaining data from state and local government agencies can be difficult at times and offers tips for researchers.[12]

Finally, it is sometimes difficult for researchers to find out exactly what some organizations' record-keeping practices are. Unless the organization publishes a description of its procedures, a researcher may not know what decisions have guided the record-keeping process. This can be a special problem when these practices change, altering in an unknown way the measurements reported.

Although the running record has its disadvantages, political scientists often must rely on it if they wish to do any empirical research on a particular topic. To illustrate some of the problems with using written records, we conclude this section with a description of PollingReport.com, one of many Web sites dedicated to providing users with national- and state-level public opinion data.

Presidential Job Approval

PollingReport.com is a popular source of public opinion polling data, as evidenced by Time.com's inclusion of the site on its list of the 50 best Web sites in 2007.[13] PollingReport.com provides national poll results, free of charge, from well-known polling organizations such as Gallup, Pew, and Quinnipiac and news organizations such as CNN, CBS, and the *Los Angeles Times*. The Web site also offers state-level poll results to paid subscribers. In this section we focus on the data available for free.

PollingReport.com organizes its poll results into categories for Elections, the State of the Union, National Security, In the News, and Issues. As shown in Figure 9-2, each of these categories offers a number of subtopics of interest. The State of the Union category, for example, includes subtopics covering each branch of the federal government, President Bush, Congress, and the Supreme Court, as well as the "Direction of the country," "National priorities," and "Consumer confidence." Likewise, the Issues category is divided into several subtopics. A great deal of useful public opinion data may be found among these many subtopics.

Let's assume that you are interested in studying public support for the president. The obvious place to start would be by finding data on President Bush's job approval ratings. By clicking on the President Bush subtopic under State of the Union, you will find several different kinds of presidential

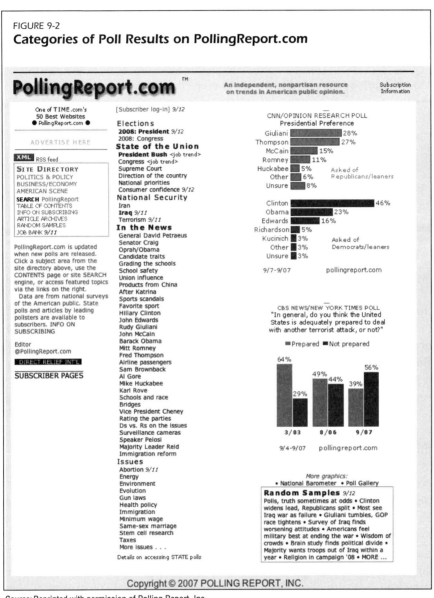

FIGURE 9-2
Categories of Poll Results on PollingReport.com

Source: Reprinted with permission of Polling Report, Inc.

support data (Figure 9-3). PollingReport.com provides data on "President Bush's Job ratings," "Favorability ratings," the "Bush Administration," the "White House leak investigation," "President Bush and history," as well as similar data for the Clinton presidency. Since you are interested in President Bush, not his administration, the "Job ratings" and "Favorability ratings" might be useful. We will explore President Bush's job ratings poll results in

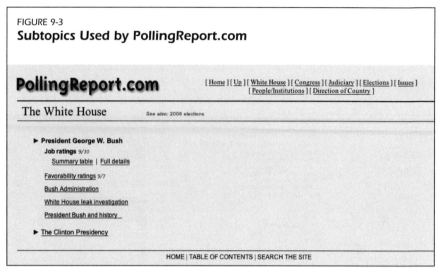

FIGURE 9-3
Subtopics Used by PollingReport.com

Source: Copyright © 2007 Polling Report, Inc. Reprinted with permission.

this example. You can click on "Summary table" to find poll results from various polling organizations that answer a question similar to Gallup's job approval rating question: "Do you approve or disapprove of the way George W. Bush is handling his job as president?" (Figure 9-4).[14] As you can see, the results span a two-year period from September 2005 to September 2007 on an irregular basis as poll data were available. Clicking on "Complete trend" at the top of the page takes you to a page with a more complete set of poll results from seven select organizations from the beginning of the Bush administration in 2001 until the present.

PollingReport.com has many advantages for students using the written record. First, and perhaps most important, PollingReport.com offers free, high-quality data with an easy-to-use Web site. The results found at Polling Report.com are from the same professional polling organizations that news organizations around the country rely on. Second, students have access to multiple surveys administered during different time periods using very similar question wording. But as valuable as PollingReport.com is, it shares some disadvantages with other examples of the running record. Perhaps most glaring is the lack of consistency and regularity in the poll results provided on the Web page. This is not an indictment of PollingReport.com but a reality of the kind of information the Web site provides. PollingReport.com can report only the data made available by other polling organizations. Although those other organizations provide a great deal of data, sometimes a large number of surveys are administered at the same time whereas no data are available for other time periods. And, although a great many organizations are listed on the Web site, it does not include all polling organizations.

FIGURE 9-4
Survey Data from PollingReport.com

PollingReport.com   [Home] [Up] [Bush: Job Ratings] [Bush: Favorability] [Bush Administration]
[Leak Probe] [Clinton] [Presidents & History]

▼Advertise on PollingReport.com▼

PollingReport.com

PRESIDENT BUSH – Overall Job Rating in recent national polls

See also: Complete trend

	Survey Dates	Approve %	Disapprove %	Unsure %	Approve minus Disapprove
USA Today/Gallup	9/7-8/07	33	62	5	-29
CBS/New York Times	9/4-8/07	30	64	6	-34
ABC/Washington Post	9/4-7/07	33	64	3	-31
FOX/Opinion Dynamics RV	8/21-22/07	33	56	11	-23
Pew	8/1-18/07	31	59	10	-28
Gallup	8/13-16/07	32	63	5	-31
Quinnipiac RV	8/7-13/07	29	64	7	-35
CBS	8/8-12/07	29	65	6	-36
AP-Ipsos	8/6-8/07	35	62	*	-27
CNN/Opinion Research Corp.	8/6-8/07	36	61	3	-25
USA Today/Gallup	8/3-5/07	34	62	4	-28
Newsweek	8/1-2/07	29	63	8	-34
NBC/Wall Street Journal	7/27-30/07	31	63	6	-32
Pew	7/25-29/07	29	61	10	-32
CBS/New York Times	7/20-22/07	30	62	8	-32
Diageo/Hotline RV	7/19-22/07	33	63	4	-30
ABC/Washington Post	7/18-21/07	33	65	2	-32
FOX/Opinion Dynamics RV	7/17-18/07	32	61	7	-29
CBS/New York Times	7/9-17/07	29	64	7	-35
Gallup	7/12-15/07	31	63	6	-32
Newsweek	7/11-12/07	29	64	7	-35
AP-Ipsos	7/9-11/07	33	65	*	-32
USA Today/Gallup	7/6-8/07	29	66	5	-37
Newsweek	7/2-3/07	26	65	9	-39
CBS	6/26-28/07	27	65	8	-38
FOX/Opinion Dynamics RV	6/26-27/07	31	60	9	-29
CNN/Opinion Research Corp.	6/22-24/07	32	66	3	-34
Newsweek	6/18-19/07	26	65	9	-39
Gallup	6/11-14/07	32	65	3	-33
NBC/Wall Street Journal	6/8-11/07	29	66	5	-37
Quinnipiac RV	6/5-11/07	28	65	7	-37
L.A. Times/Bloomberg	6/7-10/07	34	62	4	-28
FOX/Opinion Dynamics RV	6/5-6/07	34	57	9	-23
AP-Ipsos	6/4-6/07	32	66	*	-34
USA Today/Gallup	6/1-3/07	32	62	6	-30
Pew	5/30 - 6/3/07	29	61	10	-32
ABC/Washington Post	5/29 - 6/1/07	35	62	3	-27
CBS/New York Times	5/18-23/07	30	63	7	-33
Diageo/Hotline RV	5/16-20/07	32	64	4	-32
FOX/Opinion Dynamics RV	5/15-16/07	34	56	10	-22
Gallup	5/10-13/07	33	62	5	-29
AP-Ipsos	5/7-9/07	35	61	*	-26
CNN/Opinion Research Corp.	5/4-6/07	38	61	1	-23
USA Today/Gallup	5/4-6/07	34	63	3	-29
Newsweek	5/2-3/07	28	64	8	-36
WNBC/Marist RV	4/26 - 5/1/07	33	61	6	-28
Quinnipiac RV	4/25 - 5/1/07	35	60	6	-25
Diageo/Hotline RV	4/26-30/07	35	62	3	-27
NPR LV	4/26-29/07	37	59	5	-22
CBS/New York Times	4/20-24/07	32	61	7	-29

Source: Copyright © 2007 Polling Report, Inc. Reprinted with permission.

Second, even though the president's job approval rating question is one of the most frequently asked questions in national political surveys, a single organization is not likely to have a survey in the field every week of the year, so there will be gaps in the results. Relying on multiple surveys from different organizations can introduce bias to the conclusions because the results of each survey are influenced by question wording—and all organizations use slightly different wording to ask the same question about presidential job approval. Third, although PollingReport.com provides the results from the most basic measure of presidential approval, representing the whole population, the results for sometimes more interesting questions, like presidential support divided into subgroups by party identification, race, or gender, are not consistently available on the Web site.

Finally, PollingReport.com provides poll results from the Bush and Clinton administrations, but results from previous administrations are not available. If you wish to compare President Bush's approval ratings with those of previous presidents, you will have to search for those results elsewhere. These are only some of the potential problems that might be encountered with PollingReport.com or other examples of the running record. Problems like these are generally not going to prevent you from using such sources, but they can be a nuisance depending on the purpose of your research project.

Content Analysis

Sometimes researchers extract excerpts, quotations, or examples from the written record to support an observation or relationship. Those who rely on the episodic record, such as Charles Beard and James David Barber, often use the written record in this way. Alternatively, a researcher might measure the number of times the president references the economy in his State of the Union address. This use of the written record is an example of **content analysis.** We can think of document analysis, much like other forms of analysis, as taking both a qualitative and a quantitative form. In the earlier examples, we can label Beard and Barber's use of the written record as qualitative because they are using the words of others, or interpreting written documents without the use of a numeric coding scheme, to provide evidence for their arguments. Alternatively, we can label the use of State of the Union addresses as quantitative because quantitative content analysis enables us to "take a verbal, nonquantitative document and transform it into quantitative data." A researcher "first constructs a set of mutually exclusive and exhaustive categories that can be used to analyze documents, and then records the frequency with which each of these categories is observed in the documents studied." [15] This is exactly what Segal and Cover had to do with newspaper editorials to produce a quantitative measure of the political ideologies of Supreme Court justices.[16] In this section we focus primarily on quantitative content analysis for use in statistical analyses, but remember that a qualitative approach to documents can be just as useful, if not more so, depending on the purpose of a research project.

Content Analysis Procedures

The first step in content analysis is deciding what sample of materials to include in the analysis. If a researcher is interested in the political values of candidates for public office, a sample of political party platforms and campaign speeches might be suitable. If the level of sexism in a society is of interest, then a sample of television entertainment programs and films might be drawn. Or if a researcher is interested in what liberals are thinking about, liberal opinion magazines might be sampled. Actually, two tasks are involved at this stage: selecting materials germane to the researcher's subject (in other words, choosing the appropriate sampling frame) and sampling the actual material to be analyzed from that sampling frame. Once the appropriate sampling frame has been selected, then any of the possible types of samples described in Chapter 10—random, systematic, stratified, cluster, and nonprobability—could be used.

The second task in any content analysis is to define the categories of content that are going to be measured. A study of the prevalence of crime in the news, for example, might measure the amount of news content that either deals with crime or does not. Content that deals with crime might be further subdivided into the kinds of crime covered. A study of news coverage of a presidential campaign might measure whether news content concerning a particular presidential candidate is favorable, neutral, or unfavorable. Or a study might measure the personality traits of various prime-time television characters—such as strength, warmth, integrity, humility, and wisdom—and the sex, age, race, and occupation of those characters. This process is in many respects the most important part of any content analysis because the researcher must measure the content in such a way that it relates to the research topic, and he or she must define this content so that the measures of it are both valid and reliable.

The third task is to choose the recording unit. For example, from a given document, news source, or other material, the researcher may want to code (1) each word, (2) each character or actor, (3) each sentence, (4) each paragraph, or (5) each item in its entirety. To measure concern with crime in the daily newspaper, the recording unit might be the article. To measure the favorableness of news coverage of presidential campaigns in news weeklies, the recording unit might be the paragraph. And to measure the amount of attention focused on different government institutions on television network news, the recording unit might be the story.

In choosing the recording unit, the researcher usually considers the correspondence between the unit and the content categories (stories may be more appropriate than words in determining whether crime is a topic of concern, whereas individual words or sentences rather than larger units may be more appropriate for measuring the traits of political candidates). Generally, if the recording unit is too small, each case will be unlikely to possess any of the content categories. If the recording unit is too large, however, it will be difficult to measure the single category of a content variable that it possesses (in other words, the case will possess multiple values of a given content variable). The selection of the appropriate recording unit is often a matter of trial and error, adjustment, and compromise in the pursuit of measures that capture the content of the material being coded.

One good example of a content analysis of this type is a study of presidential campaign coverage in 1980 by Michael J. Robinson and Margaret A. Sheehan.[18] We discuss the procedures they followed and some of the strengths and weaknesses of their analysis.

At the beginning Robinson and Sheehan had to select the news coverage to be included in their study. Given the overwhelming amount of print and broadcast coverage that a presidential election campaign stimulates, there was no way that they could carefully analyze it all. In 1980 there were more than 1,000 daily newspapers and 6,000 broadcast stations in the United States. Consequently, they had to select, or sample, a portion of the news coverage to analyze. Six different decisions were involved in choosing the sample.

First, the researchers decided what type of medium to analyze. Primarily because of their estimates of the audience reached by different media, they chose national network television and newspaper wire service copy. In the process, they decided not to select several regional daily newspapers and the news weeklies, as had been done in a study of the 1976 campaign, and not to draw a representative sample of daily newspapers, as had been done in a study of the 1974 congressional elections.[19]

A SIMPLE COMPUTER CONTENT ANALYSIS
You can use your Internet browser to find and analyze speeches or other printed records if you are willing to keep careful records of your work. As an example, suppose you wanted to compare how presidents have defined the role of government over time. To do this you could analyze presidential inaugural addresses. When you locate a particular address you could use the browser's "Find in page" feature to look for, say, "we must" or a similar phrase. The sentence or ideas that follow may indicate the presidents' feelings about the role of government, since in saying "we" the president is usually referring to government or society. You can then make a count of the types of references to see how they have changed over time.

Second, because Robinson and Sheehan's resources were limited, they had to decide which of the media outlets to select. In other words, which television network and which wire service would be chosen? Based again on audience size, as well as professional prestige, they selected CBS and the Associated Press (AP). But AP refused to cooperate—an example of one of those disturbing yet all too frequent developments that cause the best laid research plans to go awry. Consequently, the researchers switched reluctantly to United Press International (UPI), even though it had far fewer clients and generally placed fewer stories in daily newspapers. CBS and UPI, then, became their case studies for 1980.

What products of these two media outlets should be included in the study? This was the third decision facing Robinson and Sheehan. CBS produces several versions of the nightly news, as well as morning news shows, midday news shows, news interviews, and news specials. And UPI offers several news services, among them an "A" wire, which is the national wire; a city wire; and a radio wire. The "A" wire itself has two versions: the night cycle, which runs from noon to midnight, and the day cycle, which runs from midnight to noon. The researchers decided to use the day "A" wire, for reasons of scope of coverage as well as accessibility, and the CBS nightly news (the 7:00 p.m. Eastern time edition), primarily for financial reasons and convenience.

Fourth, they had to decide which of the material from these news shows and wire copy to include. They decided to include only campaign or campaign-related stories. Thus they used any story that "mentioned the presidential campaign, no matter how tangentially; mentioned any presidential candidate in his campaign role; mentioned any presidential candidate or his immediate family in a noncampaign, official role (almost always a story about the president); or discussed to a substantial degree any campaign lower than the presidential level."[20] Just over 5,500 stories on UPI and CBS—22 percent of UPI and CBS total news coverage—met these selection criteria.

Fifth, Robinson and Sheehan had to decide what time period to include in their study. Although a presidential campaign has a fairly clear ending point, election day, the beginning date of the campaign is uncertain. The researchers decided to include weekday coverage throughout 1980 (that is, from January 1 to December 31). They gave no justification for excluding the weekend news.

Finally, Robinson and Sheehan made an important decision to exclude some of the content of both CBS's and UPI's news coverage. They decided not to include any photographs, film, videotape, or live pictures and to rely exclusively on verbal (CBS) or written (UPI) expression. They defended this decision on the grounds that it is more difficult to interpret the meaning of visuals and that the visual message usually supports the verbal message. Moreover, they thought that comparing the visual component of CBS with that of UPI would be difficult.

Having selected the news content to be analyzed, Robinson and Sheehan then decided on the unit of analysis to use when coding news content. Generally, they analyzed the story, although at times they analyzed the content sentence by sentence and word by word. Most content analyses of this type have also used the story as the unit of analysis, but it is unfortunate that Robinson and Sheehan did not explain this choice in any detail or discuss how difficult it was to tell where one story ended and another began.

Having selected the news content to be used in the measurement of campaign news coverage, Robinson and Sheehan then had to decide the content categories to be encoded and the definitions of the values for these content categories. They coded some twenty-five different aspects of each 1980 campaign story. Some of these were straightforward, such as the story's date, length, and reporter. Other categories that pertained to the central subject matter of the study were not as readily defined or measured.

The researchers were primarily concerned with five characteristics: Were CBS and UPI (1) objective, (2) equitable in providing access, (3) fair, (4) serious, and (5) comprehensive? Consequently, they needed to decide how to measure each of these attributes of news coverage.

Robinson and Sheehan measured the objectivity of the press's coverage in four ways: by the number of explicit and unsupported conclusions drawn by journalists about the personal qualities of the candidates; by the number of times the journalists expressed personal opinions concerning the issues of the campaign; by counting the number of sentences that were either descriptive, analytical, or judgmental; and by counting the number of verbs used by journalists that were either descriptive, analytical, or insinuative. Clearly, each of these content categories involved judgments by the researchers concerning what constituted an explicit and unsupported conclusion, what constituted a personal opinion, and what constituted a descriptive versus an analytical sentence. The researchers provided examples of different types of coded content, and they also gave some brief definitions of what each of the categories meant to them: for example, descriptive sentences "present the who, what, where, when of the day's news, without any meaningful qualification or elaboration," analytical sentences "tell us *why* something occurs or predicts as to whether it might," and judgmental sentences "tell us how something ought to be or ought not to be."[21]

To determine whether the press granted appropriate access to each of the presidential candidates, Robinson and Sheehan measured how much coverage (in seconds for CBS and in column inches for UPI) each of the candidates received. They did not say whether this coding procedure presented any difficulties, although they did evaluate whether the amount of access granted each candidate was justified.

Determining whether press coverage was fair was much more difficult than measuring access, since an evaluation of fairness requires that the tone of campaign coverage be measured. Establishing tone in a reliable and valid way is not easy. Robinson and Sheehan defined tone and fairness in these terms:

> *Tone pertains not simply to the explicit message offered by the journalist but the implicit message as well. Tone involves the overall (and admittedly subjective) assessment we made about each story: whether the story was, for the major candidates, "good press," "bad press," or something in between. "Fairness," as we define it, involves the sum total of a candidate's press tone; how far from neutrality the candidate's press score lies.*[22]

They evaluated content by whether it represented good press (a story that had three times as much positive information as negative information about a candidate) or bad press (a story that had three times as much negative as positive information). But they never discussed how they determined what constituted positive and negative information. Furthermore, in their effort to restrict their analysis to the behavior of journalists, Robinson and Sheehan excluded information about political events (such as the failure of the Iranian hostage rescue mission), polls, comments made by partisans, remarks of "criminals and anti-Americans" (such as Fidel Castro and the Ayatollah Khomeini), and statements made by the candidates themselves.[23] In short, their measurement of fairness depended on the wisdom of their decisions regarding the encoding of campaign stories. Some of these decisions are questionable, such as using an arbitrary three-to-one ratio to determine good press/bad press and excluding political events, polls, and the words of the candidates themselves from the analysis.

The seriousness of press coverage was measured by coding each story and, at times, each sentence, according to whether it represented policy issues, candidate issues, "horse-race coverage," or something else. Policy issues were ones that "involve major questions as to how the government should (or should not) proceed in some area of social life"; candidate issues "concern the personal behavior of the candidate during the course of his or her campaign"; and horse-race coverage focuses on "any consideration as to winning or losing." Because of the difficulty of encoding entire stories into only one of these categories, the researchers shifted to the more exacting sentence-by-sentence analysis. Some sentences did not fit into one and only one of these categories, but "the majority of sentences were fairly easy to classify as one form of news or another."[24] The seriousness of UPI and CBS campaign coverage in 1980 was then measured by comparing their amount of policy issue coverage with the policy issue coverage in other media in previous presidential election years.

Finally, to evaluate how comprehensive press coverage was in 1980, Robinson and Sheehan coded campaign stories as to the level of office covered: presidential, vice presidential, senatorial, congressional, and gubernatorial. More than 90 percent of both CBS and UPI campaign coverage was of the presidential and vice presidential races.[25]

Over the Wire and on TV represents one of the most thorough content analyses ever performed by political scientists. Certainly, in regard to the time period covered and the sheer quantity of material analyzed, it was an ambitious study. The value of the study was weakened, however, by the inadequate explanation of the content analysis procedures. The definitions of the categories used were brief and the illustrative material sketchy. Furthermore, Robinson and Sheehan dispensed with the issue of measurement quality in only one paragraph, where they reported that intercoder reliability figures among four members of the coding team averaged about 95 percent agreement.[26] However, they failed to report any details about how this reliability was measured or about the agreement scores for different content categories. Despite these shortcomings, this study exemplifies how content analysis can reveal useful information about a significant political phenomenon. It also illustrates how practical limitations—such as AP's refusal to participate, as well as financial constraints—all too often limit what researchers can actually accomplish.

Advantages and Disadvantages of the Written Record

Using documents and records, or what we have called the written record, has several advantages for researchers. First, it allows us access to subjects that may be difficult or impossible to research through direct, personal contact, because they pertain either to the past or to phenomena that are geographically distant. For example, the record keeping of the Puritans in the Massachusetts Bay Colony during the seventeenth century allowed Erikson to study their approach to crime control, and late eighteenth-century records permitted Beard to advance and test a novel interpretation of the framing of the U.S. Constitution. Neither of these studies would have been possible had there been no records available from these periods.

A second advantage of data gleaned from archival sources is that the raw data are usually nonreactive. As we mentioned in previous chapters, human subjects often consciously or unconsciously establish expectations or other relationships with investigators, which can influence their behavior in ways that might confound the results of a study. But those writing and preserving the records are frequently unaware of any future research goal or hypothesis or, for that matter, that the fruits of their labors will be used for research purposes at all. The record keepers of the Massachusetts Bay Colony were surely unaware that their records would ever be used to study how a society defines and reacts to deviant behavior. Similarly, state loan officers during the late 1700s had no idea that two hundred years later a historian would use their records to discover why some people were in favor of revising the Articles of Confederation. This nonreactivity has the virtue of encouraging more accurate and less self-serving measures of political phenomena.

Record keeping is not always completely nonreactive, however. Record keepers are less likely to create and preserve records that are embarrassing to them, their friends, or their bosses; that reveal illegal or immoral actions; or that disclose stupidity, greed, or other unappealing attributes. Richard Nixon, for example, undoubtedly wished that he had destroyed or never made the infamous Watergate tapes that revealed the extent of his administration's knowledge of the 1972 break-in at Democratic National Committee headquarters. Today many record-keeping agencies employ paper shredders to ensure that a portion of the written record does *not* endure. Researchers must be aware of the possibility that the written record has been selectively preserved to serve the record keepers' own interests.

A third advantage of using the written record is that sometimes the record has existed long enough to permit analyses of political phenomena over time. The before-and-after research designs discussed in Chapter 5 may then be used. For example, suppose you are interested in how changes in the 55-mile-per-hour speed limit (gradually adopted by the states and then later dropped by many states on large stretches of their highway systems) affected the rate of traffic accidents. Assuming that the written record contains data on the incidence of traffic accidents over time in each state, you could compare the accident rate before and after changes in the speed limit in those states that changed their speed limit. These changes in the accident rate could then be compared with the changes occurring in states in which no change in the speed limit took place. The rate changes could then be "corrected" for other factors that might affect the rate of traffic accidents. In this way an interrupted time series research design could be used, a research design that has some important advantages over cross-sectional designs. Because of the importance of time, and of changes in phenomena over time, for the acquisition of causal knowledge, a data source that supports longitudinal analyses is a valuable one. The written record more readily permits longitudinal analyses than do either interview data or direct observation.

A fourth advantage to researchers of using the written record is that it often enables us to increase sample size above what would be possible through either interviews or direct observation. For example, it would be terribly expensive and time consuming to observe the level of spending by all candidates for the House of Representatives in any given year. Interviewing candidates would require a lot of travel, long-distance phone calls, or the design of a questionnaire to secure the necessary information. Direct observation would require gaining access to many campaigns. How much easier and less expensive it is to contact the Federal Election Commission in Washington, D.C., and request the printout of campaign spending for all House candidates. Without this written record, resources might permit only the inclusion of a handful of campaigns in a study; with the written record, all 435 campaigns can easily be included.

This raises the fifth main advantage of using the written record: cost. Since the cost of creating, organizing, and preserving the written record is borne by the record keepers, researchers are able to conduct research projects on a much smaller budget than would be the case if they had to bear the cost themselves. In fact, one of the major beneficiaries of the record-keeping activities of the federal government and of news organizations is the research community. It would cost a prohibitive amount for a researcher to measure the amount of crime in all cities larger than 25,000 or to collect the voting returns in all 435 congressional districts. Both pieces of information are available at little or no cost, however, because of the record-keeping activities of the FBI and the Elections Research Center, respectively. Similarly, using the written record often saves a researcher considerable time. It is usually much quicker to consult printed government documents, reference materials, computerized data, and research institute reports than it is to accumulate data ourselves. The written record is a veritable treasure trove for researchers.

Collecting data in this manner, however, is not without some disadvantages. One problem mentioned earlier is selective survival. For a variety of reasons, record keepers may not preserve all pertinent materials but rather selectively save those that are the least embarrassing, controversial, or problematic. It would be surprising, for example, if political candidates, campaign consultants, and public officials saved correspondence and memoranda that cast disfavor on themselves. Obviously, whenever a person is selectively preserving portions of the written record, the accuracy of what remains is suspect. This is less of a problem when the connection between the record keeper's self-interest and the subject being examined by the researcher is minimal.

A second, related disadvantage of the written record is its incompleteness. Large gaps exist in many archives due to fires, losses of other types, personnel shortages that hinder record-keeping activities, and the failure of the record maker or record keeper to regard a record as worthy of preservation. We all throw out personal records every day; political entities do the same. It is difficult to know what kinds of records should be preserved, and it is often impossible for record keepers to bear the costs of maintaining and storing voluminous amounts of material.

Another reason records may be incomplete is simply because no person or organization has assumed the responsibility for collecting or preserving them. For example, before 1930, national crime statistics were not collected by the FBI, and before the creation of the Federal Election Commission in 1971, records on campaign expenditures by candidates for the U.S. Congress were spotty and inaccurate.

A third disadvantage of the written record is that its content may be biased. Not only may the record be incomplete or selectively preserved, but it also may be inaccurate or falsified, either inadvertently or on purpose. Memoranda or copies of letters that were never sent may be filed, events may be conveniently forgotten or misrepresented, the authorship of documents may be disguised, and the dates of written records may be altered; furthermore, the content of government reports may tell more about political interests than empirical facts. For example, Soviet and East European governments apparently released exaggerated reports of their economic performance for many years; scholars (and investigators) attempting to reconstruct the actions in the Watergate episode have been hampered by alterations of the record by those worried about the legality of their role in it. Often, historical interpretations rest upon who said or did what, and when. To the extent that falsifications of the written record lead to erroneous conclusions, the problem of record-keeping accuracy can bias the results of a research project. The main safeguard against bias is the one used by responsible journalists: confirming important pieces of information through several dissimilar sources.

A fourth disadvantage is that some written records are unavailable to researchers. Documents may be classified by the federal government; they may be sealed (that is, not made public) until a legal action has ceased or the political actors involved have passed away; or they may be stored in such a way that they are difficult to use. Other written records—such as the memoranda of multinational corporations, campaign consultants, and Supreme Court justices—are seldom made public because there is no legal obligation to do so and the authors benefit from keeping them private.

Finally, the written record may lack a standard format because it is kept by different people. For example, the Chicago budget office may have budget categories for public expenditures different from those used in the San Francisco budget office. Or budget categories used in the Chicago budget office before 1960 may be different from the ones used after 1960. Or the French may include items in their published military defense expenditures that differ from those included by the Chileans in their published reports. Consequently, a researcher often must expend considerable effort to ensure that the formats in which records are kept by different record-keeping entities can be made comparable.

Despite these limitations, political scientists have generally found that the advantages of using the written record outweigh the disadvantages. The written record often supplements the data we collect through interviews and direct observation, and in many cases it is the only source of data on historical and cross-cultural political phenomena.

Conclusion

The written record includes personal records, archival collections, organizational statistics, and the products of the news media. Researchers interested in historical research, or in a particular event or time in the life of a polity, generally use the episodic record. Gaining access to the appropriate material is often the most resource-consuming aspect of this method of data collection, and the hypothesis testing that results is usually more qualitative and less rigorous (some would say more flexible) than with the running record.

The running record of organizations has become a rich source of political data as a result of the record-keeping activities of governments at all levels and of interest groups and research institutes concerned with public affairs. The running record is generally more quantitative than the episodic record and may be used to conduct longitudinal research. Measurements using the running record can often be obtained inexpensively, although the researcher frequently relinquishes considerable control over the data collection enterprise in exchange for this economy.

One of the ways in which a voluminous, nonnumerical written record may be turned into numerical measures and then used to test hypotheses is through a procedure called content analysis. Content analysis is most frequently used by political scientists interested in studying media content, but it has been used to advantage in studies of political speeches, statutes, and judicial decisions.

Through the written record, researchers may observe political phenomena that are geographically, physically, and temporally distant from them. Without such records, our ability to record and measure historical phenomena, cross-cultural phenomena, and political behavior that does not occur in public would be seriously hampered.

Notes

1. Frank Way and Barbara J. Burt, "Religious Marginality and the Free Exercise Clause," *American Political Science Review* 77 (September 1983): 652–665; Samuel C. Patterson and Gregory A. Caldeira, "Getting Out the Vote: Participation in Gubernatorial Elections," *American Political Science Review* 77 (September 1983): 675–689; William Zimmerman and Glenn Palmer, "Words and Deeds in Soviet Foreign Policy: The Case of Soviet Military Expenditures," *American Political Science Review* 77 (June 1983): 358–367; and Alan P. L. Liu, "The Politics of Corruption in the People's Republic of China," *American Political Science Review* 77 (September 1983): 602–623.

2. Steven C. Poe and C. Neal Tate, "Repression of Human Rights to Personal Integrity in the 1980s: A Global Analysis," *American Political Science Review* 88 (December 1994): 853–872; Jeff Yates and Andrew Whitford, "Presidential Power and the United States Supreme Court," *Political Research Quarterly* 51 (June 1998): 539–550; Jeffrey A. Segal and Albert D. Cover, "Ideological Values and the Votes of U.S. Supreme Court Justices," *American Political Science Review* 83 (June 1989): 557–565; and Kim Fridkin Kahn and Patrick J. Kenney, "Do Negative Campaigns Mobilize or Suppress Turnout? Clarifying the Relationship between Negativity and Participation," *American Political Science Review* 93 (December 1999): 877–890.

3. Charles Beard reports that he was able to use some records in the U.S. Treasury Department in Washington "only after a vacuum cleaner had been brought in to excavate the ruins." See Charles Beard, *An Economic Interpretation of the Constitution of the United States* (London: Macmillan, 1913), 22.

4. Kai T. Erikson, *The Wayward Puritans* (New York: Wiley, 1966).

5. The records of the governor were edited by Nathaniel B. Shurtleff and printed by order of the Massachusetts legislature in 1853–1854; the records of the courts were edited by George Francis Dow and published by the Essex Institute in Salem, Massachusetts.

6. Erikson, *The Wayward Puritans,* 209–210.

7. Beard, *An Economic Interpretation.*

8. Ibid., 324. Beard's interpretation has been challenged by several historians. Among his critics are Robert E. Brown, *Charles Beard and the Constitution* (Princeton: Princeton University Press, 1956); Forrest McDonald, *We the People: The Economic Origins of the Constitution* (Chicago: University of Chicago Press, 1958); and Gordon Wood, *The Creation of the American Republic* (New York: Norton, 1972). Although Beard's interpretation continues to be controversial, the authors of one mainstream political science textbook state, "[A]lthough histor-

ical evidence does not fully support Beard's conclusions, most historians acknowledge that economic interests were very much at issue in the framing and ratification of the Constitution." Lewis Lipsitz and David M. Speak, *American Democracy*, 2d ed. (New York: St. Martin's, 1989), 76.

9. James David Barber, *The Presidential Character*, 3d ed. (Englewood Cliffs, N.J.: Prentice-Hall, 1985), 4, 5.

10. James David Barber, *The Presidential Character*, 1st ed. (Englewood Cliffs, N.J.: Prentice-Hall, 1972), ix.

11. A critique of Barber's analysis may be found in Garry Wills, *The Kennedy Imprisonment* (Boston: Little, Brown, 1982).

12. Emily Van Dunk, "Getting Data through the Back Door: Techniques for Gathering Data from State Agencies," *State Politics and Policy Quarterly* 1 (Summer 2001), 210–218.

13. *Time* Magazine, "50 Best Websites 2007." Retrieved from www.time.com/time/specials/2007/article/0,28804,1633488_1639316,00.html.

14. PollingReport.com, "Bush: Job Ratings." Retrieved from www.pollingreport.com/BushJob1.htm.

15. Kenneth D. Bailey, *Methods of Social Research*, 2d ed. (New York: Free Press, 1982), 312–313.

16. Segal and Cover, "Ideological Values."

17. Bailey, *Methods of Social Research*, 319.

18. Michael J. Robinson and Margaret A. Sheehan, *Over the Wire and on TV* (New York: Russell Sage Foundation, 1983); on their survey decisions, discussed in the following paragraphs, see pp. 17–27.

19. On the 1976 campaign, see Thomas Patterson, *The Mass Media Election: How Americans Choose Their President* (New York: Praeger, 1980). On the 1974 congressional elections, see Arthur Miller, Edie Goldenberg, and Lutz Erbring, "Type-Set Politics: Impact of Newspapers on Public Confidence," *American Political Science Review* 73 (March 1979): 67–84.

20. Robinson and Sheehan, *Over the Wire*, 20.

21. Ibid., 49–50.

22. Ibid., 92.

23. Ibid., 94–95.

24. Ibid., 144, 145, 155.

25. Ibid., 173.

26. Ibid., 22.

(Te) erms Introduced

CONTENT ANALYSIS. A procedure by which verbal, nonquantitative records are transformed into quantitative data.

EPISODIC RECORD. The portion of the written record that is not part of a regular, ongoing record-keeping enterprise.

INTERCODER RELIABILITY. Demonstration that multiple analysts, following the same content analysis procedure, agree and obtain the same measurements.

RUNNING RECORD. The portion of the written record that is enduring and covers an extensive period of time.

WRITTEN RECORD. Documents, reports, statistics, manuscripts, and other recorded materials available and useful for empirical research.

Suggested Readings

Hovey, Kendra A., and Harold A. Hovey. *CQ's State Fact Finder 2004.* Washington, D.C.: CQ Press, 2004.

Miller, Delbert C. *Handbook of Research Design and Measurement.* Thousand Oaks, Calif.: Sage Publications, 2002.

Stanley, Harold W., and Richard K. Niemi. *Vital Statistics on American Politics: 2003–2004.* Washington, D.C.: CQ Press, 2003.

Van Dunk, Emily. "Getting Data through the Back Door: Techniques for Gathering Data from State Agencies." *State Politics and Policy Quarterly* (Summer 2001): 210–218.

Webb, Eugene J., and others. *Unobtrusive Measures,* Rev. ed. Thousand Oaks, Calif.: Sage Publications, 1999.

CHAPTER 10

Survey Research and Interviewing

In the winter of 2007 Vermont, like several other states, wrestled with the question of whether or not state law should be rewritten to allow doctors to assist terminally ill individuals to end their lives. The act before the state legislature, labeled in the press as "death with dignity" or "physician-assisted suicide," provoked considerable debate. As with many issues, those on both sides frequently invoked public opinion to support their positions. The blog GreenMountainDaily.com, which backed the legislation, claimed "opinion polls virtually anywhere show strong support for such options, particularly in Vermont where a February Zogby poll showed a whopping 82 percent in favor of the concept."[1] Another advocacy group wrote, "Polls show that most Catholics support aid in dying, but the U.S. Council of Bishops strongly opposes it, as does the Vatican."[2] On the other side, the Vermont Alliance for Ethical Healthcare asserted that polls were misleading: "It is important for the public to know that physician assisted suicide does *not* (emphasis added) have the widespread approval that its supporters and the media claim. For example, biased polling techniques and inaccurate reporting have affected public perception of the facts surrounding [physician-assisted suicide]."[3]

Most readers may be familiar with this form of argumentation because even a cursory glance at the news demonstrates how pervasive polling has become in American politics. Polls are part and parcel of the efforts of many groups not just to study public opinion but also to use it for political ends. So it behooves anyone who wants to understand debates about public policies and issues to become familiar with this activity.

More generally, polling, or as we call it, survey research, is an indispensable tool in social and political research. Suppose we want to know whether or not Americans are "isolationists" or "internationalists" when it comes to foreign affairs. We might try to answer the question by making indirect or unobtrusive observations, such as reading letters to the editor in a dozen or so newspapers and coding them as pro or con involvement. Or we might observe protest demonstrations for or against various international activities to see what kinds of people seem to be participating. But these indirect methods probably would not tell us what we wanted to know. It would seem far preferable (and maybe even easier) to ask citizens up front how they felt about world affairs and the proper U.S. role in them. This was Miroslav Nincic's strategy, for he relied heavily on poll data in his article "Domestic Costs, U.S. Public Opinion, and the Isolationist Calculus."[4] And it is used by countless other political scientists who feel that the best way to measure people's preferences, beliefs, opinions, and knowledge about current events is to ask them.

In this chapter we explain two related methods of collecting data from people: (1) survey research, which involves collecting information via a questionnaire or survey instrument (a carefully structured or scripted set of questions that may be administered face-to-face, by telephone, by mail, by Internet, or other means) and (2) interviewing, which involves direct and personal communication with individuals in a less formal and less structured style—more in the nature of a constrained conversation. Although we describe both techniques in a moment, for now let us just say that these approaches range from talking to a handful of people to gathering data from 1,000 or more members of the public across an entire nation. In either approach, the researcher is trying to get at what people think and do by asking them for self-reports.

Because both methods rely on interpersonal communication, they might seem to entail no special considerations: just think of some questions and ask them. This is not the case, however. To see why, refer to the research described in Chapter 1 regarding voter turnout. As you may recall, the issue boils down to who votes and who doesn't. Political scientists have developed all sorts of hypotheses and theories to answer the question, but testing them rests on a seemingly simple and straightforward but in reality quite difficult task: determining who actually voted in any given election. You might think it would be easy to ask people, "Did you vote in the last election?" And that is precisely what most survey researchers do. The problem is that often two-thirds to three-fourths of the respondents claim to have voted. But we know from vote counts reported by election officials that these survey estimates must be too high, for voter turnout rarely exceeds 50 percent and is often much less. So the questionnaire method usually overestimates participation, or the number of people who claimed to have voted when they probably did not. Overcounting of voters calls into question conclusions based on the replies to these questions.[5]

As a result, the design and implementation of surveys and interviews have to be scrutinized. We begin with a thorough discussion of the problem of obtaining accurate information about attitudes and beliefs by asking people questions rather than by directly observing their behavior. This background puts us in a position to examine survey and interview methods carefully and thoroughly.

Fundamentals: Ensuring Validity and Reliability

Since survey and interview methods produce only indirect measures of attitudes and behavior, measurement problems, as discussed in Chapter 4, come to the fore. In particular, what is recorded on a piece of paper or an audiotape is usually not an exact, error-free measure of a object. This is particularly true when the objects are attitudes, beliefs, or self-described behavior. An observed "score" (for example, a response to a question) is composed of a true or real (but unobserved) part plus various types of error. The errors may be random or systematic. Random errors arise by chance or happenstance and (it is hoped) cancel one another out. A systematic error, by contrast, results when a measuring device consistently over- or underestimates a true value, as when a scale always reads two pounds less than a person's real weight. The goal of any research design, of course, is to minimize these errors. Stated differently, our investigative procedures have to ensure validity and reliability. A valid measure produces an accurate or true picture of an object, whereas a reliable one gives consistent results (measurements) across time and users. In the case of survey research, attaining both goals can be a daunting but surmountable problem. This is an important point for making sense of claims based on polls.

Figure 10-1 illustrates a naive view of survey research. It shows a pollster asking a man if he supports or opposes civil unions, a legal status that gives gay and lesbian couples rights equal to or similar to those enjoyed by traditionally married people. When asking such a question, most people expect

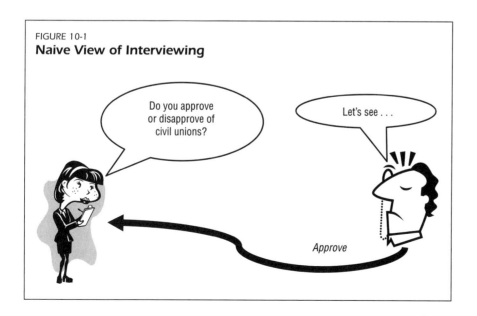

FIGURE 10-1
Naive View of Interviewing

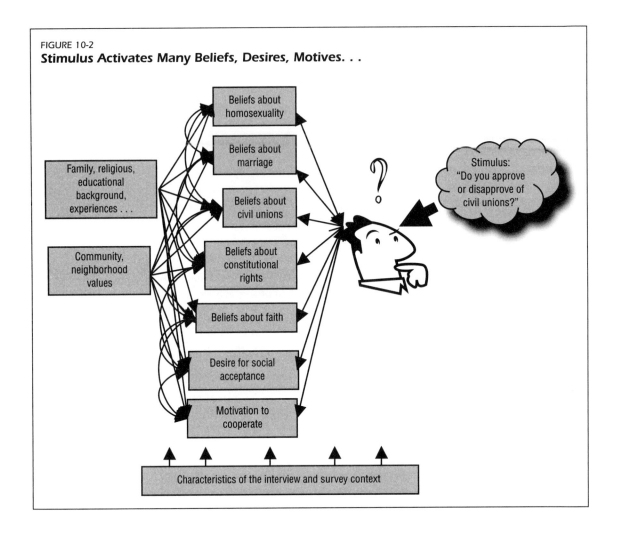

FIGURE 10-2
Stimulus Activates Many Beliefs, Desires, Motives. . .

that most of the time the responses will be a precise representation of what the respondent intends or thinks and that all parties understand the question and answers the same way. In the case of a supposedly objective scientific method such as survey research, the replies may appear to be straightforward statements that accurately express a person's real feelings. It is usually assumed, in other words, that the response is unproblematic; that *approve* means "approve" and there is nothing more to the story than that.

To see what can go wrong, consider Figure 10-2. It shows that, contrary to Figure 10-1, a fully formed attitude does not simply sit in someone's mind isolated from all other mental states. Instead, a verbalized or written opinion that is stimulated by a question is a distillation of a number of beliefs, hopes, desires, and motivations. Furthermore, one person who hears the phrase *civil union* may not be familiar with the term, whereas another, perhaps more informed about current events, may realize the question deals with marriage among homosexual couples. They may express opposing views, but the opinions can be based on unrelated beliefs about the subject. Or, if both people reply with, say, "approve," they may do so for different reasons. Even though two people may give the same response, the answers may not be comparable when this opinion is associated with other attitudes or behaviors. And consider that even if two people share a common understanding of the term, they may differ greatly in other respects: the intensity of their feelings about the matter, their willingness to cooperate with the research, or their desire to "please" the interviewer by giving a socially acceptable response. We must also factor in the interviewer's characteristics (for example, demeanor, race, gender) and the context of the research (for example, nature of the sponsoring organization, time constraints on participants). The net result is that an interview, even one carried out over the phone or using a mail survey, involves a complex set of potential interactions that can confuse the interpretation of responses.

To deal with problems of this nature, even in the simplest situations, we need to ensure that several assumptions have been met, including the following (for simplicity, let R stand for the respondent and I for the interviewer):

- The requested information must be available to R (that is, not forgotten or misunderstood).

- R must know what is to I a relevant and appropriate response.

- R must be motivated to provide I with the information.

- R must know how to provide the information.

- I must accurately record R's responses.

- The responses must reflect R's intentions, not I's.

- Other users of the data must understand the questions and answers the same way R and I do.

We are not trying to overanalyze the situation. The point is that if questionnaires and interviews are to produce any useful information, they must take into account the mental context of the respondent and the interview situation. For that reason we spend the greater portion of this chapter discussing crucial topics such as question wording, questionnaire layout, administrative protocols, efforts to balance demands for completeness versus costs, ways of motivating cooperation, and interviewer characteristics and deportment. These factors contribute to the validity and reliability of the measurements or observations obtained through questionnaires and interviews.

Survey Research

The term *survey research* has two broad meanings. In the context of research design it indicates an alternative data collection method to experiments, simulations, and formal modeling. Instead of manipulating an independent variable, for instance, to see the effect on behavior, the survey design asks people if they have been exposed to some factor and, if so, by how much, and then relates the responses to other questions about their behavior. In this chapter we use the term a bit more specifically to mean research based on direct or indirect interview methods. Simply stated, a group of individuals respond to or fill out more or less standardized questionnaires. (The questionnaires may take different forms to investigate different hypotheses, but they do not involve freewheeling or spontaneous conversations.) Also known as opinion polling, it is one of the most familiar political science research methods.

As the use of surveys has grown, so too has the amount of research on the method itself. This research tries to improve the validity and reliability of the method while keeping costs manageable. We now know more about many aspects of survey research than was known when it was first used, much to the benefit of researchers and consumers of survey research.

We begin with a review of the types of surveys and some of their important characteristics. Then we take up response quality and question wording, the heart and soul of this type of research.

Types of Surveys

A survey solicits information by posing a standard set of questions and stimuli to a sample of individuals drawn from an appropriate target population (see Chapter 7). The forms of the instrument and means of administration vary widely depending on a host of factors ranging from cost to comprehensiveness. Table 10-1 lists the main types of surveys along with a few of their properties or characteristics. As you can see from the table, surveys range from personal or face-to-face interviews to contacting subjects by mail or telephone to more or less hit-or-miss methods such as posting questions on a Web site or leaving them in a public area. As Floyd J. Fowler points out, researchers may be able to combine mail, telephone, and personal interviews in a research project to take advantage of the particular strengths of each type.[6] Often however, researchers must make compromises in choosing a survey instrument. As Don A. Dillman notes, "The use of any [of these types] requires accepting less of certain qualities to achieve others, the desirability of which cannot be isolated from a consideration of the survey topic and the population to be studied."[7]

Perhaps the most familiar surveys are those conducted personally or face-to-face. The interviewer typically follows a structured questionnaire that is ad-

TABLE 10-1

Types and Characteristics of Surveys

Type of Survey	Characteristics				
	Overall Cost[a]	Potential Completion Rate[b]	Sample-Population Congruence	Questionnaire Length[c]	Data-Processing Costs
Personal/face-to-face	High	High to medium	Potentially high	Long-medium	High
Telephone	Medium	Medium	Medium	Medium-short	High to low
Mail	Low	Low	Medium	Medium-short	Medium
E-mail	Low	Depends but low	Low	Medium-short	High to low
Internet	Low	Depends but low	Low	Medium-short	High to low
Group administration	Very low	High once group is convened	Depends on group selection process	Variable	High to low
Drop-off/pick-up	Very low	Low	Low	Short	Low

[a] Costs of design, administration, and processing.
[b] Assumes a general target population (see text): high: greater than 75%; medium: 30% to 75%; low: less than 30%.
[c] Length can refer to the number of questions or the time to complete (see text).

ministered to all members of the sample, although sometimes different forms are used with slightly different question wording and order. Not only are the same questions asked of everyone, but the manner in which they are posed is standardized to the maximum extent possible. The results are then coded or transcribed for further analysis. Moreover, for a variety of reasons, the principal investigator does not usually conduct the interviews but uses paid or volunteer assistants. Hence, this kind of research can be quite expensive.[8]

Academic and commercials polls are increasingly being conducted in whole or part by phone or mail. A mail survey, which may be preceded by an introductory letter, has to be self-contained with clear questions and instructions for completing and returning it. Motivating participants, anticipating misunderstandings, and obtaining unambiguous results demands a lot of careful planning and pre-testing. It also requires a list of addresses drawn from the population of interest. Although somewhat less expensive, phone interviews raise a number of tricky problems of their own (discussed below). Nevertheless, the basic idea is the same: pose a series of questions or stimuli and record the responses.

The explosive growth in e-mail and Internet use has led many researchers to adapt these technologies to survey research. Some researchers send questionnaires via e-mail. Just as with mail surveys the objective is to identify and contact potential participants and send them a questionnaire to be filled out and returned. In this case the task is carried out electronically. The use of e-mail has some significant advantages (for example, distribution and processing costs are reduced), but it is hard to apply to general populations. Not everyone uses e-mail or it may be difficult or impossible to establish a list of addresses that conforms with a desired target population (for example, people living in poverty). A related approach is the Web or Internet survey. Here participants are somehow directed or enticed to visit a site where they answer questions by filling out computer-generated forms. This form of research engenders a lot of suspicion—sampling is again a major issue—but some large organizations (Zogby Interactive and Harris Poll Online, to name two) encourage people to register in order to participate in their ongoing polls. The results, the sponsors claim, are both accurate and influential.[9]

Finally, it is possible, even necessary sometimes, to prepare a survey that is administered to a group (for example, a political science class or visitors to a senior center) or made available at a public location (library, museum, dormitory lounge). The finished forms are then collected or returned to the same or another convenient spot. The results are generally suspect in some people's minds and probably not publishable because they may not be representative, but the method offers considerable savings in effort and cost. For this reason they are commonly used at schools and colleges.

As might be expected, each of these types has advantages and disadvantages. The entries in Table 10-1 are merely suggestive and comparative, and have to be interpreted flexibly. A phone survey, for example, generally has to be shorter (in time necessary to complete it) than a personal interview because respondents to the former may be reluctant to tie up a phone line or may be distracted by those around them. Given an interesting topic, plenty of forewarning, and trained interviewers, however, it is possible to hold people's attention for longer periods.

Characteristics of Surveys

Cost Any type of survey research takes time and incurs at least some expenses for materials. Among the factors determining survey costs are the amount of professional time required for questionnaire design, the length of the questionnaire, the geographical dispersion of the sample, call-back procedures, respondent selection rules, and availability of trained staff.[10] Personal interviews are the most expensive to conduct because interviewers must, after being trained, locate and contact subjects, a frequently time-consuming process. For example, some well-established surveys ask interviewers to visit a household and, if the designated subject is not available, make one or more call-backs. National in-person surveys also incur greater administrative costs. Regional supervisory personnel must be hired and survey instruments sent back and forth between the researcher and interviewers. Mail surveys are less expensive but require postage and materials. Electronic surveys (for example, e-mail or Internet) do not necessitate interviewer time but must still be set up by individuals with technical skills. Although mail surveys are thought to be less expensive than telephone polls, Fowler argues that the cost of properly executed mail surveys is likely to be similar to that of phone surveys.[11] Thus, when deciding among personal interviews, telephone interviews, and mail surveys, researchers must consider cost and administrative issues.

Compared with personal interviewing, telephone surveys have several administrative advantages.[12] Despite the cost of long-distance calls, centralization of survey administration is advantageous. Training of telephone interviewers may be easier, and flexible working hours are often attractive to the employees. But the real advantages to telephone surveys begin after interviewing starts. Greater supervision of interviewers and prompt feedback to them is possible. Also, callers can easily inform researchers of any problems they encounter with the survey. Coders can begin coding data immediately. If they discover any errors, they can inform interviewers before a large problem emerges. With proper facilities, interviewers may be able to code respondents' answers directly on computer terminals. In some cases, the whole interview schedule may be computerized, with questions and responses displayed on a screen in front of the interviewer. These are known as computer-assisted telephone interviews (CATIs). The development of computer and telephone technologies gives telephone surveys a significant time advantage over personal interviews and mail surveys. Telephone interviews may be completed and data analyzed almost immediately.[13] On the down side, people tend to be home mostly in the evenings and weekends. But calls made during these hours often meet resistance. This problem used to be aggravated by an explosion in the use of telemarketing; now, however, with the advent of "do-not-call" lists the situation may not as dire as it once was.

Telephone surveys are particularly good for situations in which statistically rare subgroups must be reached or estimated. For example, a telephone survey was used to estimate the disabled population in an area (the only problem being that the number of hearing-impaired people was underestimated).[14] A large sample was required to obtain enough disabled persons for the survey. Where large samples are required, telephone surveys are one-half to one-third the cost of personal interviews.[15] Telephone interviews cut down on the cost of screening the population. In some cases telephone surveys may be used to locate appropriate households, and then the survey itself may be completed by personal interview. Telephone surveys are also best if the research must be conducted in a short period of time. Personal surveys are not as fast, and mail surveys are quite slow.

Almost needless to say, group surveys, those that are distributed to members of groups who might be expected to fill them out at a group meeting or online, and drop-off surveys, questionnaires that are left in public places like libraries, malls, or offices with collection boxes on site, are least expensive. After all, they require minimal administrative and personnel costs to gather the data. On the other hand, the questionnaires still have to be carefully constructed and tested. This is particularly true if the investigator is not in a position to provide guidance or answer questions during survey administration.

COMPLETION RATES. One of the maddening characteristics of many commercial polls is that they often do not indicate how many people refused to take part in the survey. As a typical example, a CBS poll contained the following information: "This poll was conducted among a random sample of 831 adults nationwide, interviewed by telephone March 26–27, 2007." This number, however, most likely refers to the number of complete or nearly complete questionnaires and not to the refusals to participate in the survey in the first place. This information can be important to have.

A completion rate or **response rate** refers to the proportion of persons initially contacted who actually participate. In a mail survey, for instance, the denominator is the total number of questionnaires sent out, not the number returned. Three distinguished researchers, Robert M. Groves, Robert B. Cialdini, and Mick P. Couper, succinctly summarized the significance of this quantity for the social sciences:

> *Among the alternative means of gathering information about society, survey research methods offer unique inferential power to known populations. . . . This power, however, is the cumulative result of many individual decisions of sample persons to participate in the survey. When full participation fails to obtain, the inferential value of the method is threatened.*[16]

We need to explore this point in slightly greater detail. If the response rate is low, either because individuals cannot be reached or because they refuse to participate, the researchers' ability to make statistical inferences for the population being studied may be limited. Also, those who do participate may differ systematically from those who do not, creating other biases. Increasing the size of the survey sample to compensate for low response rates may only increase costs without alleviating the problem.

Most of what we know about response rates comes from studies of personal interview, mail, and telephone surveys. It is difficult, perhaps impossible in some cases, to measure participation levels in electronic and drop-off studies. At one time, response rates were clearly superior for personal interview surveys of the general population than for other types of surveys. Response rates of 80 to 85 percent were often required for federally funded surveys.[17] Higher response rates were not uncommon. By the 1970s, however, response rates for personal interview surveys declined. In 1979 it was reported that in "the central cities of large metropolitan areas the final proportion of respondents that are located *and* consent to an interview is declining to a rate sometimes close to 50 percent." [18]

In general, the decrease in response rates for personal interview surveys has been attributed to both an increased difficulty in contacting respondents and an increased reluctance among the population to participate in surveys. There are more households now in which all adults work outside the home, which makes it difficult for interviewers to get responses. Moreover, pollsters continually worry about public resistance to their craft.[19]

In large cities, nonresponse can be attributed to several additional factors: respondents are less likely to be home or are more likely to be people who do not have a full command of English, or both; interviewers are less likely to enter certain neighborhoods after dark; and security arrangements in multiple-unit apartment buildings make it difficult for interviewers to reach potential respondents. Moreover, many individuals such as undocumented immigrants or people receiving welfare benefits are often skittish about talking to "official looking" strangers. Because of poor working conditions, it is hard to find skilled and experienced interviewers to work in large cities. In smaller cities and towns also, people have shown an increased tendency to refuse to participate in surveys.[20]

Higher refusal rates may be due to greater distrust of strangers and fear of crime as well as to the increased number of polls. For example, in one study of respondents' attitudes toward surveys, about one-third did not believe that survey participation benefited the respondent or influenced government.[21] An equal number thought that too many surveys were conducted and that too many personal questions were asked. Some survey researchers feared that the National Privacy Act, which requires researchers to inform respondents that their participation is voluntary, would lead to more refusals. However, one study found that privacy concerns and past survey experience were more frequent reasons for refusal than was being informed of the voluntary nature of participation.[22]

Some of these findings about why people do not participate in personal interview surveys raise the possibility that survey research of all types may become increasingly difficult to conduct. The increased nonresponse has reduced the advantage of the personal interview over mail and telephone surveys. In fact, Dillman, using his "total design method" for mail and telephone surveys, has achieved response rates rivaling those for personal interviews.[23] He concludes that the chance someone will agree to be surveyed is best for the personal interview but that telephone interviews are now a close second, followed by mail surveys. Other research comparing response rates of telephone and personal interview surveys have also found little difference.[24]

It is often thought that personal interviews can obtain higher response rates because the interviewer can ask neighbors the best time to contact a respondent who is not at home, thus making return visits more efficient and effective. But repeated efforts by interviewers to contact respondents in person are expensive. Much less expensive are repeated telephone calls.

Two norms of telephone usage have contributed to success in contacting respondents by phone and completing telephone interviews.[25] First, most people feel compelled to answer the phone if they are home when it rings. A telephone call represents the potential for a positive social exchange. With the increase in telephone solicitation and surveys, this norm may be revised, however. Caller ID and answering machines can be used to screen and redirect unwanted calls. Telephone surveys may increasingly become prearranged and conducted after contact has been established by some other method.

A second norm of telephone usage is that the initiator should terminate the call. This norm gives the interviewer the opportunity to introduce himself or herself. And in a telephone interview the introductory statement is crucial (see the discussion on motivation below). Because the respondent lacks any visual cues about the caller, there is uncertainty and distrust. Unless the caller can quickly alleviate the respondent's discomfort, the respondent may refuse to finish the interview. For this reason telephone interviews are more likely to be terminated before completion than are personal interviews. It is harder to ask an interviewer to leave than it is simply to hang up the phone.

One advantage of mail surveys is that designated respondents who have changed their address may still be reached, since the postal service forwards mail for about a year. It is not as easy in phone surveys to track down persons who have moved. In personal and telephone interviews it is also harder to change the minds of those who initially refuse to be interviewed, since personal contact is involved and respondents may view repeated requests as harassment.

Recontacts made by mail are less intrusive.[26] Because of the importance attached to high response rates, much research on how to achieve them has been conducted. For example, an introductory letter sent prior to a telephone interview has been found to reduce refusal rates.[27] In fact, such letters may result in response rates that do not differ significantly from those for personal surveys.[28] Researchers have also investigated the best times to find people at home. One study found that for telephone interviews, evening hours are best (6:00 to 6:59, especially), with little variation by day (weekends excluded).[29] Another study concluded that the best times for finding someone at home were late afternoon and early evening during weekdays, although Saturday until four in the afternoon was the best time period overall.[30]

Because mail surveys usually have the poorest response rates, many researchers have investigated ways to increase responses to them.[31] Incentives (money, pens, and other token gifts) have been found to be effective, and prepaid incentives are better than promised incentives. Follow-up, prior contact, type of postage, sponsorship, and title of the person who signs the accompanying letter are also important factors in improving response rates. Telephone calls made prior to mailing a survey may increase response rates by alerting respondents to the survey's arrival. Telephone calls also are a quick method of reminding respondents to complete and return questionnaires. Good follow-up procedures allow a researcher to distinguish between respondents who have replied and those who have not without violating the anonymity of respondents' answers.[32] Generally, mail surveys work best when the population is highly literate and interested in the research problem.[33]

In sum, response rates are an important consideration in survey research. When evaluating research findings based on survey research, you should check the response rate and what measures, if any, were taken to increase it. Should you ever conduct a survey of your own, a wealth of information is available to help you to achieve adequate response rates.

SAMPLE-POPULATION CONGRUENCE. **Sample-population congruence,** which refers to how well the sample subjects represent the population, is always a major concern. Here we are speaking of how well the individuals in a sample represent the population from which they are presumably drawn. Bias can enter either through the initial selection of respondents or through incomplete responses of those who agree to take part in the study. In either case a mismatch exists between the sample and the population of interest. These problems arise to varying degrees in every type of survey.

Some of the cheapest and easiest surveys, such as drop-off or group questionnaires, encounter difficulties in matching sampling frames with the target population, as Figure 10-3 suggests. Suppose, for example, you wanted to survey undergraduates at your college about abortion. One option would be to draw a sample of names and addresses from the student directory. Assuming all currently enrolled students are correctly listed there, the sampling frame (the directory) should closely match the target population, the undergraduate student body. A random sample drawn from the list would presumably be representative. If instead you left a pile of questionnaires in the library, you would have much less control over who responds. Now your "sample" might include graduate students, staff, and outside visitors. It would then be difficult to draw inferences about the student body. One solution would be to add a "filter" question (for example, "Are you a freshman, sophomore, . . . ?", to sort out nonstudents, but the potential for problems can easily be imagined.[34]

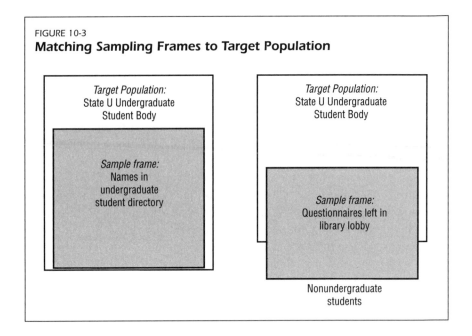

FIGURE 10-3

Matching Sampling Frames to Target Population

Recall from Chapter 7 that when all members of a population are listed, theoretically there is an equal opportunity for all members to be included in the sample. Only rarely, however, are all members included. Personal interviews based on cluster samples in which all the households of the last sampled cluster or area are enumerated and then selected at random, which gives each household an equal chance of being selected, are likely to be more representative than are mail or telephone surveys based on published lists, which are often incomplete.

Telephone surveys attempt to improve the representativeness of samples with a procedure called random digit dialing (the use of randomly generated telephone numbers instead of telephone directories; see Chapter 7) and by correcting for households with more than one number. Thus people who have unlisted numbers or new numbers may be included in the sample. Otherwise, a telephone survey may be biased by the exclusion of these households. Estimates of the number of households in the United States that do not have phones vary from 2 to 10 percent, whereas only about 5 percent of dwelling units are missed with personal interview sampling procedures.[35]

A relatively recent technical innovation has put a wrinkle into telephone interviewing: the rise of cell-phone-only users. During the competitive 2004 presidential election, polling companies and media outlets worried that unless this growing group could be included in their sampling frames, predictions about the outcome could be in error. That is, if interest lies in surveying the general public, and a small but noticeable fraction of people are excluded from the sample frame, bias might creep into the study. Apparently there was some basis for concern. According to a Pew Research Center report, in 2006 about 7 to 9 percent of Americans relied on cell phones instead of landlines or combinations of cell and landline phones. More ominous, this subpopulation differs from the general public in several significant ways. Cell-phone-only individuals tend to be younger, earn less, and hold somewhat more "liberal" views on social issues (for example, abortion, gay marriage), compared with the populace as a whole.[36] The study also pointed out that cell-phone-only users are easier to contact than landline users, but doing so is more difficult and expensive, and that responses rates are lower and refusal rates are higher among cell-phone-only users. Nevertheless, by making statistical adjustments, the Pew Center found that substantive conclusions about attitudes on political issues were not greatly distorted. Still, this is an area in need of additional research.

Sometimes researchers make substitutions if respondents cannot or will not participate. Substituting another member of a household may bias results if the survey specifically asks about the respondent rather than the respondent's household, which is why many professional and academic research designs insist that a pre-sampled individual, not a replacement, be interviewed. Substituting another household from the same block for personal surveys might be better than substituting another telephone number, because city blocks tend to be homogeneous, whereas there is no way to estimate the similarity of households reached by telephone. Substitution of respondents in mail surveys may pose a special problem since the researcher cannot control whether the intended respondent or another member of the household completed the questionnaire. The extent of bias introduced by substitution of respondents depends on the nature of the survey.

As mentioned earlier, one of the major reasons for worry about sample-population congruence is the possibility that those who are not included will differ from those who are.[37] Some evidence of this likelihood appears in the literature. African Americans, for example, have been found to be more likely to refuse telephone interviews.[38] Refusals also are more common among older, middle-class persons, urban residents, and westerners.[39]

The amount of bias introduced by nonresponses due to refusal or unavailability varies, depending on the purpose of the study and the explanatory factors stressed by the research. For example, if urbanization was a key explanatory variable and refusals were concentrated in urban areas, the study could misrepresent respondents from urban areas because the urban respondents who agreed to participate could differ systematically from those who refused. The personal interview provides the best direct opportunity to judge the characteristics of those who refuse and to estimate whether their refusals will bias the analysis.[40]

The bottom line is that you should always ascertain how well sample proportions match the population of interest on key demographic variables (for example, age, gender, ethnicity). Nearly every survey analyst takes this first step, even if the results are not always reported.

QUESTIONNAIRE LENGTH. Subject fatigue is always a problem in survey research. Thus, if a survey poses an inordinate number of questions or takes up too much of the respondents' time, the respondents may lose interest or start answering without much thought or care. Or they may get distracted, impatient, or even hostile. And keeping people interested in the research is exactly what is needed.

A survey needs to include enough questions to gather the data necessary for the research topic, but a good rule of thumb is to keep the survey as short as possible. Given that general advice, how many questions are too many? It is almost impossible to give a precise answer. This is why hypotheses have to be stated carefully. A fishing expedition will in all likelihood end up producing little useful information. Getting the length right is another reason why repeated pretesting is crucial. By trying out the questionnaire on the same types of subjects that will be included in the final study, it is possible to head off respondent weariness.

Unless the pool of participants is especially well motivated (see below), three or four dozen items may be the limit. Especially when there is only a limited possibility to interact with subjects (for example, with mail, Internet and e-mail, or drop-off surveys), the number of questions should be kept to a minimum (within the confines of the project's goals, of course). Alternatively, the questionnaire should take less than, say, 40 minutes (and, for phone surveys, much less time). This seems to be about the time needed for surveys conducted by government and academic institutions. A respondent's attention may be held longer in some situations than in others; hence, some questionnaires can be longer. Personal interviews generally permit researchers to ask more questions than do phone or mail surveys, and certainly more than can be asked on dropped-off forms, especially if experienced and personable interviewers do the interviewing. But again, researchers need to experiment before going into the field.

IS THE SAMPLE GOOD ENOUGH?

Survey analysts commonly compare their sample results with known population quantities. They do so by asking a few questions about demographic characteristics such as gender, age, and residency (e.g., urban versus rural). If these quantities are known for the target population, the sample results can be compared with the known quantities and any discrepancies noted. Suppose, for example, that you wanted to survey undergraduates at a college but had limited resources and had to rely on a relatively low-cost class or drop-off procedure (see text). You might ask students in several large introductory courses to participate in your study. Imagine that your questionnaire asks about gender, class (e.g., freshman, sophomore), residency (in-state or out-of-state), and living arrangements (e.g., dorm, off-campus apartment). The registrar or office of institutional research probably makes these data available online or in print. Finding them should not be a major problem. So just compare the sample percentages with those for the entire school. If the sample percentages are low or high in a category, you can factor that into your analysis. This method works for any group or population for which adequate measures of common traits are available.

DATA-PROCESSING ISSUES. Finally, although technology is making data collection, preparation, and analysis easier, data processing remains an important subject. After the surveys have been collected, the answers still have to be tabulated. And that requirement can be costly. Consider a written questionnaire administered to 500 people that contains 50 agree-disagree items plus 10 other questions for a total of 60. The responses will simply be marks on pieces of paper. These data need to be coded (translated) in such a way that a computer can process them. (Usually an "agree" answer would be coded as, say, a "1" and a "disagree" as a "5.") If all of the responses can be given numeric codes, there will be 60×500 or 30,000 bits of data to record. If, however, any of the questions are open ended, with respondents replying in their own words, this information has to be transcribed and coded. The task used to be done laboriously by hand. (Since the late 1990s, software has become available for this purpose, although it can be expensive and requires training to use.[41] And, as might be expected, skeptics wonder if machines can ever really decode verbatim transcripts.) If proper forms have been used, scanners can be put to work. Otherwise, the numbers or codes have to be entered manually. In the days of IBM cards and keypunches these chores were the bane of the survey researcher.

Data-processing costs are a major reason for the adoption of Internet and even telephone surveys (CATIs). In the latter case, an operator uses a monitor and software to guide the interviewer through the questionnaire and record the data. One company puts it this way:

> The most important aspect of a CATI system is that it uses computers to conduct the interviews. Because a computer controls the questionnaire, skip patterns are executed exactly as intended, responses are within range, and there are no missing data. And, because answers are entered directly into the computer, data entry is eliminated—data analysis can start immediately.[42]

The software also dials the numbers, records no-answers, and handles many other administrative details.

Response Quality

As we said at the outset, it is easy to take respondents' answers at face value. But doing so can be a mistake. A mere checkmark next to "approve" may or may not indicate a person's true feelings. Also, political scientists and other specialists often forget that not everyone shares their enthusiasm or knowledge of current events. What is exciting to one person may bore another. If you are a political science major, terms like *party identification, World Trade Organization, Senate filibuster,* and *Kyoto Treaty* may have a clear meaning. But the public as a whole may not be nearly as familiar with them.

Or they may be aware of issues such as *global warming* and *greenhouse gases* but not comprehend *volatile organic compounds* or CO_2 *emissions.* Nor will they always understand a question the way you do. You may know that *Roe v. Wade* invalidated state laws outlawing abortion, but a person in the street may only be aware that a controversial Supreme Court decision "legalized" abortion. Asking about *Roe v. Wade* might produce too many "don't know" or "no opinion" responses. Equally important, people may be reluctant to express their opinions with strangers, even if they can do so anonymously, or they may view social research as trivial or a waste of time. Finally, everyone seems to be busy; what is in your mind a short interview may be a major interference in someone else's busy life. All of these factors may affect the quality of the data obtained through a survey or an interview.

These observations lead to some important guidelines. You can apply them in your own research and, more important, should be on the lookout for how others handle (or do not handle) them.

- Motivate respondents. Good survey researchers try hard not just to induce people to take part in their studies, but to do so as enthusiastically as possible. They want more than perfunctory responses; they hope participants will be careful and thoughtful.

- Always pre-test a questionnaire with the types of respondents to be included in the study, not just your friends or colleagues. Find out ahead of time what works and what doesn't.

- Be neat, organized, professional, and courteous.

- If you are using interviewers, train them especially in the skill and art of putting subjects at ease and probing and clarifying. The more experience they have, the better. Make sure they don't betray any political, ethnic, gender, age, class, or other biases that would affect the truthfulness of responses.

- Have reasonable expectations. It is not possible to conduct "the perfect study." As desirable as a personal or mail survey may be, it may not feasible. So think about adopting an alternative and making it as rigorous as your resources allow. Regardless of the choice, keep in mind that some types of surveys have advantages over others in regard to response quality.

These guidelines pertain to **response quality,** which refers to the extent to which responses provide accurate and complete information. It is the key to making valid inferences. Response quality depends on several factors, including the respondents' motivations, their ability to understand and follow directions, their relationship with the interviewer and sponsoring organization, and, most important, the quality of the questions being asked. Indeed, this last point is so important that we discuss it in a separate section.

ENGAGING RESPONDENTS.   To engage respondents, it is important to get off on a good footing by introducing yourself, your organization, your purpose, your appreciation of their time and trouble, your nonpartisanship, your awareness of the importance of anonymity, and your willingness to share your findings. Here, for example, is how the *Washington Post* began one of its telephone surveys:

> *Hello, I'm (NAME), calling for the* Washington Post *public opinion poll. We're not selling anything, just doing an opinion poll on interesting subjects for the news.*[43]

This introduction is short and businesslike but friendly. The interviewers have no doubt rehearsed the message countless times so that they can repeat it with confidence and professionalism.

In general, interviewers are expected to motivate the respondents. Generally it has been thought that warm, friendly interviewers who develop a good rapport with respondents motivate them to give quality answers and to complete the survey. Yet some research has questioned the importance of rapport.[44] Friendly, neutral, "rapport-style" interviews in which interviewers give only positive feedback no matter what the response may not be good enough, especially if the questions involve difficult reporting tasks. Both types of feedback—positive ("yes, that's the kind of information we want") and negative ("that's only two things")—may improve response quality. Interviewers also may need to instruct respondents about how to provide complete and accurate information. This more businesslike, task-oriented style has been found to lead to better reporting than rapport-style interviewing.[45]

Interviewer style appears to make less difference in telephone interviews, perhaps because of the lack of visual cues for the respondent to judge the interviewer's sincerity.[46] Even something as simple as intonation, however, may affect data quality. Interviewers whose voices go up rather than down at the end of a question appear to motivate a respondent's interest in reporting.[47]

Despite the advantages of using interviewers to improve response quality, the interviewer-respondent interaction may also bias a respondent's answers. The interviewer may give a respondent the impression that certain answers are expected or are correct. For example, interviewers who anticipate difficulties in persuading respondents to respond or to report sensitive behavior have been found to obtain lower response and reporting rates.[48] The age, sex, or race of the interviewer may affect the respondent's willingness to give honest answers. For example, on questions about race, respondents interviewed by a member of another race have been found to be more deferential to the interviewer (that is, trying harder not to cause offense) than those interviewed by a member of their own race.[49] Education also has an impact on race-of-interviewer effects: less-educated blacks are more deferential than better-educated blacks, and better-educated whites are more deferential than less-educated whites.[50]

Interviewer bias, which occurs when the interviewer influences the respondent's answers, may have a larger effect on telephone surveys than in-person surveys.[51] Because of its efficiency, and because telephone interviewers, even for national surveys, do not need to be geographically dispersed, telephone interviewing requires fewer interviewers than does personal interviewing to complete the same number of interviews. Centralization of telephone interviewing operations, however, allows closer supervision and monitoring of interviewers, making it easier to identify and control interviewer problems. For both personal and telephone interviewers, training and practice is an essential part of the research process.

PROBING.   As just noted, politics is not on the top of everyone's mind. Consequently, it is often necessary to tease out responses. An interviewer can probe for additional information or clarification. He or she can gently encourage the respondent to think a bit or add more information rather than just provide an off-the-cuff answer. For example, suppose you want to know how people feel about presidential candidates. You could, as many polls do, list a number of qualities or characteristics that respondents can apply to the choices. But this technique assumes that you know what people are thinking. By contrast, if you simply ask a subject, "What do you think about Candidate X?", the first reply is often something like "Hmm...not much" or "She's a jerk." This may or may not be a true feeling, and in all likelihood it is not complete. Often, however, people pause a moment before responding. A trained interviewer waits a short while for the person to gather his or her thoughts. If the answer is not totally clear, the interviewer can ask for clarification. This is how the American National Election Studies, a series of major academic surveys, handle the problem. The lead-in begins, "Now I'd like to ask you about the good and bad points of the major candidates for President. . . . Is there anything in particular about [name of candidate] that might make you want to vote for him?" The questionnaire then reads:

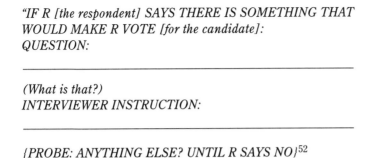

*"IF R [the respondent] SAYS THERE IS SOMETHING THAT
WOULD MAKE R VOTE [for the candidate]:
QUESTION:*

_____

*(What is that?)
INTERVIEWER INSTRUCTION:*

_____

{PROBE: ANYTHING ELSE? UNTIL R SAYS NO}[52]

Survey Type and Response Quality

The ability to obtain quality responses differs according to the type of survey used. A drawback of mail and drop-off surveys is that an interviewer is not able to probe for additional information or clarification, discussed in the previous section. Nevertheless, mail surveys may have an advantage in obtaining truthful answers to threatening or embarrassing questions because anonymity can be assured and answers given in private. A mail survey also gives the respondent enough time to finish when it is convenient; this enables the respondent to check records to provide accurate information, something that is harder to arrange in telephone and personal interviews.

Disadvantages of the mail survey include problems with open-ended questions. (As we see in the next section, an open-ended question asks for the respondent's own words, as in "Is there anything in particular that you like about the Republican Party?" The respondent can say whatever he or she thinks.) Some respondents may lack writing skills or find answering at length a burden. No interviewer is present to probe for more information, to clarify complex or confusing questions, to motivate the respondent to answer tedious or boring questions, or to control who else may contribute to or influence answers.

Personal and telephone interviews share many advantages and disadvantages with respect to obtaining quality responses, although some important differences exist. Several of the advantages of personal and telephone interviews over mail surveys stem from the presence of an interviewer. As noted earlier, an interviewer may obtain better-quality data by explaining questions, by probing for more information to open-ended questions, and by making observations about the respondent and his or her environment. For example, in a personal interview, the quality of furnishings and housing may be an indicator of income, and in a telephone interview, the amount of background noise might affect the respondent's concentration. In a personal interview, the interviewer may note that another household member is influencing a respondent's answers and take steps to avoid it. Influence by others is generally not a problem with telephone interviews, since only the respondent hears the questions. One response quality problem that does occur with telephone interviews is that the respondent may not be giving the interviewer his or her undivided attention. This may be difficult for the interviewer to detect and correct.

Numerous studies have compared response quality for personal and telephone interviews. One expected difference is in answers to open-ended questions. Telephone interviewers lack visual cues for probing. Thus telephone interviews tend to be quick paced; pausing to see if the respondent adds more to an answer is more awkward on the telephone than in person. Research findings, however, have been mixed. One study found that shorter answers were given to open-ended questions in telephone interviews, especially among respondents who typically give complete and detailed responses; another study found no difference between personal and telephone interviews in the number of responses to open-ended questions.[53] Asking an open-ended question early in a telephone survey helps to relax the respondent, reduce the pace of the interview, and ensure that the respondent is thinking about his or her answers.[54]

Response quality may be lower for telephone interviews than for face-to-face interviews because of the difficulty of asking complex questions or questions with many response categories over the phone. Research has found more acquiescence, evasiveness, and extremeness in telephone survey responses than in personal survey responses. In addition, phone respondents give more contradictory answers to checklist items and are less likely to admit to problems.[55] This finding contradicts the expectation that telephone interviews result in more accurate answers to sensitive questions because of reduced personal contact.

As we mentioned earlier, one advantage attributed to mail questionnaires is greater privacy for the respondent in answering sensitive questions. As a result, researchers using personal and telephone interviews have developed techniques to obtain more accurate data on sensitive topics.[56]

Problems often can be avoided simply by careful wording choice. For example, for questions about socially desirable behavior, a casual approach reduces the threat by lessening the perceived importance of the topic. The question "For whom did you vote in the last election?" could inadvertently stigmatize nonvoting. Here, once more, is how the American National Election Studies put the question:

> *In talking to people about elections, we often find that a lot of people were not able to vote because they weren't registered, they were sick, or they just didn't have time. How about you—did you vote in the elections this November?*

In other words, giving respondents reasons for not doing something perceived as socially desirable reduces threat and may cut down on overreports of the behavior.

A different approach to the problem of obtaining accurate answers to sensitive questions is the **randomized response technique** (RRT).[57] Examples might be, "Have you ever had an abortion?" or "Have you used marijuana in the past month?" or "Do you believe gays should have the same legal rights as other people?" The RRT technique allows respondents to answer these kinds of sensitive questions truthfully without the interviewer's knowing the question being answered. For example, the interviewer gives the respondent a card with two questions, one sensitive and one not sensitive. A device such as a coin or a box with two colors of beads is used to randomly determine which question the respondent will answer. If a coin is used, the respondent will be instructed to answer one question if the result is a heads and the other if tails shows up. The respondent flips the coin and, without showing the interviewer the outcome of the toss, answers the appropriate question. One can then use probability theory to estimate the distribution of sensitive responses (see the "How It's Done" box).

The accuracy of RRT depends on the assumption that a respondent will answer both the sensitive and nonsensitive questions truthfully. The success in obtaining accurate information depends on the respondent's ability to understand the method and follow instructions and his or her belief that the random-choice device is not rigged.[58] The technique seems to work better when the nonsensitive question deals with a socially positive activity, further reducing the stigma attached to a yes response.[59]

Research has found RRT to be superior to other methods of asking threatening questions, such as having the respondent answer a direct question and return it in a sealed envelope.[60] For example, use of RRT produced higher estimates of abortion than previous measures.[61] RRT can be used for telephone as well as personal interviews. Random-choice devices can be supplied by the respondent—thus eliminating suspicion that the device is fixed—or they can be mailed by the researcher to the respondent.[62] However, since the method can be expensive and cumbersome, it is not widely used in public opinion research. In any event, careful question wording and the use of experienced interviewers can alleviate some of these difficulties.

Question Wording

The central problem of survey and interview research is that the procedures involve a structured interaction between a social scientist and a subject. This is true even if the method used involves indirect contact such as a mail or an Internet survey. Since the whole point of survey research is to accurately

How It's Done

RANDOMIZED RESPONSE TECHNIQUE

To calculate the proportion of yes answers to the sensitive question, the expected proportion of yes answers to the nonsensitive question must be known. Thus the nonsensitive question could be "Were you born in July?" Assuming that birthdays are distributed equally among the months, one-twelfth of the respondents would be expected to say yes to the nonsensitive question. Or the proportion of persons exhibiting the nonsensitive behavior could be estimated by asking a sample population a direct question about the nonsensitive behavior. The proportion of respondents answering yes to the sensitive question can be estimated using the formula

$$R_{yes} = P(S_{yes}) + (1 - P)(N_{yes}),$$

where

R_{yes} = probability of obtaining yes answer to random question;
P = probability of respondent choosing sensitive question;
$1 - P$ = probability of respondent choosing nonsensitive question;
S_{yes} = proportion of respondents exhibiting sensitive behavior;
N_{yes} = known proportion of respondents exhibiting nonsensitive behavior.

Therefore

$$S_{yes} = R_{yes} - (1 - P)(N_{yes})/P.$$

Let's assume that out of 1,000 respondents, we get 500 yes responses and that we have estimates showing that 80 percent of our sample should answer yes to the nonsensitive question. If a balanced coin is used, P equals .5. Making the substitutions, we get

$$S_{yes} = 500/1000 - (1 - .5)(.80)/.5 = .5 - .5(.80)/.5 = .20 \text{ or } 20\%.$$

measure people's attitudes, beliefs, and behavior by asking them questions, we need to spend time discussing good and bad questions. Good questions prompt accurate answers; bad questions provide inappropriate stimuli and result in unreliable or inaccurate responses. When writing questions, researchers should use objective and clear wording. Failure to do so may result in incomplete questionnaires and meaningless data for the researcher. The basic rule is this: *the target subjects must be able to understand and in principle have access to the requested information.* Try to put yourself in the respondent's place. Would an ordinary citizen, for example, be able to reply meaningfully to the question "What's your opinion of the recently passed amendments to the Import/Export Bank Authorization?"

WHY IT MATTERS. Before proceeding with an examination of the do's and don'ts of question wording, we present a substantive example of the importance of question wording.

TABLE 10-2
Question Wording Matters: Responses to "Death with Dignity" Questions (percent)

Question	Favor/ Allow/ Yes	Oppose/ Do not Allow/No	Depends/ Other
1. "Would you favor or oppose withdrawing life support systems, including food and water, from hopelessly ill or irreversibly comatose patients if they or their families request it?"	76	15	10
2. "Do you think the law should allow doctors to honor the written instructions of their patients, even if it means allowing them to die?"	81	8	11
3. "Do you think the law should allow doctors to administer drugs that might shorten the patient's life in severe cases with no hope of recovery, if the patient requested the drug and understands the consequences?"	76	21	3
4. "When a person has a disease that cannot be cured, do you think doctors should be allowed by law to end the patient's life by some painless means if the patient and his family requests it?"	70	25	5
5. "When a person has a disease that cannot be cured, do you think doctors should be allowed by law to assist the patient to commit suicide if the patient requests it, or not?"	52	42	6
6. "If someone is terminally ill, is in great pain and wants to kill themselves, should it be legal for a doctor to help them to commit suicide or not?"	50	47	3
7. "If you were seriously ill with a terminal disease, would you consider suicide, or not?"	32	59	9
8. "If a terminally ill relative or close friend asked you to help him or her commit suicide to end his or her suffering, do you think you would do it, or not?	14	72	14

Source: John M. Benson, "The Polls—Trends," *Public Opinion Quarterly* 63 (Summer 1999): 263–277. Reproduced with permission of Oxford University Press via Copyright Clearance Center.

Note: These questions come from polls conducted by various organizations. All involved national samples of about 1,000 individuals and were conducted during roughly the same period, the early 1990s.

The chapter began by mentioning how groups use polls to advance their causes. The particular issue discussed, end-of-life legislation, travels under various names in the media: "death with dignity," "physician-assisted suicide," "the right to die," for example. And herein lies the problem. As noted earlier, people know things by different names and descriptions. So question wording is fundamental when asking people for their opinions. John Benson's review of public opinion polls shows why.[63] Table 10-2 lists several questions drawn from various surveys. Each pertains to the end-of-life issue but approaches it (via the choice of wording) from a different perspective. For instance, it appears that the public backs the idea of doctors helping people end their lives when the patients are terminally ill and the families are in agreement. (See the first four questions.) But notice that the minute the words *suicide* or *kill themselves* appear, approval drops precipitously. (See questions 5 and 6.) Finally, when asked if they would commit suicide or help someone do so, even in the case of terminal illness, the overwhelming majority say no!

The upshot, then, is that how one frames an issue partly determines the public's expressed or verbalized stances on it. In this case, although people do not favor suicide, they seem sympathetic to ending irreversible pain and suffering. Advocacy groups of course have to take into account this phenomenon, but as social scientists we have to be aware of it as well. The following sections describe in more detail some considerations when writing or evaluating questionnaires.

OBJECTIVITY AND CLARITY.   Certain types of questions make it difficult for respondents to provide reliable, accurate responses. These include double-barreled, ambiguous, and leading questions. A **double-barreled question** is really two questions in one, such as "Do you agree with the statement that the situation in Iraq is deteriorating and that the United States should increase the number of troops in Iraq?" How does a person who believes that the situation in Iraq is deteriorating but who does not wish an increase in troops answer this question? Or someone who doesn't feel the situation is worse but nevertheless believes that more troops would be advisable? And how does the researcher interpret an answer to such a question? It is not clear whether the respondent meant for his or her answer to apply to both components or whether one component was given precedence over the other.

Despite a conscious effort by researchers to define and clarify concepts, words with multiple meanings or interpretations may creep into questions. An **ambiguous question** is one that contains a concept that is not defined clearly. An example would be the question "What is your income?" Is the question asking for family income or just the personal income of the respondent? Is the question asking for earned income (salary or wages), or should interest and stock dividends be included? Similarly, the question "Do you prefer Brand A or Brand B?" is ambiguous. Is the respondent telling us which brand is purchased or which brand would be purchased if there were no price difference between the brands?

Ambiguity also may result from using the word *he*. Are respondents to assume that *he* is being used generically to refer both to males and females or to males only? If a respondent interprets the question as applying only to males and would respond differently for females, it would be a mistake for the researcher to conclude that the response applies to all people.[64]

After the 1976 debate between presidential contenders Jimmy Carter and Gerald Ford, respondents to one poll were asked to rate the performance of the two debaters as good, bad, or indifferent. This ambiguous question confused respondents, who then asked whether the ratings were to be made (1) by comparing the debate with other, unspecified political debates that they had witnessed, (2) by comparing the debate with the one between John F. Kennedy and Richard Nixon, (3) by comparing Carter and Ford with each other, or (4) by measuring the candidates against the respondents' pre-debate expectations. Respondents also were unsure whether performance meant style, substance, or something else.[65]

Researchers must avoid asking leading questions. A **leading question,** sometimes called a reactive question, encourages respondents to choose a particular response because the question indicates that the researcher expects it. The question "Don't you think that global warming is a serious environmental problem?" implies that to think otherwise would be unusual. Word choice may also lead respondents. Research has shown that people are more willing to help "the needy" than those "on welfare." Asking people if they favor "socialized medicine" rather than "national health insurance" is bound to decrease affirmative responses. Moreover, linking personalities or institutions to issues can affect responses. For example, whether or not a person liked the governor would affect responses to the following question: "Would you say that Governor Burnett's program for promoting economic development has been very effective, fairly effective, not too effective, or not effective at all?"[66] There are numerous other ways to lead the respondent, such as by characterizing one response as the preference of others, thereby creating an atmosphere that is anything but neutral.[67]

Polls conducted by political organizations and politicians often include leading questions. For example, a 1980 poll for the Republican National Committee asked, "Recently the Soviet armed forces openly invaded the independent country of Afghanistan. Do you think the U.S. should supply military equipment to the rebel freedom fighters?"[68] Before accepting any interpretation of survey responses, we should check the full text of a question to make sure that it is neither leading nor biased.

Indeed, some campaigns, parties, and political organizations have begun converting survey research into a form of telemarketing through a technique called **push polls.** Interviewers, supposedly representing a research organization, feed respondents (often) false and damaging information about a candidate or cause under the guise of asking a question. The caller may ask, for example, "Do you agree or disagree with Candidate X's willingness to tolerate terrorism in our country?" The goal, of course, is not to conduct research but to use innuendo to spread rumors and lies.

Questions should be stated in such a way that they can produce a variety of responses. If you simply ask, "Do you favor cleaning up the environment—yes or no?", almost all the responses will surely be yes in the yes category. At the same time, the alternatives themselves should encourage thoughtful replies. For instance, if the responses to the question "How would you rate President Bush's performance so far?" are "(1) great," "(2) somewhere between great and terrible," and "(3) terrible," you probably are not going to find very much variation, since the list practically demands that respondents pick choice (2). Also, an alternative should be available for each possible situation. For example, response options for the question "For whom did you vote in the 2004 presidential election?" should list George W. Bush and John Kerry, as well as other candidates (for example, Ralph Nader) and certainly should include generic "other" and "did not vote" options. (The section on "question type" discusses this topic in more depth.)

Other kinds of bias can creep into questionnaires. As Margrit Eichler demonstrates, gender bias may occur if, for example, a respondent is asked to agree or disagree with the following statement: "It is generally better to have a man at the head of a department composed of both men and women employees." This wording does not allow the respondent to indicate that a woman is preferable as the department head. It would be better to rephrase the statement as "What do you think is generally better: To have a woman or a man at the head of a department that is composed of both men and women employees?" and to give respondents the opportunity to indicate that a man is better, a woman is better, or there is no difference.[69]

Use of technical words, slang, and unusual vocabulary should be avoided, since their meaning may be misinterpreted by respondents. Questions including words with several meanings will result in ambiguous answers. For example, the answer to the question "How much bread do you have?" depends on whether the respondent thinks of bread as coming in loaves or dollar bills. The use of appropriate wording is especially important in cross-cultural research. For researchers to compare answers across cultures, questions should be equivalent in meaning. For example, the question "Are you interested in politics?" may be interpreted as "Do you vote in elections?" or "Do you belong to a political party?" The interpretation would depend on the country or culture of the respondent.

Questions should be personal and relevant to the respondent. For example, in a questionnaire on abortion, the question "Have you ever had an abortion?" could be changed to "Have you [your wife, girlfriend] ever had an abortion?" This will permit the researcher to include the responses of men as well as women.

Attention to these basic guidelines for question wording increases the probability that respondents will interpret a question consistently and as intended, yielding reliable and valid responses. Luckily every researcher does not have to formulate questions anew. We discuss archival sources of survey questions later in this chapter.

Question Type

The form or type of question as well as its specific wording is important. There are two basic types of questions: closed ended and open ended. A **closed-ended question** provides respondents with a list of responses from which to choose. "Do you agree or disagree with the statement that the government ought to do more to help farmers?" and "Do you think that penalties for drunk driving are too severe, too lenient, or just about right?" are examples of closed-ended questions.

A variation of the closed-ended question is a question with multiple choices for the respondent to accept or reject. A question with multiple choices is really a series of closed-ended questions. Consider the following example: "Numerous strategies have been proposed concerning the federal budget deficit. Please indicate whether you find the following alternatives acceptable or unacceptable: (a) raise income taxes, (b) adopt a national sales tax, (c) reduce military spending, (d) reduce spending on domestic programs."

In an **open-ended question,** the respondent is not provided with any answers from which to choose. The respondent or interviewer writes down the answer. An example of an open-ended question is, "Now I'd like to ask you about the good and bad points of the major candidates for president. Is there anything in particular about MR. KERRY that might make you want to vote FOR him?" [70]

CLOSED-ENDED QUESTIONS: ADVANTAGES AND DISADVANTAGES.   The main advantage of a closed-ended question is that it is easy to answer and takes little time. Also, the answers can be pre-coded (that is, assigned a number) and the code then easily transferred from the questionnaire to a computer. Another advantage is that answers are easy to compare, since all responses fall into a fixed number of predetermined categories. These advantages aid in the quick statistical analysis of data. In open-ended questions, by contrast, the researcher must read each answer, decide which answers are equivalent, decide how many categories or different types of answers to code, and assign codes before the data can be computerized.

Another advantage of closed-ended questions over open-ended ones is that respondents are usually willing to respond on personal or sensitive topics (for example, income, age, frequency of sexual activity, or political views) by choosing a category rather than stating the actual answer. This is especially true if the answer categories include ranges. Finally, closed-ended questions may help clarify the question for the respondent, thus avoiding misinterpretations of the question and unusable answers for the researcher.

Critics of closed-ended questions charge that they force a respondent to choose an answer category that may not accurately represent his or her position. Therefore, the response has less meaning and is less useful to the researcher. Also, closed-ended questions often are phrased so that a respondent must choose between two alternatives or state which one is preferred. This may result in an oversimplified and distorted picture of public opinion. However, a closed-ended question allowing respondents to pick more than one response (for example, with instructions to choose all responses that apply) may be more appropriate in some situations. The information produced by such a question indicates which choices are acceptable to a majority of respondents. In fashioning a policy that is acceptable to most people, policymakers may find this knowledge much more useful than simply knowing which alternative a respondent prefers.

Just as the wording of a question may influence responses, so too may the wording of response choices. Changes in the wording of question responses can result in different response distributions. Two questions from the 1960s concerning troop withdrawal from Vietnam illustrate this problem.[71] A June 1969 Gallup Poll question asked,

> *President Nixon has ordered the withdrawal of 25,000 United States troops from Vietnam in the next three months. How do you feel about this—do you think troops should be withdrawn at a faster rate or a slower rate?*

The answer "same as now" was not presented but was accepted if given. The response distribution was as follows: faster, 42 percent; same as now, 29 percent; slower, 16 percent; no opinion, 13 percent. Compare the responses with those to a September–October 1969 Harris Poll in which respondents were asked,

> *In general, do you feel the pace at which the president is withdrawing troops is too fast, too slow, or about right?*

Responses to this question were as follows: too slow, 28 percent; about right, 49 percent; too fast, 6 percent; no opinion, 18 percent.

Thus support for presidential plans varied from 29 to 49 percent. The response depended on whether respondents were directly given the choice of agreeing with presidential policy or had to mention such a response spontaneously.

Response categories may also contain leading or biased language and may not provide respondents with equal opportunities to agree or disagree. Response distributions may be affected by whether the researcher asks a **single-sided question,** in which the respondent is asked to agree or disagree with a single substantive statement, or a **two-sided question,** which offers the respondent two substantive choices. An example of a one-sided question is

> *Do you agree or disagree with the idea that the government should see to it that every person has a job and a good standard of living?*

An example of a two-sided question is

Do you think that the government should see to it that every person has a job and a good standard of living, or should it let each person get ahead on his or her own?

With a single-sided question a larger percentage of respondents tend to agree with the statement given. Forty-four percent of the respondents to the single-sided question given above agreed that the government should guarantee employment, whereas only 30.3 percent of the respondents to the two-sided question chose this position.[72] Presenting two substantive choices has been found to reduce the proportion of respondents who give no opinion.[73]

Closed-ended questions may provide inappropriate choices, thus leading many respondents to not answer or to choose the "other" category. Unless space is provided to explain "other" (which then makes the question resemble an open-ended one), it is anybody's guess what "other" means. Another problem is that errors may enter into the data if the wrong response code is marked. With no written answer, inadvertent errors cannot be checked. A problem also arises with questions having a great many possible answers. It is time consuming to have an interviewer read a long list of fixed responses that the respondent may forget. A solution to this problem is to use a response card. Responses are typed on a card that is handed to the respondent to read and choose from.

OPEN-ENDED QUESTIONS: ADVANTAGES AND DISADVANTAGES. Unstructured, free-response questions allow respondents to state what they know and think. They are not forced to choose between fixed responses that do not apply. Open-ended questions allow respondents to tell the researcher how they define a complex issue or concept. As one survey researcher in favor of open-ended questions points out, "Presumably, although this is often forgotten, the main purpose of an interview, the most important goal of the entire survey profession, is to let the respondent have his say, to let him tell the researcher what he means, not vice versa. If we do not let the respondent have his say, why bother to interview him at all?"[74]

Sometimes researchers are unable to specify in advance the likely responses to a question. In this situation, an open-ended question is appropriate. Open-ended questions are also appropriate if the researcher is trying to test the knowledge of respondents. For example, respondents are better able to *recognize* names of candidates in a closed-ended question (that is, pick the candidates from a list of names) than they are able to *recall* names in response to an open-ended question about candidates. Using only one question or the other would yield an incomplete picture of citizens' awareness of candidates.

Paradoxically, a disadvantage of the open-ended question is that respondents may respond too much or too little. Some may reply at great length about an issue—a time-consuming and costly problem for the researcher. If open-ended questions are included on mail surveys, some respondents with poor writing skills may not answer. This may bias responses. Thus the use of open-ended questions depends on the type of survey. Another problem is that interviewers may err in recording a respondent's answer. Recording answers verbatim is tedious. Furthermore, unstructured answers may be difficult to code, interpretations of answers may vary (affecting the reliability of data), and processing answers may become time consuming and costly. For these reasons, open-ended questions are often avoided—although unnecessarily, in Patricia Labaw's opinion:

> *I believe that coding costs have now been transferred into dataprocessing costs. To substitute for open questions, researchers lengthen their questionnaires with endless lists of multiple choice and agree/disagree statements, which are then handled by sophisticated dataprocessing analytical techniques to try to massage some pattern or meaning out of the huge mass of pre-coded and punched data. I have found that a well-written open-ended question can eliminate the need for several closed questions, and that subsequent data analysis becomes clear and easy compared to the obfuscation provided by data massaging.*[75]

Question Order

The order in which questions are presented to respondents may also influence the reliability and validity of answers. Researchers call this the **question-order effect.** In ordering questions, the researcher should consider the effect on the respondent of the previous question, the likelihood of the respondent's completing the questionnaire, and the need to select groups of respondents for certain questions. In many ways, answering a survey is a learning situation, and previous questions can be expected to influence subsequent answers. This presents problems as well as opportunities for the researcher.

The first several questions in a survey are usually designed to break the ice. They are general questions that are easy to answer. Complex, specific questions may cause respondents to terminate an interview or not complete a questionnaire because they think it will be too hard. Questions on personal or sensitive topics usually are left to the end. Otherwise, some respondents may suspect that the purpose of the survey is to check up on them rather than to find out public attitudes and activities in general. In some cases, however, it may be important to collect demographic information first. In a study of attitudes toward abortion, one researcher used demographic information to infer the responses of those who terminated the interview. She found that older, low-income women were most likely to terminate the interview on the abortion section. Since their group matched those who completed the interviews and who were strongly opposed to abortion, she concluded that termination expressed opposition to abortion.[76]

One problem to avoid is known as a **response set,** or straight-line responding. A response set may occur when a series of questions have the same answer choices. Respondents who find themselves agreeing with the first several statements may skim over subsequent statements and check "agree" on all. This is likely to happen if statements are on related topics. To avoid the response set phenomenon, statements should be worded so that respondents may agree with the first, disagree with the second, and so on. This way the respondents are forced to read each statement carefully before responding.

Additional question-order effects include saliency, redundancy, consistency, and fatigue.[77] Saliency is the effect that specific mention of an issue in a survey may have in causing a respondent to mention the issue in connection with a later question: the earlier question brings the issue forward in the respondent's mind. For example, a researcher should not be surprised if respondents mention crime as a problem in response to a general question on problems affecting their community, if the survey had earlier asked them about crime in the community. Redundancy is the reverse of saliency. Some respondents, unwilling to repeat themselves, may not say crime is a problem in response to the general query if earlier they had indicated that crime was a problem. Respondents may also strive to appear consistent. An answer to a question may be constrained by an answer given earlier. Finally, fatigue may cause respondents to give perfunctory answers to questions late in the survey. In lengthy questionnaires, response set problems often arise due to fatigue.[78]

Lee Sigelman has used survey research techniques to explore the effects of question order on the results of different presidential popularity polls.[79] He found that placing the presidential popularity item early in the survey elicited more "no opinion" answers than occurred when it was asked toward the end of the interview. This finding is explained by the tendency of respondents to respond in a safe or socially desirable way early in an interview before their critical faculties have been fully engaged and before they begin to trust the interviewer. Since presidential popularity is usually measured in terms of all respondents, including those with no opinions, the percentage of people approving or disapproving of a president will be deflated by early placement of the question.

Another study tested the assumption that specific questions create a saliency effect that influences answers to more general questions.[80] People were found to express significantly more interest in politics and religion when these questions followed specific questions on political and religious matters. However, respondents' evaluations of the seriousness of energy and economic problems were not affected by previous questions about these problems. Perhaps interest is more easily influenced by question order than is evaluation because evaluation questions require more discriminating, concrete responses than do interest questions. The study also suggests that if specific questions about behavior are asked first, they give respondents concrete, behavioral references for answering later on a related, more general question.[81] For example, the answer to "How actively do you engage in sports?" may depend on whether the respondent had first been asked about participation in specific sporting activities.

The learning that takes place during an interview may be an important aspect of the research being conducted. The researcher may intentionally use this process to find out more about the respondent's attitudes and potential behavior. Labaw refers to this as "leading" the respondent and notes it is used "to duplicate the effects of information, communication and education on the respondent in real life."[82] The extent of a respondent's approval or opposition to an issue may be clarified as the interviewer introduces new information about the issue.

In some cases, such education *must* be done to elicit needed information on public opinion. For example, one study set out to evaluate public opinion on ethical issues in biomedical research.[83] Because the public is generally uninformed about these issues, some way had to be devised to enable respondents to make meaningful judgments. The researchers developed a procedure of presenting "research vignettes." Each vignette described or illustrated a dilemma actually encountered in biomedical research. A series of questions asking respondents to make ethical judgments followed each vignette. Such a procedure was felt to provide an appropriate decision-making framework for meaningful, spontaneous answers and a standard stimulus for respondents. A majority of persons, even those with less than a high school education, were able to express meaningful and consistent opinions.

If there is no specific reason for placing questions in a particular order, researchers may vary questions randomly to control question-order bias. Computerized word processing of questionnaires makes this an easier task.[84]

Question order also becomes an important consideration when the researcher uses a **branching question,** which sorts respondents into subgroups and directs these subgroups to different parts of the questionnaire, or a **filter question,** which screens respondents from inappropriate questions. For example, a marketing survey on new car purchases may use a branching question to sort people into several groups: those who bought a car in the past year, those who are contemplating buying a car in the next year, and those who are not anticipating buying a car in the foreseeable future. For each group, a different set of questions about automobile purchasing may be appropriate. A filter question is typically used to prevent the uninformed from answering questions. For example, respondents in the 1980 National Election Study were given a list of presidential candidates and asked to mark those names they had never heard of or didn't know much about. Respondents were then asked questions only about those names that they hadn't marked.

Branching and filter questions increase the chances for interviewer and respondent error.[85] Questions to be answered by all respondents may be missed. However, careful attention to questionnaire layout, clear instructions to the interviewer and the respondent, and well-ordered questions will minimize the possibility of confusion and lost or inappropriate information.

Questionnaire Design

The term **questionnaire design** refers to the physical layout and packaging of the questionnaire. An important goal of questionnaire design is to make the questionnaire attractive and easy for the interviewer and the respondent to follow. Good design increases the likelihood that the questionnaire will be completed properly. Design may also make the transfer of data from the questionnaire to the computer easier.

Design considerations are most important for mail questionnaires. First, the researcher must make a favorable impression based almost entirely on the questionnaire materials mailed to the respondent. Second, because no interviewer is present to explain the questionnaire to the respondent, a mail questionnaire must be self-explanatory. Poor design increases the likelihood of response error and nonresponse. Whereas telephone and personal interviewers can and should familiarize themselves with questionnaires before administering them to a respondent, the recipient of a mail questionnaire cannot be expected to spend much time trying to figure out a poorly designed form.

Using Archived Surveys

Now that you have a better idea of what surveys are and how they are properly designed and administered, it is important to understand the costs and benefits of designing and administering your own survey to collect data versus using survey questions written by or data collected by someone else. Because of the high costs involved in designing and administering a survey and the concerns about validity and reliability of the data, most students who want to design their own survey would be remiss if they did not at least consult existing surveys. In this section we explain how you can search for survey questions and data in archives and what you can expect to find.

Advantages of Using Archived Surveys

As explained previously, designing and administering surveys can be a complex and costly venture. Although some students can rely on funding from their universities for research (usually a couple hundred dollars to defray costs), most students will not have access to such funds for research projects. Without funding, most students wishing to collect survey data to analyze in a research paper will turn to the least expensive options available. The most popular source of survey data for students is a sample of undergraduate students. These efforts generally involve less expensive collection measures like in-class group surveys, or surveys conducted via e-mail or the Internet. Given the limitations on available resources, these are acceptable choices and students designing and administering their own surveys will undoubtedly learn a great deal about the pitfalls of survey research when choosing this option. This type of firsthand experience can be invaluable in fully understanding survey design and administration, and the experience cannot be replicated simply by reading a textbook on the subject.

One of the biggest drawbacks of students designing and administering their own surveys is that the questions, the survey form, and the administration will likely be of quite low quality without considerable input from an instructor or advisor. Although this is not a problem if a survey is intended to be a learning exercise, students hoping to collect high-quality data from their own survey might be disappointed with the results. To be able to make valid conclusions based on the data, students should consider instead using survey questions written by professionals. Such questions are widely available for free to students through publicly available archives. In Chapter 8 we discussed the availability of pre-processed or pre-analyzed data in regard to document analysis. The data to which we referred are useful to students because someone else has already worked with the raw data, or answers to survey questions, and produced results in the form of tables, figures, or statistical output. In this chapter, we focus instead on the survey questions and the raw survey data, or the unvarnished answers to survey questions. These data can be more difficult to work with, but they will lend greater flexibility to students in analyzing data and making conclusions.

There are many advantages to drawing upon professional surveys for use in your own. First and foremost, imagine that you are ready to embark on a research project for which you will need survey data. You would be safe in assuming that using data someone else collected would save a great deal of time, effort, and resources. The key, of course, is finding data that will allow you to test your hypotheses and answer your research questions. Fortunately, myriad data archives are publicly available, with surveys and sample data collected from many different populations and about many topics. Second, using a professionally designed survey should lead to better data, collected from answers to well-written questions. Having taken a data analysis course, and read this chapter, students should have a good idea of what to look for in a survey design to determine the quality of the questions and, subsequently, the data. Third, using a professional survey can help convince readers that the results reported in a research report are valid because the questions used to collect the data have been used by others, and potentially used in published, peer reviewed work.

Publicly Available Archives

For archived surveys and data, it is best to start with sources available through your school library before attempting your own general search. Searching for surveys should be done carefully. Using a search engine like Google or Yahoo to locate survey questions and data is likely to be a difficult process; you will find thousands of Web sites with information or data that are useless for your purpose. In addition, unlike finding an article published on the Internet, which represents just one piece of information in a larger project, relying on someone else's survey questions or data should be done very carefully because the data will make up a critical part of the research project. It is much better to begin with a trusted source for survey questions and data. For example, the University of Michigan library maintains a Web page of useful data sources for political science (www.lib.umich.edu/govdocs/stpolisc.html). Although the page includes links to many different kinds of data, for this chapter we are most interested in the section on public opinion. The section includes links to over thirty Web sites that feature useful survey questions and data. When visiting the University of Michigan page, you will notice that some of the links are active only for University of Michigan students, but you might be able to access the sites through your own library's subscription. Whether you use the University of Michigan's page as a starting point, or a page at your own library, you will find three different sources of surveys: archives, organizations, and individuals' Web sites.

The Inter-university Consortium for Political and Social Research (ICPSR) or the Public Opinion Poll Question Database (POPQD) are archives maintained by major research universities, the University of Michigan and the University of North Carolina, respectively, that offer a large number of data sets collected by different individuals and organizations. The value of these archives is that you are likely to find a variety of surveys and data covering a multitude of topics. Organizations like the Pew Research Center or Polling Report.com offer a more narrow selection of surveys, but many of the included surveys have been using the same questions for years and allow analysis over time. Finally, you will also find individuals' Web pages that archive their own data, or have links to useful data. The best individual Web pages are those maintained by survey professionals such as university professors or employees of survey research firms.

FIGURE 10-4
Search Tool on the Roper Center Site

ROPER
CENTER
PUBLIC OPINION ARCHIVES
Providing access to the world's voice since 1947

[Search]
⊙ Site   ○ Datasets   Advanced Search

University of
Connecticut

| QUICK LINKS | DATA ACCESS | MEMBERSHIP | EDUCATION | RESEARCH | ABOUT THE CENTER |

? ✳ 📖

Search for Datasets

Browse the Center's on-line catalog of studies. Researchers interested in secondary analysis of survey data files can determine studies relevant to their areas of interest. Opinion **data are available to members using the** Roper*Express* service. Non-members may contact the Center to place orders.

Search   [   ]   [Search] [Reset]
                                                        Search Help

Date Range   [   ] to [   ]

Country   [All Countries   ▾]

Survey Organization   [   ]

Type of Sample   [   ]

☐ Search Roper*Express* Studies Only

© 2007 Roper Center, University of Connecticut. All Rights Reserved.

Source: The Roper Center public opinion data archives; Catalogue of Dataset Holdings. Reprinted with permission.

As an example, imagine that you are looking for survey questions to use in a research project on public opinions about abortion among college students, and that you will be designing a survey to administer to a random sample of students at your college or university. You could start your search for questions on the University of Michigan Web page mentioned earlier, selecting one of the listed archives and searching for keywords that describe the types of questions for which you are looking. Let's start by examining the data available through a data archive listed on the page, an archive maintained by the Roper Center at the University of Connecticut. Figure 10-4 is a screen shot of the search tool on the Roper Center site that will allow you to search for a survey that might include questions you would like to include in your own survey. By typing the words "abortion and college students" in the search field you will find two surveys matching the search terms. The two surveys (Figure 10-5) were administered in 2006 and 1985 to samples of undergraduate students by survey researchers at Harvard and the Gallup Organization.

Clicking on the most recent of the two studies, you will see that although the documentation for the study is publicly available at no charge, the data are available only to students whose college or university maintains a sub-

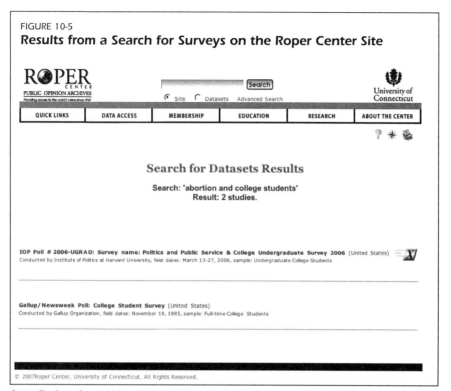

FIGURE 10-5
Results from a Search for Surveys on the Roper Center Site

Source: The Roper Center public opinion data archives; Catalogue of Dataset Holdings. Reprinted with permission.

scription to the Roper Center archive. For our purpose in this example—finding survey questions for use in your own project—this archive will work well with or without the subscription. First, as shown in Figure 10-6, the search results page provides information about the study, including the title, the survey firm, and the dates during which the data were collected in the field. You also will find the type of sample used—in this case it was a survey of 1,200 undergraduates attending a four-year college or university. And you will find the major topics covered by the sample and the number of variables. With eighty-seven variables covering a variety of topics, this survey looks like it will be a good fit with your survey question needs.

By clicking on the PDF or Word versions of the study documentation, you will be able to read and save a copy of the documentation file that includes seventy-five survey questions. In particular, as you will see in Figure 10-7, the survey includes two questions (questions 13 and 14) pertaining to abortion. If these questions about abortion, and the other questions in the survey, are appropriate for your research project, the search is over and you can use some of these questions when writing your own survey. Recall here an important point from our discussion in earlier chapters about conceptualization and

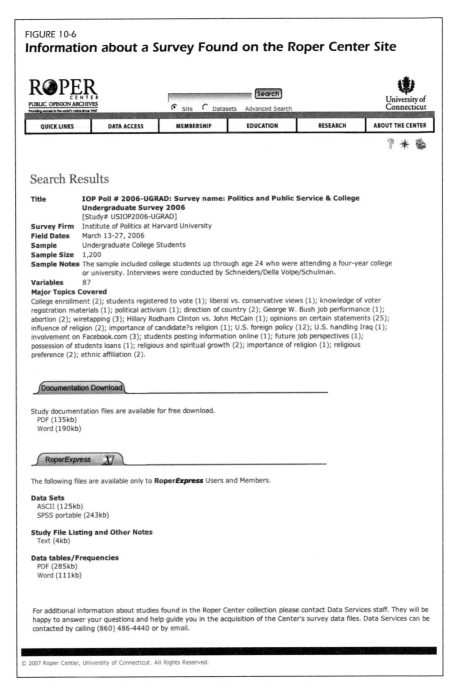

FIGURE 10-6

Information about a Survey Found on the Roper Center Site

Source: The Roper Center public opinion data archives; Catalogue of Dataset Holdings. Reprinted with permission.

FIGURE 10-7

Questions from a Survey Located on the Roper Center Site

8.   When it comes to most political issues, do you think of yourself as a liberal, a conservative or a moderate? [IF MODERATE, ASK: Do you lean liberal or conservative?]

Liberal	1
Moderate-Liberal	2
Moderate	3
Moderate-Conservative	4
Conservative	5
Don't know/Refused	6

9.   Have you ever seen voter registration materials on your campus?

Yes	1
No	2
Don't know	3

10.   Do you consider yourself to be politically engaged or politically active?

Yes	1
No	2
Don't know	3

11.   All in all, do you think that things in the nation are generally headed in the right direction, or do you feel that things are off on the wrong track?

Right direction	1
Wrong track	2
Don't know	3

12.   In general, do you approve or disapprove of the job that George W. Bush is doing as president?

Approve	1
Disapprove	2
Not sure	3

13.   Currently, there are at least three abortion cases that could go to the Supreme Court in the near future. Generally speaking, do you favor keeping Roe v Wade and abortion law as is – or do you favor setting a new legal precedent that would allow restrictions on abortion?

Keep Roe v Wade as is	1
Favor new legal precedent that would allow restrictions	2
Don't know	3

14.   Do you think abortions should be legal under any circumstances, legal only under certain circumstances, or illegal in all circumstances?

Legal under any circumstances	1
Legal only under certain circumstances	2
Illegal in all circumstances	3
Don't know	4

operationalization. If you are planning to use survey questions that you did not write to measure concepts in your own work, you must make sure that the questions measure the same concepts you intend to measure. If the concept you wish to measure differs from the concept the question measures, you will be introducing a weakness into your data. Remember also that you must provide a full citation of the study from which you copied the survey questions. If these questions do not meet all your needs, you can save the questions that you felt were suitable, and make sure to save a full citation for your project, and then turn your attention to another study and continue your search for survey questions.

If you are interested in using the data collected by a survey, you can return to the search results page and clink on the data links and save the data. Whenever you use data collected by someone else, you must save not only the data but also the corresponding codebook, which explains how the data are organized, how they should be read and analyzed, and whether any special considerations must be taken into account before you begin your analysis of the data. Because data come in a variety of formats, you should consult the codebook for instructions on use. In general, most data are provided in a text format that can be opened by the analytical package of your choice (for example, Excel, STATA, SPSS). Text files are usually organized with one case or observation per line and come in one of two formats: delimited or fixed width. Delimited files use a character such as a space or tab between columns of data in the text file, whereas fixed-width files require you to indicate where the column breaks should be, based on the description in the documentation or codebook.

Interviewing

Interviewing is simply the act of asking individuals a series of questions and recording their responses. The interaction may be face-to-face or over the phone.[86] In some cases the interviewer asks a predetermined set of questions; in others the discussion may be more spontaneous or freewheeling; and in still others both structured and unstructured formats are used. The key is that an interview, like a survey, depends on the participants sharing a common language and understanding of terms. And whereas a formal questionnaire, once constructed, limits opportunities for empathetic understanding, an in-depth interview gives the interviewer a chance to probe, to clarify, to search for deeper meanings, to explore unanticipated responses, and to assess intangibles such as mood and opinion intensity.

Perhaps one of the finest examples of the advantages of extended interviews is Robert E. Lane's study of 15 "urban male voters."[87] Although the sample seems small, Lane provides evidence that it is representative of working- and middle-class men living in an Atlantic seaboard town he calls "Eastport." More important for his purposes, his method—a series of extended and taped individual interviews lasting a total of ten to fifteen hours per subject—allowed him to delve into the political consciousness of his subjects in a way no cut-and-dried survey could.

Among many other topics, Lane explored these men's attitudes toward "equality" and a hypothetical movement toward an equalitarian society. Of course, he could have written survey-type questions that would have asked respondents if they agreed or disagreed with this or that statement. Instead, he let his subjects speak for themselves. And what he found turned out to be very interesting and unexpected.

> *The upper working class, and the lower middle class, support specific measures embraced in the formula "welfare state," which have equalitarian consequences. But, so I shall argue, many members of the working class do not want equality. They are afraid of it. In some ways they already seek to escape from it.*[88]

Why did he come to this startling conclusion? Because during his long interviews he uncovered several latent patterns in the men's thinking, patterns that would have been difficult to anticipate and virtually impossible to garner from a standardized questionnaire. For example, when asked about the desirability of greater equality of opportunity and income, one man, Sullivan, a railroad fireman, said:

> *I think it's hard. . . . Supposing I came into a lot of money, and I moved into a nice neighborhood—class—maybe I wouldn't know how to act then. I think it's very hard, because people know that you just—word gets around that you . . . never had it before you got it now. Well, maybe they wouldn't like you . . . maybe you don't know how to act.*[89]

Lane termed this response a concern with "social adjustment" and found that others shared the sentiment. He discovered another source of unease: those in the lower classes would not necessarily deserve a "promotion" up the social ladder. Thus, Ruggiero, a maintenance worker, believed "There's laziness, you'll always have lazy people," while another man said:

> *But then you get a lot of people who don't want to work; you got welfare. People will go on living on that welfare—they're happier than hell. Why should they work if the city will support them.*[90]

The research uncovered similar fears that Lane's subjects experienced when envisioning an equalitarian society. They believed such a society would be unfair to "meritorious elites," would entail the loss of "goals" (if everyone is equal, why work?), and would cause society to "collapse."

Our quick review of Lane's research should not be interpreted as an argument that his is the definitive study. One could, in fact, interpret some of the men's statements quite differently. But the men of Eastport, like all citizens, had mixed, frequently contradictory thoughts, and only after hours of conversation and considerable analysis of the transcripts could Lane begin to classify and make sense of them.

The Ins and Outs of Interviewing

Interviewing, as we use the term, differs substantially from the highly structured, standardized format of survey research.[91] There are many reasons for this difference. First, a researcher may lack sufficient understanding of events to be able to design an effective, structured survey instrument, or schedule of questions. The only way for researchers to learn about certain events is to interview participants or eyewitnesses directly. Second, a researcher is usually especially interested in an interviewee's own interpretation of events or issues and does not want to lose the valuable information that an elite "insider" may possess by unduly constraining responses. As one researcher put it, "A less structured format is relatively exploratory and stresses subject rather than researcher definitions of a problem."[92]

Finally, some people, especially elites or those in positions of high standing or power, may resent being asked to respond to a standardized set of questions. In her study of Nobel laureates, for example, Harriet Zuckerman found that her subjects soon detected standardized questions. Because these were people used to being treated as individuals with minds of their own, they resented "being encased in the straightjacket of standardized questions."[93] Therefore, those who interview elites often vary the order in which topics are broached and the exact form of questions asked from interview to interview.

In this method, eliciting valid information may require variability in approaches.[94] Interviewing is not as simple as lining up a few interviews and chatting for a while. The researcher using the in-depth interview technique must consider numerous logistical and methodological questions. Advance preparation is extremely important. The researcher should study all available documentation of events and pertinent biographical material before interviewing a member of an elite group. Advance preparation serves many purposes. First, it saves the interviewee's time by eliminating questions that can be answered elsewhere. The researcher may, however, ask the interviewee to verify the accuracy of the information obtained from other sources. Second, it gives the researcher a basis for deciding what questions to ask and in what order. Third, advance preparation helps the researcher to interpret and understand the significance of what is being said, to recognize a remark that sheds new light on a topic, and to catch inconsistencies between the interviewee's version and other versions of events. Fourth, the researcher's serious interest in the topic impresses the interviewee. At no time, however, should the researcher dominate the conversation to show off his or her knowledge. Finally, good preparation buoys the confidence of the novice researcher who is interviewing important people.

The ground rules that will apply to what is said in an interview should be made clear at the start.[95] When the interview is requested, and at the beginning of the interview itself, the researcher should ask whether confidentiality is desired. If he or she promises confidentiality, the researcher should be careful not to reveal a person's identity in written descriptions. A touchy problem in confidentiality may arise if questions are based on previous interviews. It may be possible for an interviewee to guess the identity of the person whose comments must have prompted a particular question.

A researcher may desire and promise confidentiality in the hope that the interviewee will be more candid.[96] Interviewees may request confidentiality if they fear they may reveal something damaging to themselves or to others. Some persons may want to approve anything written based on what they have said. In any event, it often is beneficial to the researcher to give interviewees a chance to review what has been written about them and the opportunity to clarify and expand on their comments. Sometimes a researcher and an interviewee may disagree over the content or interpretation of the interview. If the researcher has agreed to let an interviewee have final say on the use of an interview, the agreement should be honored. Otherwise the decision is the researcher's—to be made in light of the needs of the investigation.

ASK THE RIGHT QUESTIONS
The importance of thoroughly researching a topic before conducting elite interviews cannot be stressed enough. In addition to the guidelines discussed in the text, ask yourself this question: Can the information be provided only (or at least most easily) by the person being interviewed? If you can obtain the answers to your questions from newspapers or books, for example, then it is pointless to take up someone's time going over what is (or should be) already known. If, however, the subject believes that only she or he can help you, then you are more likely to gain her or his cooperation. Looking and acting professional is absolutely essential. So, for example, do not arrive at the interview wearing a ball cap or without paper and pen.

Sometimes, gaining access to influential people is difficult. They may want further information about the purpose of the research or need to be convinced of the professionalism of the researcher. Furthermore, many have "gatekeepers" who limit access to their bosses. It is advisable to obtain references from people who are known to potential interviewees. For example, suppose you want to talk to a few state senators. Try getting your own representative to make a few phone calls or write an introductory letter. Sometimes a person who has already been interviewed will assist a researcher in gaining access to other elites. Having a letter of recommendation or introduction from someone who knows the subject can be extremely helpful in this regard.

Two researchers encountered particular access problems in their study of the 1981 outbreaks of civil disorder in several British cities when they attempted to interview community activists.[97] These activists, whom the researchers termed the "counter-elite" or the "threatened elite," were reluctant to cooperate, even hostile. They feared that the findings might be abused, that the research was for the benefit of the establishment and part of a system of oppression, and that cooperation would jeopardize their standing in their community. Unlike conventional elites, they did not assume that social science research was useful.

Whom to interview first is largely a theoretical decision. Interviewing persons of lesser importance in an event or of lower rank in an organization first allows a researcher to become familiar with special terminology used by an elite group and more knowledgeable about a topic before interviewing key elites. It also may bolster a researcher's experience and confidence. Lower-level personnel may be more candid and revealing about events because they are able to observe major participants and have less personal involvement. Talking to superiors first, however, may indicate to subordinates that being interviewed is permissible. Moreover, interviewing key elites first may provide a researcher with important information early on and make subsequent interviewing more efficient. Other factors, such as age of respondents, availability, and convenience, may also affect interview order.

A tape recorder or handwritten notes may be used to record an interview. There are numerous factors to consider in choosing between the two methods. Tape recording allows the researcher to think about what the interviewee is saying, to check notes, and to formulate follow-up questions. If the recording is clear, it removes the possibility of error about what is said. Disadvantages include the fact that everything is recorded. The material must then be transcribed (an expense) and read before useful data are at hand. Much of what is transcribed will not be useful—a problem of elite interviewing in general. A tape recorder may make some interviewees uncomfortable, and they may not be candid even if promised confidentiality; there can be no denying what is recorded. Sometimes the researcher will be unfamiliar with recording equipment and will appear awkward.

Many researchers rely on handwritten notes taken during an interview. It is important to write up interviews in more complete form soon after the interview, while it is still fresh in the researcher's mind. Typically this takes much longer than the interview itself, so enough time should be allotted. Only a few interviews should be scheduled in one day; after two or three, the researcher may not be able to recollect individual conversations distinctly. How researchers go about conducting interviews will vary by topic, by researcher, and by respondent.

Although interviews are usually not rigidly structured, researchers still may choose to exercise control and direction in an interview. Many researchers conduct a semistructured or flexible interview—what is called a **focused interview**—when questioning elites. They prepare an interview guide, including topics, questions, and the order in which they should be raised. Sometimes alternative forms of questions may be prepared. Generally the more exploratory the purpose of the research, the less topic control exercised by the researcher. Researchers who desire information about specific topics should communicate this need to the person being interviewed and exercise enough control over the interview to keep it on track.

Interviewing is difficult work. A researcher must listen, observe nonverbal behavior, think, and take notes all at the same time. Maintaining appropriate interpersonal relations is also required. A good rapport between the researcher and subject, although it may be difficult to establish, facilitates the flow of information. How aggressive should a researcher be in asking questions? The issue is often debated. Although aggressive questioning may yield more information and allow the researcher to ferret out misinformation, it also may alienate or irritate the interviewee. Zuckerman used the tactic of rephrasing the interviewee's comments in extreme form to elicit further details, but in some cases the Nobel laureates expressed irritation that she had not understood what they had already said.[98]

Establishing the meaningfulness and validity of the interview data is important. The data may be biased by the questions and actions of the interviewer. Interviewees may give evasive or untruthful answers. As noted earlier, advance preparation may help an interviewer recognize remarks that differ from established fact. Examining their plausibility, checking for internal consistency, and corroborating them with other interviewees also may determine the validity of an interviewee's statements. John P. Dean and William Foote Whyte argue that a researcher should understand an interviewee's mental set and how it might affect his or her perception and interpretation of events.[99] Raymond L. Gordon stresses the value of being able to empathize with interviewees to understand the meaning of what they are saying.[100] Lewis Dexter warns that interviews should be conducted only if "interviewers have enough relevant background to be sure that they can make sense out of interview conversations or . . . there is reasonable hope of being able to . . . learn what is meaningful and significant to ask."[101]

Despite the difficulties, interviewing is an excellent form of data collection, particularly in exploratory studies or when thoughts and behavior can be described or expressed only by those who are deeply involved in political processes. Interviewing often provides a more comprehensive and complicated understanding of phenomena than other forms of research designs, and it provides researchers with a rich variety of perspectives.

Conclusion

In this chapter we discussed two ways of collecting information directly from individuals—through survey research and interviewing. Whether data are collected over the phone, through the mail, or in person, the researcher attempts to elicit information that is consistent, complete, accurate, and instructive. This goal is advanced by being attentive to questionnaire design and taking steps to engage and motivate respondents. The choice of an in-person, telephone, or mail survey can also affect the quality of the data collected. Interviews of elite populations require attention to a special set of issues and generally result in a less structured type of interview.

Although you may never conduct an elite interview or a public opinion survey of your own, the information in this chapter should help you evaluate the research of others. Polls, surveys, and interview data have become so prevalent in American life that an awareness of the decisions made and problems encountered by survey researchers is needed to independently judge conclusions drawn from such data.

Notes

1. GreenMountainDaily.com, "Death with Dignity" in Trouble: Newton, Religion and the "Passion Gap," March 14, 2007. Retrieved March 24, 2007, from www.greenmountaindaily.com/showDiary.do?diaryId=1040.

2. Death with Dignity Vermont, Inc., "A Supportive Voice from the Pulpit." Retrieved March 26, 2007, from www.deathwithdignityvermont.org/spong_interview.htm.

3. Vermont Alliance for Ethical Healthcare, "Physician Assisted Suicide Puts Our Lives at Risk." Retrieved March 26, 2007, from www.vaeh.org.

4. Miroslav Nincic, "Domestic Costs, U.S. Public Opinion, and the Isolationist Calculus," *International Studies Quarterly* 41 (1997): 593–610.

5. See Robert F. Belli and others, "Reducing Vote Overreporting in Surveys: Social Desirability, Memory Failure, and Source Monitoring," *Public Opinion Quarterly* 63 (Spring 1999): 90–108; and Janet M. Box-Steffensmeirer, Gary C. Jacobson, and J. Tobin Grant, "Question Wording and the House Vote Choice: Some Experimental Evidence," *Public Opinion Quarterly* 64 (Autumn 2000): 257–270.

6. Floyd J. Fowler, *Survey Research Methods,* rev. ed. (Newbury Park, Calif.: Sage, 1988), 61.

7. Don A. Dillman, *Mail and Telephone Surveys: The Total Design Method* (New York: Wiley, 1978), 40.

8. Indeed, a large (1,000 or more respondents) national survey using probability sampling of the sort explained in Chapter 7 might cost more than $100,000.

9. The Harris organization says, for example, "By participating, you'll not only have your say in matters that affect you, you'll also be able to see the results from the surveys you complete. You can then compare your opinions and experiences to many others'—people who are like, and unlike, you. Occasionally, you may participate in surveys whose results will be published in national or international media." Harris Interactive Inc. "Harris Poll Online." Retrieved March 27, 2007, from www.harrispollonline.com/main.asp. Zogby stresses its accuracy: "The strength of the results of the 2004 Presidential election has validated the new method. Zogby Interactive accurately predicted the winner in 85 percent of the states that it polled, while by state, the poll was within 4 points on average." Zogby International, "Welcome to the Zogby Interactive." Retrieved March 27, 2007, from http://interactive.zogby.com/index.cfm.

10. Fowler, *Survey Research Methods,* 68.

11. Ibid.

12. Robert M. Groves and Robert L. Kahn, *Surveys by Telephone: A National Comparison with Personal Interviews* (New York: Academic Press, 1979); and James H. Frey, *Survey Research by Telephone* (Beverly Hills, Calif.: Sage, 1983).

13. Frey, *Survey Research by Telephone,* 24–25.

14. Howard E. Freeman and others, "Telephone Sampling Bias in Surveying Disability," *Public Opinion Quarterly* 46 (1982): 392–407.

15. Ibid.

16. Robert M. Groves, Robert B. Cialdini, and Mick P. Couper, "Understanding the Decision to Participate," *Public Opinion Quarterly* 56 (Winter 1992): 474.

17. Earl R. Babbie, *Survey Research Methods* (Belmont, Calif.: Wadsworth, 1973), 171.

18. Groves and Kahn, *Surveys by Telephone,* 3.

19. See, for example, Burns W. Roper, "Evaluating Polls with Poll Data," *Public Opinion Quarterly* 50 (Spring 1986): 10–16.

20. Charlotte G. Steeh, "Trends in Nonresponse Rates, 1952–1979," *Public Opinion Quarterly* 45 (1981): 40–57.

21. Laure M. Sharp and Joanne Frankel, "Respondent Burden: A Test of Some Common Assumptions," *Public Opinion Quarterly* 47 (1983): 36–53. Note that another survey found that people had generally favorable beliefs about polls (Roper, "Evaluating Polls with Poll Data"), but even the author of this study worries that the public may grow weary and distrustful of polling.

22. Theresa J. DeMaio, "Refusals: Who, Where, and Why," *Public Opinion Quarterly* 44 (1980): 223–233.

23. Dillman, *Mail and Telephone Surveys.*

24. See Theresa F. Rogers, "Interviews by Telephone and in Person: Quality of Responses and Field Performance," *Public Opinion Quarterly* 39 (1975): 51–64; and Groves and Kahn, *Surveys by Telephone.* Response rates are affected by different methods of calculating rates for the three types of surveys. For example, nonreachable and ineligible persons may be dropped from the total survey population for telephone and personal interviews before response rates are calculated. Response rates to mail surveys are depressed because all nonresponses are assumed to be refusals, not ineligibles or nonreachables. Telephone response rates may be depressed if nonworking but ringing numbers are treated as nonreachable but eligible respondents. Telephone companies vary in their willingness to identify working numbers. If noneligibility is likely to be a problem in a mail survey, ineligibles should be asked to return the questionnaire anyway so that they can be identified and distinguished from refusals.

25. Frey, *Survey Research by Telephone,* 15–16.

26. Herschel Shosteck and William R. Fairweather, "Physician Response Rates to Mail and Personal Interview Surveys," *Public Opinion Quarterly* 43 (1979): 206–217.

27. Don A. Dillman, Jean Gorton Gallegos, and James H. Frey, "Reducing Refusal Rates for Telephone Interviews," *Public Opinion Quarterly* 40 (1976): 66–78.

28. Fowler, *Survey Research Methods,* 67.

29. Gideon Vigderhous, "Scheduling Telephone Interviews: A Study of Seasonal Patterns," *Public Opinion Quarterly* 45 (1981): 250–259.

30. M. F. Weeks and others, "Optimal Times to Contact Sample Households," *Public Opinion Quarterly* 44 (1980): 101–114.

31. See J. Scott Armstrong, "Monetary Incentive in Mail Surveys," *Public Opinion Quarterly* 39 (1975): 111–116; Arnold S. Linsky, "Stimulating Responses to Mailed Questionnaires: A Review," *Public Opinion Quarterly* 39 (1975): 82–101; James R. Chromy and Daniel G. Horvitz, "The Use of Monetary Incentives in National Assessment of Households Survey," *Journal of the American Statistical Association* 73 (1978): 473–478; Thomas A. Heberlein and Robert Baumgartner, "Factors Affecting Response Rates to Mailed Questionnaires," *American Sociological Review* 43 (1978): 447–462; R. Kenneth Godwin, "The Consequences of Large Monetary Incentives in Mail Surveys of Elites," *Public Opinion Quarterly* 43 (1979): 378–387;

Kent L. Tedin and C. Richard Hofstetter, "The Effect of Cost and Importance Factors on the Return Rate for Single and Multiple Mailings," *Public Opinion Quarterly* 46 (1982): 122–128; Anton J. Nederhof, "The Effects of Material Incentives in Mail Surveys: Two Studies," *Public Opinion Quarterly* 47 (1983): 103–111; Charles D. Schewe and Norman G. Cournoyer, "Prepaid vs. Promised Monetary Incentives to Questionnaire Response: Further Evidence," *Public Opinion Quarterly* 40 (1976): 105–107; James R. Henley Jr., "Response Rate to Mail Questionnaire with a Return Deadline," *Public Opinion Quarterly* 40 (1976): 374–375; Thomas A. Heberlein and Robert Baumgartner, "Is a Questionnaire Necessary in a Second Mailing?" *Public Opinion Quarterly* 45 (1981): 102–108; and Wesley H. Jones, "Generalized Mail Survey Inducement Methods: Population Interactions with Anonymity and Sponsorship," *Public Opinion Quarterly* 43 (1979): 102–111.

32. For detailed instructions on improving the response rate to mail surveys, see Dillman, *Mail and Telephone Surveys.*

33. Fowler, *Survey Research Methods,* 63.

34. And, of course, you still would not have a probability sample. See Chapter 7.

35. Groves and Kahn, *Surveys by Telephone,* 214; and Frey, *Survey Research by Telephone,* 22.

36. Scott Keeter, "The Cell Phone Challenge to Survey Research" (The Pew Research Center, Washington, D.C., May 15, 2006). Retrieved March 2, 2007, from http://people-press.org/reports/display.php3?ReportID=276.

37. For research estimating the amount of bias introduced by nonresponse due to unavailability or refusal, see F. L. Filion, "Estimating Bias Due to Nonresponse in Mail Surveys," *Public Opinion Quarterly* 39 (1975): 482–492; Michael J. O'Neil, "Estimating the Nonresponse Bias Due to Refusals in Telephone Surveys," *Public Opinion Quarterly* 43 (1979): 218–232; and Arthur L. Stinchcombe, Calvin Jones, and Paul Sheatsley, "Nonresponse Bias for Attitude Questions," *Public Opinion Quarterly* 45 (1981): 359–375.

38. Carol S. Aneshensel and others, "Measuring Depression in the Community: A Comparison of Telephone and Personal Interviews," *Public Opinion Quarterly* 46 (1982): 110–121.

39. DeMaio, "Refusals," 223–233; and Steeh, "Trends in Nonresponse Rates," 40–57.

40. Dillman, *Mail and Telephone Surveys.*

41. See, for example, Matthew B. Miles and A. Michael Huberman, *Qualitative Data Analysis: An Expanded Sourcebook,* 2d ed. (Thousand Oaks, Calif.: Sage Publications, 1995); and Renata Tesch, *Qualitative Research: Analysis Types and Software Tools* (Bristol, Penn.: The Falmer Press, 1990).

42. SawTooth Technologies, "WinCati for Computer-Assisted Telephone Interviewing." Retrieved March 11, 2007, from www.sawtooth.com/products/cati/index.htm#whatis.

43. *Washington Post* Virginia Governor Poll #2, October [computer file]. ICPSR04522-v1. Horsham, Penn.: Taylor Nelson Sofres Intersearch [producer], 2005. Ann Arbor, Mich.: Inter-university Consortium for Political and Social Research [distributor], 2007-03-09. Retrieved March 31, 2007, from www.icpsr.umich.edu/access/index.html.

44. See Willis J. Goudy and Harry R. Potter, "Interview Rapport: Demise of a Concept," *Public Opinion Quarterly* 39 (1975): 529–543; and Charles F. Cannell, Peter V. Miller, and Lois Oksenberg, "Research on Interviewing Techniques," in Samuel Leinhardt, ed., *Sociological Methodology 1981* (San Francisco: Jossey-Bass, 1981), 389–437.

45. Rogers, "Interviews by Telephone and in Person," 51–65.

46. Ibid.; and Peter V. Miller and Charles F. Cannell, "A Study of Experimental Techniques for Telephone Interviewing," *Public Opinion Quarterly* 46 (1982): 250–269.

47. Arpad Barath and Charles F. Cannell, "Effect of Interviewer's Voice Intonation," *Public Opinion Quarterly* 40 (1976): 370–373.

48. Eleanor Singer, Martin R. Frankel, and Marc B. Glassman, "The Effect of Interviewer Characteristics and Expectations on Response," *Public Opinion Quarterly* 47 (1983): 68–83; and Eleanor Singer and Luane Kohnke-Aguirre, "Interviewer Expectation Effects: A Replication and Extension," *Public Opinion Quarterly* 43 (1979): 245–260.

49. Patrick R. Cotter, Jeffrey Cohen, and Philip B. Coulter, "Race-of-Interviewer Effects in Telephone Interviews," *Public Opinion Quarterly* 46 (1982): 278–284; and Bruce A. Campbell,

"Race of Interviewer Effects among Southern Adolescents," *Public Opinion Quarterly* 45 (1981): 231–244.

50. Shirley Hatchett and Howard Schuman, "White Respondents and Race-of-Interviewer Effects," *Public Opinion Quarterly* 39 (1975): 523–528; and Michael F. Weeks and R. Paul Moore, "Ethnicity of Interviewer Effects on Ethnic Respondents," *Public Opinion Quarterly* 45 (1981): 245–249.

51. See Singer, Frankel, and Glassman, "The Effect of Interviewer Characteristics and Expectations on Response"; Groves and Kahn, *Surveys by Telephone;* Dillman, *Mail and Telephone Surveys;* and John Freeman and Edgar W. Butler, "Some Sources of Interviewer Variance in Surveys," *Public Opinion Quarterly* 40 (1976): 79–91.

52. Adapted from American National Election Study (ANES) 2004, "HTML Codebook Produced July 14, 2006." Retrieved March 10, 2007, from http://sda.berkeley.edu.

53. See Groves and Kahn, *Surveys by Telephone;* and Lawrence A. Jordan, Alfred C. Marcus, and Leo G. Reeder, "Response Styles in Telephone and Household Interviewing," *Public Opinion Quarterly* 44 (1980): 210–222.

54. Dillman, *Mail and Telephone Surveys.*

55. Jordan, Marcus, and Reeder, "Response Styles"; Groves and Kahn, *Surveys by Telephone.* See also Rogers, "Interviews by Telephone and in Person."

56. For example, see Seymour Sudman and Norman M. Bradburn, *Asking Questions: A Practical Guide to Questionnaire Design* (San Francisco: Jossey-Bass, 1982), 55–86; Jerald G. Bachman and Patrick M. O'Malley, "When Four Months Equal a Year: Inconsistencies in Student Reports of Drug Use," *Public Opinion Quarterly* 45 (1981): 542.

57. RRT was first proposed by S. L. Warner in "Randomized Response," *Journal of the American Statistical Association* 60 (1965): 63–69.

58. Frederick Wiseman, Mark Moriarty, and Marianne Schafer, "Estimating Public Opinion with the Randomized Response Model," *Public Opinion Quarterly* 39 (1975): 507–513.

59. S. M. Zdep and Isabelle N. Rhodes, "Making the Randomized Response Technique Work," *Public Opinion Quarterly* 40 (1976): 531–537.

60. Ibid.

61. Iris M. Shimizu and Gordon Scott Bonham, "Randomized Response Technique in a National Survey," *Journal of the American Statistical Association* 73 (1978): 35–39.

62. Robert G. Orwin and Robert F. Boruch, "RRT Meets RDD: Statistical Strategies for Assuring Response Privacy in Telephone Surveys," *Public Opinion Quarterly* 46 (1982): 560–571.

63. John M. Benson, "The Polls—Trends," *Public Opinion Quarterly* 63 (Summer 1999): 263–277.

64. Margrit Eichler, *Nonsexist Research Methods: A Practical Guide* (Winchester, Mass.: Allen and Unwin, 1988), 51–52.

65. Doris A. Graber, "Problems in Measuring Audience Effects of the 1976 Debate," in George F. Bishop, Robert G. Meadow, and Marilyn Jackson-Beeck, eds., *The Presidential Debates: Media, Electoral and Policy Perspectives* (New York: Praeger, 1978), 116.

66. Charles H. Backstrom and Gerald Hursh-Cesar, *Survey Research,* 2d ed. (New York: Wiley, 1981), 142, 146.

67. Ibid., 141.

68. Republican National Committee, *1980 Official Republican Poll on U.S. Defense and Foreign Policy.*

69. Eichler, *Nonsexist Research Methods,* 43–44.

70. Quoted from 1996 American National Election Study, "HTML Codebook Produced March 30, 2000." Retrieved November 22, 2000, from http://csa.berkeley.edu:7502/archive.htm.

71. John P. Dean and William Foote Whyte, "How Do You Know if the Informant Is Telling the Truth?" in Lewis Anthony Dexter, ed. *Elite and Specialized Interviewing* (Evanston, Ill.: Northwestern University Press, 1970), 127.

72. Raymond L. Gordon, *Interviewing: Strategy, Techniques, and Tactics* (Homewood, Ill.: Dorsey, 1969), 18.

73. Dexter, *Elite and Specialized Interviewing,* 17.

74. Patricia J. Labaw, *Advanced Questionnaire Design* (Cambridge, Mass.: Abt Books, 1980), 132.

75. Ibid., 132–133.

76. Ibid., 117.

77. Norman M. Bradburn and W. M. Mason, "The Effect of Question Order on Responses," *Journal of Marketing Research* 1 (1964): 57–64.

78. A. Regula Herzog and Jerald G. Bachman, "Effects of Questionnaire Length on Response Quality," *Public Opinion Quarterly* 45 (1981): 549–559.

79. Lee Sigelman, "Question-Order Effects on Presidential Popularity," *Public Opinion Quarterly* 45 (1981): 199–207.

80. Sam G. MacFarland, "Effects of Question Order on Survey Responses," *Public Opinion Quarterly* 45 (1981): 208–215.

81. Ibid., 213, 214.

82. Labaw, *Advanced Questionnaire Design,* 122.

83. Glen D. Mellinger, Carol L. Huffine, and Mitchell B. Balter, "Assessing Comprehension in a Survey of Public Reactions to Complex Issues," *Public Opinion Quarterly* 46 (1982): 97–109.

84. William D. Perrault Jr., "Controlling Order-Effect Bias," *Public Opinion Quarterly* 39 (1975): 544–551.

85. Donald J. Messmer and Daniel T. Seymour, "The Effects of Branching on Item Nonresponse," *Public Opinion Quarterly* 46 (1982): 270–277.

86. Occasionally the investigator may obtain the information from some form of written communication.

87. Robert E. Lane, "The Fear of Equality," *American Political Science Review* 53 (March 1959): 35. The complete results of Lane's work are found in his *Political Ideology: Why the Common Man Believes What He Does* (New York: Free Press, 1963).

88. Lane, "The Fear of Equality," 35.

89. Ibid., 46.

90. Ibid., 44–45.

91. There are exceptions to this general rule, however. See John Kessel, *The Domestic Presidency* (Belmont, Calif.: Duxbury, 1975). Kessel administered a highly structured survey instrument to Richard Nixon's Domestic Council staff.

92. Joseph A. Pika, "Interviewing Presidential Aides: A Political Scientist's Perspective," in George C. Edwards III and Stephen J. Wayne, eds., *Studying the Presidency* (Knoxville: University of Tennessee Press, 1982), 282.

93. Harriet Zuckerman, "Interviewing an Ultra-Elite," *Public Opinion Quarterly* 36 (1972): 167.

94. Gordon, *Interviewing,* 49–50.

95. Dom Bonafede, "Interviewing Presidential Aides: A Journalist's Perspective," in Edwards and Wayne, eds., *Studying the Presidency,* 269.

96. Richard F. Fenno Jr., *Home Style: House Members in Their Districts* (Boston: Little, Brown, 1978), 280.

97. Margaret Wagstaffe and George Moyser, "The Threatened Elite: Studying Leaders in an Urban Community," in George Moyser and Margaret Wagstaffe, eds., *Research Methods for Elite Studies* (London: Allen and Unwin, 1987), 186–188.

98. Zuckerman, "Interviewing an Ultra-Elite," 174.

99. Dean and Whyte, "How Do You Know if the Informant Is Telling the Truth?," 127.

100. Gordon, *Interviewing,* 18.

101. Dexter, *Elite and Specialized Interviewing,* 17.

Terms Introduced

Ambiguous question. A question containing a concept that is not defined clearly.

Branching question. A question that sorts respondents into subgroups and directs these subgroups to different parts of the questionnaire.

Closed-ended question. A question with response alternatives provided.

Double-barreled question. A question that is really two questions in one.

Filter question. A question used to screen respondents so that subsequent questions will be asked only of certain respondents for whom the questions are appropriate.

Focused interview. A semistructured or flexible interview schedule used when interviewing elites.

Interviewer bias. The interviewer's influence on the respondent's answers; an example of reactivity.

Interviewing. Interviewing respondents in a nonstandardized, individualized manner.

Leading question. A question that encourages the respondent to choose a particular response.

Open-ended question. A question with no response alternatives provided for the respondent.

Push poll. A poll intended not to collect information but to feed respondents (often) false and damaging information about a candidate or cause.

Question-order effect. The effect on responses of question placement within a questionnaire.

Questionnaire design. The physical layout and packaging of a questionnaire.

Randomized response technique (RRT). A method of obtaining accurate answers to sensitive questions that protects the respondent's privacy.

Response quality. The extent to which responses provide accurate and complete information.

Response rate. The proportion of respondents selected for participation in a survey who actually participate.

Response set. The pattern of responding to a series of questions in a similar fashion without careful reading of each question.

Sample-population congruence. The degree to which sample subjects represent the population from which they are drawn.

Single-sided question. A question with only one substantive alternative provided for the respondent.

SURVEY INSTRUMENT. The schedule of questions to be asked of the respondent.

TWO-SIDED QUESTION. A question with two substantive alternatives provided for the respondent.

Suggested Readings

Aldridge, Alan, and Kenneth Levine. *Surveying the Social World.* Buckingham, England: Open University Press, 2001.

Braverman, Marc T., and Jana Kay Slater. *Advances in Survey Research.* San Francisco: Jossey-Bass, 1998.

Converse, J. M., and Stanley Presser. *Survey Questions: Handcrafting the Standardized Questionnaire.* Beverly Hills, Calif.: Sage Publications, 1986.

Dillman, Don A. *Mail and Electronic Surveys.* New York: Wiley, 1999.

Frey, James H., and Sabine M. Oishi. *How to Conduct Interviews by Telephone and in Person.* Thousand Oaks, Calif.: Sage Publications, 1995.

Nesbary, Dale. *Survey Research and the World Wide Web.* Needham Heights, Mass.: Allyn and Bacon, 1999.

Newman, Isadore, and Keith A. McNeil. *Conducting Survey Research in the Social Sciences.* Lanham, Md.: University Press of America, 1998.

Patten, Mildred L. *Questionnaire Research: A Practical Guide,* 2d ed. Los Angeles: Pyrczak, 2001.

Rea, Louis M., and Richard A. Parker. *Designing and Conducting Survey Research.* San Francisco: Jossey-Bass, 1997.

Sapsford, Roger. *Survey Research.* Thousand Oaks, Calif.: Sage Publications, 1999.

Tanur, Judith M., ed. *Questions about Questions.* New York: Russell Sage Foundation, 1992.

CHAPTER 11

Statistics:

First Steps

Many political science students wonder why they spend a semester or more studying empirical methods and statistics. After all, aren't topics such as current events, politics in general, law, foreign affairs, voting, and legislatures more interesting? Why bother with something as formal as data collection and analysis? We give two reasons, both equally compelling. First, for better or worse, you need to understand a few basic statistical concepts and methods in order to have more than a superficial understanding of political science research, which most of its practitioners regard as a partly quantitative discipline. Second, good citizenship requires an awareness of statistical concepts. Why? Because, to one degree or another, many issues and policies involve quantitative arguments. Statistical arguments underlie innumerable policy and legal debates.[1]

This chapter takes readers the first steps down the road to understanding applied statistics. Besides a bit of introductory material, the chapter covers two main topics:

- Methods for describing and exploring data. These tools include tables, summary statistics, and graphs. They all work toward the same end: extracting information hidden in numbers.

- Statistical inference, or the process of drawing defensible conclusions about populations from samples.

We proceed slowly because the concepts, though not excessively mathematical, do require thought and effort to comprehend. But it will be worth the effort because the knowledge will make you not just a better political science student but also a better citizen.

The Data Matrix

Most of the statistical reports you come across in both the mass media and scholarly publications usually show only the final results of what has been a long process of gathering, organizing, and analyzing a large body of data. But knowing what goes on behind the scenes is as important as understanding the empirical conclusions. Conceptually, at least, the first step is the arrangement of the observed measurements into a **data matrix,** which is simply an array of rows and columns that stores the values. Separate rows hold the data for each case or unit of analysis. If you read across one row, you see the specific values that pertain to that individual case. Each column contains the values on a single variable for all the cases. The column headings list the variable names. To find out where a particular case stands with regard to a particular variable, just look for the row for that case and read across to the appropriate column. Table 11-1 is a data matrix that gives the raw data for 10 different variables for 36 nations in a sample. Table 11-2 contains descriptions of the variables. We turn to these data for numeric examples throughout the remaining chapters.

Data Description and Exploration

As presented in Table 11-1, the data are not very helpful, partly because they overwhelm the eye and partly because it is hard to see even what an average value is, much less the degree of variability or range of values. Nor does a matrix reveal many patterns in the data or tell us much about what causes low or high scores. Still, its creation is an essential initial step in data analysis. Moreover, it represents pictorially how text and numbers are stored in a computer. We can instruct a computer program, for example, to sum all the numerical values in column 2 (IMR) and divide by 36 to obtain a mean infant mortality rate.

To go from raw data to meaningful conclusions, you begin by summarizing and exploring the information in the matrix. Several kinds of tables, statistics, and graphs can be used for this purpose. Different statistical procedures assume different levels of measurement. Recall from Chapter 4 that we distinguished four broad types of measurement scales:

■ Nominal: Variable values are unordered names or labels. (Examples: ethnicity, gender, country of origin)

■ Ordinal: Variable values are labels having an implicit but unspecified or measured order. Numbers may be assigned to categories to show ordering

TABLE 11-1
Aggregate Data for 36 Nations

Country	IMR	Phone	Ag	GNP	Rights	Press	VA	Status	Docs	Region
Austria	4	462.0	4.3	35,500	1	21	1.24	1	337.8	Eur
Belgium	4	464.4	1.6	31,800	1	11	1.31	1	449.4	Eur
Canada	5	642.7	2.1	35,200	1	17	1.32	1	213.9	NAm
Greece	4	578.4	14.6	23,500	1	28	0.95	1	438.0	Eur
Iceland	2	650.1	7.2	38,100	1	9	1.38	1	361.6	Eur
Ireland	5	499.4	8.8	43,600	1	15	1.41	1	278.6	Eur
Israel	5	437.2	2.3	26,200	1	28	0.61	1	382.0	ME
Italy	4	447.5	4.4	29,700	1	35	1.00	1	420.3	Eur
Japan	3	460.0	3.2	33,100	1	20	0.94	1	197.6	Asia
Luxembourg	4	*	1.5	68,800	1	11	1.34	1	266.2	Eur
Netherlands	4	484.4	3.0	31,700	1	11	1.45	1	314.9	Eur
New Zealand	5	461.1	8.6	26,000	1	12	1.39	1	236.6	Asia
Norway	3	*	4.0	47,800	1	10	1.45	1	313.3	Eur
Portugal	4	402.5	11.1	19,100	1	14	1.32	1	342.3	Eur
Spain	4	415.2	6.0	27,000	1	22	1.12	1	329.5	Eur
Sweden	3	715.4	2.7	31,600	1	9	1.41	1	328.4	Eur
United States	6	606.0	1.8	43,500	1	17	1.19	1	256.4	NAm
Bangladesh	54	6.1	51.8	2,200	4	68	−0.50	5	25.7	Asia
Burkina Faso	96	6.1	92.2	1,300	5	40	−0.37	5	5.9	Afr
Cambodia	98	*	68.5	2,600	6	62	−0.94	5	15.6	Asia
Colombia	17	171.4	18.3	8,400	3	63	−0.32	5	135.0	LA
Cuba	6	67.8	12.8	3,900	7	96	−1.87	5	590.6	LA
Dominican Republic	26	106.5	14.2	8,000	2	38	0.20	5	187.6	LA
Ecuador	22	122.2	23.2	4,500	3	41	−0.16	5	147.6	LA
Egypt	28	135.5	30.8	4,200	5	68	−1.15	5	53.5	ME
Haiti	84	17.1	60.2	1,800	7	66	−1.41	5	25.0	LA
Kenya	79	9.2	73.6	1,200	3	61	−0.12	5	13.9	Afr
Malawi	79	7.5	81.3	600	4	54	−0.45	5	2.2	Afr
Malaysia	10	173.8	15.9	12,700	4	69	−0.41	5	70.2	Asia
Morocco	36	43.8	33.1	4,400	4	63	−.76	5	51.5	ME
Mozambique	100	*	80.3	1,500	3	45	−0.06	5	2.7	Afr
Nicaragua	30	37.7	17.2	3,000	3	42	−0.01	5	37.4	LA
Paraguay	20	*	32.5	4,700	3	56	−0.19	5	110.7	LA
Rwanda	118	2.7	90.1	1,600	6	84	−1.32	5	4.7	Afr
Sri Lanka	12	51.0	44.3	4,600	3	56	−0.26	5	54.5	Asia
Tanzania	76	*	78.7	800	4	51	−0.31	5	2.3	Afr

Note: See Table 11-2 for definitions of variable labels and sources of data; asterisks indicate missing data.

TABLE 11-2
Data Definitions and Sources

Variable Label	Definition
IMR	Infant mortality rate (IMR). Deaths of children between birth and the age of one per 1,000 live births.[a]
Phone	Main telephone lines per 1,000 people.[b]
Ag	Agricultural labor force as a percentage of the total labor force, 2004.[c]
GNP	Gross national product per capita, 2005.[d]
Rights	Index of political rights. Measures the degree of freedom in the electoral process, political pluralism, participation, and functioning of government. Scale = 1–7 (1 represents the most freedom and 7 the least freedom).[e]
Press	Index of press freedom. Measures the degree to which each country permits the free flow of information. Scale = 1–100 (1–30 indicates that the country has a "free" media; 31–60, "partly free" media; and 61–100, "not free."[f]
VA	Voice and accountability. Measures the extent to which a country's citizens are able to participate in selecting their government, as well as freedom of expression, freedom of association, and a free media. Negative (low) values mean less public control, 0 is about average, and positive values indicate greater degrees of accountability.[g]
Status	Developmental status of the nation: 1 = developed; 5 = developing.[h]
Docs	Physicians per 100,000 people; latest available data for years 1990–2004.[i]
Region	LA: Latin America (including Caribbean); NAm: North America; Asia (including New Zealand); Eur: Europe; ME: Middle East (including Egypt and Morocco); Afr: sub-Saharan Africa.

a. United Nations Children's Fund (UNICEF). 2006. *The State of the World's Children 2007: The Double Dividend of Gender Equality,* Table 1. New York: UNICEF. Retrieved from www.unicef.org/sowc07/.
b. Food and Agriculture Organization of the United Nations (FAO). 2006. FAOSTAT Online Statistical Service. Rome: FAO. Retrieved from http://faostat.fao.org.
c. Central Intelligence Agency, *The World FactBook,* 2005.
d. Development Data Group, The World Bank. 2006. *2006 World Development Indicators Online.* Washington, DC: The World Bank. Retrieved from http://publications.worldbank.org/ecommerce/catalog/product?item_id=631625.
e. Freedom House. 2006. *Freedom in the World 2006: The Annual Survey of Political Rights and Civil Liberties.* Retrieved from www.freedomhouse.org/template.cfm?page=15&year=2005.
f. Freedom House. 2005. *Freedom of the Press 2005: A Global Survey of Media Independence.* Retrieved from www .freedomhouse.org/template.cfm?page=16&year=2005.
g. Daniel Kaufmann, Aart Kraay, and Massimo Mastruzzi. 2006. "Aggregate and Individual Governance Indicators for 1996–2005," *World Bank Policy Research Working Paper 4012.* Daniel Kaufmann, Aart Kraay, and Massimo Mastruzzi. 2006. "Governance Matters V: Governance Indicators for 1996–2005," *World Bank Policy Research Working Paper,* and "A Decade of Measuring the Quality of Governance," World Resources Institute.
h. Based on the classification used by Earth Trends Environmental Information, World Resources Institute. Available online at http://earthtrends.wri.org/.
i. World Health Organization (WHO). 2006. *World Health Report 2006,* Annex Table 4. Geneva: WHO. Retrieved from www.who.int/whr/2006/annex/en/index.html. Also available in WHO. *Global Atlas of the Health Workforce* (www.who .int/globalatlas/default.asp).

or ranking, but strictly speaking, arithmetical operations (for example, addition) are inappropriate. (Example: scale of ideology)

■ Interval: Numbers are assigned to objects such that interval differences are constant across the scale, but there is no true or meaningful zero point. (Examples: temperature,[2] intelligence scores)

■ Ratio: In addition to having the properties of interval variables, these scales have a meaningful zero value. (Examples: income, percentage of the population with a high school education)

What is appropriate for interval scales, for example, may not be useful or may be misleading when applied to ordinal scales.[3] In the discussion that follows, we clarify which techniques apply to which kinds of variables.

Frequency Distributions, Proportions, and Percentages

An empirical **frequency distribution** is a table that shows the number of observations having each value of a variable. The number of observations, also called the frequency or count, in the kth category of a variable is often represented by the small letter f_k. In addition, a frequency distribution is usually accompanied by a number called a **relative frequency,** which simply transforms the raw frequency into a proportion or percentage. A *proportion*—the ratio of a part to a whole—is calculated by dividing the number of observations (f_k) having a specific property, or individuals who gave a particular response, by the total number of observations (N). A *percentage,* or "parts per 100," is found by multiplying a proportion by 100 or equivalently moving the decimal two places to the right. Percentages are especially popular because the quantities communicate information that is often more meaningful and easier to grasp than plain frequencies.

Table 11-3 gives a simple example of a frequency distribution. In particular, it shows the number of cases (nations) in each level or category of the variable "region" (see Table 11-1). We see, to take one example, that 6 countries have been placed in the category "Africa." This table also gives the relative frequencies in the form of proportions and percentages. As an

TABLE 11-3
Frequency Distribution of Region

Region	Frequency (f_k)	Relative Frequency Proportion (p_k)	Relative Frequency Percent
Africa	6	.167	16.7
Asia	6	.167	16.7
Europe	12	.333	33.3
Latin America	7	.194	19.4
Middle East	3	.083	8.3
North America	2	.056	5.6
Total	36	1.00	100

Source: See Table 11-1.

TABLE 11-4
Frequency Distribution of Party Identification

Response Category	Frequency of Response	Proportion Giving Response	Percent Giving Response	Cumulative Percent
Strong Democrat	197	.165	16.5	16.5
Weak Democrat	186	.156	15.6	32.1
Independent-leaning Democrat	208	.174	17.4	49.5
Independent	116	.097	9.7	59.2
Independent-leaning Republican	140	.117	11.7	70.9
Weak Republican	149	.125	12.5	83.5
Strong Republican	197	.165	16.5	99.9
Total	1,193	1.00	99.9[a]	99.9[a]

Source: 2004 National Election Study.

Note: Question: "Generally speaking, do you usually think of yourself as a Republican, a Democrat, an independent, or what? Would you call yourself a Strong [Democrat/Republican] or a Not Very Strong [Democrat/Republican]? Do you think of yourself as closer to the Republican Party or to the Democratic party?"

[a] Does not add to 100 because of rounding.

example, note that the portion of the sample classified as "Africa" is .167 or 16.7 percent.

Since there are not many cases ($N = 36$) in our example, we could almost have counted by hand the number of countries in the various regions of the world. But if we were investigating a major national survey, such as the 2004 National Election Study (NES), we would find far too many observations to put on paper and scan by eye.

Table 11-4 shows the distribution of 1,193 responses to a question on the 2004 NES asking people about party identification: "Generally speaking, do you usually think of yourself as a Republican, a Democrat, an independent, or what?"[4] (Note incidentally that Table 11-4 does not contain any "don't know" or "other" responses, which have been treated as missing data.) This frequency distribution also includes **cumulative proportions** (or percentages), which tell the reader what portion of the total is at or below a given point. Notice, for instance, that about half (49.5 percent) of the respondents think of themselves as a Democrat of one kind or another.

A glance at the table shows how helpful a summary distribution can be. You can, for example, read the number and percentage (relative frequency) of people in the sample who give any specific response. For example, 197 respondents out of a total of 1,193 called themselves "strong Democrats." That is, $(197/1193) \times 100 = 16.5$ percent. The frequency distribution reveals that Americans are roughly evenly divided among the party identification categories and that, perhaps surprisingly to someone who feels that political independence has become an important aspect of politics in the United States, only about 116 out of 1,193 respondents or less than 10 percent (9.7 percent) think of themselves as pure independents. Most people either identify with or lean toward one of the two parties.

PROBABILITIES AND ODDS.   Let's look at the frequency distribution once more for further information and interpretations. Start with the number of people who say they think of themselves as "strong Republicans." If you look back at Table 11-4, you find that 197 people gave that response. Knowing that 197 people in a sample of American adults pick "strong Republican" tells us very little. So we calculated the proportion $p = 197/1193 = .165$. We can also interpret this number as a probability. (We will continue to use the same symbol, p.) So you might prefer to read "$p = .165$" as saying, "the probability of a randomly selected adult American identifying as a strong Republican is about .165" or "there are about 16 or 17 chances in a hundred that a person will select that label." Later on we use p to estimate a corresponding population probability or proportion, P.

In this text we define *probability* in commonsense terms as the likelihood that, in the long run, an observation will have a specific characteristic or an event will occur. This is known as a "relative frequency approach." The proper definition and conceptualization of probability is a bone of contention in statistics and social science, and we do not discuss it further here. Suffice it to say that (1) a probability must lie between 0 and 1.0, inclusive, and (2) for a random process, such as a coin toss or responses to a poll question, the sum of the probabilities of all the individual outcomes must equal 1.0. (This just means something must happen.).

A related concept is the odds. **Odds** refer to the extent or degree to which one response, event, or characteristics exceeds another response, event, or characteristic. A couple of examples may clarify this point. The 2004 NES questionnaire asked respondents about their positions on a law that would prohibit late-term or partial-birth abortions.[5] From the entire sample of 1,122 cases (exclusive of missing data), 705 people said they would favor such a law and 417 said they would oppose it. What are the odds that a person supports prohibition of this procedure? They are 705 to 417, or about 1.7 to 1. (Divide 417 *into* 705.) Stated slightly differently, Americans in 2004 were about 1.7 times as likely to back a law forbidding this kind of procedure as they were to disagree with such a provision. Or, for every 1 person against the law, 1.7 individuals favored it.

Odds, in essence, compare the frequency of one response versus the frequency of another. Accordingly, the way to find odds is to divide the number or frequency of the first response by the number of the second.[6] Thus odds can run from zero to infinity. When no one makes a given response or has a specific trait, its odds are nil. However, if there are no cases in the second group,

How It's Done

Proportions, Percentages, and Odds

Consider a nominal or ordinal variable, Y, with K values or categories and N observations. Let f_k be the frequency or number of observations in the kth category or class. (k goes from 1 to K.)

$$\sum_{k=1}^{k}$$ means add frequencies or proportions starting at 1 and stopping at k.

The *proportion* or *relative frequency* of cases in the kth category is $\dfrac{f_k}{N} = p_k$.

The cumulative proportion in the kth category is $\displaystyle\sum_{k=1}^{k} p_k = p_1 + p_2 + \ldots + p_k = P_k$

The *percentage* of cases in the kth category is $\dfrac{f_k}{N}(100)$.

Let j and k be two different categories of Y with frequencies f_j and f_k, respectively.

The *odds* of being in class j as opposed to k is $\dfrac{f_j}{f_k} = O_{jvsk}$.

the odds in favor of the first are infinite. Numbers in between are read as "such and such is this much more likely than that." (Odds can be a difficult concept, so try using that phrase if someone gives you the "odds" of an event.)[7]

A variable that has more than two classes or categories will have several possible odds. Go back to the distribution of partisanship (see Table 11-4), for which there are seven categories. You could calculate, for example, the following two odds: the odds of a person's being a strong Democrat as opposed to a strong Republican (they are, by the way, 197/197 = 1.0) and the odds of a person's being a strong Democrat rather than an independent (this time, they're 197/116 = 1.7). (Did you notice that a sample participant's likelihood of being a strong Democrat is the same as the odds of being a strong Republican, namely 1 to 1; but the odds of being a strong Democrat [or Republican] are nearly twice those of being an independent. Remember this when people tell you the parties are losing favor with the public.) Given that a distribution with more than two categories will have several odds,[8] how do you choose which to report? There is no easy answer. The best bet is to select one category to serve as a reference or baseline and find the odds of being in that category rather than in each of the other categories. Frequently the nature of the problem or hypothesis will suggest a reference category.

Odds are *not* the same as probabilities. The probability of being for a ban on late-term abortion is 705/1122 = .63. (If you pick 100 people at random, about 63 will favor the ban. This is roughly 1.7 times as many who will oppose it, 417/1122 = .37 or 37 out of 100.)

Why bother spending so much time on these quantities? Besides having descriptive uses in their own right, probabilities and odds underlie a lot of advanced analysis techniques. (Chapter 13 takes up logistic regression, a procedure that has become perhaps one of the most widely applied tools in political science. It makes explicit use of odds and the logarithm of odds, which is known as the logit.) These techniques can be combined or manipulated in various ways into more advanced methods that we take up in ensuing chapters.

MISSING DATA, PERCENTAGES, AND PROPORTIONS. Let's conclude this section by nailing down a very important point. When calculating and describing proportions or percentages or odds, it is essential to have a firm grasp on the *base,* or denominator, of all these numbers. Missing data such as nonresponses can cause difficulties in statistical analyses if not handled appropriately. Not everyone is willing to identify with a party or to publicly take a side on a controversial issue. This nonrecorded information is usually referred to as missing data. Look at Table 11-5. It shows that 152 people out of a total of 1,212 (12.5 percent) did not offer a substantive response to the question about working women or their responses were for one reason or another not recorded. We thus have two totals. The first is the sample size—often designated by capital *N*—that is, the total of all the cases in the study, regardless of whether or not information is available for each and every person. Second, there is a subtotal of "valid" or recorded responses for each item. Look at the fourth column, where you will find substantive, or valid, responses. (The total is 311 + 138 + 610 = 1,059.) The percentages of the valid responses are *not* the same as the total percentages.

TABLE 11-5
The Effect of Missing Data on Percentages

Response	Frequency	Total Percent	Valid Percent	Cumulative Percent
Agree	311	25.7	29.4	29.4
Neither agree nor disagree	138	11.4	13.1	42.4
Disagree	610	50.3	57.6	100
Valid Responses	1,059	—	100	
Missing	152	12.5		
Total	1,212	100		

Source: 2004 National Election Study.
Note: Question: "A working mother can establish just as warm and secure a relationship with her children as a mother who does not work. (Do you agree, neither agree nor disagree, or disagree with this statement)?"

TABLE 11-6

The Effect of Missing Data

	Raw Frequency	Percent of Total *N*	Percent of "Valid" *N*
Approve	160	16	80
Disapprove	40	4	20
Subtotal	200	—	100
Missing	800	80	
Total	1,000	100	

(Pay attention to the last column, which contains cumulative percentages. About 42 percent of the sample either "agree" *or* "neither agree nor disagree" with the statement.)

When discussing your results, and especially when calculating percentages or proportions, it is essential to keep the two totals separate and to use the one most appropriate to your study's purposes. In this case, 29.4 percent *of those (1,059) respondents with substantive or valid responses* agreed that a working woman can establish "just as warm and secure a relationship" with the family as a stay-at-home mom. In the complete data set the value was 25.7 percent, a difference of about 2 percentage points. Here the numbers differ only slightly, but such will not always be the case. (By the way, are you surprised that more than half of the subjects in this survey disagreed with the statement about working mothers?)

The differences in percentages between those with a valid response and the complete data set may be considerable. Why might this be important? Imagine that someone tells you that a survey proves that 80 percent of Americans favor stem cell research. You might assume that since only 20 percent are opposed, there must be overwhelming sentiment for this scientific activity. But suppose 1,000 people took part in the poll, and only 200 gave "approve" or "disapprove" responses. All the others were recorded as "don't know what stem cell research is," "no opinion," or "refused." In this case the "80 percent" in favor could be potentially misleading. Table 11-6 shows the hypothetical distribution responses.

We see that only 20 percent of the *total* sample took any position. For the overwhelming number of respondents, there is no interpretable or meaningful response, so you might hesitate to make sweeping generalizations about the public, given these data. The lesson: Always be aware of the base of a percentage calculation.[9] If you choose to present only the valid responses, you should also tell your audience which categories have been excluded and the number of cases in each.

Descriptive Statistics

Frequency distributions, like those displayed in Tables 11-3, 11-4, 11-5, and 11-6, help us make sense of a large body of numbers and consequently are a good first step in describing and exploring the data. They have, however, a couple of shortcomings. First, and perhaps most obvious, it would be nice to have one, two, or at most a few numerical indicators that would in some sense describe the crucial aspects of the information at hand rather than keeping track of many relative frequencies, proportions, or percentages. Another problem with the frequency distributions is that they aren't much help in describing quantitative (interval and ratio) variables, for which there is often just one observation for each observed value of the variable. If you refer to Table 11-1, for instance, you can see that the 36 nations have different gross national products (GNPs). For these reasons analysts turn to descriptive statistics.

A **descriptive statistic** is a number that, because of its definition and formula, describes certain characteristics or properties of a batch of numbers. These descriptive indicators have two important applications:

■ As the name suggests, they provide a concise summary of variables. If interpreted carefully and in conjunction with knowledge of the subject matter, they can help answer questions such as "Are women more liberal than men?" or "Are democracies less stable than other kinds of polities?"

■ The word *statistic* in the term reminds us that these measures underlie statistical inference, the process of estimating unknown or unmeasured population characteristics from a sample. For example, if a sample shows that 10 percent more women than men identify themselves as strong Democrats, can we say the difference holds for the adult population of Americans as a whole? What is the best guess of the size of the gender gap? On what grounds does the answer rest? Assuming certain assumptions have been met and the data come from a probability sample, nearly all of the statistics we present in this and later chapters have the properties of not only describing data but also of being mathematically justifiable estimators of population parameters.

What kinds of statistics are used? In this section we describe statistics for measuring central tendency, variation, and the occurrence or rates of events or properties. In later chapters we introduce statistics that describe the association or connection between two variables. We further organize the statistics by level of measurement.

Measures of Central Tendency

Formally speaking, a measure of **central tendency** locates the middle or center of a distribution, but in a more intuitive sense it describes a typical case. A measure of central tendency applied to Table 11-1 can tell you the average or typical GNP or rate of infant mortality across the 36 nations.

THE MEAN. The most familiar measure of central tendency is the **mean,** called the *average* in everyday conversation. A simple device for summarizing a batch of numbers,[10] it is calculated by adding the values of a variable and dividing the total by the number of values. For example, if we want the mean of the variable "physicians per 100,000 population" for the 19 developing nations in Table 11-1, we just add the values and divide by 19:

$$\bar{Y} = \frac{\begin{array}{c}(25.7+5.9+15.6+135.0+590.6+187.6+147.6+53.5+25.0\\+13.9+2.2+70.2+51.5+2.7+37.4+110.7+4.7+54.5+2.3)\end{array}}{19} = 80.9$$

The mean, denoted $\bar{Y}$ (read as "Y bar"), is thus 80.9 physicians for every 100,000 people. In plain English, there about 81 doctors for every 100,000 people in a typical developing nation. Incidentally, the corresponding mean for the 17 developed nations is 321.6, quite a difference.

Another example: The average of the "press freedom" index scores for the six African nations in Table 11-1 is

$$\bar{Y} = \frac{(40+61+54+45+84+51)}{6} = \frac{335}{6} = 55.83$$

The mean is appropriate for interval and ratio (that is, truly quantitative) variables, but it is sometimes applied to ordinal scales in which the categories have been assigned numbers, as in the previous example. After all, everyone uses the mean to get grade point averages (GPAs), which are usually based on assigning a value of 4.0 for an A and so forth. Another substantive example is the mean political ideology in the 2004 NES data. The questionnaire asks respondents to place themselves on a seven-point liberalism-conservatism scale, for which the responses are coded 1 for "extremely liberal," 2 for "liberal," 3 for "slightly liberal," 4 for "moderate," 5 for "slightly conservative," 6 for "conservative," and 7 for "extremely conservative." (These integers have no inherent meaning; they are just a way of coding the data.) The mean of the 907 cases with valid (that is, nonmissing) values is 4.28. Since the center of the scale, 4, represents a middle-of-the-road position, a mean scale score of 4.28 suggests that the sample is very slightly conservative. As we will say countless times, numbers do not speak for themselves. A mean ideology score of 4.28 does lean a tad in the conservative direction, but is there a functional or practical consequence? Just how conservative is public opinion? These are questions statistics cannot answer. Only a political scientist or journalist can. The result agrees with many academic studies showing that Americans are not particularly ideological in their politics.

Of course, summarizing even quantitative data with a single figure has a potential disadvantage because the message contained in one number provides incomplete information. How many

How It's Done

The Mean

The mean is calculated as follows:

$$\bar{Y} = \frac{\sum_{i=1}^{N} Y_i}{N}, \quad \text{where the symbol } \sum_{i=1}^{N} Y_i \text{ means}$$

summing Y values starting with 1 and continuing until all N of them have been added.

times have you heard a claim like "my GPA doesn't reflect what I got out of college?" Or suppose a classmate has a GPA of 3.0 (on a 4.0 grade scale). It is impossible to learn from that indicator alone whether she excelled in some courses and struggled in others or whether she consistently received B's.

Furthermore, although the mean is widely known and used, it can mislead the unwary. Here's a simple illustration. Suppose you have been told that Community A has a lower crime rate than Community B. You hypothesize that the gap stems partly from differences in economic well-being. To test this supposition you take a random sample of 10 households per community, obtain the family income of each household, and compute the means for both neighborhoods. The results appear in Table 11-7. The mean income of Community A is $37,500; the mean for Community B is $20,500. Since Community A has a higher average (look at the bottom row of the table), you might believe the hypothesis holds water.

On closer inspection, however, note that the incomes are identical in each community except for the last one. These two families have substantially different earnings. Concentrating on just the mean income of the communities and ignoring any large values would give you the erroneous impression that people in Community A are financially much better off than people in Community B. In reality only one family in A is much better off than others in B. Could this one family explain the difference? Possibly, but before deciding, look at how the mean is calculated. For both communities, the first nine numbers add to 10,000 + 10,000 + 12,000 + . . . + 30,000 = 175,000. For Community A the last income, $200,000, brings this total to $375,000, which when divided by 10 (the number of cases), gives a mean of $37,500. But for Community B the last addition is $30,000, for a total of $205,000. When the total is divided by 10, we get a mean of $20,500. This example illustrates how one (or a few) extreme or very large (or small) values can affect or skew the numerical magnitude of the

TABLE 11-7

Hypothetical Incomes in Two Communities

Community A	Community B
$10,000	$10,000
10,000	10,000
12,000	12,000
18,000	18,000
20,000	20,000
22,000	22,000
25,000	25,000
28,000	28,000
30,000	30,000
200,000	30,000
$\bar{Y}$ = $37,500	$\bar{Y}$ = $20,500

TAKE IT EASY ON THE READER
When presenting the results of a statistical analysis, you can help your readers by doing a couple of things:

- Whenever possible, make sure the measurement units are clear. Instead of writing, for example, "the mean is 37,500," state what the number refers to: "the mean is 37,500 dollars." If the numbers appear in a table, use a typographical symbol such as "$" for dollars or "%" for percent. Thus write $37,000, not simply 37,000. (And, of course, in a printed or published report, use commas as needed.) This won't always be possible because a variable may be measured on a rather abstract scale as with an index of liberalism-conservatism. But the point is to make your results as reader friendly as possible.

- Don't present a lot of unnecessary decimal places. Suppose a mean liberalism-conservatism score is 2.0884163. You lose nothing by rounding and keeping just one or two decimals (for example, 2.1 or 2.09) in the finished report. Again, the lesson is straightforward: spare the reader.

How It's Done

The Trimmed Mean

1. Order the data from lowest to highest values.
2. Decide how many observations to drop from the low and high ends of the variable. Call this integer t.

The trimmed mean is calculated as follows:

$$\overline{Y}_{trimmed} = \left(\frac{Y_{t+1} + Y_{t+2} \cdots Y_{N-2t} + Y_{N-t}}{N-2t} \right) = \frac{1}{N-2t} \sum_{i=t+1}^{N-t} Y_i.$$

mean (and other statistics). For this reason, other measures of central tendency, known as **resistant measures,** which are not sensitive to one or a few extremes values, are frequently used.

THE TRIMMED MEAN. Because the mean is susceptible or sensitive to a few very large or small values—in statistical argot, it is "not resistant to outliers"—analysts sometimes modify its calculation by dropping or excluding some percentage of the largest and smallest values. The result is called the **trimmed mean.** This tactic automatically removes the influence of the discrepant values. Return to the previous example, in which we want to compare the incomes of communities A and B. If we drop the first and last observations from each community, we have 8 cases each and the means are both $20,625. Seen this way, the communities have the same standard of living. You can see that both will be the same by noting that the calculation of the trimmed mean turns out to be the same for both communities:

$$\overline{Y}_{trimmed} = \frac{(10,000 + 12,000 + 18,000 + 20,000 + 22,000 + 25,000 + 28,000 + 30,000)}{8} = 20,625$$

Most software programs will produce trimmed means along with the standard arithmetic mean. The usual default is to exclude the highest and lowest 5 percent of values, but some programs let the user choose the percentage. (In our example, we effectively dropped 20 percent, but only because we had such an artificially low number of cases.)

To further demonstrate the effects of trimming, let's return to the variable "physicians per 100,000 population." For the 19 less developed countries, the mean was 80.9 doctors. The raw data (ordered from smallest to largest) are shown in Table 11-8. If we drop the highest and lowest values (2.2 and 590.6), in effect trimming the data by about 5 percent at each end, and recalculate the mean, we find

$$\overline{Y}_{trimmed} = \frac{\begin{array}{c}(2.3 + 2.7 + 4.7 + 5.9 + 13.9 + 15.6 + 25.0 + 25.7 + 37.4 \\ + 51.5 + 53.5 + 54.5 + 70.2 + 110.7 + 135.0 + 147.6 + 187.6)\end{array}}{17} = 55.5$$

This mean is considerably smaller than the original (unadjusted) mean (80.9). What happened? Notice that one country (Cuba) has a recorded physician rate of 590.6, a value that lies far above the next closest (187.6, in the Dominican Republic). By removing this value (and the lowest one, which is not really out of line with the others), we show what might be considered the typical value of developing nations. Moreover, this finding encourages us to inquire about the Cuban figure. Is it a mistake? Is there a recording error? Or has the rate been noted correctly, and there is something about Cuba's health care system that needs to be taken into account? (This would be a good topic for discussion and might lead to some interesting insights about public perceptions of our southern neighbor.)

Finally, we also said that the mean for the developed countries is 321.6; the trimmed mean for them is 321.3, practically the same as the unadjusted mean, a fact that suggests there are no extremely large or small rates among them. To wrap up, we can conclude that (1) by this indicator of health care, the developing nations trail far behind the more industrialized ones (321 versus 55.5 physicians per 100,000) and (2) Cuba has a rate equal to or better than most of the developed countries, including the United States.

How valid is this procedure? It might seem like a case of lying with statistics. Yet statisticians regard it as standard practice because they feel the technique resists the effects of a few extreme values. (As we explain later, variation of the data is also reduced.)

Inasmuch as trimming adjusts the arithmetic mean by dropping a small percentage of cases, it is, like the arithmetic mean, appropriate for interval or ratio scales but typically would not be applied to ordinal scales.

TABLE 11-8

Physicians per 100,000 Population

Country	Number of Physicians
Malawi	2.2
Tanzania	2.3
Mozambique	2.7
Rwanda	4.7
Burkina Faso	5.9
Kenya	13.9
Cambodia	15.6
Haiti	25.0
Bangladesh	25.7
Nicaragua	37.4
Morocco	51.5
Egypt	53.5
Sri Lanka	54.5
Malaysia	70.2
Paraguay	110.7
Colombia	135.0
Ecuador	147.6
Dominican Republic	187.6
Cuba	590.6

Source: See Table 11-1.

THE MEDIAN. A measure of central tendency that is fully applicable to ordinal as well as interval and ratio data is the median. The **median** (frequently denoted M) is a (not necessarily unique) value that divides a distribution in half. That is, half the observations lie above the median and half below it. Stated differently, to find the median we need to locate the middle of the distribution.

You can find the middle of an *odd* number of observations by arranging them from lowest to highest and counting the same number of observations from the top and bottom to find the middle. If, for example, the data consist of seven values, you will count in three observations and the middle observation will be the fourth. The seven Latin American nations in Table 11-1 have press freedom index scores (ordered smallest to largest) of 38, 41, 42, 56, 63, 66, and 96. The fourth in line is 56, which is the median index for these countries: three values lie below 56 and three above it, so the median divides the distribution in half.

If you have lots of observations, an easy way to find the middle one is to apply the following formula:

$$mid_{obs} = \frac{(N+1)}{2}$$

(For the previous example, this formula yields $(7 + 1)/2 = 4$, as it should.)

If, however, the number of observations is even, a modification is required because the middle observation number will contain a .5. (For example: $(6 + 1)/2 = 3.5$.) What to do? Simply use the observations above and below mid_{obs} and average their values. For the 12 European countries in Table 11-1, the median press freedom index scores are (again, from smallest to largest) 9, 9, 10, 11, 11, 11, 14, 15, 21, 22, 28, and 35. The formula for determining the middle two observations gives $(12 + 1)/2 = 6.5$. Thus we average the values for the sixth and seventh observations, 11 and 14, to find the median:

$$M = \frac{(11+14)}{2} = 12.5$$

Briefly, the median press freedom index for these 12 European nations is 12.5. Since low scores represent greater freedom than do high ones, we see that on average—as indicated by the median—there is more freedom of the press in Europe than in Latin America.

The median, like the trimmed mean, is a resistant measure in that extreme values (outliers) do not overwhelm its computation. Figure 11-1 shows the calculation of the median for the two hypothetical communities discussed earlier (see Table 11-7). Recall that the means of the two differed quite a bit: $\bar{Y}_A$ = \$37,000 versus $\bar{Y}_B$ = \$20,500. But the medians are identical, \$21,000. This reflects the fact that the incomes and hence the standards of living in the two areas are essentially the same. Consequently, if A's crime rate is lower than B's, the cause may be something other than wealth.

Returning to a more realistic case, we can compare the means, trimmed means, and medians of the developed and developing countries for the number of physicians per 100,000 people (Table 11-9). For practical purposes, all three statistics for the developed nations are the same. No matter how you

How It's Done

The Median

This procedure is practical if the number of cases, *N,* is not large (say, less than 30 to 40):

1. Sort the values of the observations from lowest to highest.
2. If the number of cases is an odd number, locate the middle one and record its value. This is the median.
3. If the number of cases is an even number, locate the two middle values. Average these two numbers. The result is the median.

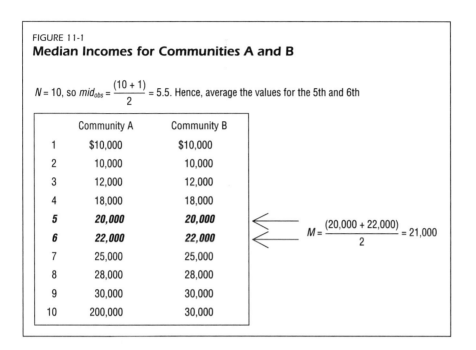

FIGURE 11-1
Median Incomes for Communities A and B

$N = 10$, so $mid_{obs} = \dfrac{(10 + 1)}{2} = 5.5$. Hence, average the values for the 5th and 6th

	Community A	Community B
1	$10,000	$10,000
2	10,000	10,000
3	12,000	12,000
4	18,000	18,000
5	**20,000**	**20,000**
6	**22,000**	**22,000**
7	25,000	25,000
8	28,000	28,000
9	30,000	30,000
10	200,000	30,000

$$M = \frac{(20,000 + 22,000)}{2} = 21,000$$

look at it, these statistics suggest that there are more than 320 doctors for every 100,000 people in the developed countries. By contrast, the developing areas are much worse off. And pay attention to the discrepancy between the mean and the two resistant measures (the trimmed mean and the median), a gap that points to the existence of one or more outlying values in the data.

THE MODE.   A common measure of central tendency, especially for nominal and categorical ordinal data, is the **mode,** or modal category. It is simply the category with the greatest frequency of observations. As an example, start with Table 11-4, which shows the distribution of responses to a party identification question from the 2004 NES. The modal (most frequent) answer was "independent-leaning Democratic," with 208 responses.

TABLE 11-9
Comparison of Measures of Central Tendency

	Physicians per 100,000 Population	
Measure of Central Tendency	Developed	Developing
Mean	321.6	80.9
Trimmed mean	321.3	55.5
Median	328.4	37.4

The mode has less utility in describing interval and ratio data per se but is helpful in descriptions of the *shape* of distributions of all kinds of variables. When one category or range of values has many more cases than all the others, we describe the distribution as being unimodal, which is to say it has a single peak. But there can be more than two dominant peaks or spikes in the distribution. In these cases we speak of multimodal distributions. Look carefully at the distribution of party identification in Table 11-4 (third column). You will see a more or less even distribution of observations; the relative frequencies vary roughly between 10 percent and 20 percent. These data don't really have a pronounced peak. Yes, there is a mode (independent-leaning Democrat), but it's barely larger than the other categories. The term *rectangular* is typically used to describe this type of distribution. Graphs of distributions shown later in this chapter illustrate these properties.

Measures of Variability or Dispersion

We come now to a key concept in statistics: variation or the differences among the units on a variable. Naturally we want to know what a typical case in our study looks like. But equally important, we need to take stock of the variability among the cases and understand why this variation arises.

GENERAL REMARKS. Go back to Table 11-1. Consider the IMR variable, the data in the second column. The values at the top of the table, for the mostly European and North American countries, lie between 2 and 6 infant deaths per 1,000 live births; the figures for the other countries range between 6 and 118. The former seem more or less alike, whereas the latter differ considerably among themselves. Or examine the index of political rights ("Rights"). Even a quick glance indicates that the developed countries all have the same score, 1, while those for the less developed nations vary between 2 and 7. Here we have two examples of variation or differences between units of analysis such as people or countries or organizations.

In this section we do a couple of things:

- Briefly discuss the causes or reasons for observed variation in a data set
- Describe a few statistical measures that summarize and quantify the amount of variation

Why do things differ among themselves? This question, we contend, lies at the heart of social and political research. As a matter of fact, hard-nosed social scientists might argue that explaining variation is one of the very purposes of empirical research. If men's and women's political attitudes differ on the whole, one naturally wants to know why. Is it just biology? Or, and more

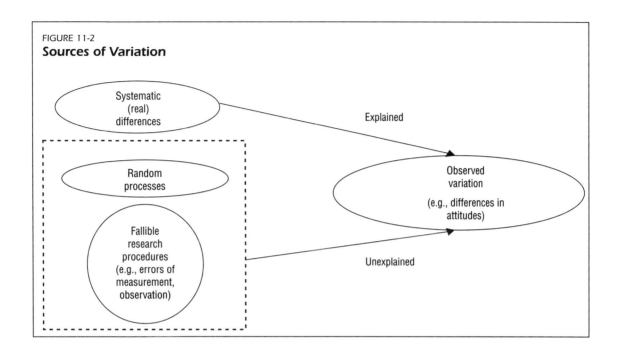

FIGURE 11-2
Sources of Variation

likely, do the sexes have different socialization, educational, employment, and life experiences that cause them to disagree about politics? Of course, it's possible that there are no real differences and that any observed diversity in attitudes and behavior results purely from observational errors or happenstance. Or perhaps it is a combination of systematic *and* random factors. When we discuss modeling relationships in Chapters 12 and 13, we will see in more detail how to sort out these issues. For now, we list three possible sources of variation (see Figure 11-2):

■ Systematic differences. We might think of these as causal variables. Knowing what they are makes variability in the data intelligible or understandable. If we decide, for example, that male-female differences can be traced to childrearing and schooling practices, we have a substantive explanation of the differences in attitudes and behavior between men and women.

■ Random processes. Quintessential examples of a random process include the toss of a (fair) coin, the throwing of two dice, and the spin of a roulette wheel. The essence of randomization is that the outcomes just occur; there is no other explanation for them. Maybe some women just happen to be slightly more liberal than some men, and there is nothing more to say. Obviously, this account is not very satisfying, which is why we search for other possibilities. In the end, we may be forced to conclude that some differences can't be explained and that we have to attribute variation to randomness.

■ Fallible research procedures. Another possibility is that variation stems at least in part from mistakes or shortcomings in operational definitions, measurement, observation, recording, analysis, or any of a multitude of decisions and acts that go into an empirical study. Most of what we know about public opinion comes from surveys, and, as we have seen throughout this book, surveys are subject to errors of one kind or another.

Throughout this and later chapters, we assume that the errors cancel each other out. They are, that is, random. If, for example, one person's score on a variable is accidentally recorded as one unit too high, there will be another score in the study that is one unit too low. Needless to say, this is a big assumption and as often as not it is violated. Indeed, it is easy for unmeasured or unobserved biases of one sort or another to creep into research. Social scientists and statisticians are fully aware of the problem and devote considerable time and effort to dealing with it. However, the complexity of the voluminous literature on the subject goes far beyond the scope of this book, and we simply assume that our data are subject only to random error. In this section we simply describe a few methods for measuring the amount of *observed* variation.

PROPERTIES OF MEASURES OF VARIATION. Ideally a single summary number would clearly express the exact amount of variation in a variable. That way we could precisely or objectively compare variability among groups of objects or apportion it among known and unknown causes. In fact, many single indices of variation exist, but many lack a simple, commonsense interpretation. Let's use the symbol V for some as-yet undefined statistical measure of variation. In the limiting case, if all the units in a batch of data have the same value, V is zero. All measures have at least this property. And this trait seems reasonable only because it is impossible for there to be less than zero variation; hence, V will always be equal to or greater than zero. But does V have an upper limit? Nearly all of the measures considered herein do not have an upper bound, which means that they can take on any positive value subject to constraints of the measurement scales. Unfortunately, the substantive interpretation of the numerical value of a measure of variation is not always obvious. Suppose you are told V equals 2.0. What exactly does this mean? The number will be pretty abstract unless you understand V's definition and the theoretical or practical content of the variable it describes. Even then the meaning may not be intuitively satisfying, so it is often helpful to interpret measures of variability with the help of other statistics and graphs.

The properties of measures of variation or **dispersion** described in this book can be summed up in three statements:

1. If there is *no* variability (all the scores have the same value), the measure will equal 0.

2. The measure will always be a positive number (there can't be less than no variation).

3. The greater the variability in the data, the larger the numerical value of the measure.

THE RANGE.   For interval- and ratio-level scales, the **range** is a particularly simple measure of gross variation: it is just the largest (maximum) value of a variable minus the smallest (minimum) value:

$$\text{Range} = \text{maximum} - \text{minimum}.$$

Look carefully at the GNP figures in Table 11-1. The largest value is $68,800 (Luxembourg) and the smallest is $600 (Malawi). The range in incomes is therefore $68,800 − $600 = $68,200. In plain language, an enormous disparity exists in well being between the richest and poorest countries in this sample. Now consider the political rights index for the developed nations: the highest and lowest scores are both 1, so the range is 0. This example illustrates the property that when all units have the same value, a measure of variation will be 0.

INTERQUARTILE RANGE.   Finding the range, of course, requires sorting or ordering the data from largest to smallest or at least finding the maximum and minimum values. With a large data set, this chore can be arduous, but computers make it easy. Yet however it is done, getting the data in ascending or descending order can be handy because another measure of variation, the interquartile range, is easily computed from data ordered in this way. (And, as we pointed out earlier, the median is found by sorting the numbers.)

Imagine that, after ordering your data, you divide them into four equal-sized batches. The first bunch would contain 25 percent of the cases, the next would have 25 percent, the next 25 percent, and the last the remaining 25 percent. The division points defining these groups are called quartiles, abbreviated Q. Now, find the range as before but using the third and first quartiles as the maximum and minimum. Their difference is the **interquartile range** (IQR):

$$\text{IQR} = Q3 - Q1,$$

where Q3 stands for the third quartile (sometimes it's called the 75th percentile because 75 percent of the cases lie below it) and Q1 is the first (or

FIGURE 11-3
Calculating the Interquartile Range

Obs.	Percent in agriculture	Percent in agriculture ordered	
1.	4.3	1.5	
2.	1.6	1.6	
3.	2.1	1.8	
4.	14.6	2.1	
5.	7.2	2.3	← Q1 = (2.1+2.3)/2 = 2.2
6.	8.8	2.7	
7.	2.3	3.0	
8.	4.4	3.2	
9.	3.2	4.0	← M = Q2 = 4.0
10.	1.5	4.3	
11.	3.0	4.4	
12.	8.6	6.0	
13.	4.0	7.2	
14.	11.1	8.6	← Q3 = (7.2+8.6)/2 = 7.9
15.	6.0	8.8	
16.	2.7	11.1	
17.	1.8	14.6	

Interquartile range (IQR) = 7.9 − 2.2 = 5.7

25th percentile). Since the interquartile range, a single number, indicates the difference or distance between the third and second quartiles, the middle 50 percent of observations lie between these values.

Another way to think about the computation of the IQR is to obtain the median, which divides the data in half. Then, find the medians in *each* half; these medians will be the first and third quartiles. *Remember: Whenever you have an even number of cases, the median will be the average of the two middle cases.* Figure 11-3 shows how the IQR is calculated for the 17 developed nations that have values on agricultural workers.

Thus 50 percent of the developed nations have agricultural workforces that constitute between 2.2 percent and 7.9 percent of their populations. There does not seem to be much difference among them in this regard. Not surprisingly, perhaps, much more diversity is found in the developing world. In our sample the IQR for agricultural labor is 60.4 percent. This relatively large number indicates that some of these countries have turned away from agriculture, whereas many others have not. A number of early studies on political development argue that democracy requires at least a modest transition from a rural to an industrial economy.[11]

Quartiles, the range, and the interquartile range (as well as the median) have the property that we have been calling resistance: extreme or outlying values do not distort the picture of the majority of cases, which is a major advantage, especially in small samples. Although these measures of dispersion are useful, they do not play as important a role in more advanced statistics as do three common measures of variation: the mean absolute deviation, variance, and the standard deviation, which we now introduce.

Deviations from Central Tendency

These statistics evaluate deviations from means. A deviation from the mean is simply a data point's distance from the mean.[12] Let's take a hypothetical case. Suppose the mean of a variable is 50. If a particular observation has a value of, say, 49, it deviates from the mean by just one unit. Another case that has a score of 20 departs from the mean by 30 units. Now, suppose we find out how far each observation's score is from the mean and total these deviations. We would then have the beginnings of a measure of variation. Artificial data demonstrate the point (Table 11-10). We have three samples each containing just three observed values.

The first data set clearly has no variation. The numbers in the second set are almost the same, while those in the third exhibit considerable diversity. The measures of variation to be discussed combine the deviations (the third set of columns) into a single indicator that summarizes the data.

THE MEAN ABSOLUTE DEVIATION.   First consider a quantitative (ratio or interval) variable. The **mean absolute deviation** measures variation or, as it is sometimes called, dispersion by totaling the absolute value of the deviation of each score from the mean. This sum is then divided by the number of cases (N). Table 11-11 shows a sample calculation for the infant mortality rate of the 17 developed nations. We find the mean of the IMR scores—it turns out to be 4.059—and then subtract each score from this mean. These calculations give

---

TABLE 11-10
Deviations from the Mean

Observed Values	Mean	Deviations from the Mean	Interpretation
50 50 50	50	0 0 0	No deviation implies no variation.
49 50 51	50	−1 0 1	Small deviations imply little variation.
20 50 80	50	−30 0 30	Large deviations imply considerable variation.

TABLE 11-11
Calculation of the Mean Absolute Deviation (MAD)

Country	IMR (Y)	Deviations (IMR–Mean)	Absolute Value of Deviations (|IMR–Mean|)
Austria	4	–0.059	0.059
Belgium	4	–0.059	0.059
Canada	5	0.941	0.941
Greece	4	–0.059	0.059
Iceland	2	–2.059	2.059
Ireland	5	0.941	0.941
Israel	5	0.941	0.941
Italy	4	–0.059	0.059
Japan	3	–1.059	1.059
Luxembourg	4	–0.059	0.059
Netherlands	4	–0.059	0.059
New Zealand	5	0.941	0.941
Norway	3	–1.059	1.059
Portugal	4	–0.059	0.059
Spain	4	–0.059	0.059
Sweden	3	–1.059	1.059
United States	6	1.941	1.941
Total	69	0	11.412

$N = 17$

Mean IMR $= \dfrac{69}{17} = 4.059$

MAD $= \dfrac{11.4127}{17} = 0.671$

Source: See Table 11-1.

us the deviation of each value from the mean (see column 3). *The sum of deviations from a mean is always zero—an important mathematical property of the mean.* (Check this by adding the numbers in column 3 for yourself.) But because we are interested only in the amount of deviation and not in the direction or sign of the deviation, we can add the absolute values of the deviations (column 4). Taking the absolute value of a difference simply amounts to ignoring minus signs if present. Then we divide this sum by the number of scores to find the mean deviation of scores from the mean. The larger the mean deviation, the greater the dispersion of scores around the mean.

How It's Done

The Mean Absolute Deviation

The mean absolute deviation is calculated as follows:

$$MAD = \frac{\sum_{i=1}^{N} |Y_i - \overline{Y}|}{N},$$ where the vertical bars or lines,

| |, indicate that you take the absolute value (that is, disregard minus signs) of the difference.

According to this statistic, the average departure or deviation from the mean IMR of the 17 developed nations is about 0.67 or almost one death per 1,000 births. As noted previously, this figure may not have a lot of intrinsic meaning on its own. But we can compare it to mean absolute deviation for the 19 developing countries, which is 33.1 infant deaths. Compared with the more developed nations, the developing countries have considerably more heterogeneity on this variable.

THE VARIANCE.   The **variance** is the average or mean of squared deviations, or the average of all the squared differences between each score and the mean. Denoted σ^2, the variance closely resembles the mean absolute deviation, but instead of working with the absolute values of the deviations, they are squared. More precisely, the variance is the sum of the squared deviations from the mean divided by N. So, to calculate the variance for the infant mortality rate for developed nations, simply square the number in column 4 in Table 11-11 for each country, add up the results, and divide by 17. This turns out to be 11.1702. When calculating the variance for a sample, $N - 1$ appears in the denominator for technical reasons. Since N and $N - 1$ are close when N is large, you can think of the variance and sample variance as a kind of average of squared deviations.

The variance and sample variance follow the rules of a measure of variation: the greater the dispersion of data about the mean, the higher the value of the variance. If all the values are the same, it equals zero. And it is always nonnegative. The variance is a fundamental concept in mathematical and applied statistics, as we shall see shortly.

THE STANDARD DEVIATION FOR A POPULATION AND A SAMPLE.   The most commonly computed and calculated measure of variation is the **standard deviation,** which we denote by σ for a population and by $\hat{\sigma}$ for a sample. The standard deviation is simply the square root of the variance.

The standard deviation (and also the variance), like the mean, is sensitive to extreme values. Here's a quick example to illustrate this property (see Table 11-12). Here we slightly alter the three hypothetical data sets given in Table 11-10.

The first group of numbers (50, 50, 50) have no variation—their deviations from the mean are 0—and, in keeping with our conception of a measure of variation, the standard deviation, σ,

How It's Done

The Variance and Sample Variance

The variance is calculated as follows:

$$\sigma^2 = \frac{\sum_{i=1}^{N}(Y_i - \overline{Y})^2}{N}, \text{ where } Y_i \text{ stands for the } i\text{th}$$

value, $\overline{Y}$ is the mean of the variable, and N is the sample size.

The sample variance is calculated as follows:

$$\hat{\sigma}^2 = \frac{\sum_{i=1}^{N}(Y_i - \overline{Y})^2}{N-1}.$$

How It's Done

The Standard Deviation and the Sample Standard Deviation

The standard deviation is calculated as follows:

$$\sigma = \sqrt{\dfrac{\sum\limits_{i=1}^{N}(Y_i - \overline{Y})^2}{N}},$$ where Y_i stands for the ith

value, $\overline{Y}$ is the mean of the variable, and N is the population size.

The sample standard deviation is calculated as follows:

$$\hat{\sigma} = \sqrt{\dfrac{\sum\limits_{i=1}^{N}(Y_i - \overline{Y})^2}{N-1}}.$$

is 0. The next set (49, 50, 51) has almost no variability, and its standard deviation works out to .82. Inasmuch as we haven't identified a measurement scale, this number does not have much theoretical or practical meaning. But it is quite small (less than one unit) and indicates very little dispersion among the values. The last group (49, 50, 80) has one outlying value, at least compared to the other two, and the standard deviation is more than 17 units. Is there that much variation? According to σ, the answer is yes. But the last value (80) contributes virtually everything to the total and hence inflates the magnitude of the statistic.

It's time to examine a more substantive example. We observed earlier that the physicians rate for the developing nations contained an outlying or very large value, Cuba, with a rate of 590.6 physicians per 100,000 population. For the sample of 19 developing countries, $\hat{\sigma}$ is 135.1, whereas if Cuba is dropped from the calculation $\hat{\sigma}_{(-cuba)}$ becomes 56.5, about half of its previous value. If you look merely at the standard deviation, there would appear to be a great deal of variation in physician availability across the developing world. But in fact, except for Cuba, there is more homogeneity than meets the eye. The corresponding value of $\hat{\sigma}$ for developed nations is 74.6; these countries have considerably less variation, compared to the developing countries with Cuba included, which indicates that they share a more similar level of health care than do the developing nations. If Cuba is excluded, however, there would appear to be less variation among developing nations than developed ones. You would not know this unless you had taken the time to examine the data for outliers or had noticed that the trimmed mean

TABLE 11-12

Calculation of the Standard Deviation for a Population

Data Set 1	Squared Deviations	Data Set 2	Squared Deviations	Data Set 3	Squared Deviations
50	$(50-50)^2 = 0$	49	$(49-50)^2 = 1$	49	$(49-50)^2 = (-1)^2 = 1$
50	$(50-50)^2 = 0$	50	$(50-50)^2 = 0$	50	
50	$(50-50)^2 = 0$	51	$(51-50)^2 = 1$	80	$(80-50)^2 = (30)^2 = 900$
$\overline{Y} = 50$	Sum = 0	$\overline{Y} = 50$	Sum = 2	$\overline{Y} = 50$	Sum = 901
	$\sigma = \sqrt{\dfrac{0}{3}} = 0$		$\sigma = \sqrt{\dfrac{2}{3}} = .82$		$\sigma = \sqrt{\dfrac{901}{3}} = 17.333$

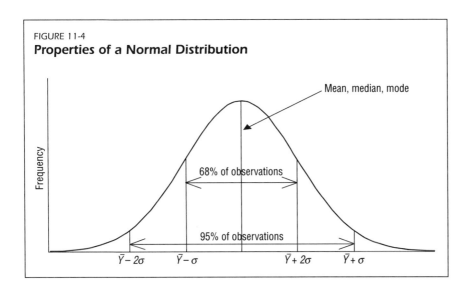

FIGURE 11-4
Properties of a Normal Distribution

Frequency

Mean, median, mode

68% of observations

95% of observations

$\bar{Y} - 2\sigma$ $\bar{Y} - \sigma$ $\bar{Y} + 2\sigma$ $\bar{Y} + \sigma$

and median were different from the mean for developing nations. The lesson in all of this is that you should not rely on a single number to summarize or describe your data. Rather, use as much information as is reasonable, and take your time interpreting each variable. Computers spit out results, but they never interpret them.

MORE ON THE INTERPRETATION OF THE STANDARD DEVIATION. The significance of the standard deviation in statistics is illustrated by considering a common situation. Suppose a large set of data have a distribution approximately like the one shown in Figure 11-4. What we see there is a "bell-shaped" distribution (often called a **normal distribution**) with the following features:

■ The bulk of observations lie in the center, where there is a single peak.

■ More specifically, in a normal distribution, half (50 percent) of the observations lie *above* the mean and half lie *below* it.

■ The mean, median, and mode have the same numerical values.

■ Fewer and fewer observations fall in the tails of the distribution.

■ The spread of the distribution is symmetric: one side is the mirror image of the other.

If data have such a distribution, the frequency or proportion of cases lying between any two values of the variable can be described by "distances" between the mean and standard deviations. Following what Alan Agresti calls the "empirical rule," the spread of observations can be described this way:[13]

■ Approximately 68 percent of the data lie between $\bar{Y} - \sigma$ and $\bar{Y} + \sigma$. Read this as "68 percent of the observations are between plus and minus one standard deviation of the mean." For example, if the mean of a variable is 100 and its standard deviation is 10, then about 68 percent of the cases will have scores somewhere between 90 and 110.

■ Approximately 95 percent of the cases will fall between $\bar{Y} - 2\sigma$ and $\bar{Y} + 2\sigma$. In the first example, 95 percent or so would be between 80 and 120.

■ Almost all of the data will be between $\bar{Y} - 3\sigma$ and $\bar{Y} + 3\sigma$.

This feature of the standard deviation and normal distribution has an important practical application. For all suitably transformed normal distributions, the areas between the mean and various distances above and below it, measured in standard deviation units, are precisely known and have even been tabulated (see Appendix A). Figure 11-4 illustrates this information for one and two standard deviation units above and below the mean.

How do we know these percentages? Because mathematical theory proves that normal distributions have this property. Of course, if data are not perfectly normally distributed, the percentages will only be approximations. Yet many naturally occurring variables do have nearly normal distributions or they can be transformed into an almost normal distribution (for instance, by converting each number to logarithms).

For a substantive example of the connection between areas under the normal curve and standard deviations, examine Figure 11-5, which graphs the distribution of infant mortality rates for the 17 developed nations in our previous example. (Bear in mind that all we are doing now is exploring the data.) As you can see, these data have a bell-shaped form. (Figure 11-5 also has a superimposed graph of a normal distribution.) The mean is just about 4, and the standard deviation is almost 1. Using the empirical rule, we would then expect approximately 68 percent of the countries to have values between 3 (that is, 4 − 1) and 5 (that is, 4 + 1). If you go back to Table 11-1 and count, you should find that the actual value is about 71 percent, which is close, given the small number of cases and the fact that the distribution may not be perfectly normal. (As an exercise, check to see what percentage of the countries fall between the mean and plus and minus two standard deviations. Does your calculation agree with the rule?)

MEASURES OF VARIATION FOR CATEGORICAL DATA. So far we have constructed measures of variability by subtracting numbers such as individual scores from their mean or the maximum minus minimum values. These procedures assume the numbers have mathematical meaning. But what about nominal or ordinal scales? Since the values of these variables are category names or maybe numbers assigned arbitrarily, arithmetical operations do not seem jus-

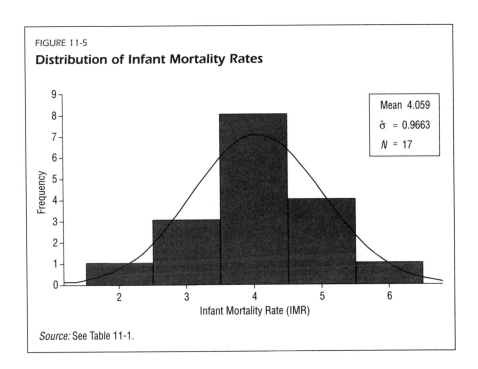

FIGURE 11-5

Distribution of Infant Mortality Rates

Mean 4.059
$\hat{\sigma} = 0.9663$
$N = 17$

(y-axis: Frequency; x-axis: Infant Mortality Rate (IMR))

Source: See Table 11-1.

tified. Still, considerable research has gone into defining and studying indices of variation for nonquantitative variables. The only problem is that these measures are not widely available in textbooks or social science software. Consequently, we describe two such measures that provide insight into the variation of nominal data and that can be calculated by hand, as long as a variable does not have an extraordinary number of classes or categories.

Consider an unordered nominal variable such as religious preference or ethnicity. Each category will have a certain number of observations or cases. The data for this summary measure are frequencies in the categories. Look ahead to Table 11-14, which we will analyze in a moment. It contains the frequencies of the responses to the following question asked in the 2004 NES: "What racial or ethnic group or groups best describe you?" Participants could place themselves in a dozen or so categories, but the choices have been combined into six groups: white, black, Hispanic, Asian, Native American, and other. The table lists the frequencies and proportions in each category. The numbers in these ethnic groups constitute the "data" to be used in calculating the measures.

Many measures of variation can be calculated from these frequencies,[14] but here we discuss just two, the **index of diversity**[15] and the **index of qualitative variation.**[16] The index of diversity (ID) can be interpreted as "the probability that two individuals selected at random would be in different categories."[17] (Given that the index can be interpreted as a probability, this

379

How It's Done

Indices of Categorical Variation

Assume a variable has K categories and a total of N observations. The frequency or count in the kth category is f_k. The proportion in the kth category is p_k.

The index of diversity is calculated as follows:

$$IQV = \frac{1 - \sum_{k=1}^{K}\left(\frac{f_k}{n}\right)^2}{1 - \left(\frac{1}{K}\right)} = \frac{1 - \sum_{k=1}^{K} p_k^2}{1 - \left(\frac{1}{K}\right)}$$

The index of qualitative variation is calculated as follows:

$$IQV = 1 - \frac{\sum_{k=1}^{K}\left(\frac{f_k}{N}\right)^2}{1 - (1/K)} = 1 - \frac{\sum_{k=1}^{K} p_k^2}{1 - (1/K)}$$

TABLE 11-13

Frequencies and Proportions Used to Calculate ID and IQV for Two Hypothetical Categorical Variable Distributions

Category	Distribution A	
	Frequency f_k	Proportion p_k
M	100	1.0
N	0	0
O	0	0
P	0	0
Total	100	1.0
ID = 0		
IQV = 0		

Category	Distribution B	
	Frequency f_k	Proportion p_k
M	25	.25
N	25	.25
O	25	.25
P	25	.25
Total	100	1.0
ID = .75		
IQV = 1.0		

measure of variability is "bounded" between 0, for no diversity, and 1, for complete diversity). Other ways of viewing it are as an index of inequality, heterogeneity, uniformity, disagreement, segregation, and cohesion.[18] The index of qualitative variation (IQV) is a bounded measure in the sense that its value will always range between 0, for no variability, and 1, for the maximum variability given the number of cases. The "How It's Done" box at left gives the formulas for both measures.

Consider the distributions A and B shown in Table 11-13. Each has four categories ($K = 4$) and a total of $N = 100$ observations. In distribution A, all of the observations fall in the first category (M) so that the proportions in this distribution are $100/100 = 1.0$, $0/100 = 0$, $0/100 = 0$, and $0/100 = 0$. Plainly, there is no variation,[19] and indeed the statistic ID works out to be 0, as desired. Hence, there is no chance that two people will be in different categories (or, if you prefer, it's certain they will share the same category).

$$ID = 1 - (1.0^2 + 0^2 + 0^2 + 0^2) = 0.$$

In distribution B, by contrast, the 100 observations are spread evenly among the four categories and the ID works out to be

$$ID = 1 - (.25^2 + .25^2 + .25^2 + .25^2) = .75.$$

This means that two individuals or units selected randomly would have a probability of .75 of being in different categories. Or conversely, they have a 25 percent probability of being in the same category.

Some social scientists see a slight problem with the ID index. Since we are interpreting the statistic as a probability, its maximum value should be 1.0. But it attains the maximum only if every observation is in a category by itself, that is, if $K = N$. The more categories a variable has, the greater the possible diversity as measured by ID. In other words, more variability is possible in a variable having 10 categories than in one having 2 categories. (This is because the upper limit of ID, which we want to be 1.0, can be attained only as the number of values equals the number of observations; that is, ID = 1 if and only if $K = N$.) Here's where the IQV comes in. It is the index of diversity divided by a correction factor that equals the maximum that ID can reach given the number of categories, K. This adjustment ensures that the measure can attain an upper limit of 1.0 even if there are only a few categories. The correction factor is $(K - 1)/K$, or equivalently, $1 - (1/K)$. If a variable has four categories, the maximum possible value of ID is $(4 - 1)/4 = 1 - (1/4) = .75$. Given the way the index is constructed and the fact that there are four categories, the highest probability that two observations drawn at random will fall into different categories is .75.

Going back to the data in Table 11-13, we find that the IQV for distribution A is 0, which indicates no variation, as desired:

$$IQV = \frac{1-(1.0^2 + 0^2 + 0^2 + 0^2)}{1-(1/4)} = \frac{0}{.75} = 0.$$

But for distribution B the IQV turns out to be 1.0.

$$IQV = \frac{1-(.25^2 + .25^2 + .25^2 + .25^2)}{1-(1/4)} = \frac{.75}{.75} = 1.0.$$

This means that, *given the number of categories,* there is a 100 percent chance that two observations drawn at random will be in different categories. Although the IQV may be cited more commonly than the ID, the interpretation of the IQV may be a bit more difficult to grasp because of the qualifier "given the number of categories." The probability refers to what's possible in the situation. Agresti and Agresti suggest that when comparing the variability of two distributions with an unequal number of classes, IQV may be preferred. In general, for most practical applications, either one will suffice.[20]

Let's take up the distribution of ethnicity as presented in Table 11-14. The index of diversity is calculated as follows:

$$ID = 1 - (.707^2 + .151^2 + .07^2 + .03^2 + .01^2 + .04^2) = 1 - .53 = .47.$$

Hence, in this particular sample, there is a less than 50 percent chance of two people being in different ethnic groups, the result of the preponderance of respondents being white. The IQV works out to be .56.

$$IQV = \frac{1-(.707^2 + .151^2 + .07^2 + .03^2 + .01^2 + .04^2)}{1-(1/6)} = \frac{.470}{.833} = .56.$$

You might want to use one or both of these measures to compare two or more distributions. Table 11-15 shows a two-variable frequency distribution, party identification by sex. We will let you check the calculations, but the ID

TABLE 11-14
Calculation of Categorical Variation

Ethnic Group	Frequency (f_k)	Relative Frequency (proportion)
White	850	.707
Black	182	.151
Hispanic	81	.07
Asian	31	.03
Native American	11	.01
Other	47	.04
Total	1,202	1.01

$ID = .47$

$IQV = .56$

Source: 2004 National Election Study.
Note: Question: What racial or ethnic group or groups best describe you? Proportions do not add to 1.0 because of rounding errors.

TABLE 11-15
Two-Variable Frequency Distribution

Party Identification Response Category	Gender	
	Male	Female
Strong Democrat	11.9%	20.9%
	(69)	(128)
Weak Democrat	15.0	16.2
	(87)	(99)
Independent-leaning Democrat	18.6	16.5
	(108)	(101)
Independent	10.2	9.3
	(59)	(57)
Independent-leaning Republican	14.5	9.1
	(84)	(56)
Weak Republican	13.9	11.1
	(81)	(68)
Strong Republican	16.0	17.0
	(91)	(104)
Total	100.1	100.1
	(581)	(613)

Source: 2004 National Election Study.
Note: Numbers in parentheses are frequencies or the number of cases. Percentages do not add to 100 because of rounding.

for men is .85, which tells us there is about as much variation on partisanship among men as possible. The IQV confirms the point because its value is .99. The corresponding values for women are .84 and .98. This result means that the dispersion among women in the sample is also high and that the distributions do not differ in this respect. The table does show, however, that females are slightly more apt to call themselves strong Democrats.

We close by noting that measures of variation for ordinal variables have been proposed.[21] But their underlying logic is not grasped as readily by introductory students as the previous ones, and we omit them here.

We conclude this section with Table 11-16, which summarizes the descriptive statistics discussed in this chapter.

TABLE 11-16
Summary of Descriptive Statistics

Statistic	Symbol (if any)	Description (what it shows)	Resistant to Outliers	Most Appropriate For
		Measures of Central Tendency		
Mean	$\bar{Y}$	Arithmetic average: identifies center of distribution	No	Interval, ratio scales
Trimmed mean	$\bar{Y}_{trimmed}$	Arithmetic mean with a certain percentage of cases dropped from each end of the distribution	Yes	Interval, ratio scales
Median	M	Identifies middle value: 50% of observations lie above, 50% below	Yes	Interval, ratio, ordinal, scales; ranks
Mode	Mode	Identifies the category (or categories) with highest frequencies	No	Categorical (nominal, ordinal) scales
		Measures of Variation		
Range	Range	Maximum–minimum	NA	Interval, ratio scales
Interquartile range	IQR	Middle 50% of observations	Yes	Interval, ratio scales; ranks
Mean deviation	MAD	Average of absolute value of deviations from the mean	No	Interval, ratio scales
Standard deviation	σ	Square root of average of squared deviations	No	Interval, ratio scales
Variance	σ^2	Average of squared deviations	No	Interval, ratio scales
Index of diversity	ID	Probability that two observations will be in the same category	NA	Categorical (nominal, ordinal) scales
Index of qualitative variation	IQV	Probability that two observations will be in the same category	NA	Categorical (nominal, ordinal) scales

Note: NA = not applicable.

Graphs for Presentation and Exploration

In statistics the maxim "a picture is worth a thousand words" has a special place. We have already discussed the difficulty of using a large data matrix and the need to condense the information in it to a few descriptive numbers. But even those numbers can be uninformative, if not misleading. A major development since the 1980s has been the emphasis on graphic displays to explore and analyze data.[22] These visual tools may lead you to see aspects of the data that are not revealed by tables or a single statistic. And they assist with developing and testing models.

In particular, for any data matrix, a well-constructed graph can answer several different questions at one time:

■ Central tendency. Where does the center of the distribution lie?

■ Dispersion or variation. How spread out or bunched up are the observations?

■ The shape of the distribution. Does it have a single peak (one concentration of observations within a relatively narrow range of values) or more than one?

■ Tails. Approximately what proportion of observations is in the ends of the distribution or in its tails?

■ Symmetry or asymmetry (also called skewness). Do observations tend to pile up at one end of the measurement scale, with relatively few observations at the other end? Or does each end have roughly the same number of observations?

■ Outliers. Are there values that, compared with most, seem very large or very small?

■ Comparison. How does one distribution compare to another in terms of shape, spread, and central tendency?

■ Relationships. Do values of one variable seem related to those of another?

Figure 11-6 illustrates some ways a variable (Y) can be distributed. Panel A displays a symmetric, unimodal (one-peak) distribution that we previously called bell-shaped or normal. Panel B depicts a rectangular distribution. In this case, each value or range of values has the same number of cases. Panel C shows a distribution that, although unimodal, is **negatively skewed** or skewed to the left. In other words, there are a few observations on the left or low end of the scale, with most observations in the middle or high end of the scale. Finally, panel D represents the opposite situation: there are comparatively few observations on the right or high end of the scale. The curve is skewed to the right or **positively skewed.** These are ideal types. No empiri-

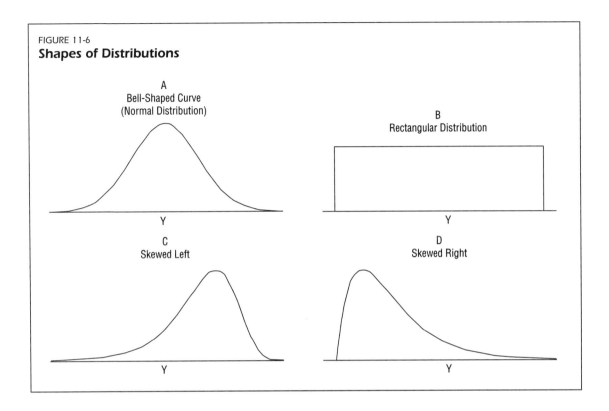

FIGURE 11-6
Shapes of Distributions

A
Bell-Shaped Curve
(Normal Distribution)

Y

B
Rectangular Distribution

Y

C
Skewed Left

Y

D
Skewed Right

Y

cal distribution will look exactly like any one of them. Yet if you compare these shapes with the graphs of your own data, you can quickly approximate the kind of distribution you have.

Why is it important to look at the shape of a distribution? For one thing, many statistical procedures assume that variables have an approximately normal distribution. And, if they do not, they can sometimes be mathematically changed or transformed into a bell-shaped configuration.

We can somewhat arbitrarily divide graphs into two general varieties, aimed at two audiences.

■ Presentation graphs. Some graphs are intended to be end products. Everyone has seen bar graphs and pie diagrams. The mass media use them to summarize information or even to prove a point. These pictorial devices commonly appear in the media and in books on policy and politics. They are appropriate for summarizing data to be published or otherwise publicly disseminated.

■ Exploratory graphs. Graphic analysis works in the background to assist in the exploration of data. As a matter of fact, one of the most vigorous research activities in applied statistics is the search for newer and better ways to visualize quantitative and qualitative information. Beginning with techniques developed in the 1970s by John Tukey and others, research on the visualization of data has become a growth industry.[23] Sometimes these exploratory graphs appear in published research literature, but they are mainly for the benefit of the analyst and are intended to tease out aspects of data that numerical methods do not supply. These diagrams amount to all-in-one devices that simultaneously display various aspects of data, such as central tendency, variation, and shape.

Presentation Graphs: Bar Charts and Pie Diagrams

Numerical information can be presented in many ways, the most common of which are bar charts and pie diagrams. A **bar chart** is a series of bars in which the height of each bar represents the proportion or percentage of observations that are in the category. A **pie diagram** is a circular representation of a set of observed values in which the entire circle (or pie) stands for all the observed values and each slice of the pie is the proportion or percentage of the observed values in each category. Figure 11-7 presents a bar chart and a pie diagram of the party identification data from Table 11-4. In a presentation, you would include one *or* the other.

Both types of graphs are most useful when the number of categories or values of a variable is relatively small and the number of cases is large. They most commonly display percentages or proportions. Be careful about constructing a bar chart or pie diagram when you have, say, a dozen or more categories. By the same token, these graphs will not reveal much if you have fewer than 10 or 15 cases. And you have to make sure the graphic elements (bars and slices) are proportional to the data. Unless there is a substantive reason to do otherwise, keep miscellaneous and missing value categories out of the picture.

Presentation graphs are no longer constructed by hand. Computers have greatly simplified the task, churning out dozens of graphs with a few key strokes. Yet this computational power will create problems if it is overused. Remember, a graph is supposed to provide the viewer with a visual description of the data. Many computer programs create wonderful-looking charts. But many of them add so many extra features, such as three-dimensional bars, exploded slices, cute little icons, or colorful fills, that the data easily get lost in the ink. It is usually best to keep lines and areas as simple as possible so that readers can easily see the point

STUDY GRAPHS FIRST

Participants in debates frequently mobilize graphs to bolster their arguments. In most instances, it's a good idea to scan a graph *before* reading the author's claims about what it says. The application of data analysis to real-world problems is at least as much a matter of good judgment as it is an exact science. A researcher, who will have a lot invested in demonstrating a point, may see great significance in a result, whereas your independent opinion is "big deal." You can maintain this independence by first drawing your own conclusions from the visual evidence and then checking them with the writer's assessments. But if you don't study the information for yourself, you become a captive, not a critic, of the research.

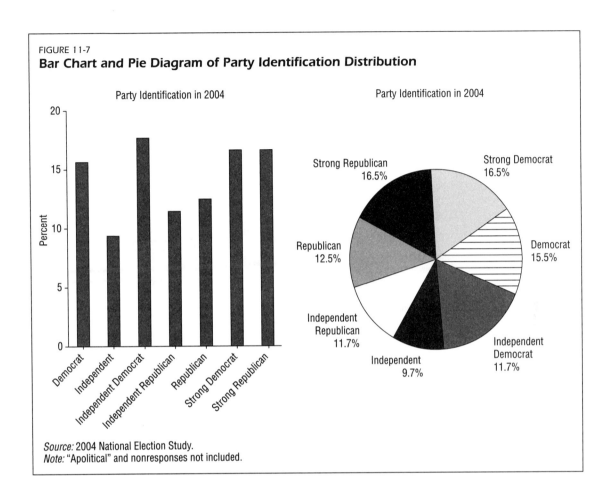

FIGURE 11-7

Bar Chart and Pie Diagram of Party Identification Distribution

Party Identification in 2004

Party Identification in 2004

Source: 2004 National Election Study.
Note: "Apolitical" and nonresponses not included.

being conveyed.[24] Also, be sure to title charts and graphs and provide clear labels for variables and both axes (the vertical axis is referred to as the y-axis and the horizontal axis is the x-axis), where appropriate: do not use abbreviations that mean nothing to a general reader. Also, use annotations (text and symbols) judiciously to highlight features of the data you believe the viewer should study. In other words, make your graphs as self-contained as possible.

Exploratory Graphs

You could include in a report or paper some of the graphs described in this section, but they are especially helpful in revealing different aspects of data.

HISTOGRAMS. A **histogram** is a type of bar graph in which the height and area of the bars are proportional to the frequencies in each category of a nominal variable or in intervals of a continuous variable. If the variable is continuous,

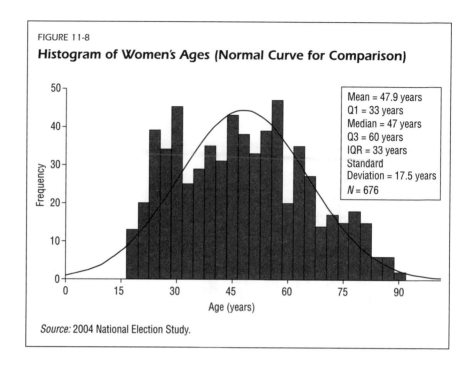

FIGURE 11-8

Histogram of Women's Ages (Normal Curve for Comparison)

Mean = 47.9 years
Q1 = 33 years
Median = 47 years
Q3 = 60 years
IQR = 33 years
Standard
Deviation = 17.5 years
N = 676

Source: 2004 National Election Study.

such as age or income, construct a histogram by dividing the variable into intervals, or bins, counting the number of cases in each one, and then drawing the bar heights to reveal the proportion of total cases falling in each bin. If the variable is nominal or ordinal with a relatively few discrete categories, just draw bar sizes so as to reflect the proportion of cases in each class.

A histogram, like other descriptive graphic devices, reduces a data matrix in such a way that you can easily see important features of the variable. For example, you can quickly see modes if there are any, the shape of the distribution (that is, whether or not it is skewed), and even extreme values. It helps to annotate your exploratory graphs with summary statistics so that you have everything required for analysis in one place.

Figure 11-8 shows a histogram of women's ages from the 2004 NES. (Several midpoints of the intervals are shown on the x-axis.) It is important to examine this variable because (1) we want to make sure the sample distribution approximates the population and (2) if discrepancies exist, we can identify and adjust for them. Although a couple of spikes can be observed in this distribution, the overall appearance is very roughly normal or bell-shaped. The graph succinctly sums up the more than 600 female respondents in the sample. Notice among other things that the middle of the distribution, as measured by both the mean and the median, is just about 47. The lower end of the scale is truncated because females less than 18 years of age were ineligible for participation in the study. The "location" statistics (Q1, median, and Q3) are instructive: 50 percent of the women are somewhere between 33 and 60 years old, and fully 25 percent are older than 60 years.

If you have to construct a histogram by hand (unlikely in this computer-dominated era), draw a horizontal line and divide it into equal-sized intervals or bins for the variable. The y-axis shows the frequency or proportion of observations in each interval. Simply count the number of units in each group and draw a bar proportionate to its share of the total. Suppose you have a total of 200 cases, of which 20 fall in the first interval. The bar above this interval should show that 20 observations or 20/200 = 10 percent of the observations are in it.

Drawing a histogram can be time consuming, but most statistical programs, even elementary ones, do the job easily. This capability comes in handy because an investigator may want to try creating histograms on a given data set using many different numbers of intervals. Histograms are helpful, as we have indicated, because they summarize both the spread of the values and their average magnitude. They are, however, quite sensitive to the delineations or definitions of the cut points or bins. (By *sensitive* we mean that the shape of the distribution can be affected by the number and width of the intervals. Some programs do not give the user much control over the intervals, so be cautious when using them.)

DOT PLOTS. A dot plot displays all of the observed values in a batch of numbers as dots above a horizontal line that represents the variable. Since it shows the entire data set, the number of observations should be relatively small, for example, less than 75. The great advantage of this plot is that it presents the main features of a distribution. To construct a simple dot plot, draw a horizontal line that stands for the variable. Below the line, print a reasonable number of values of the variable. Finally, using the data scores to position the dots, draw one dot above the line for each observation. If two or more observations have the same value, simply stack them one on top of the other. (In most cases, especially if there are a lot of distinct values, the dots will have to be drawn approximately over the scale in order to create legible stacks.)

Figure 11-9 shows a simple dot plot of infant mortality rates in the developed countries. It does not reveal anything that Figure 11-5 does not, except that we can now identify individual countries. (Interestingly or not, the United States has the highest rate.)

BOXPLOT. Perhaps the most useful graphic device for summarizing and exploring interval and ratio level data is the boxplot. It does not display individual points but does explicitly and simultaneously let you see five resistant

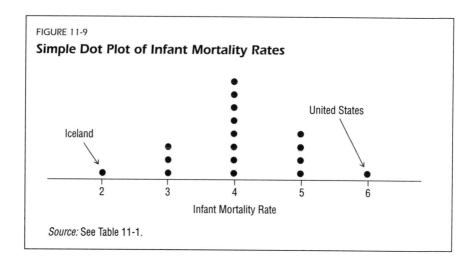

FIGURE 11-9

Simple Dot Plot of Infant Mortality Rates

Source: See Table 11-1.

descriptive statistics: the minimum, Q1, the median, Q3, and the maximum. Moreover, it makes obvious the interquartile range (IQR) and shows outlying or deviant values. What is more, it can be annotated in various ways to reveal even more information. Boxplots are sometimes called box-and-whisker plots because they appear to have a whisker at each end of a box.

Constructing a boxplot (even with paper and pencil) is relatively simple:[25]

1. Find the maximum and minimum, the first and third quartiles, the interquartile range, and the median.

2. Draw a horizontal line to indicate the scale of the variable. Mark off intervals of the variable. Be sure to fully label the scale.

3. Above the line, say about half an inch or so, draw a small vertical line to indicate the median. It should correspond to the appropriate value on the scale.

4. Next, draw short vertical lines of the same size above the scale to indicate Q1 and Q3.

5. Sketch a rectangle with the two quartiles (Q1 and Q3) at the ends. The median will be in the box somewhere. The height of the rectangle does not matter.

6. Next, calculate 1.5 times the interquartile range.

7. Go that distance *below* Q3 (the left end of the box) and place a mark. Draw a line from Q3 to your mark. This is the first (lower) whisker and marks the minimum value that is not an outlier.

8. Do the same for the upper end. Draw a line from Q3 to the point 1.5 times the IQR. This is the maximum value that is not an outlier.

9. Place points or symbols to indicate the actual location of extreme values. These should be labeled with the observation name or number.

10. If desired, and N is not large, individual observations can be marked and labeled.

The boxplot in Figure 11-10 has been annotated to show its main features. The variable is the "voice and accountability scale," a composite index that purports to measure the freedom and power citizens have to elect and hold their political leaders accountable. (Negative scores indicate less freedom.) The notations on the graph (for example, "median," "Q1") are not usually included because viewers supposedly understand this kind of graph. We include them merely for instructional purposes. The boxplot tells us that the average median VA value is –.4 and that 50 percent of the developing countries have scores between roughly –.9 and –.2. In addition, Cuba is far below average, and few of the nations have positive scores, the highest being the Dominican Republic. It takes a bit of getting used to, but you can actually detect the shape of the distribution: it is skewed slightly to the left or lower end of the scale. (You could check this by constructing either a histogram or a dot plot.)

TIME SERIES PLOT. Political scientists frequently work with variables that have been measured or collected at different points in time. In a time series

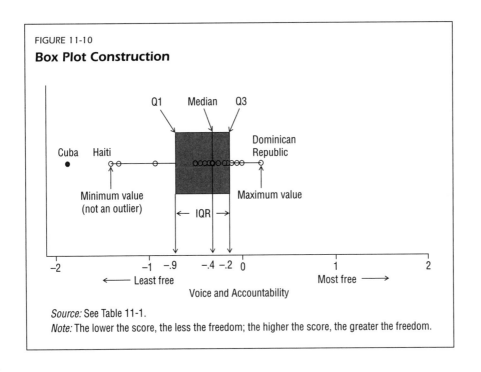

FIGURE 11-10
Box Plot Construction

Source: See Table 11-1.
Note: The lower the score, the less the freedom; the higher the score, the greater the freedom.

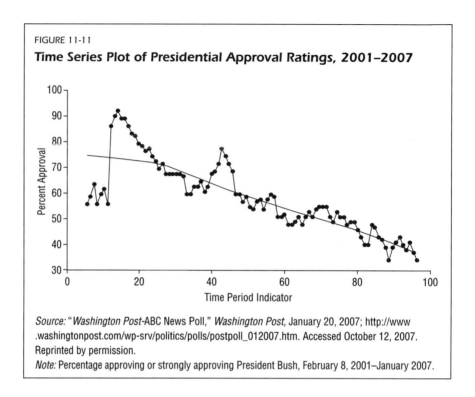

FIGURE 11-11

Time Series Plot of Presidential Approval Ratings, 2001–2007

Source: "*Washington Post*-ABC News Poll," *Washington Post,* January 20, 2007; http://www
.washingtonpost.com/wp-srv/politics/polls/postpoll_012007.htm. Accessed October 12, 2007.
Reprinted by permission.
Note: Percentage approving or strongly approving President Bush, February 8, 2001–January 2007.

plot the x-axis marks off time periods or dates, whereas the y-axis represents
the variable. These sorts of plots frequently appear in newspapers and magazines and could well be described as a type of presentation graph. However,
they are also helpful in exploring trends in data in preparation for explaining
or modeling them.

The time series plot in Figure 11-11 shows the decline in President George W.
Bush's approval ratings from the beginning of his first term in January 2001 to
January 2007. The graph contains the results of 96 polls and, to simplify the pre-
sentation, we assumed that they were taken during equally spaced intervals.

One of the characteristics of time series data is that they appear to
"seesaw" around a trend: there is frequently some or a lot of random variation
in addition to the trend itself. You can observe this in Figure 11-11, which
contains innumerable peaks and valleys. The pattern in the figure seems
clear, but with other data the variation may obscure or confuse the picture.
Hence, some software optionally applies a "smoothing" process that in a
sense averages out the ups and downs. (The math underlying this procedure
goes way beyond the scope of this book.) The smoothed data can then be
plotted as a line that better reveals the trend, if one exists. In the figure, the
downward sloping line amid the swarm of points represents the smoothed
data. It demonstrates that, despite a few reversals, the president's approval

TABLE 11-17
Typical Presentation and Exploratory Graphs

Type of Graph	What Is Displayed	Most Appropriate Level of Measurement	Number of Cases	Comments
Bar chart or pie diagram	Relative frequencies (percentages, proportions)	Categorical (nominal, ordinal)	3–10 categories	Common presentation graphic
Dot plot	Frequencies, distribution shape, outliers	Quantitative (interval, ratio)	$N \leq 50$ cases	Displays actual data values
Histogram	Distribution shape	Quantitative	$N \geq 50$ cases	Essential exploratory graph for interval or ratio variables with a large number of cases
Boxplot	Distribution shape, summary statistics, outliers	Quantitative	$N \geq 50$ cases	Can display several distributions; actual data points, an essential exploratory tool
Time series plot	Trends	Quantitative (percentages, rates)	$10 \leq N \leq 100$	Common in presentation and exploratory graphics

Note: Entries are guidelines, not hard and fast rules.

ratings have plunged steadily during his terms of office. (This situation, incidentally, is not unusual in presidential politics, a fact that you could check by obtaining and plotting approval ratings for other presidents.)

Table 11-17 summarizes the kinds of graphs we have discussed and offers a few tips on their proper use.

Statistical Inference

Chapter 7 discussed sampling. The nub of the statistical problem is this: A social scientist believes that some characteristics of an unmeasured population are worth investigating, particularly if they might help resolve substantive controversies. The trouble is that a scholar or policy analyst cannot study and measure every member or unit in the population. Instead it is necessary to take a sample and make inferences on the basis of sample measurements.

As obvious and mundane as that statement appears, it points to an extremely important fact about both scholarship and practical politics: a great deal of what is known or thought to be known about the world is based on sampling and the inferences drawn from the samples. Therefore, even if you have no intention of becoming an empirical social scientist, as a citizen and consumer of political arguments you should understand the concepts and logic behind statistical inference.

Before continuing, let's clarify some statistical vocabulary. These concepts have been discussed in various places earlier in the book, but it will help to review them:

■ A population refers to any well-defined set of objects such as people, countries, states, organizations, and so on. The term doesn't simply mean the population of the United States or some other geographical area.

■ A sample is a subset of the objects drawn in some known manner. If the selection is made so that each case is chosen independently of all the others and each has an equal chance of being drawn, then the subset is considered to be a simple random sample, also called a random sample or even just a sample. We use those three labels interchangeably from here on out because when we make statistical inferences we always assume the data derive from a simple random sample. (Violations of this assumption don't always spell disaster in empirical research. As long as the probabilities of being included in the sample are known, you can adapt most procedures and formulas accordingly. But that discussion goes far beyond the scope of this book.)

■ Populations have interesting numerical features that are called parameters. A population parameter may be the proportion of people in the United States who identify as independents. A parameter, in short, is a number that describes a specific feature of a population.

■ A sample statistic or estimator, as we call it, is a numerical summary of a sample that corresponds to a population parameter of interest and is used to estimate the population value. An example is the mean income in a sample that can estimate the average income of the population from which the sample derives.

■ Statistical presentations rely heavily on symbols and ours is no exception, although we attempt to keep them to a minimum. With that in mind, N stands for the number of cases or observations in the sample. Also, in line with what we have been doing, we let capital letters stand for variables. We may even embellish them to symbolize aspects of these variables just as $\overline{Y}$ (read "Y bar") is the mean or average of the variable Y.

■ We distinguish sample statistics from population statistics, which we call parameters, by using Greek letters for the latter. Thus, for example, the Greek letter mu (μ) designates the population mean.

■ In many books, Greek letters with a caret or hat ($\wedge$) over them denote sample estimators, but no consensus exists on the best typographical style. Hence, we use a mixture of symbols. The context and the text itself will, we believe, make clear what statistic is under discussion.

Two Kinds of Inference

Statistical inference means many things to many people, but for us it involves two core activities:

■ Hypothesis testing. Many empirical claims can be translated into specific statements about a population that can be confirmed or disconfirmed with the aid of probability theory. To take a simple example, the assertion that there is *no* gender gap in American politics can be restated as "Hypothesis: men's mean liberalism scores equal women's mean liberalism scores."

■ Point and interval estimation. The goal here is to estimate unknown population parameters from samples and to surround those estimates with confidence intervals. Confidence intervals suggest the estimate's reliability or precision.

We discuss each activity in turn.

Hypothesis Testing

The problem with depending on samples is that we cannot be sure whether any particular result arises because of a real underlying phenomenon in the population or because of sampling error. As a consequence, we are forced to make inferences. A common inferential tool is hypothesis testing. These tests involve translating propositions into exact statistical hypotheses and using probability to test them. A statistical hypothesis asserts that a population parameter theta (θ) equals a particular value such as 1 or 0. This assertion is evaluated in light of a sample statistic. That's really about all there is to it.

Although hypotheses are simple to state, hypothesis testing entails a few concepts that can confuse beginners. So we start with an intuitive explanation. Imagine playing a game of chance with someone you don't know very well. Suppose you and your opponent agree to toss a coin. If heads comes up, he wins a dollar; if tails comes up, you win the dollar. Naturally, you assume that the coin is fair, which effectively means the probability or chances of obtaining a head on a single flip is one-half. In addition, the outcome of any one toss is independent of the ones before and after it, and the probability of heads stays constant. The game consists of the opponent tossing the coin 10 times. Your main concern is where you stand at the end of the game. If the coin and opponent are fair, you would expect to break even, maybe winning or losing a couple of dollars, but not much more.

The game starts, and to your consternation your opponent gets 9 out of 10 heads. You are down eight dollars. (Why eight dollars?) But you're no sucker, and you walk away before paying, saying, "You're a cheat!" Your decision leaves your adversary fuming, and he yells back, "The game was fair. You're a chiseler." As far-fetched as the idea sounds, this experience can be construed as a parable of a statistical hypothesis test.

Let's reconstruct the game and its conclusion in statistical terms. In the abstract, a hypothesis test involves several steps, some of which are supposed to be made before any data are collected. The following list is a reconstruction or idealization of the test procedure, and it will make more sense when viewed in context of the cases to follow.

1. Starting with a specific verbal claim or proposition, recast it as a hypothesis about a population parameter: "My opponent is a gentleman. He plays fair and so the chances of getting a heads are the same as the chances of tails, namely, one-half."

2. More precisely, state a **null hypothesis.** This is a statement that a population parameter equals a specific value. (*Note:* a null hypothesis never sets a range of values.) Letting P stand for probability, you can often state a null hypothesis in this way: H_0: $P = .5$, which reads "the null hypothesis is that the probability of 'producing' heads is one-half." After all, the very essence of a fair coin is that the probability of heads is one-half. In many research reports, the null hypothesis (H_0) is that something (for example, a mean, a proportion) equals zero.[26] Hence, the word *null*—because 0 represents no effect, such as no difference. But keep in mind that a hypothesis can be an assertion that a population parameter equals any *single* number such as .5 or 100.

3. Specify an **alternative hypothesis,** such as the parameter does *not* equal the number in the null hypothesis or is greater or less than it is. In the present case, the alternative hypothesis is that the coin is *not* fair, which when translated into statistical lingo is H_A: $P \neq .5$. This particular alternative simply asserts that the probability of heads is not one-half but something greater or lesser. Such an open-ended hypothesis is called two-sided, which means that a sample result's deviating too far in any direction from the null hypothesis will be cause for rejection. If only outcomes greater (or less) than the null hypothesis matter, a one-sided hypothesis can be formulated: H_A: $P > .5$ or H_A: $P < .5$. So, depending on your alternative hypothesis, if the opponent throws "too many" heads *or* tails, you may challenge the fairness of the game.

4. Identify the sample estimator that corresponds to the parameter in question. In the example at hand, the population parameter—the probability of heads—must be estimated from the "data," namely, a series of 10 flips of a coin. A logical sample statistic is $p,$ the sample proportion of heads in 10 tries.

5. Determine how this sample statistic is distributed in repeated random samples. That is, specify the sampling distribution of the estimator. Here, we want to know the likelihood of various outcomes in 10 tosses of a fair coin. In other words, we want the chances of getting 10 heads in 10 flips ($p = 1.0$), 9 heads ($p = .9$), 8 heads ($p = .8$), and so on down to 0 heads ($p = 0$). A sampling distribution provides this kind of information. Briefly stated, it is a

function that shows the probabilities of obtaining possible values for a statistic *if* a null hypothesis is true. We use it to assess the possibility that a particular sample result could have occurred by chance.

6. Make a decision rule based on some criterion of probability or likelihood. It is necessary to determine probable and improbable values of the statistic if the null hypothesis is true. In the social sciences, a result that occurs with a probability of .05 or less (that is, 1 chance in 20) is considered unusual and consequently is grounds for rejecting a null hypothesis. Other thresholds (for example, .01, .001) are common, and you are free to adopt a different decision rule.

7. In light of the decision rule, define a critical region, according to the following logic: Assume the null hypothesis is true. Draw a sample from a population and calculate the sample statistic. Under the null hypothesis, only a certain set of outcomes of a random process or sample should arise regularly. If you toss a fair coin 10 times, you expect heads to come up *about* half the time. If you see 10 heads (or 10 tails) in 10 tosses, you may suspect something is amiss. The critical region consists of those outcomes deemed so unlikely to occur *if* the null hypothesis is true that, should one of them appear, you have cause to reject the null hypothesis. The demarcation points between probable and improbable results are called critical values. They define areas of "rejection" and "nonrejection." If a result falls at or beyond a critical value, the null hypothesis is rejected in favor of the alternative; otherwise, H_0 is not rejected. In this game, you might think, "If my opponent tosses 9 or more heads *or* 9 or more tails, I'm going to reject the hypothesis that the coin is fair. Those outcomes are just too unlikely to happen if the coin is unbiased." According to this rule, the critical values are 1 and 9 because should one of them (or one even more "extreme," that is, 0 or 10) occur, you will reject the hypothesis of fairness (Figure 11-12). The other possible outcomes—2 heads, 3 heads, . . . 8 heads—are not in the critical region and hence provide no reason for rejection.

8. In theory, these choices are made in advance of any data collection (for example, before the game begins). The probability claims that back your conclusions will be invalid if you first obtain the data and *then* make your decision rules.

9. Collect a random sample and calculate the sample estimator. (Start the game.)

10. Calculate the observed test statistic. A test statistic converts the sample result into a number that can be compared with the critical values specified by your decision rule and critical values. We always show you how to

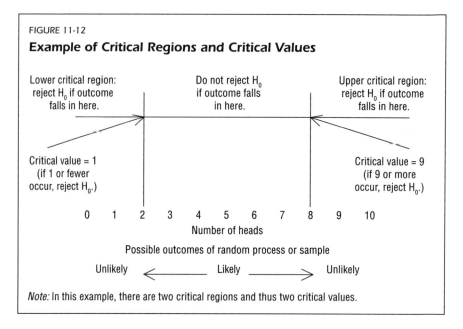

FIGURE 11-12

Example of Critical Regions and Critical Values

Note: In this example, there are two critical regions and thus two critical values.

obtain the observed test statistic, and anyway, most of the time a computer does the number crunching for you. What is important is that you understand what the test does and doesn't say.

11. Examine the observed test statistic to see if it falls in the critical region. If it does, reject the null hypothesis. If not, do not reject the hypothesis. In the coin-flipping example, if the final tally is, let's say, 8 heads, you may be suspicious, but you don't reject H_0. However, suppose the outcome is 9 heads, 1 tail. This outcome does lie in the critical region, and accepting the alternative hypothesis ($H_A: P \neq .5$) is the decision.

12. Make a practical or theoretical interpretation of the findings and take appropriate action. At most, the result of a hypothesis test is a bit of information. You have to couple the statistical result with other evidence to know whether you have uncovered something of practical or theoretical value. Recall from Chapter 2 that inference is not the same as deductive proof. So the results of a statistical test, no matter how carefully carried out, prove very little by themselves.

Let's now explore aspects of these steps in more detail, beginning with the notion of what inference entails.

Types of Errors in Inference. To paraphrase dictionary definitions, *inference* refers to reasoning from available information or facts to reach a conclusion. The end product, however, is not guaranteed to be true. It might in

TABLE 11-18
Types of Inferential Errors

Decision is to	In the "real" world, the null hypothesis is	
	True	False
Accept H_0	Correct decision	Type II error
Reject H_0	Type I error	Correct decision

fact turn out to be incorrect. This possibility certainly arises in the case of statistical inference, in which values of unknown population characteristics such as means or probabilities are estimated and hypotheses about those characteristics are tested.[27]

This discussion implies that errors can arise in statistical inferences. Take hypothesis testing, the essence of which is making a decision about a null hypothesis. We can make two kinds of mistakes, as illustrated in Table 11-18.

The first type of mistake is to reject a true null hypothesis. Going back to the coin-tossing dispute, suppose that (1) the coin is fair (that is, H_0: P = .5 is true), (2) ahead of time you have decided to reject H_0 if 0, 1, 9, or 10 heads occurs (that is, your critical region), (3) your opponent obtains 9 heads in 10 flips, and (4) you do reject the hypothesis and accuse the person of being unfair. Given that the coin is fair ($P = .5$), you have made a mistake. Statisticians call this mistake a **type I error**: the incorrect or mistaken rejection of a true null hypothesis.

As you might have surmised, the probability of committing a type I error is the "size" of the critical region. It is designated by small Greek letter alpha (α). Thus α = the probability of a type I error given the decision rule, critical value, and sample size.

Before we explain further, let's look at another possibility (see Table 11-18), namely, failing to reject a null hypothesis that is false. Suppose you are a trusting soul and, prior to the game, establish a small critical region: to wit, you will reject H_0 only if the opponent tosses 10 out of 10 heads. You play the game, and the outcome is 9 heads, which this time does not equal or exceed your critical value. The null hypothesis is *not* rejected. But what if the person really is a cheat? What if, in other words, P does *not* equal .5 but instead equals, say, .9. This time, H_0 should have been discarded. In this case, where the result does not fall in the critical region and you do not reject H_0, you have made a mistake of a different sort: a **type II error,** that is, failing to reject a false null hypothesis.

The probability of committing a type II error (normally designated with the lowercase Greek letter beta, β) depends on how far the true value of the population parameter is from the hypothesized one. Assuming a fixed α level (the probability of making a type I error) and a fixed sample size, N, suppose the null hypothesis is H_0: P = .5. If the true P is .6, it may be hard to reject the null hypothesis even though it should be rejected, and β will be relatively large. But if P = .9, it will be easier to reject H_0, and β will be relatively small. We cannot go into more detail because the necessary background information would take us too far afield, except to note that for a fixed α, the probability of a type II error (β) decreases as the sample size increases.[28] In a life-or-death showdown, you might be better off having the opponent toss the coin many more times than 10.

Now that we have identified two possible inferential errors, we need to consider how to measure and control the probability of their occurrence.

SAMPLING DISTRIBUTIONS.   On several occasions in this chapter and previous ones, we have mentioned sampling distributions.[29] Yet it is worth explaining the concept once more because it is so important to statistical inference.

Here again is the general idea. Given an indefinitely large number of independent samples of size N, a sampling distribution is a mathematical function that indicates the probability of different values of the estimator occurring.[30] As another way of understanding the concept, think of a sampling distribution as a picture of the variability of sample results. Grasping this idea takes a bit of imagination because a sampling distribution is a theoretical concept rather than something observed in the course of research.

Imagine that you take an endless number of independent samples of size N from a fixed population that has a mean of μ and a standard deviation σ. Each time, you calculate the sample mean, $\bar{Y}$, and the standard deviation, $\hat{\sigma}$. In the end, you will have literally thousands of $\bar{Y}$'s. If you drew a graph of their distribution, much like the histogram shown in Figure 11-8, the result would be what is called a sampling distribution. How do we know? Statistical theory demonstrates that the distribution of sample means is normal with these two characteristics: the mean of the sampling distribution would be μ, the population mean; and its standard deviation would be $\sigma/\sqrt{N}$. The standard deviation of a sampling distribution is called the standard error of the mean (on computer-generated reports, it is usually denoted by "s.e." or "se. mean," or some such abbreviation). If the population standard deviation is unknown, as it almost always is, the standard error has to be estimated by using the sample standard deviation, $\hat{\sigma}$, and calculating it as $\hat{\sigma}/\sqrt{N}$.

The expression for the standard error implies that as N gets larger and larger, the standard error decreases in numerical value. The lesson here is that as N increases, you would expect the sample estimate to get closer and closer to the

---

How It's Done

Calculating the Standard Error of the Mean

The standard error is calculated as follows:

$$\hat{\sigma}_{\bar{Y}} = \frac{\hat{\sigma}}{\sqrt{N}},$$

where $\hat{\sigma}$ is the sample standard deviation and N is the sample size.

true value (μ). It might not be exactly on target, but with large samples the average of the estimates should congregate around this value. The standard error is a measure of the imprecision in the estimators, and we use it again and again.

All of this discussion is theoretical, but you can grasp the gist of it by considering a simple example. Let's return to the coin-tossing contest. There are 10 flips of a supposedly fair coin per game. At the end, the proportion (p) of heads is counted. Suppose you were to compete a billion times more, each time recording the proportion of heads that come up in the 10 tosses. How often would you see 10 out of 10 heads, 9 out of 10 heads, 8 out of 10 heads, and so forth?

TABLE 11-19
Binomial Sampling Distribution ($N = 10$)

Outcome (p)	Probability of Outcome p
0	.001
.1	.010
.2	.044
.3	.117
.4	.205
.5	.246
.6	.205
.7	.117
.8	.044
.9	.010
1.0	.001

Statistics proves that a distribution, called the binomial, answers those questions. The binomial is an equation that takes as input the hypothesized probability of heads (the general term is the hypothesized probability of a success) and the number of tosses (more generally, the number of trials), and gives the relative frequency or probability of each possible outcome. We do not give the formula here,[31] but Table 11-19 shows an example of the binomial distribution as it pertains to 10 tosses of a supposedly fair coin. Interpret the table as follows: in 10 flips of a coin for which the probability of heads is $P = .5$, the chances of getting exactly no heads or $p = 0$ is .001, or 1 in a thousand.[32] Similarly, the likelihood of 1 head in 10 trials is about .01 or 1 in a hundred. At the other end of the scale, the probability of getting all heads is also quite remote, .001. (By the way, since this sampling distribution assumes that $P = .5$, you might expect to get on average 5 heads half of the time. The table, though, shows that the probability of obtaining *exactly* 5 heads in 10 flips is roughly .25. As counterintuitive as the probability seems, it is a fact.)

The lesson is that a sampling distribution such as the binomial shows you the probabilities of various possible sample outcomes. Therefore, when you observe a particular result you can decide whether it is unusual and thus whether to reject a hypothesis.

To use a sampling distribution, you have to first decide on how large an error in estimation you are willing to tolerate. In other words, you have to determine the areas of rejection and nonrejection of H_0 and find critical values that define those areas, as explained next. Then you compare these numbers with the observed test statistic derived from the one sample you have. If the observed value is greater than or equal to the critical value, reject the null hypothesis. If not, do not reject it.

The binomial is merely one of many sampling distributions. Different sample estimators require different distributions for inference. In this book we talk mainly about four others: the normal distribution, the t distribution, the chi-square distribution, and the F distribution. This list may seem intimidating, but all four work pretty much the same way as the binomial.

CRITICAL REGIONS AND CRITICAL VALUES.   Refer to Table 11-19, which shows a binomial distribution, as you read this section. Remember that a critical region is a set of possible values of a sample result, the actual occurrence of any one of which will cause the null hypothesis to be rejected. When setting a critical region and the corresponding critical points that define it, we have a couple of choices. For one, we determined the size of the regions by selecting the probability of making a type I error, α. Assume the choice is to reject if *either* 0 *or* 10 heads appear in the sample of 10 coin tosses. Go back to the table and add up the probabilities of getting either 0 or 10 heads in 10 attempts. You should see that it is .001 + .001 = .002. This is the size of the critical region. More generally, the size of a critical region is the total probability of a sample result's landing in it under the null hypothesis. So in the present situation, under the decision rule, the size of the critical region is $\alpha = .002$.

This probability is quite small, so you might consider it a conservative test. We use the term *conservative* because only two rare events will lead to rejection of the null hypothesis. You might gain a clearer understanding of this point by thinking about changing the decision rule, that is, by changing the size of the critical region. If you were hypersuspicious, you might tolerate almost no deviation from what you thought was a fair game. Hence, the rejection area could have been 0, 1, 2, 8, 9, or 10 heads. If any of those numbers comes up, you question the opponent's honesty. Now, what is the size of the critical region or, put differently, what is the probability of making a type I error (falsely accusing the opponent of cheating)? Again, add the probabilities of the outcomes in your rejection region: .001 + .010 + .044 + .044 + .010 + .001 = .110. This means you have about 11 chances out of 100 of incorrectly rejecting the null hypothesis. Is it too big a risk?

Unhappily there is no right or wrong answer because statistical analysis cannot be divorced from practical or substantive considerations. Instead, it is always necessary to ask, "What are the costs of a mistaken decision?" If rejection of your opponent's integrity leads to gunplay, perhaps you want to be conservative by choosing critical values that minimize a type I error. On the other hand, if there is little or no down side to a false accusation of unfairness—maybe you don't know the person well and can leave the relationship without pain—you might be less rigorous or demanding and choose the .11 level.

Whatever the verdict, it is made with a certain probability of a type I error and this probability is called the level of significance.

LEVELS OF SIGNIFICANCE   All of this background comes to a head in the notion of **statistical significance,** perhaps one of the most used (and abused) terms in empirical analysis. To claim that a result of a hypothesis test is statistically significant is to assert that a (null) hypothesis has been rejected with a specified probability of making a type I error: falsely rejecting a null hypothesis. The three commonest levels of significance in political science are .05, .01, and .001, but you should not be chained to these standards.

Before explaining, let's decode the following statement: "The result is significant at the .05 level." Translated into common usage, the statement means that a researcher has set up a null hypothesis, drawn a sample, calculated a sample statistic or estimator, and found that its particular value is unlikely given the null hypothesis. How unlikely? It would occur by chance at most only 5 percent of the time. Because the outcome is deemed unlikely in the circumstances, the null hypothesis is rejected. This might be a mistake, but the probability of falsely making a mistake is only .05. That's the long and short of "level of significance." Once you get used to the vocabulary, the idea becomes less daunting.

The popular and scholarly literatures contain a variety of ways to phrase this idea. For example, let's assume that you have established a less-than-stringent critical region and will reject the null hypothesis of a fair coin if your adversary gets 8, 9, or 10 heads in 10 chances. In this new circumstance, the critical value is 8 because if the result of the contest is 8 *or more* heads, the null hypothesis is rejected and the size of the critical region is .044 + .010 + .001 = .055 (see Table 11-19). One way of expressing this decision rule is to say the "test is being conducted at the .055 level." Furthermore, if you do reject the null hypothesis, it is common to express the finding as "significant at the .055 level." Sometimes, the alpha symbol is used in the declaration, as in "the hypothesis is rejected at α = .055." These statements all mean the same thing: your sample result is *so* unlikely given the truth of the null hypothesis (and other assumptions including independent random trials, accurate counting, and so forth) that you are willing to reject the null hypothesis and entertain the alternative, H_A.

READING TABLES OF STATISTICAL RESULTS
The text mentions a number of forms for presenting the results of a hypothesis test. In most scholarly presentations, the format is far more succinct. Authors give a table of sample results (for example, p = .8) and place a typographical symbol (for example, an asterisk "*") next to them. Then, at the bottom of the table (or somewhere else), a brief explanation will appear, such as "* prob < .05," which means a null hypothesis is rejected at the .05 level. If a null hypothesis is not rejected, you often see "NS" for "not significant" or the estimate does not have any accompanying mark.

Conversely, with the same decision rule, if the opponent gets 7 heads, then the result does not equal or exceed the critical value (8) or lie in the critical region. Now you might write, "not significant at the .055 level" or "failed to reject at α = .055."

More generally, the media often report that something is "statistically significant" or is "not statistically significant" and let the matter drop there. What lies behind such statements is that someone has conducted a statistical test and rejected (or did not reject) the null hypothesis. The sample size and α level almost never get reported, so the claim of (in)significance by itself does not reveal much information. Before we go into more detail about whether or not statistical significance should be considered a big deal, we need to wrap up some loose ends of hypothesis testing.

ONE- OR TWO-SIDED TESTS. You may or may not have noticed that in the previous section we pulled a fast one. In the very first coin-tossing scenario, a decision rule was to reject if the observed number of heads turned out to be 0 *or* 10. In effect, we were using *both* ends of the sampling distribution, which meant that a low *or* high number of heads would count against the null hypothesis. But for the last example we changed the rejection criterion to consider only 8, 9, or 10 heads as evidence against the hypothesis. This time just one (the upper) end of the binomial distribution (the high values) constituted the critical region. The former case is called a *two-sided test of significance,* whereas the latter is called (sensibly enough) a *one-sided test.*

The difference, of course, is that both small and large values of the sample result count in a two-sided test, whereas in a one-sided test either the low or high results (but not both!) make up the critical region. More precisely, a two-sided test requires two critical values and two critical regions, one for small values and one for large ones. With a one-tailed test, only one critical value and region are relevant.

So which kind of test do you use? The substantive goals of the research ultimately determine the response. One thing is certain: try to make a test of significance as informative as possible. If previous research or common sense suggests that a population parameter will be larger than the one in the null hypothesis, use a one-sided test. For example, assume you watch your opponent in action against other players and his coin generally comes up heads. As an alternative to H_0: $P = .5$, you might establish the alternative hypothesis H_A: $P > .5$ and thus use the upper tail of the distribution for the critical region, say, 8 or more heads. This is a one-tailed test. That way if the final result is 7 or fewer (even 0 or 1) heads, you won't reject the null hypothesis. But if you cannot predict in which direction the population parameter might be—that is, if you have no reason to expect large or small P values—you have to use a two-sided test.

COMPARING OBSERVED TEST STATISTICS TO CRITICAL VALUES. We made the coin-toss example simple in order to explain different aspects of hypothesis testing. One simplification was the sampling distribution, which was based on the probability of success (heads) equaling .5 and 10 trials (tosses). Although we did not supply the formula for the binomial distribution, we produced a table that listed every possible sample outcome and the probability of their occurrence under the null hypothesis. We could thus directly link a realized result (say, 9 heads) to a probability and decide whether or not that result was probable. Imagine that we bumped up the number of trials (tosses) to 1,000. Now, a similar table of the binomial would have to have a thousand rows, one for each possible result (for example, 0 heads out of 1,000 tries, 1 head, . . . , 999 heads, 1,000 heads). Besides being impracticable, this strategy is unnecessary.[33]

In the applications covered in this book, we use sampling distributions that have tabular summaries. To find critical values, all we have to do is decide on a level of significance and look up the critical values that define the critical region. Suppose, for instance, we want a two-tailed test of some hypothesis at the .01 level. This means we want the total area of the critical region to be .01. To find critical values that create an area that size, we use a tabulation of an appropriate sampling distribution.

In the previous example, the observed test statistic was p, the observed proportion of heads, and we compared it directly to critical values to reach a decision about whether or not to reject the null hypothesis. In most instances, however, it is not possible to compare a sample value directly to a critical value obtained from some distribution. Instead we need to convert the estimator into an *observed test statistic,* which is a function of the sample data. The observed statistic is then compared to the critical value so that a decision about rejection can be reached. Many test statistics take the following general form:

$$\text{Observed test statistic} = \frac{\text{Sample estimate} - \text{Hypothesized population parameter}}{\text{Standard error}}$$

To remind you, a standard error is the standard deviation of a sampling distribution. When needed, we always provide the required formula, and statistical software always computes standard errors along with other hypothesis test information. In any event, the observed test statistic is compared to a critical value, and the decision to reject or not reject the null hypothesis depends on the outcome of the comparison:

- If the observed statistic's *absolute value* is greater than or equal to the critical value, reject the null hypothesis in favor of the alternative.

- Otherwise, do not reject the null hypothesis.

Figure 11-13 summarizes the steps in hypothesis testing. It is time now to put these ideas to work on a "real" problem.

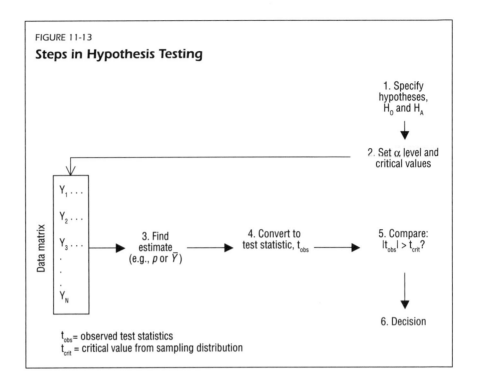

FIGURE 11-13

Steps in Hypothesis Testing

Significance Tests of a Mean

Suppose someone tells you the "average American has left the middle of the road and now tends to be somewhat conservative." You, however, are not so sure. You think that the average American is not conservative. This situation could be tested using responses to a seven-point ideology scale with 1 representing "extremely liberal," 4 "moderate" or "neither liberal or conservative," and 7 "extremely conservative." A 5 on this scale represents a slightly conservative position. So, the above claim can be interpreted as saying that the mean ideology score of voting-age individuals is 5. The null hypothesis is, thus, H_0: $\mu = 5$, where μ is the population mean ideology score. Given the way the problem has been set up, the alternative hypothesis is H_A: $\mu < 5$.

So, the statistic for this test is the sample mean of liberalism-conservatism scores. For this example, let's set $\alpha = .05$ for the level of significance. The next step is to specify an appropriate sampling distribution and gather the data for the test. When testing a hypothesis about a mean, you will use the sample size to determine the appropriate sampling distribution.

SMALL-SAMPLE TEST OF A MEAN.   If the sample is small—less than or equal to about 25—statistical theory asserts that the appropriate sampling distribution for a test about a mean is the t distribution. A graph of this distribution is a symmetrical (bell-shaped) curve with a single mode. It resembles the normal distribution described in conjunction with the standard deviation but is a bit "fatter" in that it has more area in its tails. The shape of the *t* distribution depends on the sample size (*N*) and is thus a "family" of distributions. But as *N* gets larger, the *t* distribution approaches the shape of the normal distribution; at *N* = 30 or 40, they are essentially indistinguishable.

To use the *t* distribution for a small-sample hypothesis test of a mean, follow these steps:

1. Determine the size of the sample to be collected. This number, *N*, should be less than or equal to 30. Otherwise, use the normal distribution, as explained in the next section.
2. Find the degrees of freedom (*df*) by calculating *N* – 1. The degrees of freedom is for now an abstract concept that we do not explain. But in this situation it is always the sample size minus one.[34]
3. Choose the level of significance and directionality of the test, a one- or two-tailed test at the α level.
4. Given these choices, find the critical value(s). Looking at a tabulation of a *t* distribution (see Appendix B for an example), go down the rows of the table until you locate the degrees of freedom. Move across the row until you find the column that corresponds to the area of the size of the designated level of significance. The number at the intersection of the degrees of freedom row and area under the curve column is the critical value. Call this number t_{crit}.

At this point you would now collect the sample data, find the sample mean, and compute the observed test statistic. In keeping with the general formula given above, the observed *t* is found by subtracting the hypothesized mean from the observed mean and dividing by the estimated standard error.

The observed value, labeled t_{obs}, is then compared to the critical value and a statistical decision is made according to the following rule:

If $|t_{obs}| \geq t_{crit}$, reject H_0; otherwise, do not reject.

In plain English, if the absolute value of the observed test statistic is greater than or equal to the critical value, reject the null hypothesis; if the observed *t* is smaller than the critical value, do not reject it.

Let's test the statistical hypothesis that the population mean liberalism-conservatism score is 5. Suppose we use a small sample of 25 observations from the 2004 NES. Table 11-20 outlines

How It's Done

Calculating the Observed *t*

The observed *t* is calculated as follows:

$$t_{obs} = \frac{(\overline{Y} - \mu)}{\hat{\sigma}/\sqrt{N}},$$ where $\overline{Y}$ is the sample mean, μ is the hypothesized population mean, $\hat{\sigma}$ is the sample standard deviation, and *N* is the sample size.

TABLE 11-20
Example Setup of a *t* Test for the Mean

Null hypothesis	$H_0: \mu = 5$
Alternative hypothesis	$H_A: \mu < 5$
Sampling distribution[a]	*t* distribution
Level of significance	$\alpha = .05$
Test	one-tailed
Critical value (*df* = 25 − 1 = 24)	$t_{crit(.05)} = 1.711$

[a] t Distribution required for small-sample ($N \leq 30$) test of the mean.

the test guidelines. It shows that we are using a one-sided test at the .05 level of significance. As a reminder, this phrase means that if we observe a sample result greater in absolute value than the critical value, 1.711, we will reject the null hypothesis that the population mean liberalism-conservatism score is 5 and entertain the alternative that it is less than 5.

Figure 11-14 shows how the critical value was found from the *t* distribution shown in Appendix B. (For the sake of brevity, many table entries have been deleted.)

The level of significance, .05, is found by looking in the second column. The degrees of freedom for this problem is calculated as 25 − 1 = 24, so we use the 24th row. The intersection of this row and the third column leads to the critical value, 1.711. Therefore, if the observed test statistic equals or is greater than 1.711, we reject the null hypothesis in favor of the alternative. Otherwise, we do not reject.

So far all of these decisions and choices have presumably been made before the data have been examined. (In fact, hypothesis testing is so standardized and automated that rarely would anyone go through all of these steps. Instead, they plug the numbers into a computer program and simply interpret the results as they come out. Still, it is not considered correct to

FIGURE 11-14
Finding the *t* Value

Level of significance = *total* size of critical region for one-sided test.
(Probability of type I error (α level) ——

Alpha Level for One-Tailed Test

	.05	.025	.01	.005	.0025	.001	.005

Alpha Level for Two-Tailed Test

Degrees of Freedom (*df*)	.10	.05	.02	.01	.005	.002	.001
. . .	. . .	. . .	. . .	. . .	. . .	. . .	. . .
21	1.721	2.080	2.518	2.831	3.135	3.527	3.819
22	1.717	2.074	2.508	2.819	3.119	3.505	3.792
23	1.714	2.069	2.500	2.807	3.104	3.485	3.767
24	1.711	1.064	2.492	2.797	3.091	3.467	3.745
25	1.708	2.060	2.485	2.787	3.078	3.450	3.725
. . .	. . .	. . .	. . .	. . .	. . .	. . .	. . .

df = 25 − 1 = 24 ———————— Critical value for $\alpha = .05$ with 24 *df*.

Source: Excerpt from "Critical Values from *t* Distribution," appendix table B, p. 576.

change decision rules *after* the data have been examined.) The next step is to calculate the test statistic from the sample and compare it to the critical value.

Assume we drew a sample of 25 cases from the 2004 NES data and found the sample mean liberalism-conservatism score to be 4.44, with a standard deviation of 1.23. This observed mean is slightly below the hypothesized average. The big question is, Does this result give us any grounds for rejecting the hypothesis that the population mean is 5? To find out, we must convert the sample mean to an observed *t,* which we can compare with the critical value:

$$t_{obs} = \frac{(\overline{Y}-\mu)}{\frac{\hat{\sigma}}{\sqrt{N}}} = \frac{(4.44-5)}{\frac{1.23}{\sqrt{25}}} = -2.28.$$

The denominator is the standard error. It is what we use to scale the difference between the sample mean and the hypothesized mean to fit to a tabulated sampling distribution such as the *t.* Notice that the test statistic can be negative, but we are only interested in its absolute value (that is, disregarding the negative sign). And since $|t_{obs}| = 2.28$ is greater than 1.711 (the critical value), we reject the null hypothesis. The implication is that Americans are not as conservative as hypothesized, and at this point the best estimate of the true mean is 4.44, a very slightly conservative value. (To cement your understanding of this section, use the sample information provided and the decision rule to test the hypothesis that H_0: $\mu = 5$. We have supplied all the necessary information. Then, for further practice, pick a different decision criterion, say, $\alpha = .01$ for the level of significance.[35])

Computer programs now perform most statistical analyses. Although the advantages in saved time and effort are obvious, it is essential to understand what the computer output is telling us. That's why we spend so long going over the ideas behind hypotheses testing. But whether as a student or in another capacity, you are likely to be a consumer of software-generated reports. Figure 11-15 illustrates the results of a small-sample *t* test cranked out by a popular software package. Once the abbreviations have been explained, the meaning of the numbers should present no problem. The only extra information is the "P" in the last column; it stands for "p-value." Instead of indicating that the result is significant at, say, the .05 level, this program, like most others of its kind, gives the probability of getting a *t* statistic *at least as large as the one actually observed if the null hypothesis is true.* In the present case, in which the sample mean is 4.44, the evidence is that a population value of 5 is not very likely. More precisely, the probability of a sample mean this far or farther from the hypothesized value is only about .016 or 16 chances in 100.

Finally, given the ubiquity of computers, you might as well take advantage of their services and follow this rule: whenever the p-value is available, report

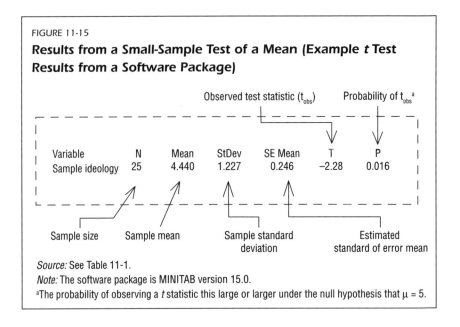

FIGURE 11-15

Results from a Small-Sample Test of a Mean (Example *t* Test Results from a Software Package)

Observed test statistic (t$_{obs}$)   Probability of t$_{obs}$[a]

Variable	N	Mean	StDev	SE Mean	T	P
Sample ideology	25	4.440	1.227	0.246	−2.28	0.016

Sample size   Sample mean   Sample standard deviation   Estimated standard of error mean

Source: See Table 11-1.

Note: The software package is MINITAB version 15.0.

[a]The probability of observing a *t* statistic this large or larger under the null hypothesis that $\mu = 5$.

it and not an arbitrary level of significance. Why follow this advice? Compare these two assertions:

■ The result is significant at the .05 level.

■ The p-value is .016.

The first statement tells us only that the probability of the result (or one more extreme) is less than .05. But is it .04? Or .02? The second statement indicates that under the null hypothesis the probability of the result (or one more extreme) is .016.

LARGE-SAMPLE TEST OF A MEAN. In the previous example, we drew a sample of a sample. This time let's rely on a larger sample. The *t* distribution, we noted, is a collection of distributions, the shape and areas of which depend on the sample size. But as *N* grows larger, the distributions get closer and closer to the normal distribution. More specifically, an important theorem in statistics states that, given a random sample of 30 or more cases, the distribution of the mean of a variable (Y) is approximately a normal distribution with a mean equal to μ (the mean of the population from which the sample was drawn) and a standard deviation of $\hat{\sigma}_{\bar{Y}}$.[36] As noted previously, the latter symbol (read "sigma sub Y bar") is called the standard error of the mean. It measures how much variation (or imprecision) there is in the sample estimator of the mean. For us the theorem boils down to the fact that we can test large-sample means with a **standard normal distribution.**

The standard normal distribution is specified by an equation, the graph of which is a unimodal, symmetrical (bell-shaped) curve. This particular distribution has a mean of 0 and a standard deviation of 1. (Figure 11-5 displays a graph of the standard normal distribution.) Recall from the previous discussion of the standard deviation and the empirical rule that the areas between the mean and any point along the horizontal axis have been tabulated. (Appendix A contains such a table.) So if you travel up from the mean a certain distance, you can easily find how much of the area under the curve lies beyond that point. Thinking of these areas as probabilities, you can thus establish critical regions and critical values. The values along the horizontal scale, called **z scores**, are multiples of the standard deviation. For example, $z = 1.96$ means 1.96 standard deviations above the mean.

Although this is our first application of the standard normal distribution to assess a hypothesis such as H_0: $\mu = 5$, the general logic and methods explained previously apply here. We need only use a different table and compute a slightly different test statistic to be compared with critical values. More concretely, we need to establish null and alternative hypotheses, choose a decision rule, find appropriate critical regions, draw the sample, calculate the observed test statistic, and make a decision.

Figure 11-16 shows you how to use the tabulated distribution to find critical values. Suppose, as we have, that we have a large sample ($N > 30$) and want to test H_0: $\mu = 5$ at the .01 level with a two-tailed test. We need two critical regions, but their total area must equal .01 to give the desired level of significance. This requirement means that the size of each tail region must be $.01/2 = .005$. The numbers appearing in the body of the table show the area or proportion of the distribution lying above the z scores defined by the row and column headings. For example, scan down the column under "z" until you come to the row marked "2.5." This corresponds to a z value of 2.5. Now move across the row until you come to the entry ".0049." Then move up the column to the top row under "Second Decimal Place of z." There you should see ".08." The combination of the row label (2.5) and column label (.08) gives 2.58 (just add 2.5 and .08). The area at and above this z score is .0049 or about .005. That's the size of the region we want, so the critical value for the test is 2.58. In probability language, 2.58 creates a critical region for which (assuming H_0 is true) the probability of a sample result's landing in it is .005.

The table gives values only for the upper half of the distribution, but the normal is symmetric, so −2.58 (note the minus sign) defines an area at the lower end equal to about .005. Consequently, if we get an observed test statistic that is greater than or equal to either −2.58 *or* +2.58, we will reject the null

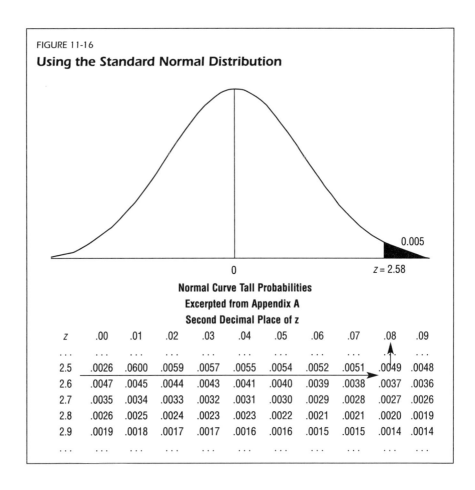

FIGURE 11-16

Using the Standard Normal Distribution

0.005

0

$z = 2.58$

Normal Curve Tail Probabilities
Excerpted from Appendix A
Second Decimal Place of z

z	.00	.01	.02	.03	.04	.05	.06	.07	.08	.09
. . .	. . .	. . .	. . .	. . .	. . .	. . .	. . .	. . .	. . .	. . .
2.5	.0026	.0600	.0059	.0057	.0055	.0054	.0052	.0051	.0049	.0048
2.6	.0047	.0045	.0044	.0043	.0041	.0040	.0039	.0038	.0037	.0036
2.7	.0035	.0034	.0033	.0032	.0031	.0030	.0029	.0028	.0027	.0026
2.8	.0026	.0025	.0024	.0023	.0023	.0022	.0021	.0021	.0020	.0019
2.9	.0019	.0018	.0017	.0017	.0016	.0016	.0015	.0015	.0014	.0014
. . .	. . .	. . .	. . .	. . .	. . .	. . .	. . .	. . .	. . .	. . .

hypothesis at the .01 level. Once again, .01 can be thought of as the probability of making a type I error (or incorrectly rejecting a true null hypothesis).

To help secure the procedure in your mind, let's find the critical values for (1) a two-tailed test at the .05 level and (2) a one-tailed test at the .002 level.

1. Since we want a two-tailed test, we have to divide .05 in half and look for the z value in the table that marks off the $.05/2 = .025$ proportion of the distribution. Look in Appendix A for ".0250," the size of the critical region. When you have found it, look at the row and column labels (you may want to use a straight edge). The row should be "1.9" and the column ".06" so that the critical value is $1.9 + .06 = 1.96$. This value is compared to the observed z to arrive at a decision.

2. We need be concerned with only one end of the distribution. Therefore, search the bottom of the table for ".002." Again, when it has been located, the conjunction of row and column labels should be "2.88." You would com-

pare this number to the observed z to make the decision.

A concrete example may clarify the procedure. Once more let's assess the possibility that Americans are as a whole slightly conservative in their political views; that is, that $\mu = 5$. But to vary the presentation, we conduct the test at the .01 level and use a two-sided test, so the alternative hypothesis is H_A: $\mu \neq 5$. To carry out the test with the 2004 NES data, we have to find the sample mean ideology score ($\bar{Y}$) and the sample standard deviation ($\hat{\sigma}$). With these numbers in hand, we can calculate the observed test statistic, which in this case is called the observed z. Its formula has the same outward appearance as that of the observed t. We just subtract the sample mean from the hypothesized mean and divide by the estimated standard error, which you may recall is $\hat{\sigma}_{\bar{Y}} = \hat{\sigma}/\sqrt{N}$.

TABLE 11-21
Large-Sample Test of a Mean

Null hypothesis	H_0: $\mu = 5$
Alternative hypothesis	H_A: $\mu \neq 5$
Sampling distribution	Standard normal
Level of significance	$\alpha = .01$
Size of each critical region	.005
Critical value	$z_{crit} = 2.58$
Sample size	920
Sample mean ($\bar{Y}$)	4.27
Estimated population standard deviation ($\hat{\sigma}$)	1.47
Estimated standard error ($\hat{\sigma}_{\bar{Y}}$)	.048
Observed test statistic	$z_{obs} = -15.21$

Table 11-21 summarizes the test criteria, sample values, and decision for testing the hypothesis that the average population liberalism-conservatism score is 5.

The observed z is calculated as follows:

$$z_{obs} = \frac{(\bar{Y} - \mu)}{\dfrac{\hat{\sigma}}{\sqrt{N}}} = \frac{(4.27 - 5)}{\dfrac{1.47}{\sqrt{920}}} = \frac{-.73}{.048} = -15.21.$$

The absolute value of this result greatly exceeds the chosen critical value (2.58), so the null hypothesis would be rejected. As a matter of fact, the observed z exceeds *any* value in the tabulated standard normal. Consequently, we would conclude that the probability of making a type I error is vanishingly small. Figure 11-17 shows how most software represents the probability (as 0.000). This does not mean that there is *no* possibility the null hypothesis is true; it only suggests a very small likelihood that cannot be presented conveniently.

What are we to make of this highly significant result? It could be presented with great fanfare. We might declare that, based on our statistical evidence, Americans can in no way be construed as being even slightly conservative. After all, the findings, which are based on a reputable scholarly survey, are significant at considerably below the .001 level. The next section, however, argues that such a claim may be unwarranted. Besides, the sample mean is 4.27, a value ever so slightly in the conservative direction. And how much does a somewhat arbitrary ideology scale reveal about attitudes and how much substantive or practical importance can we place on scale score differences of 1 or

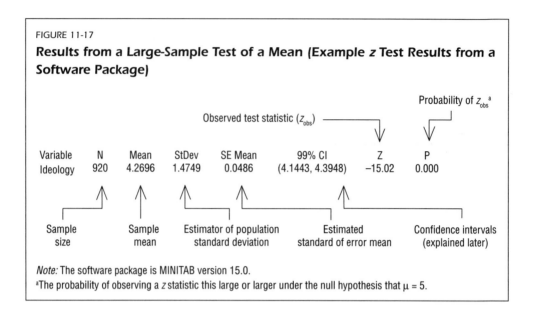

FIGURE 11-17

Results from a Large-Sample Test of a Mean (Example *z* Test Results from a Software Package)

Probability of z_{obs} [a]

Observed test statistic (z_{obs})

Variable	N	Mean	StDev	SE Mean	99% CI	Z	P
Ideology	920	4.2696	1.4749	0.0486	(4.1443, 4.3948)	−15.02	0.000

Sample size

Sample mean

Estimator of population standard deviation

Estimated standard of error mean

Confidence intervals (explained later)

Note: The software package is MINITAB version 15.0.
[a]The probability of observing a *z* statistic this large or larger under the null hypothesis that $\mu = 5$.

even .5? The soundest conclusion seems to be that the public is in the middle of the political road or just a bit to the right.

STATISTICAL SIGNIFICANCE AND THEORETICAL IMPORTANCE. We declared at the outset of the discussion on inference that people depend on the knowledge generated from samples. An integral part of that knowledge is the concept of statistical significance. Policy makers, politicians, journalists, academics, and lay people frequently try to prove a point by claiming something to the effect that "studies indicate a statistically significant difference between A and B" or that "there is no statistically significant association between X and Y." Hypothesis testing has become a common feature of both social and scientific discourse.

As empirical political scientists, we are happy that people resort to data and statistics to justify their positions. (Chapter 2 discussed the value of scientific epistemology.) Nevertheless, great confusion exists about what "significance" really entails. We have given you a lot of background about what goes into hypothesis testing and the assertions that something is or is not significant. Keep in mind, though, that these tests rest on specific assumptions and procedures, and making meaningful generalizations from samples depends on how thoroughly these assumptions and procedures are understood.

Sometimes when a person says that "findings are statistically significant," the implication is that a possibly earth-shattering discovery has been produced. But several factors affect the decision to reject or not reject a null hypothesis (Figure 11-18). The first thing that bears on a result is the objective or real situation in the world. There may indeed be differences between men and women and between developed and developing nations. A generic expression for this idea is that effects exist, and social scientists, like scientists in general, want to discover and measure these effects. But the hard truth is that a significance test does not prove that a meaningful effect has been uncovered. Too many other factors can cloud the interpretation of a hypothesis test.

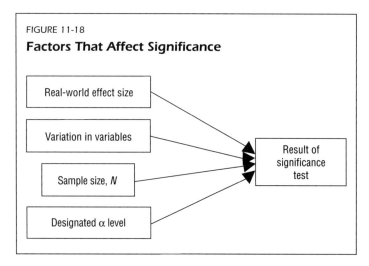

FIGURE 11-18
Factors That Affect Significance

- Real-world effect size
- Variation in variables
- Sample size, N
- Designated α level

→ Result of significance test

A major factor is the sample size. All other things being equal, the larger the sample, the easier it is to find significance, that is, to reject a null hypothesis. Why? The sample size does its work through the standard error, the measure of a sample estimator's precision or, loosely speaking, its closeness to the population value it estimates. If N is relatively small, sample estimates of a parameter will jump all over the place. But when N is relatively large, sample estimates tend to be close to one another. This variation shows up in the magnitude of the standard error, which in turn goes into the formulas for observed test statistics.

To demonstrate the point, suppose a null hypothesis is H_0: $\mu = 100$. Further, assume the sample mean is $\bar{Y} = 105$ and $\hat{\sigma} = 50$. (We will use the z statistic for illustrative purposes.) Let the sample size increase from 25 to 100 to 400 to 900. The observed z values are as follows:

$$z_{obs1} = \frac{(\bar{Y}-\mu)}{\frac{\hat{\sigma}}{\sqrt{N}}} = \frac{(105-100)}{\frac{50}{\sqrt{25}}} = \frac{5}{\frac{50}{5}} = \frac{5}{10} = .5.$$

$$z_{obs2} = \frac{(105-100)}{\frac{50}{\sqrt{100}}} = \frac{5}{\frac{50}{10}} = \frac{5}{5} = 1.0.$$

$$z_{obs3} = \frac{(105-100)}{\frac{50}{\sqrt{400}}} = \frac{5}{\frac{50}{20}} = \frac{5}{2.5} = 2.0.$$

$$z_{obs4} = \frac{(105-100)}{\frac{50}{\sqrt{900}}} = \frac{5}{\frac{50}{30}} = \frac{5}{1.6667} = 3.0.$$

The lesson is that a relatively small departure of a sample mean from the hypothesized mean becomes more and more statistically significant as the sample size increases, when everything else remains the same. This kind of argument backs up an old saying popular among statistics teachers: "You can *always* prove something if you take a large enough sample." A less cynical view is that "statistical significance is not the same thing as substantive significance."

Even more disturbing perhaps is the fact that you can influence whether something is found to be statistically significant by changing the decision rule or level of significance, or by making other decisions.[37] Consequently, when you are confronted with statistical arguments, the key questions should always be, "Does the reported effect have theoretical or practical or substantive import, and what research choices went into finding it?"

Confidence Intervals and Confidence Levels: Reporting Estimates of Population Parameters

Recall from Chapter 7 that, according to sampling theory, if we take many samples to obtain estimates of a population parameter, our estimates will be normally distributed and cluster around the true value of the population parameter. Sampling distributions tell us the probability that our estimates fall within certain distances of the population parameter. This probability is known as the **confidence level.** The **confidence interval** refers to the range of likely values associated with a given probability or confidence level. Thus for every confidence level, a particular confidence interval exists.

The general form of the confidence interval is as follows:

Estimated parameter value ± standard error × critical value.

Let's return to the question of ideology in America. Look back at Figure 11-17. You will see the sample mean, $\bar{Y} = 4.2696$; the standard deviation $\hat{\sigma} = 1.4749$; and the standard error of the mean, $\hat{\sigma}_{\bar{Y}} = 1.4749/\sqrt{920}$. Let's start out with a 99 percent confidence level. Since we have a large sample, we use the table of z scores to find the critical value. We need to find the critical value associated with 1 percent ($\alpha = .01$) of the distribution in the tails. The table reports the percentage of estimates likely to fall in one tail. So we need to divide .01 by 2, which is .005. Then we look in Appendix A for this value. Looking at Figure 11-16, we find .049, which corresponds with a z score of 2.58. This is our critical value. Substituting 2.58 into the above equation, we find that our confidence interval is $4.2696 \pm 2.58 \times .0486$, or between 4.144 and 4.395. Thus we can say that we are 99 percent confident that the actual mean in the population is between the values of 4.144 and 4.395. Figure 11-17 shows the confidence interval for the 99 percent confidence level. If, for instance, you want

How It's Done

The Construction of Confidence Intervals: The Range of Values of an Estimated Population Parameter for a Given Level of Statistical Significance or Confidence Level

Sample estimator ± (estimated standard error × critical value)

Function of sample
standard deviation
($\hat{\sigma}$) and sample size, *N*.

Critical value derived from
sampling distribution for
desired level of significance, α.

the 95 percent interval, then α is .05. The *z* score defining the upper α/2 = .05/2 = .025 tail area of the standard normal distribution is 1.96.

The same procedure is followed using the *t* distribution for small samples. Based on our earlier example, the mean number of physicians per 100,000 people for the sample of 17 developed nations is 321 with a standard deviation of 72.4. To report the range of values that might include the unknown population mean of the number of physicians for a 95 percent confidence level, we would find the critical value for α = .05 for a two-tailed test with 16 degrees of freedom. The critical value is 2.120. To calculate the interval, we must first calculate the standard error for the sample, which is $72.4/\sqrt{17}$ or 17.56. Thus the 95 percent confidence interval is 321 ± 17.56 × 2.120, or between 283.77 and 385.23. You interpret these two numbers as saying, "We are 95 percent sure that the average number of physicians per 100,000 people of the developed world lies somewhere between roughly 283.77 and 358.23 physicians. We are not 100 percent positive, but the evidence is overwhelmingly in that direction."

Here's a very important and useful tip. You can learn a lot from a confidence interval that you can't get from just one statistic: the sample mean. Confidence intervals, in a sense, suggest a span of plausible values for a parameter. By the same token, they indicate values that are implausible. For example, the interval in the previous example strongly suggests that the mean number of physicians per 100,000 persons of all the developed nations is not, say, 400; it's even less likely to be 600. Why? These values fall outside the interval. Or, to anticipate the next section, suppose someone claims that the mean for these nations is a meager 200. This argument can be considered to be a statistical hypothesis and can be tested. Since the hypothesized value falls beneath the lower confidence interval, you have reason to doubt it.

Now flip around the argument. Suppose another research team estimates the average number of physicians to be 295. Although this value lies considerably below our estimate of 321, is it really inconsistent? Remember, we are

TABLE 11-22

Confidence Intervals Calculated for Four Different Confidence Levels

Sample mean = 3789
$N = 19$

Percent confidence	Confidence Interval		
	Upper limit	Lower limit	Interval width
80	4730.9	2847.1	1883.8
90	5017.0	2561.0	2456.0
95	5276.9	2301.1	2975.9
99	5827.2	1750.8	4076.4

pretending that the data come from random samples, so both estimates inevitably have some imprecision. Our confidence intervals, however, go from about 283 to slightly more than 358. Hence, they include the alternative estimate. On just these grounds, we can't argue that the other estimate is wrong, and neither can they say that ours is wrong.

There is a rule of thumb in statistics as in life: the more certain you need to be, the more information you have to have. In the case of confidence intervals, the higher the assurance you have to have, the wider your interval will be for a given sample size. Table 11-22 illustrates this point. In it we calculate 80 percent, 90 percent, 95 percent, and 99 percent confidence intervals for the mean GNP of the developing nations. The first column gives the different degrees of confidence requested. The last column shows the widths of the resulting intervals; you might loosely interpret these numbers as the "margins of error" in the estimate. Notice that as you go down the table, this error margin increases. Stated differently, if you can be content with a ballpark guess, say 80 percent intervals, then the difference between the upper and lower limits is 4730.9 – 2847.1 = 1883.8, or about $1,884. But if you want to be as close as possible, or 99 percent certain, the interval becomes twice as wide, a whopping $4,076. This example is based on just 19 cases. To tell the story one more time, the estimator of the population mean is not wrong or invalid; but it is imprecise.

If you want narrower interval widths while still being, say, 95 percent confident, then you have to increase the sample size. Table 11-23 tells you what you get for larger and larger samples. As N increases, the interval widths shrink. If you could somehow take a very large sample of developing nations—you cannot, of course, but just imagine—you could increase the estimator's precision from almost $3,000 to less than $500. The down side is the expense of collecting the extra data.

These statements, incidentally, may remind you of something mentioned earlier: the larger the sample size, the smaller the standard error. This notion

TABLE 11-23
Confidence Intervals for Various Sample Sizes at 95 Percent Level of Confidence

| | Sample mean = 3789 | | |
Sample size (N)	Upper limit	Lower limit	Interval width
19	5276.9	2301.1	2975.8
50	4644.7	2933.3	1711.3
100	4393.1	3184.0	1210.1
1,000	4023.8	3554.2	469.69

comes into play here because confidence intervals are partly a function of the standard error. If the standard error decreases, so does the width of the corresponding confidence interval.

The bottom line is that statistical inference demands a balance between exactitude and sample sizes. If you want or need to be more exact, you need a bigger sample. But whether or not you need to be more or less exact is *not* a matter of statistics; it is essentially a substantive or practical question. Our feeling is that in an exploratory study in which the investigator is entering new territory, precision may not be as important as making sure the sample is drawn correctly and the measurements are made and recorded carefully and accurately. Only when a lot is riding on the accuracy of estimates will huge sample sizes be essential.

USING CONFIDENCE INTERVALS TO TEST FOR SIGNIFICANCE. We have stressed that a confidence interval gives you a range of likely values of the population parameter besides the one actually observed. This range turns out to be handy because you can tell at a glance whether or not a *hypothesized* value falls in that interval.

Let's return to the question of ideology in America. Look back at Figure 11-17, which reports the results of a test of the hypothesis that H_0: $\mu = 5$. You should see the sample mean, $\bar{Y} = 4.2696$; the standard deviation $\hat{\sigma} = 1.4749$; and the standard error of the mean, $\hat{\sigma}_{\bar{Y}} = 1.4749/\sqrt{920} = .0486$. The 99 percent confidence intervals are also reported; they range from about 4.14 to 4.49. This interval does not contain the hypothesized value, so you have reason to reject H_0.

In many instances, confidence intervals provide at least as much information as a statistical hypothesis test. We say "at least as much" because a hypothesis test just indicates whether or not a null hypothesis has been rejected. A confidence interval also supplies a set of possible values for the parameter.

Conclusion

The goal of this chapter has been to provide an introduction to statistical analysis. After presenting two examples as grist for our discussion, we emphasized the importance of getting organized before tackling a quantitative investigation. Then we showed several statistical and graphical methods for reducing, summarizing, and exploring data. We then started down the road to understanding statistical inference, including hypothesis testing and estimation. We leave you with a couple of guidelines for improving your research and evaluating that of others:

■ No single summary statistic does or can say everything important about even a small amount of data. Consequently, you should rely on several summary measures and graphs, not just one.

■ Look at each variable individually. How much variation is there? What form does its distribution have? Are there any "problem" observations? Does it seem to have an association with other variables? What kinds?

■ If this sounds like a lot of work, think carefully *before* collecting any data. Just because a variable is in a collection doesn't mean you have to include it in your study. Ask what will be just enough to support or refute a hypothesis.

■ Most readers probably will not be in a position to analyze much more than 6 to 10 variables. True, computers make number crunching easy. But it is very hard to take in and discuss in substantive terms a mass of tables, graphs, and statistics. You are probably better off studying a small data set thoroughly than analyzing a big one perfunctorily.

■ Try to understand the principles of statistical inference and think continually about the topic or phenomenon being studied and the practical or real-world meaning of the results. Do not get hung up on technical jargon.

■ A well-thought-out and carefully investigated hypothesis that the data do not support can be just as informative and important as a statistically significant result. It is not necessary to report only "positive" findings; in fact, it's misleading. Chapter 2 discussed the roles that replication and falsification play in science. If you are studying a claim that lots of people believe and discover that your data do not support it, you will have made a positive contribution to knowledge.

Notes

1. "Statistics and the law" has become an important multidisciplinary field, and lawyers who are not already familiar with the subject spend considerable time catching up in workshops and law journals.

2. One metric, the Kelvin scale of temperature, does have an absolute zero, the point at which atoms do not move and heat and energy are absent. The zero points on the other temperature scales are arbitrary.

3. Occasionally, however, it is useful to treat the numbers assigned to the categories of an ordinal or ranking scale as if they were really quantitative.

4. The full question is "Generally speaking, do you usually think of yourself as a Republican, a Democrat, an independent, or what? Would you call yourself a Strong [Democrat/Republican] or a Not Very Strong [Democrat/Republican]? Do you think of yourself as closer to the Republican Party or to the Democratic Party?"

5. The survey asked, "There has been discussion recently about a law to ban certain types of late-term abortions, sometimes called partial birth abortions. Do you favor or oppose a law that makes these types of abortions illegal?"

6. The order of the division affects the interpretation, but in an obligingly nice way. If you want the odds of *opposing to favoring* laws outlawing late-term abortions, reverse the denominator, and interpret accordingly: The chances of an arbitrarily chosen person opposing such a law is only about .6 the chances of that person approving it.

7. Try translating "the odds are three and a half in favor of the Broncos." You should come up with something like "Denver is more than three times [three and a half times, to be precise] as likely to beat its opponent as it is to lose." Or "the chances of the opponent's winning are less than a third."

8. If a variable has K categories, there are $K(K-1)/2$ pairs of odds; of these $(K-1)$ will be independent. For example, given a variable with $K = 4$ levels, there will be $(4)(3)/2 = 6$ pairs of odds. However, they will contain redundant information in the sense that some will be combinations of others. The number of odds containing nonredundant information is $4 - 1 = 3$. If the category labels are A, B, C, and D, you could examine A vs. B, A vs. C, and A vs. D (A is the reference class) or some other combination.

9. Unfortunately, it is often impossible to make the distinction in published work, especially in the mass media, because the authors simply do not provide the necessary information. This is why we encourage researchers to report total frequencies along with percentages.

10. The mean also has important functions in applied and theoretical statistics that are beyond the scope of this book.

11. See, for example, Seymour Martin Lipset, "Some Social Requisites of Democracy: Economic Development and Political Legitimacy," *American Political Science Review* 53 (March 1959): 69–105; Philip Cutright, "National Political Development: Measurement and Analysis," *American Sociological Association* 28 (April 1963): 253–264.

12. Incidentally, instead of using the means, we could explore deviations from the median. But we will not pursue that topic here.

13. Alan Agresti and Barbara Finlay, *Statistical Methods for the Social Sciences* (Upper Saddle River, N.J.: Prentice Hall, 1997), 60.

14. Allen R. Wilcox lists many of these measures in "Indices of Qualitative Variation and Political Measurement," *Western Political Quarterly* 26 (June 1973): 325–343.

15. The present treatment is based on Wilcox, "Indices of Qualitative Variation and Political Measurement" (Wilcox calls it by another name), and Alan Agresti and Barbara F. Agresti, "Statistical Analysis of Qualitative Variation," *Sociological Methodology* 9 (1978): 204–237.

16. It is thought that this measure was introduced by John H. Mueller and Karl F. Schuessler in *Statistical Reasoning in Sociology* (Boston: Houghton Mifflin, 1961).

17. Ibid., 206. The authors were referring to the population measure, but there is no harm in using the same interpretation for sample data.

18. Wilcox, "Indices of Qualitative Variation and Political Measurement," 326.

19. There may be variation *within* the group, but it is unobserved.

20. Agresti and Agresti, "Statistical Analysis of Qualitative Variation," 259. Wilcox points out that by rearranging the terms in the definition, both measures may be given other interpretations. Indeed, he maintains that they are analogous to variance. Wilcox, "Indices of Qualitative Variation and Political Measurement," 329.

21. Julian Blair and Michael G. Lacy provide a measure that uses cumulative proportions, not cell proportions. Its meaning, however, is not as simple as those for ID or IQV. Julian Blair and Michael G. Lacy, "Statistics of Ordinal Variation," *Sociological Methods and Research* 28 (February 2000): 251–280.

22. The literature on this topic is vast. For guidelines for presenting accurate and effective visual presentations, Edward Tufte is indispensable. His *The Visual Display of Quantitative Information* (Cheshire, Conn.: Graphics, 1983) is a classic. For an introduction to graphic data exploration, see William Jacoby, *Statistical Graphics for Univariate and Bivariate Data* (Thousand Oaks, Calif.: Sage Publications, 1997).

23. John Tukey, *Exploratory Data Analysis* (New York: Addison-Wesley, 1977).

24. See Tufte, *Visualizing Data,* for numerous examples of informative and misleading graphs.

25. Individuals and software draw boxplots in slightly different ways, though all follow the same basic principles given here. These guidelines are based on MINITAB, the software used for most of the analyses and graphics in this book.

26. Frequently a null hypothesis is known ahead of time to be false. Suppose, for example, that you tested the hypothesis that the mean income of political liberals in America is zero (H_0: $\mu = 0$). This proposition is nonsensical, and testing it would provide no information because any sample result would reject it. Yet surprisingly many studies in effect conduct such tests.

27. We are implicitly assuming that the "probability of heads" is a property or attribute of objects such as coins and hence can be regarded as a sort of population parameter. But the meaning of probability remains a wildly controversial subject in statistics and mathematics. Is it, for example, an inherent (physical?) trait of something, or is it a subjective phenomenon? We're just going to slide around this controversy, however.

28. The probability of detecting and thus rejecting a false null hypothesis is called the power of the test and equals $1 - \beta$, where β is the probability of a type II error. Power is an extremely important issue in statistics. Many commonly used inferential tests may have relatively low power. An excellent introduction to the topic is Jacob Cohen, *Statistical Power Analysis for the Behavioral Sciences,* 2d ed. (Hillsdale, N.J.: Erlbaum, 1988), and Jacob Cohen, "A Power Primer," *Psychological Bulletin* 112 (July 1992): 155–159.

29. See Chapter 7.

30. Building statistical inference on the idea of repeated samples initially makes students uneasy since it is indeed a difficult concept. Even more interesting is that it bothers many researchers in the field. For a readable introduction to this debate, see Bruce Western, "Bayesian Analysis for Sociologists," *Sociological Methods and Research* 28 (August 1999): 7–11.

31. The probability of getting exactly Y successes in N trials, $\Pr(Y)$, given that the probability of a success is P, is

$$\Pr(Y) = \frac{N!}{Y!(N-Y)!} P^Y (1-P)^{N-Y}.$$

The symbol ! stands for "factorial" so that $Y!$ means $(Y)(Y-1)(Y-2)(Y-3) \ldots$. (For example, $4! = (4)(3)(2)(1) = 24$.) The same definition applies to $N!$, and $0!$ is defined to equal 1. As an example, suppose, you toss a coin $N = 10$ times, and the probability of heads on each toss is $P = .5$. You get 1 head. What is the probability of getting 1 head in 10 tries?

$$\Pr(1) = \frac{1!}{1!(10-1)!} .5^1 (1-.5)^{10-1} = \frac{1!}{(1!9!)} .5^1 .5^9 = 01.$$

32. Actually it's a little less, .000977, to be a bit more precise, but we have rounded the results.

33. When samples are large, the testing procedures described in this book are sometimes based on approximations. Unfortunately, if N is not large, the approximations in many cases are not very good, at least by the standards of most statisticians. Consequently, what are known as "exact" methods come into play. In a nutshell, one strategy is literally to calculate the probability of every possible outcome. But this computation can be a daunting challenge and often involves more theory than is needed at this point.

34. We need degrees of freedom because the t distribution is really a family of distributions, each based on a degree of freedom. When N gets larger and larger, the t distribution becomes approximately normal, and we can use it when $N \geq 25$.

35. We get $t_{obs} = 1.788$, which with 24 degrees of freedom is barely greater than the critical value of 1.711. Hence, we would reject this null hypothesis and perhaps conclude that the public on average sits just barely to the right of center. If you test the hypothesis at the .01 level, you need a different critical value. With 24 degrees of freedom, we see that the critical value at the

α = .01 level is 2.492. According to this standard, we would *not* reject the null hypothesis. The bottom line here seems to be that the typical American is more or less in the middle of the road when it comes to politics.

36. This is known as the central limit theorem, which is discussed briefly in Agresti and Finlay, *Statistical Methods for the Social Sciences,* 103–105.

37. A point that the figure illustrates but that we do not go into here is that the amount of variation in the data also affects all kinds of statistical results, not just hypothesis tests. It seems logical to regard variation as a property of the world and to acknowledge that all you can do is measure it. Yet, in reality, research design—that is, the choices social scientists consciously or unconsciously make—affects observed variation. An excellent and accessible discussion of this sort of problem is Charles Manski, *Identification Problems in the Social Sciences* (Cambridge, Mass.: Harvard University Press, 1995).

Terms Introduced

ALTERNATIVE HYPOTHESIS. A statement about the value or values of a population parameter. A hypothesis proposed as an alternative to the null hypothesis.

BAR GRAPH. A graphic display of the data in a frequency or percentage distribution.

CENTRAL TENDENCY. The most frequent, middle, or central value in a frequency distribution.

CONFIDENCE INTERVAL. The range of values into which a population parameter is likely to fall for a given level of confidence.

CONFIDENCE LEVEL. The degree of belief or probability that an estimated range of values includes or covers the population parameter.

CUMULATIVE PROPORTION. The total proportion of observations at or below a value in a frequency distribution.

DATA MATRIX. An array of rows and columns that stores the values of a set of variables for all the cases in a data set.

DESCRIPTIVE STATISTIC. The mathematical summary of measurements for one variable.

DISPERSION. The distribution of data values around the most frequent, middle, or central value.

FREQUENCY DISTRIBUTION (f). The number of observations per value or category of a variable.

HISTOGRAM. A type of bar graph in which the height and area of the bars are proportional to the frequencies in each category of a nominal variable or intervals of a continuous variable.

INDEX OF DIVERSITY. A measure of variation for categorical data that can be interpreted as the probability that two individuals selected at random would be in different categories.

INDEX OF QUALITATIVE VARIATION. A measure of variation for categorical data that is the index of diversity adjusted by a correction factor based on the number of categories of the variable.

INTERQUARTILE RANGE. The middle 50 percent of observations.

MEAN. The sum of the values of a variable divided by the number of values.

MEAN ABSOLUTE DEVIATION. A measure of dispersion of data points for interval- and ratio-level data.

MEDIAN. The category or value above and below which one-half of the observations lie.

MODE. The category with the greatest frequency of observations.

NEGATIVELY SKEWED. A distribution of values in which fewer observations lie to the left of the middle value and those observations are fairly distant from the mean.

NORMAL DISTRIBUTION. A distribution defined by a mathematical formula and the graph of which has a symmetrical, bell shape; in which the mean, mode, and median coincide; and in which a fixed proportion of observations lies between the mean and any distance from the mean measured in terms of the standard deviation.

NULL HYPOTHESIS. A statement that a population parameter equals a single or specific value. Often a statement that the difference between two populations is zero.

ODDS. The ratio of the frequency of one event to the frequency of a second event.

PIE DIAGRAM. A circular graphic display of a frequency distribution.

POSITIVELY SKEWED. A distribution of values in which fewer observations lie to the right of the middle value and those observations are fairly distant from the mean.

RANGE. The distance between the highest and lowest values or the range of categories into which observations fall.

RELATIVE FREQUENCY. Percentage or proportion of total number of observations in a frequency distribution that have a particular value.

RESISTANT MEASURE. A measure of central tendency that is not sensitive to one or a few extreme values in a distribution.

STANDARD DEVIATION. A measure of dispersion of data points about the mean for interval- and ratio-level data.

STANDARD NORMAL DISTRIBUTION. Normal distribution with a mean of 0 and a standard deviation and variance of 1.

STATISTICAL SIGNIFICANCE. The probability of making a type I error.

TRIMMED MEAN. The mean of a set of numbers from which some percentage of the largest and smallest values has been dropped.

TYPE I ERROR. Error made by rejecting a null hypothesis when it is true.

TYPE II ERROR. Error made by failing to reject a null hypothesis when it is not true.

VARIANCE. A measure of dispersion of data points about the mean for interval- and ratio-level data.

Z SCORE. The number of standard deviations by which a score deviates from the mean score.

Suggested Readings

Abelson, Robert P. *Statistics as Principled Argument.* Hillsdale, N.Y.: Lawrence Erlbaum, 1995.

Agresti, Alan. *An Introduction to Categorical Data Analysis.* New York: Wiley, 1996.

Agresti, Alan, and Barbara Finlay. *Statistical Methods for Social Sciences.* Upper Saddle River, N.J.: Prentice Hall, 1997.

Cleveland, William S. *Visualizing Data.* Summit, N.J.: Hobart Press, 1993.

Jacoby, William. *Statistical Graphics for Univariate and Bivariate Data.* Thousand Oaks, Calif.: Sage Publications, 1997.

Lewis-Beck, Michael. *Data Analysis: An Introduction.* Thousand Oaks, Calif.: Sage Publications, 1995.

Tufte, Edward R. *The Visual Display of Quantitative Information.* Cheshire, Conn.: Graphics, 1983.

Velleman, Paul, and David Hoaglin. *Applications, Basics, and Computing of Exploratory Data Analysis.* Pacific Grove, Calif.: Duxbury Press, 1983.

CHAPTER 12

Investigating Relationships between Two Variables

This chapter takes up the investigation of relationships between two variables. Generally speaking, a statistical relationship between two variables exists if the values of the observations for one variable are associated with or connected to the values of the observations for the other. For example, if as people get older they vote more frequently, then the values of the dependent variable (voting or not voting) are related to the values of the independent variable (age). Therefore, the observed values for the two variables are related. Knowing that two variables are related lets us make predictions. If we know the value of one variable, we can predict (subject to error) the value of the other. But many other questions arise. How strong is the relationship? What is its direction or shape of the relationship? Is it a causal one? Does it change or disappear if other variables are brought into the picture? If the relationship has been detected in a sample, can we conclude that the relationship holds for the population?

We start this chapter with general remarks about two-variable relationships and then describe several methods for measuring and interpreting such relationships and for assessing their statistical significance. We use both numerical and graphical techniques for this purpose.

The Basics of Identifying and Measuring Relationships

Determining how the values of one variable are related to the values of another is one of the foundations of empirical social science inquiry. In making such determinations, we consider the following features of relationships:

■ The level of measurement of the variables. Different kinds of measurement necessitate different procedures. Table 12-1 summarizes these procedures.[1]

■ The form of the relationship. We can ask if changes in the variable X move in lockstep with increases (or decreases) in the variable Y or if a more sophisticated connection exists.

TABLE 12-1
Levels of Measurement and Statistical Procedures: A Summary

Type of Dependent Variable	Type of Independent Variable	Procedure
Quantitative	Dichotomous	Difference of means/proportions
Quantitative	Categorical (nominal or ordinal)	One-way analysis of variance (ANOVA)
Categorical (nominal or ordinal)	Categorical (nominal and/or ordinal)	Cross-tabulation analysis: measures of association
		Log-linear models
		Association models
Quantitative	Quantitative and/or categorical	Linear regression
Dichotomous	Quantitative and/or categorical (nominal and/or ordinal)	Logistic regression

Note: Dichotomous variables have two categories.

- The strength of the relationship. It is possible that some levels of X will always be associated with certain values of Y. More commonly, though, the values have only a tendency to covary, and the weaker the tendency, the less the strength of the relationship.

- Numerical summaries of relationships. Social scientists strive to boil down the different aspects of a relationship to a single number that reveals the type and strength of the association. These numerical summaries, however, depend on how relationships are defined.

- Conditional relationships. The variables X and Y may seem to be related in some fashion, but appearances can be deceiving. For example, in a spurious relationship (see Chapter 5), the perceived relationship between two variables may disappear once a third factor is brought into the analysis. A major activity in statistical analysis is studying how the inclusion of additional variables affects the form and strength of relationships.

Let us examine several of these topics in more depth.

Types of Relationships

A relationship between two variables, X and Y, can take one of several forms.

- General association. This type of association exists when the values of one variable, X, tend to be associated with specific values of the other variable, Y. This definition places no restrictions on how the values relate; the only requirement is that knowing the value of one variable helps us to know or predict the value of the other. For example, if religion and party identification are associated, then certain members of certain sects should tend to

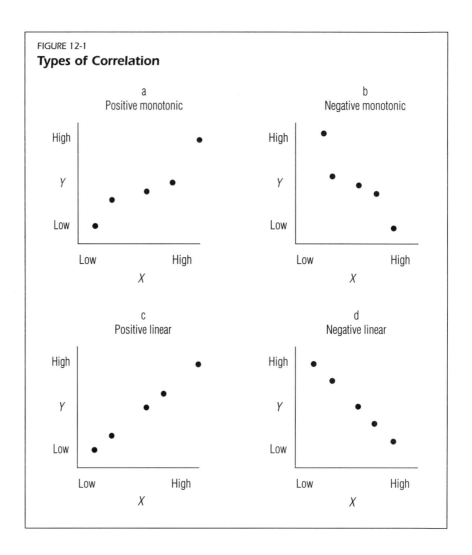

FIGURE 12-1

Types of Correlation

a
Positive monotonic

b
Negative monotonic

c
Positive linear

d
Negative linear

identify with certain political parties. Discovering that a person is Catholic should say something about his or her partisanship. If the values of X and Y are not connected in some way, then we assert that they are independent of one another. (Statistical independence is discussed later in this chapter.)

■ Positive monotonic correlation. When the values of both variables have an explicit ordering (ordinal and quantitative variables), a particular kind of relationship may hold (Figure 12-1). In one type of relationship, high values of one variable are associated with high values of the other, and conversely, low values are associated with low values. On a graph, X-Y values drift upward from left to right (see Figure 12-1a). A line drawn through the graph will be curved but never goes down once it is on its way up.

■ Negative monotonic correlation. High values of X are associated with low values of Y, and low values of X are associated with high values of Y. A graph of X-Y pairs drifts downward from left to right and never turns back up (see Figure 12-1b). It's the same idea as positive correlation, except that the points generally slope downward.

■ Positive linear correlation. This type of correlation is a particular type of monotonic relationship in which plotted X-Y values fall on (or at least close to) a straight line (see Figure 12-1c). Perhaps you recall from algebra that the equation for a straight line that slopes upward from left to right is $Y = a + bX$, where a is a constant and b is a positive constant.

■ Negative linear correlation. In this type of correlation, the plotted values of X and Y fall on a straight line that slopes downward from left to right. An example can be seen in the graph of the equation $Y = c + dX$, where d is negative (see Figure 12-1d).

Relationships between variables may have other forms, as when values of X and Y increase together until some threshold is met when they decline. Since these curvilinear patterns of association are hard to analyze, we do not address them in this book. The important point is that the first step in data analysis is the examination of plots to determine the approximate form of relationships.

The Strength of Relationships

Virtually no relationship has a perfect form. In other words, there are degrees of association, and so it makes sense to talk about their strength. Figure 12-2 provides an intuitive illustration of what is meant by the **strength of a relationship.** Observe that in Figure 12-2a the values of X and Y are tied together tightly. You could even imagine a straight line passing through or very near most of the points.

In Figure 12-2b, by contrast, the X-Y points seem spread out, and no simple line or curve would connect them. The values tend to be associated—as X increases, so does Y—but the connection is rather weak. Notice, for instance, that although as X increases, Y does as well, some cases seem to have roughly the same values on X but very different Y scores.

We always encourage you to graph relationships, but these visual devices almost always need to be supplemented by numerical indices that describe the form and strength of relationships. These statistics are generally known as measures of association.

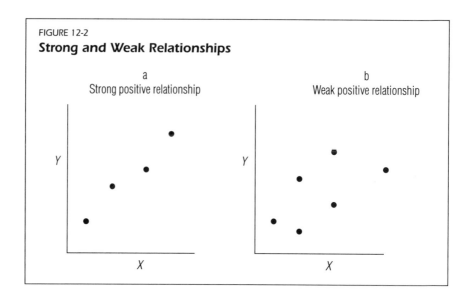

FIGURE 12-2
Strong and Weak Relationships

a
Strong positive relationship

b
Weak positive relationship

Numerical Summaries: Measures of Association

Measures of association describe in a single number or coefficient the kind and strength of relationship that exists between the values of two variables. The remainder of this chapter describes several such indicators, and most social science software programs crank them out automatically. These numbers are frequently used to support theoretical or policy claims (though attempts to do so are sometimes misguided), much as the results of statistical tests are used (see Chapter 11). It is imperative, then, to recognize what the numbers do (and do not) say about possibly complex relationships. The coefficients described in this chapter (1) assume a particular level of measurement—nominal, ordinal, interval, or ratio—and (2) rest on a specific conception of association. Stated differently, each coefficient measures a specific type of association, and to interpret its numerical value you have to grasp the kind of association it is measuring. Two variables can be associated strongly according to one coefficient and weakly (or not all) according to another. Therefore, whenever we describe a measure, such as the correlation coefficient, we need to explain the kind of relationship for which the measure is intended.

Before getting to specifics, however, here are some important properties of commonly used coefficients:

■ Null value. Zero typically indicates no association, but there are exceptions to this rule of thumb.

■ Maximum values. Some coefficients do not have a maximum value; in theory they can be very large. Many, however, are bounded: normally their lower and upper limits are 0 and |1.0|, where the brackets "| |" mean "absolute value." When a coefficient attains a bound, the variables are said to be perfectly associated according to the coefficient's definition.

- Strength of the relationship. Subject to lower and upper boundaries, a coefficient's absolute numerical value increases with the strength of the association. For example, a coefficient of .6 would indicate a stronger relationship than one of .3. (But the relationship would not necessarily be twice as strong. It all depends on how the statistic is defined.)

- Level of measurement. As indicated earlier, nominal, ordinal, and quantitative (ratio and interval) variables each require their own type of coefficient. You can, of course, treat ordinal scales as numeric and calculate a statistic intended for quantitative data—plenty of people do—but since satisfactory alternatives exist for different levels of measurement, you should be able to find one or two that will fit your data.

- Symmetry. The numerical magnitudes of some indices depend on which variable, X or $Y,$ is considered to be independent. These are asymmetric measures. The value of a coefficient calculated with Y as dependent may very well differ from the same indicator using X as the dependent variable. A symmetric measure keeps the same value no matter which variable is treated as dependent or independent.

- Standardized versus unstandardized. The measurement scale affects the numerical value of most coefficients of association. If income, for example, is measured in dollars in one study and in Euros in another, two coefficients may have different numerical values (50 versus 10, for instance), even though the strength of the relationship is exactly the same in both situations. To make the coefficients comparable, statisticians sometimes transform, or standardize, the variables so that they all have variances of 1. Once the variables have been standardized, coefficients applied to them can be compared. Often the adjustment is made by recalibrating an "unstandardized" coefficient into a standardized one. We discuss the procedure later in the chapter.

We now take up specific methods for assessing or measuring two-variable relationships.

Cross-tabulations of Nominal and Ordinal Variables

Let's return to the example considered in Chapter 11, namely, the gender gap. One way of tackling the problem is to compare men's and women's political attitudes and behavior. Suppose we hypothesize that in the United States women are more "liberal" than men. To investigate this hypothesis, we use the 2004 National Election Study data that contained questions on current events, party identification, and candidate preference. To be more specific, let us investigate the relationship between gender and vote in the 2004 presidential election.

Table 12-2 presents a frequency distribution of partisanship similar to the one shown in Chapter 11, but it does so by categories of a second variable, gender. We can construct a table showing each case's values for both variables by putting the independent variable across the top and the dependent variable down the side and creating a grid of boxes or cells, one for each combination of the variables. (This is the conventional format, but the independent variable could instead be located down the side.) Such an arrangement is called a cross-tabulation or a contingency table. A **cross-tabulation** displays the joint distribution of values of the variables by listing the categories for one of the variables along one side and the categories for the other variable across the top. Each case is then placed in the cell of the table that represents the combination of values that corresponds to its scores on the variables. In the table the counts or frequencies have been placed below the relative frequencies in parentheses to save space.

TERMINOLOGY

There is no standard term for what we defined as a cross-tabulation. Statistics books and scholars use a variety of words or phrases to label these displays: *cross-classification, contingency table, bivariate frequency distribution,* or most commonly *categorical data.* Here we treat all of these terms as synonyms. Moreover, cross-tabulations can consist of nominal, ordinal, or combinations of nominal and ordinal variables.

Cross-tabulations are one of the cornerstones of data analysis. Consider the last row, labeled "Total." In the second column you see in parentheses the number 581. This is the number of males in the sample who gave one of the responses listed in the table. (Those who replied "other" or "apolitical" have been treated as missing and are excluded.) Looking down the column for the men, you can determine that 11.9 percent of the 581 identified as strong Democrats, 15.0 percent as weak Democrats, 18.6 percent as independents who lean toward the Democratic Party, and so forth. In other words, this column provides the relative frequency distribution for males. The last column does the same for females. (As an example, 20.9 percent of the 613 women in the sample thought of themselves as strong Democrats.) By comparing the two frequency distributions, we can begin to address the hypothesis of a gender gap. About 21 percent of the women, as opposed to only about 12 percent of the men, were strong Democrats. If you compare the percentages down the rows, you will detect a slight tendency for more women than men to identify as Democrats. About 53.6 percent of women are Democrats to one degree or another, whereas the corresponding percentage for men is 45.5 percent.

The data provide some but not over-whelming evidence that women are a bit more Democratic than men and hence perhaps slightly more liberal as well. This interpretation, of course, rests on a number of assumptions: that we have a representative sample, that the party identification question correctly measures partisan feelings, that people's responses have been recorded and tallied correctly, and so on. It is also possible that the gender gap applies to just certain kinds of women, those who are young or unmarried or poor, for example. When we take up modeling in more detail, we can investigate this idea further.

The procedure for constructing a cross-tabulation is simple and requires just three steps: (1) Separate the cases into groups based on their values for the independent variable, (2) for each grouping on the independent variable compute the frequencies or percentages falling in each level of the dependent variable, and (3) decide whether the frequency or percentage distributions differ from group to group and, if so, how and by how much. Of course, for a large quantity of data, such as obtained from a poll, a computer will probably have to do the work. Cross-tabulation analysis is such a basic and common procedure that finding suitable software to carry it out is easy. The only hitch with these programs is that the user must correctly specify the percentages to calculate.

When the categories of the independent variable are arrayed across the top of the table—that is, they are the column labels—the percentages must add to 100 down the columns. (Ignore slight rounding errors, such as 99.9% or 100.1%.) These are called column percentages. You might think of the respondents in each column as a subsample. We want to know how the men differ among themselves on partisanship, and so the column totals must be used as the bases (denominators) for the percentage calculations. Thus, for the 581 men, the percentage identifying as strong Democrats (11.9 percent) plus the percentage identifying as weak Democrats (15.0 percent) plus the percentage identifying as independent-leaning Democrats (18.6 percent), and so forth

TABLE 12-2

Cross-tabulation of Gender by Party Identification

Party Identification Response Category	Gender	
	Male	Female
Strong Democrat	11.9%	20.9%
	(69)	(128)
Weak Democrat	15.0%	16.2%
	(87)	(99)
Independent-leaning Democrat	18.6%	16.5%
	(108)	(101)
Independent	10.2%	9.3%
	(59)	(57)
Independent-leaning Republican	14.5%	9.1%
	(84)	(56)
Weak Republican	13.9%	11.1%
	(81)	(68)
Strong Republican	16.0%	17.0%
	(93)	(104)
Total	100.1%	100.1%
N = 1,194	(581)	(613)

Source: 2004 National Election Study.
Note: Numbers in parentheses are frequencies or the number of cases. Percentages do not add to 100% because of rounding error.

through all the responses categories must equal 100 percent. The same is true for women.

Suppose you used a computer to calculate percentages by row totals, that is, to obtain row percentages. Table 12-3 suggests what might result and possible difficulties in interpretation. For example, if you were not careful, you might conclude that a huge gender difference exists on "strong Democrat" (35 percent versus 65 percent). But this is not what the numbers mean. There are 197 strong Democrats in the sample (look in the last column), of which 35 percent are men and 65 percent women. It *would* be appropriate to say that strong Democrats tend to be composed overwhelmingly of women whereas about half of independents are male and half are female. Still, if you believe that one variable (for example, party identification) depends on another variable (say, gender) and you want to measure the effect of the latter on the former, make sure the percentages are based on the independent variable category totals.

CATEGORIES WITH "TOO FEW" CASES

A widely accepted rule of thumb asserts that percentages based on 20 or fewer observations are not reliable indicators and should not be reported or should be reported with "warning signs." Suppose, for example, a survey contained only 15 respondents in a category of the independent variable such as Asian Americans. If you try to find the percentage of this group that identify as, say, strong Republicans, the resulting estimate will be based on such a small number (15) that many readers and analysts may not have confidence in it. Two possible solutions come to mind. First, use a symbol (for example, "†") to indicate "too few cases." Alternatively, the category could be combined with another one to increase the total frequency. The text gives some examples.

A First Look at the Strength of a Relationship

Do the data examined in the preceding section support the hypothesis about a gender gap? As we indicated, a careful examination of the column percentages suggests that the hypothesis has only minimal support. Why? Because scrutiny of the partisanship distributions by gender did not show much difference. Yet it would be desirable to have a more succinct summary, one that would reveal the strength of the relationship between sex and party identification.

The strength of an association refers to how different the observed values of the dependent variable are in the categories of the independent variable. In the case of cross-tabulated variables, the strongest relationship possible between two variables is one in which the value of the dependent variable for every case in one category of the independent variable differs from that of every case in another category of the independent variable. We might call such a connection a perfect relationship, because the dependent variable is perfectly associated with the independent variable. That is, there are no exceptions to the pattern. If the results can be applied to future observations, a perfect relationship between the independent and dependent variables enables a researcher to predict accurately a case's value on the dependent variable if the value on the independent variable is known.

TABLE 12-3

Row Percentages Are Not the Same as Column Percentages

Party Identification Response Category	Gender		
	Male	Female	Total
Strong Democrat	35.0%	65.0%	100%
	(69)	(128)	(197)
Weak Democrat	46.8%	53.2%	100%
	(87)	(99)	(186)
Independent-leaning Democrat	51.7%	48.3%	100%
	(108)	(101)	(209)
Independent	50.9%	49.1%	100%
	(59)	(57)	(116)
Independent-leaning Republican	60.0%	40.0%	100%
	(84)	(56)	(140)
Weak Republican	54.4%	45.6%	100%
	(81)	(68)	(149)
Strong Republican	47.2%	52.8%	100%
	(93)	(104)	(197)

Source: 2004 National Election Study.
Note: Numbers in parentheses are frequencies.

A weak relationship would be one in which the differences in the observed values of the dependent variable for different categories of the independent variable are slight. In fact, the weakest observed relationship is one in which the distribution is identical for all categories of the independent variable—in other words, one in which no relationship appears to exist.

To get a better handle on strong versus weak relationships as measured by a cross-tabulation, consider the hypothetical data in Tables 12-4 and 12-5. Assume we want to know if a connection exists between people's region of residency and attitudes about continuing the war in Iraq. (The hypothesis might be that southerners and westerners are more favorable than citizens in other parts of the country.) The frequencies and percentages in Table 12-4 show no relationship between the independent and dependent variables. The relative frequencies (that is, percentages) are identical across all categories of the independent variable. Another way of thinking about non-relationships is to consider that knowledge of someone's value on the independent variable does not help predict his or her score on the dependent variable. According to Table 12-4, 48 percent of the easterners responded "keep troops in Iraq," but so did 48 percent of the westerners, and for that matter, so did 48 percent of the inhabitants of the other regions. The conclusions are that (1) slightly more than half of the respondents in the survey wanted American troops

TABLE 12-4

Example of a Nil Relationship between Region and Opinions about Keeping Troops in Iraq

Opinion	Region			
	East	Midwest	South	West
Favor keeping troops in Iraq	48%	48%	48%	48%
	(101)	(103)	(145)	(97)
Favor bringing troops home	52%	52%	52%	52%
	(109)	(111)	(158)	(106)
Total	100%	100%	100%	100%
N = 930	(210)	(214)	(303)	(203)

Note: Hypothetical responses to the question "Do you favor keeping a large number of U.S. troops in Iraq until there is a stable government there OR do you favor bringing most of our troops home in the next year?"

TABLE 12-5

Example of a Perfect Relationship between Region and Opinions about Keeping Troops in Iraq

Opinion	Region			
	East	Midwest	South	West
Favor keeping troops in Iraq	0%	0%	100%	100%
	(0)	(0)	(303)	(203)
Favor bringing troops home	100%	100%	0%	0%
	(210)	(214)	(0)	(0)
Total	100%	100%	100%	100%
N = 930	(210)	(214)	(303)	(203)

Note: Hypothetical responses to the question: "Do you favor keeping a large number of U.S. troops in Iraq until there is a stable government there OR do you favor bringing most of our troops home in the next year?"

brought home and that (2) there is no difference among the regions on this point. Consequently, the hypothesis that region affects opinions would not be supported by this evidence.

Now look at Table 12-5, in which there is a strong—one might say nearly perfect—relationship between region and opinion. Notice, for instance, that 100 percent of the easterners and midwesterners favor bringing the troops home, whereas 100 percent of the southerners and westerners have the opposite view. Or, stating the situation differently, knowing a person's region of residence lets us predict his or her response.

Most observed contingency tables, like that shown in Table 12-2, fall between these extremes. The relationship between two variables may be slight (but not nil), strong (but not perfect), or moderate. Deciding which is the case requires the analyst to examine the relative frequencies carefully and determine if a substantively important pattern exists. The question "Is there a relationship between X and Y?" can rarely be answered with an unequivocal yes or no. Instead, the answer rests on judgment. If you think a relationship exists, then make the case by describing differences among percentages between categories of the independent variable. If not, then explain why you think any observed differences are more or less trivial. (The authors, for example, believe that Table 12-2 reveals a very weak relationship. We may be wrong, but because we have given our reasons, it is incumbent on our critics to explain why they disagree.) Later in this chapter we present additional methods and tools that help measure the strength of relationships.

The Direction of a Relationship

In addition to assessing the strength of a relationship, you can also examine its direction. Directionality is an issue only when analyzing ordinal or ordered categorical variables. The direction of a relationship shows which values of the independent variable are associated with which values of the dependent variable. This consideration is especially important when the variables are ordinal or have ordered categories such as "high," "medium," and "low," or "strongly agree" to "strongly disagree," or the categories can reasonably be interpreted as having an underlying order such as "least" to "most" liberal. Look back at Figure 12-1 if you need help visualizing this notion.

Table 12-6 displays the relationship between a scale of political conservatism (call it X) and a measure of opinions about gun control (Y). Both variables have

---

TABLE 12-6
Attitudes toward Gun Control by Liberalism

Make It Easier or Harder to Buy a Gun (Y)[a]	Liberalism Scale (X)			
	Least Conservative	Medium (middle of the road)	Most Conservative	Total
Least favorable to guns	66.5%	43.5%	28.2%	43.2%
(make it *much* harder to buy)	(72)	(226)	(50)	(348)
Medium	14.5%	17.0%	7.9%	14.6%
(make it harder)	(16)	(88)	(14)	(118)
Most favorable to guns	20.0%	39.5%	63.8%	42.2%
(plus "same as now")	(22)	(205)	(113)	(340)
Total	100%	100%	100%	100%
	(110)	(519)	(177)	(806)

Source: 2004 National Election Study.
Note: Respondents who refused to choose an ideological position are coded as missing data and excluded from the analysis.
[a] Question: "How much easier/harder should the federal government make it to buy a gun?" Respondents who replied "somewhat easier" and "a lot easier" are classified as "same as now" because there were too few cases to compute reliable percentages.

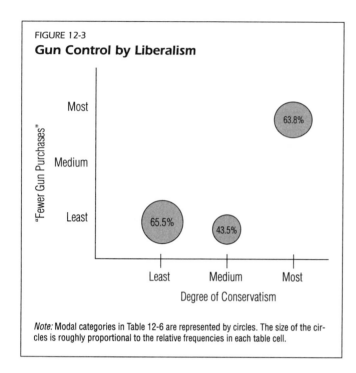

FIGURE 12-3
Gun Control by Liberalism

"Fewer Gun Purchases" (y-axis): Most, Medium, Least

63.8% (Most conservative, Most)
65.5% (Least conservative, Least)
43.5% (Medium, Least)

Degree of Conservatism (x-axis): Least, Medium, Most

Note: Modal categories in Table 12-6 are represented by circles. The size of the circles is roughly proportional to the relative frequencies in each table cell.

an inherent order. The ideology variable can be thought of as running from least to most conservative, whereas responses to the question about firearms might be considered to go from least to most favorable.

Study the numbers in the table for a moment. Start with the "least conservative" category. About two-thirds of respondents (66.5 percent) are also most supportive of restricting gun purchases. That is, "least" values of conservatism tend to be associated with "least favorable to guns." Now, look at the last column, "most conservative." A clear majority of those respondents (63.8 percent) are in the "most favorable to firearms" category of Y, the dependent variable. Here, we have a case of low values tending to be linked to low values and high values to high values. The middle group is more or less split between being for and against making it more difficult for people to buy firearms.

Until you are familiar with cross-tabulations, they can be tricky to read. You might try sketching the largest (modal) frequencies in each category of the independent variable as a circle or square. These "dingbats" can then be graphed on an X-Y coordinate system (see Figure 12-3; the tick marks are labeled with category values from Table 12-6). The pattern revealed here approximates a monotonic positive correlation: as X increases, Y also increases, but not in a straight-line, lockstep fashion.

We can generalize this way: When low values of one variable are associated with low values of another, and high values of the first are associated with high values of the second, a positive (monotonic) relationship exists. When low values of one variable are associated with high values of another, and high values of the first are associated with low values of the second, there is a negative (monotonic) relationship (see Figure 12-1).

The association between these two variables, although not perfect by the standards set forth earlier, is quite strong. Why? As a preview of things to come, try this thought experiment: Suppose you were asked to predict how Americans would respond to a question about making gun control tougher. In the absence of any other information, you might take as a first approximation the marginal distribution of responses to the question in Table 12-6. (The marginal totals are in the last column on the right of the table.) Thus you could reply, "Well most citizens are either for stricter controls (43.2 percent) or for leaving things as they are (42.2 percent) with a smattering of people (14.6 percent) in between." But suppose that you also knew people's political leanings. This knowledge would help you improve your predictions because the most liberal individuals are apt to want stronger controls whereas the most conservative (least liberal) respondents by and large favor leaving matters as they stand. So knowing a person's ideology enhances your predictive power. This idea—the proportional reduction in error—underlies several measures of association discussed in this chapter.

Coefficients for Ordinal Variables

So far we have examined the relationship between two categorical variables by inspecting percentages in the categories of the independent variable. Assessing the strength and the type (direction) of a relationship in a cross-tabulation requires that you look at relative frequencies (percentages) cell by cell. That is not at all a bad practice. To fathom these messages, we have used rough sketches and visual inspection of the tables themselves. However, if the analysis involves many tables or tables that have many cells, we need another way to summarize the information. Here we introduce four correlation coefficients for ordinal variables.

These statistics combine the data from a table into a single summary number that measures the strength and the direction of an association. Among the most common statistics are **Kendall's tau b, Kendall's tau c, Somer's d,** and **Goodman and Kruskal's gamma**—named after the individuals who developed them. Most computer programs calculate these and other coefficients as well. Each summarizes the contents of a two-way frequency distribution in similar but not identical ways.

Given that software can be used to generate these statistics, we concentrate on their numerical meaning rather than on how to calculate them. Even so, a bit of background is useful. Each coefficient compares pairs of cases by determining whether those pairs are concordant, discordant, or tied. These can be slippery concepts. Table 12-7 contains data for nine individuals (cases). In a *concordant pair,* one individual is higher than the other on *both* variables. For example, Alex and Ernesto are concordant because Alex is higher on *Y* and *X.* Alex also is concordant with Fay, Hera, Ike, and Jasmine. There are additional concordant pairs such as Dawn and Hera, Ernesto and Ike. In a *discordant pair,* one case is lower on one of the variables but higher

TABLE 12-7

Table with Concordant, Discordant, and Tied Pairs

Variable Y	Variable X		
	High	Medium	Low
High	Alex	Dawn	Gus
Medium		Ernesto	Hera
Low	Carl	Fay	Ike
			Jasmine

439

on the other. Gus, for example, has a higher score on Y but a lower score on X compared with Ernesto, Fay, or Carl. Therefore, these pairs violate the expectation that as one variable increases, so does the other. A *tied pair* is a pair in which both cases have the same value on one or both variables. Table 12-7 contains several tied pairs. For example, Alex and Dawn are tied on Y (both are in the "high" category), Alex and Carl are tied on X (but not Y), and Ike and Jasmine are tied on X and Y.

The ordinal coefficients of association (tau b, tau c, Somer's d measures, and gamma) use the number of pairs of different kinds to summarize the relationships in a table. In a population, they measure the probability of a randomly drawn pair of observations being concordant minus the probability of their being discordant with respect to Y and X:

$$\text{Measure} = p_{\text{concordance}} - p_{\text{discordance}},$$

where p means probability. The measures differ only in that the probabilities are conditional on the presence or absence of ties. Gamma, for example, is defined as

$$\gamma = p_{C|no\ ties} - p_{D|no\ ties}.$$

In plain language, gamma is the probability that a randomly drawn pair will be concordant on Y and X, given that it is not tied, minus the corresponding probability of discordance. An excess of concordant pairs over discordant pairs suggests a positive relationship; if discordant pairs are more likely, then the correlation will be negative.

In samples the basic comparison made is between the number of concordant and discordant pairs. If both types of pairs are equally numerous, the value of the statistic will be 0, indicating no relationship. If concordant pairs are more numerous, the coefficient will be positive; if discordant pairs outnumber concordant pairs, then the statistic will be negative. The degree to which concordant or discordant pairs predominate or are more frequent than the other affects the magnitude of the statistic. Hence, if only the main diagonal in the table were filled with observations, all the pairs would be concordant and the statistic would be +1—a perfect, positive relationship (Table 12-8a). If only the minor (opposite) diagonal were filled with observations, all the pairs would be discordant and the statistic would be –1—a perfect negative relationship (Table 12-8b).

Gamma can attain its maximum (1 or –1) even if not all of the observations are on the main diagonal because it ignores all tied pairs. The others measures (tau b, for example) discount the strength of the relationship by the number of ties in the table.[3] Hence, in Table 12-9, gamma would still be 1.0, whereas the coefficients would be slightly less.

A real contingency table will contain many pairs of all sorts, and counting them can be a nuisance. So we leave their calculation to the computer. The formulas for these measures have the same form: one quantity divided by another. The numerator is always the number of concordant minus discordant pairs $(C - D)$. The denominators differ, however, in how they handle ties. Gamma ignores tied pairs altogether, whereas the others incorporate them in different ways.[4] To help you understand these measures, we list a few of their properties:

- Theoretically, all vary between –1 and 1, where 1 indicates a perfect positive (monotonic) correlation and –1, a perfect negative (monotonic) correlation.

- In practice you will most likely never see one of these coefficients attain these bounds. Indeed, even for strongly related variables, the numerical values will usually be far from 1 or –1. If a measure reaches, say, .4 or .5 in absolute value, there is an association worth investigating.

- Since 0 means no correlation, values in the range of –.1 to .1 suggest a weak relationship.

- All ordinal measures of correlation will have the same sign in a given table.

- The absolute value of gamma ($\hat{\gamma}$) will always be greater than or equal to the absolute value of any of the others. The relationships among tau b, tau c, and Somer's d are harder to generalize because they are affected differently by the structure of the table (that is, the number of rows and columns).

- Somer's d is an asymmetric measure because its value depends on which variable is considered dependent. Therefore, this measure really has two

TABLE 12-8
Perfect Positive and Negative Relationships

a.
Every Pair Concordant
(Perfect Positive Relationship)

Variable Y	Variable X		
	High	Medium	Low
High	Arthur		
Medium		Candy	
Low			Ed

b.
Every Pair Discordant
(Perfect Positive Relationship)

Variable Y	Variable X		
	High	Medium	Low
High			Faith
Medium		Guy	
Low	Hilary		

TABLE 12-9
Perfect Monotonic Relationship

Variable Y	Variable X			
	Very High	Medium High	Medium Low	Very Low
Very High	Abe, Albert			
Medium High		Bertha		
Medium Low			Claudio	
Very Low				Darby, Dot
Gamma ($\hat{\gamma}$) = 1.0				

How It's Done

COMPUTING ORDINAL MEASURES OF ASSOCIATION

Let C = number of concordant pairs,

D = the number of discordant pairs,

T_X = the number of pairs tied *only* on X,

T_Y = the number of pairs tied *only* on Y,

T_{XY} = the number of pairs tied on *both* X and Y,

m = the minimum of I or J, where I and J are the numbers of categories of Y and X, respectively.

Gamma: $\hat{\gamma} = \dfrac{(C-D)}{(C-D)}$

Tau b: $\hat{\tau}_b = \dfrac{C-D}{\sqrt{(C+D+T_Y)}\ \sqrt{(C+D+T_X)}}$

Tau c: $\hat{\tau}_c = \dfrac{(C+D)}{N^2 \left[\dfrac{(m-1)}{2m}\right]}$

Somer's d: $d_{YX} = \dfrac{(C+D)}{C+D+T_Y}$

Somer's d: $d_{YX} = \dfrac{(C+D)}{C+D+T_X}$

possible versions. One, d_{YX}, has Y as the dependent variable; the other, d_{XY}, treats X as dependent.

- A single measure by itself cannot assess how and how strongly one variable is related to another. After the software calculates all of the measures, you should spend time visually inspecting the relative frequencies in the table.[5]

- Tests of statistical significance are available, but they are too sophisticated to present and explain here. Many software packages compute them along with the sample estimates. These tests, incidentally, assume large sample sizes ($N > 100$), which should not present a problem given the kinds of categorical data to which they are applied.

- These are a specific type of measure of association, namely correlation, whether linear or monotonic.

The last point is worth emphasizing. None of the coefficients is appropriate if the relationship "curves" in the sense that as X increases so does Y up to a certain point, after which an increase in X is accompanied by a decrease in Y. Consider Table 12-10, which contains four observations. A perfect association is seen here: you tell me a person's value on X, and I will predict exactly her score on Y. Yet the number of concordant pairs (3) equals the number of discordant ones (3), so their difference is 0. This difference ($C - D$) appears in the numerator of all the coefficients, so they would all be nil, implying no relationship. But an association does exist; it's just not a correlation. The moral of the story: always look at your data from several angles.

TABLE 12-10

Perfect but Not Monotonic Relationship

Variable Y	Very High	Medium High	Medium Low	Very Low
		Variable X		
Very High				Doris
Medium High	Adele			
Medium Low		Barbara		
Very Low			Connie	

THE IMPORTANCE OF SCRUTINY. A well-known psychologist and statistician, Robert Abelson, titled a book *Statistics as Principled Argument*. His point is that statistics (either the numbers or the methods) do not speak for themselves. It is always necessary to make a case for a point of view. Here is an example.

Gamma is one of the most widely reported ordinal coefficients. Its numerical value is always greater than or equal to the tau and Somer's measures, which raises the possibility that an investigator wanting to find a large relationship might think gamma gives "the most association for the money." One difficulty, however, is that its computation uses only concordant and discordant pairs and ignores all tied pairs. Yet, in some tables the tied pairs greatly outnumber the concordant and discordant ones. Look at this simple crosstabulation of Y by X.

		X		
Y		1	2	Total
A		550	608	1,158
B		20	7	27
C		12	4	16
	Total	582	619	1,201

By examining row totals you will see that the vast majority of observations are in the first row. Stated differently, the marginal total is heavily skewed or concentrated in one category. As a consequence, there are not many concordant or discordant pairs compared to ties. The next table summarizes the numbers of pairs. If someone reported just gamma (–.52), the conclusion might be that a strong Y-X relationship exists. But we see that about 96 percent of the data have been ignored and the other coefficients indicate virtually no relationship.

Type of Pair	Number	Proportion
Concordant	6,130	.01
Discordant	19,540	.02
Tied (on Y, on X, and Y and X)	694,930	.96
Total	720,600	.99

Somer's $d_{YX} = -.04$, tau b $= -.10$, gamma $= -.52$.

Note: Proportions do not add to 1.00 due to rounding error.

There are a couple important lessons here. First, always pay attention to the shape of each variable's distribution, a point made emphatically in Chapter 11. (Try imagining what a histogram of variable Y in the preceding table would look like.) Social science data sets almost always contain skewed marginal totals on at least a few variables. (The data in this example come from a table analyzed later on in the chapter.) Second, try not to rely on just one method such as ordinal coefficients to make a substantive claim.

TABLE 12-11
Vote by Conservatism

	Conservatism Scale						
	1	2	3	4	5	6	7
2004 Vote	Least Conservative (Very Liberal)			Medium (Middle)			Most Conservative
Kerry	92%	95%	85%	55%	21%	5%	4%
Bush	8	5	15	45	79	95	96
Total	100%	100%	100%	100%	100%	100%	100%
	(13)	(80)	(93)	(198)	(110)	(146)	(22)

Gamma = .854, tau c = .739, tau b = .586, Somer's d_{vote} = .463.

Source: 2004 National Election Study.
Note: "Other" and "don't know" responses are excluded from the analysis.

TWO EXAMPLES.   Hypothetical data help establish the basic ideas of these ordinal measures of association, but they do not give us much practice understanding actual survey results. Therefore we provide two tables based on data from the 2004 National Election Study (NES) survey. The first (Table 12-11) is a cross-tabulation of vote preferences in the 2004 presidential election by self-placement on a seven-point liberalism-conservatism scale. The voting variable has only two categories (Bush and Kerry), but any dichotomous variable (a variable with two categories) can be considered ordinal. You can construe the other variable as measuring the degree of conservatism. Since there are $7 \times 2 = 14$ relative frequencies to scrutinize, measures of (monotonic) correlation may help us decide how closely ideology predicts candidate preference.

You should be able to detect a clear-cut pattern: as conservatism increases across the table, the propensity to vote for Bush also increases. (Examine the percentages. Notice, by the way, that the "least" and "most" conservative categories have relatively few cases in them. We might have combined those cases with the adjacent categories to improve the precision or reliability of the cell proportion estimates. Inasmuch as we are concerned with the coefficients, we did not bother.) All the measures are large by the standards of categorical data analysis. As we mentioned earlier, these numbers seldom get close to their maximums (1.0 or –1.0), and absolute values over .4 to .5 indicate a strong correlation. So taken together, the measures suggest that ideology is highly correlated with voting. Tau c is relatively large compared with tau b and Somer's d because it is designed for tables that have a very unequal number of rows and columns. Overall, the conclusion is that liberalism-conservatism predicts voting. Note, however, that since the data show only covariance and not time

TABLE 12-12

Opinion about Legalized Abortion by Age

Attitude on Abortion Legalization	Less than 30 years	31–50 Years	More than 50 years
1. "Abortion should never be permitted by law."	11%	14%	16%
2. "The law should permit abortion only in case of rape."	37	26	35
3. "The law should permit abortions for reasons other than rape."	13	19	18
4. "By law, a woman should always be able to obtain an abortion."	39	41	30
Total	100%	100%	100%
$N = 1,047$	(218)	(394)	(435)

Gamma = −.108, tau b = −.73, tau c = −.108, Somer's $d_{abortion}$ = −.077.

Source: 2004 National Election Study.
Note: Question: "Which one of the opinions on this page best agrees with your view? You can just tell me the number of the opinion you choose."

order or the operation of other variables, we cannot say this is a causal connection.

To wrap up this section on ordinal measures, let us look at another example, the possible effect of age on attitudes regarding abortion (Table 12-12). Age has been recoded from an interval variable to an ordinal variable.[6] The question asking about when, if ever, abortion should be legalized is inherently ordinal because the responses can be regarded as going from least to most restrictive.

If a pattern exists in these responses, it is hard to detect. Most respondents, regardless of age, do not respond "never," but older individuals do seem less inclined to respond "always." Middle-aged people seem most opposed to lenient abortion laws; about 60 percent chose responses 3 or 4. Yet a trend is hard to discern one way or another. The ordinal coefficients reflect this lack of a clear-cut association. Most of them lie near zero, which, let us stress, only signifies no correlation. (Is it surprising that age does not seem to be associated with positions on this issue? It might be an interesting topic for discussion. Should other factors be included in the analysis?)

A Coefficient for Nominal Data

When one or both of the variables in a cross-tabulation are nominal, ordinal coefficients are not appropriate because the identification of concordant and discordant pairs requires that the variables possess an underlying ordering (one value higher than another). For these tables, different measures of association are needed. Some of the most useful measures rest on a **proportional-reduction-in-error (PRE)** interpretation of association. The basic idea is this: You are asked to predict a randomly selected person's category or response level on a variable following two rules. Rule 1 requires you to make the guess in the absence of any other prior information (for example, predict the individual's position on gun control). Rule 2 lets you know the person's score on a second variable, which you now take into account in making the prediction (for example, suppose you know the individual's gender). Since you are guessing in both situations, you can expect to make some errors, but if the two variables are associated, then the use of Rule 2 should lead to fewer errors than following the first. How many fewer errors depends on how closely the variables are related. If there is no association at all, the expected number of errors should be roughly the same and the reduction will be minimal. If, however, the variables are perfectly connected (that is, a one-to-one connection exists between the categories of the two variables), then you would expect no errors by following Rule 2. A PRE measure gives the proportional reduction in errors as follows:

$$PRE = \frac{(E_1 - E_2)}{E_1},$$

where E_1 is the number of errors made using Rule 1 and E_2 is the number made under Rule 2. Suppose that for a particular group of subjects the number of Rule 1 errors (E_1) made predicting variable scores on Y is 500. Think about these possibilities. First, X has no association with Y. Then even if the individuals' X scores are used, the expected number of errors will still be 500 and the proportional reduction in errors will be $(500 - 500)/500 = 0$. This is the lower limit of a proportion and it indicates no association. Second, suppose the categories of X are uniquely associated with those of Y, so that if you know X, you can predict Y exactly. The expected number of errors under Rule 2 (E_2) will be 0. Consequently, PRE = $(500 - 0)/500 = 1.0$, the upper boundary for the measure. This means perfect association (according to this definition). In the third and last situation, assume that Y and X have a moderate relationship. The expected number of errors following Rule 2 might be, say, 200. Now we have

$$PRE = \frac{(500 - 200)}{500} = \frac{300}{500} = .6.$$

In this case, there is a 60 percent reduction in prediction errors if you know the value of X, a result that suggests a modest but not complete association.

GOODMAN-KRUSKAL'S LAMBDA. Many coefficients of association (for example, gamma) can be defined in such a way as to lead to a PRE interpretation. We describe only one, however: **Goodman and Kruskal's lambda.** Lambda is a PRE coefficient. As we did earlier, imagine predicting a person's score on a variable in the absence of any other information (Rule 1). What exactly would be the best strategy? If you did not know anything, you might ask what proportion of the population had characteristic A, what proportion had characteristic B, and so on for all of the categories of the variable of interest. Let's say that B was the most common (modal) category. Then, without other information, guessing that each individual was a B would produce fewer prediction errors than if you picked any other category. Why? Well, suppose there were 10 A's, 60 B's, and 30 C's in a population of 100. Select a person at random and guess his or her category. If you picked, say, A, you would be wrong, on average, 60 + 30 = 90 times out of 100 guesses (90 percent incorrect). If, however, you chose C, you would be mistaken 10 + 60 = 70 times (70 percent errors). Finally, if you guessed the modal (most frequent) category, B, you would be wrong, on average, 10 + 30 = 40 times. By choosing B (the mode), you do indeed make some incorrect predictions but many fewer than if you picked any other category. In sum, Rule 1 states that, lacking any other data, your best long-run strategy for predicting an individual's class is to choose the modal one, the one with the most observations.

Now suppose you knew each case's score or value on a second variable, X. Say you realized a person fell in (or had property) M of the second variable. Rule 2 directs you to look only at the members of M and find the modal category. Assume that category C is most common among those who are M's. Given that the observation is an M, guessing C would (over the long haul) lead to the fewest mistakes. So Rule 2 simply involves using Rule 1 *within* each level of X.

The key to understanding lambda, a PRE-type measure of association, lies in this fact: if Y and X are associated, then the probability of making an error of prediction using Rule 1 will be greater than the probability of making an error with Rule 2. How much greater? The measure of association, lambda (λ), gives the proportional reduction in error:

$$\lambda = \frac{(p_{\text{error 1}} - p_{\text{error 2}})}{p_{\text{error 1}}},$$

where $p_{\text{error 1}}$ is the probability of making a prediction error with Rule 1 and, similarly, $p_{\text{error 2}}$ is the likelihood of making an error knowing X. If the values of X are systematically connected to those of Y, the errors made under Rule 2 will be less probable than those made under Rule 1. In this case, lambda will be greater than 0. In fact, if no prediction errors result from Rule 2, the probability $p_{\text{error 2}}$ will be 0 and

$$\lambda = \frac{(p_{error\ 1} - 0)}{p_{error\ 1}} = \frac{p_{error\ 1}}{p_{error\ 1}} = 1.0.$$

But, of course, if X and Y are unrelated, then knowing the value of X will tell you nothing about Y, and in the long run the probability of errors under both rules will be the same. So $p_{error\ 1} = p_{error\ 2}$ and

$$\lambda = \frac{(p_{error\ 1} - p_{error\ 2})}{p_{error\ 1}} = \frac{(0)}{p_{error\ 1}} = 0.$$

The upshot is that lambda lies between 0 (no association) and 1.0 (a perfect association, as defined by the prediction rules). A value of .5 would indicate a 50 percent reduction in errors, which in most situations would be quite a drop and hence suggest a strong relationship. A value of, say, .10—a 10 percent reduction—might signal a weak to nonexistent association. Note that correlation is not an issue here. If an X-Y link of whatever kind exists, lambda should pick it up.

AN EXAMPLE.   Many statistical programs report lambda as part of their cross-tabulation analyses. Here we provide an intuitive demonstration of how the calculation proceeds.

Suppose we have a sample of 1,000 primary voters and have to predict how each voted in the 2004 Democratic presidential primary. Table 12-13 helps illustrate the situation. It assumes that we are aware of the marginal totals or how many people overall voted for each candidate. The table also shows how many whites, blacks, Hispanics, and Asians are in the sample. But it does not give the particular cell values (for example, how many whites voted for Kerry).[7] Our task is to fill in those quantities, making as few errors as possible.

TABLE 12-13

Relationship between Race and 2004 Democratic Primary Votes: Predicting without the Help of an Independent Variable

Dependent Variable: 2004 Democratic Primary Vote	Independent Variable: Race				
	White	Black	Hispanic	Asian	Total
John Kerry					410
John Edwards					295
Howard Dean					295
Total	600	300	50	50	1,000

Note: Hypothetical data.

Assume we only know that overall 41 percent of the sample chose John Kerry, whereas John Edwards and Howard Dean each got 29.5 percent. Now take the first person in the sample, Mary Smith. Not knowing anything else about her we might guess that she voted for Kerry. After all, since a plurality of voters did prefer him, there is a reasonable chance she did as well. Next take Larry Mendes, the second person in the sample. Again, in the absence of other data, we might reasonably guess that he too voted for Kerry. Our reason for this prediction is the same: most voters picked Kerry, so this is still our best inference. Why? Because if we predicted, say, Dean, in the long run we would make more prediction errors. (Only about 30 percent of the Democrats voted for Dean, whereas 41 percent chose Kerry. Hence, if we guessed Dean for any particular member of the sample, we would be more likely to make a mistake than if we predicted Kerry.) The bottom line is that without any information about the respondent in the sample, the best guess of his or her vote—in the sense of making fewest prediction errors—is to pick the modal category of the dependent variable.

Return to our hypothetical sample of 1,000. Assume that we predict that all 1,000 participants in the study voted for Kerry. Undoubtedly not all did, and so we will surely make some mistakes. How many? Given the percentages above, we would in the long run expect that .41 × 1,000 = 410 guesses would be accurate and (.295 × 1,000) + (.295 × 1,000) = 590 would be wrong. This latter total (590) is the number of prediction errors made without having knowledge of X or $p_{without}$.

Now suppose we do know something about the primary electorate. In particular, election returns tell us that overall 50 percent of white voters supported Kerry, 40 percent were for Edwards, and 10 percent for Dean; that 23 percent of blacks chose Kerry, 10 percent Edwards, and 67 percent Dean; that the corresponding candidate preferences among Hispanics were 30 percent, 20 percent, and 50 percent; and finally, that the Asian vote split 50 percent for Kerry, 30 percent for Edwards, and 20 percent for Dean. These percentages appear in Table 12-14. (It is a full cross-tabulation of race by vote, whereas Table 12-13 is incomplete to reflect our lack of knowledge.)

How can we use these additional data? Look again at the first person in our sample, Mary Smith. Assume that she is white. We know that most whites (50 percent) selected Kerry. (See the first column at the top of Table 12-14.) Accordingly, we would be wise to predict that she too was a Kerry voter. Naturally, we may be wrong, but on average, if we have a white respondent, we make fewer prediction errors by guessing that he or she voted for Kerry than predicting anyone else. Here's the explanation. There are 600 whites in the sample. If we assume that all of them voted for Kerry (Kerry is the modal category, remember), then we will be right half of the time and wrong the other half. How many prediction errors will we make? Let's do the arithmetic: the

TABLE 12-14
2004 Democratic Primary Votes: Predicting with the Help of an Independent Variable

Dependent Variable: 2004 Democratic Primary Vote	Independent Variable: Race				Total	(N)
	White	Black	Hispanic	Asian		
John Kerry	50%	23%	30%	50%	41.0%	(410)
John Edwards	40	10	20	30	29.5	(295)
Howard Dean	10	67	50	20	29.5	(295)
Total	100%	100%	100%	100%	100.0%	
(N)	(600)	(300)	(50)	(50)		(1,000)
Predictions knowing X:						
John Kerry	600	0	0	50	65.0%	(650)
John Edwards	0	0	0	0	0.0	(0)
Howard Dean	0	300	50	0	35.0	(350)
Total	600	300	50	50	100.0%	(1,000)
Errors made using X:						
John Kerry	0	69	15	0		
John Edwards	240	30	10	15		
Howard Dean	60	0	0	10		
Total	300	99	25	25	449	

Note: Hypothetical data.

expected number of correct guesses will be $600 \times .5 = 300$. But we also will be making $600 \times .5 = 300$ errors. (We predict all whites vote for Kerry, but in actuality only half do—see Table 12-14 again—so 50 percent of our guesses are wrong.) How these whites voted is shown in the third panel (errors made using X).

Now move on to black voters. The second column at the top of Table 12-14 indicates that 67 percent of them voted for Dean. So if we draw a black person from our sample and make a prediction about the individual's candidate selection, we would, by the previous logic, make fewest errors by predicting that he or she voted for Dean, the modal choice among blacks. Again, there would be correct and incorrect predictions. Suppose our sample has 300 blacks. If we predict all of them voted for Dean, we would expect to make $.67 \times 300 = 201$ correct predictions and about $.33 \times 300 = 99$ incorrect ones. As before, the bottom panel in Table 12-14 shows the prediction errors for blacks.

If we follow the same reasoning for the other two groups, Hispanics and Asians, we will make accurate and inaccurate predictions using ethnicity as a basis of predicting. More specifically, the prediction errors for Hispanics will be $.5 \times 50 = 25$; for Asians it will also be $.5 \times 50 = 25$. (You should be able to verify these numbers by looking at the total number of each ethnic group and its modal choice. Then, multiply the total by the percentage for the modal choice.)

To recap, when we did not take race into account in predicting the votes of our 1,000 sample respondents, we made a total of $p_{without}$ = 590 errors. But using knowledge of ethnicity, we committed p_{with} = 300 + 99 + 25 + 25 = 449 errors. That is a lot of mistakes, but it is less than 590. In fact, lambda indicates how much less because it is a proportional-reduction-in-error (PRE) measure. Putting the errors in the formula, we find that

$$\lambda = \frac{590 - 449}{590} = .24.$$

The interpretation of this number is that knowledge of X (race) reduced the expected number of false predictions of Y (primary candidate preference) by about 24 percent.

The proportional or percentage reduction in error of prediction aspect of lambda's interpretation is an attractive feature. Say we have a dependent variable, Y, and two independent variables, X and Z. Suppose, in addition, that the lambda between Y and X is .25, whereas the lambda between Y and Z is .5. Then, other things being equal, we might have grounds for arguing that Z has the stronger relationship with Y. We probably would not want to push the claim too far by saying that Z is "twice as important" as X in understanding or predicting Y, because the value of lambda, as with the other measures we have discussed, depends partly on the marginal distributions of the variables.

Lambda has other properties as well. First, if the best prediction (mode) for each category of the independent variable is the same as the overall mode of the dependent variable, lambda will always be 0, even if the column percentages for the categories differ markedly across the rows. In that case, inspection of the column percentages would seem to indicate a relationship between the two variables even though the calculation of lambda indicates no relationship. Second, whenever there are more categories for the dependent variable than there are for the independent variable, lambda cannot take on a value of 1, even if the cases are clustered as much as the marginals permit. Finally, note that lambda is an asymmetric measure of association. As in the case of the regression coefficient to be considered later, treating Y as the dependent variable—the one being predicted—typically leads to a different numerical value than if X is considered the dependent variable.

Association in 2 × 2 Tables: The Odds Ratio

Many variables in the social sciences are dichotomous, which means, as we pointed out before, they have only two categories. Examples of dichotomous variables include the following: "Did you vote? Yes or no?," "Economic status: developed, developing," and "Gender: male, female." Tables relating dichotomous variables consist of two rows and two columns. They are so ubiquitous that statisticians have devoted enormous efforts to devising methods for analyzing them. One measure, the **odds ratio,** is perhaps the most fundamental: it is the ratio of one odds to another. (Odds were explained in Chapter 11.) Useful in its own right to describe associations, it is also a fundamental part of a procedure called logistic regression. Logistic regression, by the way, has become the most popular and widely used statistical procedure in mainstream political science. In addition, the odds ratio has uses that go far beyond applications in political science. It can be helpful in analyzing multiple-variable frequency distributions. (We dealt with a two-variable distribution when comparing men and women in terms of party identification and attitudes on social issues. A multivariable distribution contains more than two variables.) Analyses of these sorts of data are extremely common in sociology and other disciplines.

As an example, let us look at the male-female split on the issue of capital punishment. Data from the 2004 NES appear here in an abbreviated frequency table showing the numbers of men's and women's answers to the question "Do you [favor/oppose] the death penalty for persons convicted of murder?"[8]

Opinion	Male	Female
Favor	430	395
Oppose	139	197

Take men first. The odds of them being in favor of (rather than being opposed to) the death penalty are 430 to 139 or $430/139 = 3.1$. Read this as "the odds of a male favoring capital punishment to opposing it are 430 to 139, or more simply, about 3.1 to 1." A possibly better phrasing is "men are about three times as likely to support the death penalty as to oppose it." Note that this statement applies only to men. We need another odds for women. If it turns out to be more or less the same, then there isn't much difference between the sexes on the death penalty. If, however, the women's odds do not come close to the men's, then a relationship exists between gender and opinions on this topic.

So how do the men's odds compare to the women's? Reading from the table, we see that women's odds in favor are 395 to 197 = $395/197 = 2.0$. In words, the odds of females being in favor are twice those of being against. As a group, they too support capital punishment but not to the same extent.

As a way to compare the sexes on this issue, divide the male odds by female odds to form a ratio. It does not matter which is the denominator and which is the numerator, *except* for how the result is interpreted or understood. In this case we have

$$\text{Odds Ratio (OR)} = \frac{\text{odds men favor DP}}{\text{odds women favor DP}} = \frac{\dfrac{430}{139}}{\dfrac{395}{197}} = \frac{(430)(197)}{(395)(139)} = 1.54.$$

The odds of men favoring the death penalty are about one and a half times greater than the odds of women favoring it. In everyday language, there is a slight tendency for men to be more supportive of capital punishment than women. This result does not by itself tell the whole story about either support for the death penalty or gender differences. But it provides a bit more evidence. We claimed it didn't matter which set of odds got divided. Here's what happens when we take the odds of women favoring capital punishment to those of men:

$$\text{Odds Ratio (OR)} = \frac{\text{odds women favor DP}}{\text{odds men favor DP}} = \frac{\dfrac{395}{197}}{\dfrac{430}{139}} = \frac{(395)(139)}{(430)(197)} = .65.$$

Read the preceding value as "the odds of a female favoring the death penalty are only about two-thirds the odds of a male doing so." This statement is equivalent to the first interpretation. In fact, one is just the reciprocal of the other: .65 = 1/1.54. Both statements assert that men are slightly more "pro" capital punishment than women.

You may have noticed that the ratio ended up being calculated by multiplying the diagonal cells together and then dividing. For that reason the odds ratio is frequently called the cross product ratio.[9]

To shed more light on the meaning of the odds ratio, we match it up with some of the characteristics of measures of association listed in the previous section.

- The odds ratio compares chances or likelihoods of something being chosen or happening.

- In practice it is applied to discrete or categorical variables.

- Unlike most measures, the odds ratio has a null value of 1.0, not 0. If an odds ratio equals 1.0, the two individual odds are the same, and the groups (for example, men and women) do not differ in their response propensities.

- The odds ratio's boundaries are 0 and (plus) infinity. In other words, the odds ratio will always be a positive number.

- The farther from 1.0, in either direction, the stronger the association.

■ The inverse of an odds ratio gives the strength of the relationship when the comparison groups or categories have been switched. This point is easier to demonstrate than to explain. Imagine that in the case of capital punishment and gender the odds ratio came out to be 4.0. That would mean men are four times more likely to approve the death penalty than women are. The inverse, 0.25, means women are only one-fourth as likely to favor it. The two statements tell the same story. In this sense it is a symmetric measure.

■ The odds ratio is standardized. It is always expressed in units of odds, not the measurement scale of a variable.

■ The odds ratio can be applied to investigate patterns of association in tables larger than 2×2 and in multi-way cross-tabulations having more than three variables. Moreover, it acts as a sort of dependent variable in many forms of what are called log-linear and association models.[10]

These methods are now the standard for analyzing cross-tabulated data.

Testing a Cross-tabulation for Statistical Significance

Apart from the hypothetical data, the examples presented so far use data from the 2004 NES, a sample of non-institutionalized adult Americans. As samples go, this one is quite large, with slightly more than 1,200 cases. Nevertheless, this total represents only a tiny fraction of the population, and we can always ask, "Do observed relationships reflect true patterns, or did they arise from chance or what is called sampling error?" Chapter 11 introduced methods for answering that sort of question. Here we apply the technique to cross-tabulations.

STATISTICAL INDEPENDENCE.   At this point it is useful to introduce a technical term, **statistical independence,** which plays a large role in data analysis and provides another way to view the strength of a relationship. Suppose we have two nominal or categorical variables, X and Y. For the sake of convenience we can label the categories of the first $a, b, c, \ldots$ and those of the second $r, s, t, \ldots$. Let $P(X = a)$ stand for the probability that a randomly selected case has property or value a on variable X and $P(Y = r)$ stand for the probability that a randomly selected case has property or value r on Y. These two probabilities are called marginal probabilities and refer simply to the chance that an observation has a particular value on a variable (a, for instance) irrespective of its value on another variable. And, finally, $P(X = a, Y = r)$ stands for the joint probability that a randomly selected observation has both property a *and* property r simultaneously. The two variables are statistically independent if and only if the chances of observing a combination of categories is equal to the marginal probability of one category times the marginal probability of the other:

$$P(X = a, Y = r) = [P(X = a)][P(Y = r)] \text{ for all } a \text{ and } r.$$

If, for instance, men are as likely to vote as women, then the two variables—gender and voter turnout—are statistically independent, because, for example, the probability of observing a male nonvoter in a sample is equal to the probability of observing a male times the probability of picking a nonvoter.

In Table 12-15 we see that 100 out of 300 respondents are men and that 210 out of the 300 respondents said they voted. Hence, the marginal probabilities are $P(X = m) = 100/300 = .33$ and $P(Y = v) = 210/300 = .7$. The product of these marginal probabilities is $.33 \times .7 = .23$. Also, note that because 70 voters are male, the joint probability of being male *and* voting is $70/300 = .23$, the same as the product of the marginal probabilities. Since the same relation holds for all other combinations in this data set, we infer that the two variables in Table 12-15 are statistically independent.

Now suppose we had the data shown in Table 12-16. There the sample consists of 300 respondents, half of whom voted and half of whom did not. The marginal probabilities of voting and not voting are both $150/300 = .5$. Also, the marginal probabilities of being upper and lower class equal .5. If the two variables were statistically independent, the probability that an upper-class respondent voted would be $.5 \times .5 = .25$. Similarly, the predicted probability (from these marginal totals) that a lower-class individual did not vote would be $.5 \times .5 = .25$. But we can see from observed cell frequencies that actual proportions of upper- and lower-class voters are .67 and .33, respectively. Since the observed joint probabilities do not equal the product of the marginal probabilities, the variables are not statistically independent. Upper-class respondents are more likely to vote than are lower-class individuals.

TABLE 12-15
Voter Turnout by Gender

Turnout (Y)	Gender (X)		Total
	Male (m)	Female (f)	
Voted (v)	70	140	210
Did not vote (nv)	30	60	90
Total	100	200	300

Note: Hypothetical data. Cell entries are frequencies.

TABLE 12-16
Voter Turnout by Social Class

Turnout (Y)	Social Class (X)		Total
	Upper (u)	Lower (l)	
Voted (v)	100	50	150
Did not vote (nv)	50	100	150
Total	150	150	300

Note: Hypothetical data. Cell entries are frequencies.

In this context, the test for statistical significance is really a test that two variables in a population are statistically independent. The hypothesis is that in the population the variables are statistically independent, and we use the observed joint frequencies in a table to decide whether or not this proposition is tenable. Generally speaking, the stronger a relationship is, the more likely it is to be statistically significant, because it is unlikely to arise if the variables are really independent. However, even weak relationships may turn out to be statistically significant in some situations. In the case of cross-tabulations, the determination of statistical significance requires the calculation of a statistic called a chi-square, a procedure we discuss next.

CHI-SQUARE TEST FOR INDEPENDENCE.   Table 12-17 displays a cross-tabulation of attitudes on gun control by gender. (The gender hypothesis discussed at the outset of Chapter 11 proposed that women will be somewhat more inclined to back tougher laws.) By examining the cell proportions and the measures of association, you can surmise that a modest relationship exists between the two variables. (You might reinforce your understanding of the coefficients by interpreting them yourself.[11]) But is the relationship statistically significant?

Whether or not a relationship is statistically significant usually cannot be determined just by inspecting a cross-tabulation alone. Instead, a statistic called **chi-square** (χ^2) must be calculated. This statistic essentially compares an observed result—the table produced by sample data—with a hypothetical table that would occur if, in the population, the variables were statistically independent. Stated differently, the chi-square measures the discrepancy between frequencies actually observed and those we would expect to see if there was no population association between the variables. When each observed cell frequency in a table equals the frequency expected under the null hypothesis of independence, chi-square will equal zero. You might think of the situation this way: "chi-square = 0" implies statistical independence, which means no association. Chi-square increases as the departures of observed and expected frequencies grow. There is no upper limit to how big the difference can become, but if it passes a certain point—a critical value—there will be reason to reject the hypothesis that the variables are independent.

How is chi-square calculated? The observed frequencies are already present in the cross-tabulation. (They have been highlighted in Table 12-17.) Expected frequencies in each cell of the table are found by multiplying the row and column marginal totals and dividing by the sample size. As an example, consider the first cell in Table 12-17, the one showing the number of men who think buying a gun should be made a lot more difficult. That cell is in the first row, first column of the table, so multiply the row total, 522, by the column total, 582, and divide by 1,201, the total sample size in this table. The result is $(522 \times 582)/1,201 = 252.96$. This is the expected frequency in the first cell of the table; it is what we would expect to get in a sample of 1,201 (with 522 males and 582 "a lot more difficult" responses) *if there is statistical independence in the population.* This is substantially more than the number we actually have, 190, so there is a difference.

TABLE 12-17
Opinion on Gun Control by Gender

Buying a gun should be . . .	Male	Female	Total
A lot more difficult	32.6%	53.6%	
	190	**332**	**522**
Somewhat more difficult	13.6	12.6	
	79	**78**	**157**
About the same	48.3	32.0	
	281	**198**	**479**
Somewhat easier	3.4	1.1	
	20	**7**	**27**
A lot easier	2.1	.6	
	12	**4**	**16**
Total	100.0%	100.0%	
	582	**619**	**1,201**

Lambda = .134; gamma = −.369; Somer's d = −.241; χ^2_{obs} = 62.20, 4 *df* (significant at .001 level).

Source: 2004 National Election Study.
Note: Cell entries are percentages with observed frequencies highlighted in boldface. Item: "How much easier/harder should the federal government make it to buy a gun—self-placement."

Let's try another cell. There are 332 women who responded that gun purchases should be made a lot more difficult. (This is more than half the females in the sample.) How many would we expect to find if the variables were unrelated? Again, multiply the row total (522) by the column total (619) and divide by 1,201 to get $(522 \times 619)/1,201 = 269.04$. (Notice incidentally that the sum of the expected values in this row equals the observed row total. That will always be the case. Following this procedure to get expected values, we will always reproduce the marginal frequency totals.) This is many fewer than we observed. Once more the data show a discrepancy between the expectation under statistical independence and what occurred. So the top row has one cell with an excess frequency and another with too few.

Table 12-18 contains all of the expected frequencies for Table 12-17. It may be hard to discern, but if you look closely you will see that the *expected* frequencies for men and women are relatively close to each other. That should make sense, because if gender is not related to opinions, then you would expect men and women to respond in similar ways. The *observed* frequencies indicate, however, that such is not the case.

The chi-square statistic combines the squared differences between observed and expected frequencies to produce a total.

As you may recall from Chapter 11, a statistical hypothesis test entails several steps. First,

How It's Done

CALCULATING THE OBSERVED CHI-SQUARE STATISTIC

For a cross-tabulation with I rows and J columns and a total of N observations, let f_{ij} be the observed frequency in row i and column j.

The expected frequency (e_{ij}) in row I and column j is calculated as follows:

$$e_{ij} = \frac{(row\ i\ total)(column\ j\ total)}{N}.$$

The observed chi-square is $\chi^2_{obs} = \sum_{ij} \frac{(f_{ij} - e_{ij})^2}{e_{ij}},$

where $\sum_{ij}$ means sum overall the cells in the table.

TABLE 12-18

Expected Values under Hypothesis of Independence

Buying a gun should be . . .	Male	Female	Total
A lot more difficult	252.96	269.04	522
Somewhat more difficult	76.08	80.92	157
About the same	232.12	246.88	479
Somewhat easier	13.08	13.92	27
A lot easier	7.75	8.25	16
Total	100.0%	100.0%	
	582	619	1,201

Source: See Table 12-17.

specify the null and alternative hypotheses, state a sample statistic and an appropriate sampling distribution, set the level of significance, find critical values, calculate the observed test statistic, and make a decision. Assuming variable Y has I rows and variable X has J columns, a chi-square test of the statistical independence of Y and X has the same general form.

- Null hypothesis: statistical independence between X and Y.

- Alternative hypothesis: X and Y are not independent. The nature of the relationship is left unspecified.

- Sampling distribution: chi-square. The chi-square is a family of distributions, each of which depends on **degrees of freedom** (*df*). The degrees of freedom equals the number of rows minus one times the number of columns minus one or $(I - 1)(J - 1)$.

- Level of significance: the probability (α) of incorrectly rejecting a true null hypothesis.

- Critical value: the chi-square test is always one-tailed. Choose the critical value of chi-square from a tabulation to make the critical region (the region of rejection) equal to α.

- The observed chi-square is the sum of the squared differences between observed and expected frequencies divided by the expected frequency.

- Reject the null hypothesis if the observed chi-square equals or exceeds the critical chi-square. That is, reject if $\chi^2_{obs} \geq \chi^2_{crit}$; otherwise, do not reject.

We can use the gender-gap example to illustrate these steps. The null hypothesis is simply that the two variables are independent. The alternative is that they are not. (Yes, this is an uninformative alternative in that it does not specify how gender and attitudes on gun control might be related. This lack of specificity is a major criticism of the common chi-square test. But this is nevertheless a first step in categorical data analysis.) For this test we will use a level of significance of $\alpha = .01$. As the guidelines indicate, the chi-square is a family of distributions. To find a critical value, it is first necessary to find the degrees of freedom, which is done by multiplying the number of columns in the table minus one ($J - 1$) times the number of rows in the table minus one ($I - 1$), or in this case $(2 - 1)(5 - 1)$, or 4. *Note: I* and *J* refer to the variables' number of categories; *do not* count the "total" row and column.

Then we look in a chi-square table (see Appendix C) to find the value that marks the upper 1 percent (the .01 level) of the distribution. Read down the first column (*df*) until the degrees of freedom is located (4, in this case), and then go across to the column for the desired level of significance. With 4 *df,* the critical value for the .01 level is 13.28. This means that if our observed chi-square is greater than or equal to 13.28, we will reject the hypothesis of statistical independence. Otherwise, we will not reject it.

The observed chi-square for Table 12-17 is 62.20 with 4 *df. (Always report the degrees of freedom.)* This value greatly exceeds the critical value (13.28), and so we would reject the independence hypothesis at the .01 level. Indeed, if you look at the chi-square distribution table, you will see that 62.20 is much larger than the highest listed critical value, 18.47, which defines the .001 level. (We report this fact in the table.) So this relationship is significant at the .001

level. Remember, though, that our finding of a "significant" relationship means simply that we have rejected the null hypothesis. We have not necessarily produced a momentous finding. This statement leads to our next point.

Large values of chi-square occur when the observed and expected tables are quite different and when the sample size upon which the tables are based is large. A weak relationship in a large sample may attain statistical significance, whereas a strong relationship found in a small sample may not. Keep this point in mind. If N (the total sample size) is large, the magnitude of the chi-square statistic will usually be large as well, and we will reject the null hypothesis, even if the association is quite weak. This point can be seen by looking at Tables 12-19 and 12-20. In Table 12-19 the percentages and the chi-square of 1.38 suggest that there is virtually no relationship between the categories X and Y. In Table 12-20, which involves only a larger sample size, the same basic relationship holds. The entries in Table 12-19 have simply been multiplied by 10 so that now the sample size is 3,000 instead of 300. But even though the strength of the relationship between X and Y is still the same as before, namely, quite small, the chi-square statistic (13.8) is now statistically significant (at the .05 level).

The lesson to be drawn here is that when dealing with large samples (say, N is greater than 1,500), small, inconsequential relationships can be statistically significant.[12] As a result, we must take care to distinguish between statistical and substantive importance.

Despite our warnings about sample sizes, the chi-square test is reliable only for relatively large samples. Stating exactly how large is difficult because the answer depends on the table's number of rows and columns (or more formally, its degrees of freedom). Many times, as in Table 12-17, a sample will be large but the table will contain cells with small frequencies. Few respondents in the NES chose the option "a lot easier," and the entire row contains only sixteen cases. A rule of thumb directs analysts to be cautious if any cell contains expected frequencies of five or less, and many cross-tabulation programs flag these "sparse cells." If you run across this situation, the interpretation of the chi-square value remains the same but perhaps should be advanced with less certainty. Moreover,

TABLE 12-19

Relationship between X and Y Based on Sample of 300

Variable Y	Variable X			Total
	a	b	c	
A	30	30	30	90
B	30	30	36	96
C	40	40	34	114
Total	100	100	100	300

$\chi^2 = 1.38$, 4df.

Note: Hypothetical data.

TABLE 12-20

Relationship between X and Y Based on Sample of 3,000

Variable Y	Variable X			Total
	a	b	c	
A	300	300	300	900
B	300	300	360	960
C	400	400	340	1,140
Total	1,000	1,000	1,000	3,000

$\chi^2 = 13.8$, 4df.

Note: Hypothetical data.

if the total sample size is less than 20 to 25, alternative procedures are preferable for testing for significance.[13]

EXTENDING CHI-SQUARE ANALYSIS. Many students (and even some social scientists) merely compute a chi-square statistic and a few measures of association and move on from there. This can be a mistake because cross-tabulations by their nature contain information that may not be apparent at first glance. So we encourage you to carefully inspect these tables. Refer back to Table 12-17. After looking it over, you might notice that most of the "action" takes place in the first three rows. There the gender differences are most pronounced. On the whole, men seem willing to keep gun-purchase laws the same as now, while a majority of women prefer stricter regulations. At the other end of the scale, few individuals of either sex favor relaxing the rules. One way to highlight this contrast is to compute the odds of responding "a lot harder" to saying "about the same." Scan down the row totals and observe the sharp drop in frequencies in the bottom two categories.

Although we cannot go into the details, there are several ways to examine the "associations within the association." If we use a slightly different chi-square statistic, the likelihood-ratio chi-square (χ^2_{LR}),[14] the total chi-square for Table 12-17 is 63.2 with 4 df. This total chi-square can be partitioned or divided into components, each of which is associated with a subtable derived from the complete cross-tabulation. As we said, most of the observations in Table 12-17 lie in the first three rows; the last two are relatively sparse. Suppose we temporarily drop the last two rows and analyze the resulting 3×2 table shown here:

Position on Buying a Gun	Gender		Total
	Men	Women	
A lot more difficult	34.5%	54.6%	
	(190)	(332)	
	[247.9]	[274.1]	522
Some what more difficult	14.4	12.8	
	(79)	(78)	
	[74.6]	[82.4]	157
About the same	51.1	32.6	
	(281)	(198)	
	[227.5]	[251.5]	479
Total	100%	100%	100%
	(550)	(608)	(1,158)

$\chi^2_{LR} = 50.2$; 2 df.

Source: Formed from first three rows of Table 12-17.
Note: Numbers in parentheses are observed frequencies; numbers in brackets are expected frequencies.

Without showing the computation, we find that the likelihood-ratio chi-square for this table is 50.2 with 2 *df*. Next, a second table can be formed by adding the first three rows together—that is, using the total frequencies 550 and 608 (see the next to last row of the previous table)—as the first row of a new table that also includes the two excluded rows from Table 12-17.

In brief, we have created two subtables from a main cross-tabulation and found two likelihood-ratio chi-square statistics. The sum is 50.2 (from the first subtable) + 12.5 (from the second subtable) = 62.7, which is the same as the likelihood statistic for the whole table (63.2) except for rounding error. (Moreover, each of the smaller tables has 2 *df* and their total, 4, equals the degrees of freedom for Table 12-17.) We see, then, that most of the total chi-square is derived from the upper part of the table (the first three rows). In a sense, this approach may present a better picture of the gender gap than the entire cross-tabulation does. As indicated previously, few Americans seem to want to make buying firearms easier. Including those responses just clouds the differences between men and women on the firearms issue. It appears from an examination of the percentages (relative frequencies) that on this issue the sexes do differ.

The statistician Alan Agresti describes a general way to partition an $I \times J$ table (that is, one with I rows and J columns):

> In order for the chi-square components to be independent, certain rules must be followed in choosing the component tables. For example, for an $(I \times J)$ table, one can compare rows 1 and 2, then combine rows 1 and 2 and compare them to row 3, then combine rows 1 through 3 and compare them to row 4, and so forth.[15]

Other partitioning schemes are legitimate. The first I' (I' < I) rows can be combined, as we did, or if more than two columns are available, subtables can be created by combing columns, rows, or both. Ideally this should be done in a way that throws light on the research topic.[16]

Given what we said about the gender differences involving mainly disagreements about keeping gun laws the same or making them stricter, we might want a summary measure of association for just that portion of the table. Relationships within the body of a table may be analyzed with the odds

Position on Buying a Gun	Gender		Total
	Men	Women	
Much harder to same	94.5%	98.2%	
	(550)	(608)	
	[561.2]	[596.8]	1,158
Somewhat easier	3.4	1.1	
	(20)	(7)	
	[13.1]	[13.9]	27
A lot easier	2.1	.6	
	(12)	(4)	
	[7.7]	[8.3]	16
Total	(582)	(619)	1,201

$\chi^2_{LR} = 12.5; 2 \ df.$

Note: "Much harder to same" formed by combining the first three rows of Table 12-17 ("A lot more difficult," "Somewhat more difficult," and "About the same"). Numbers in parentheses are observed frequencies; numbers in brackets are expected frequencies.

ratio. In general, a cross-tabulation with I rows and J columns will contain a basic set of $(I-1)(J-1)$ independent odds ratios.[17] Table 12-17 has five rows and two columns; thus four basic odds ratios can be created and compared. Various computing algorithms are available for finding these sets that we cannot go into here. For this example, let's compare men's and women's odds of responding "make harder" to their odds of responding "the same." For men, the odds are 190/281 = .68 to 1, whereas for women, the odds are 332/198 = 1.68 to 1. Quite a different response pattern is revealed by the ratio of these odds:

$$OR = \frac{(190/281)}{(332/198)} = \frac{.68}{1.68} = .4.$$

This result tells us that the odds of a male wanting stricter rules are only about 40 percent of the corresponding odds for women. Now we see that within the boundaries of the current gun control debate, in which almost no one seems to want more relaxed gun-buying laws, the sexes appear to be a bit polarized. The overall chi-square alerts us to a relationship, but an examination of the odds ratio sheds more light on it.

The procedures described here represent the tip of the iceberg, in terms of the possibilities available for analyzing data. These methods rest on solid mathematical and statistical foundations, and social scientists and statisticians have developed a vast array of tools for analyzing contingency tables. In fact, this is one of the most active branches of applied statistics and is one well worth looking into.[18]

Analysis of Variance and the Difference of Means

Cross-tabulation is the appropriate analysis technique when both variables are nominal- or ordinal-level measures. When the independent variable is nominal or ordinal and the dependent variable is interval or ratio, however, a contingency table would have far too many columns or rows for a straightforward and meaningful analysis. Moreover, a tabular analysis would not take advantage of the information packed in numerical scales. Therefore, another analysis technique is appropriate, the **analysis of variance.**

In a nutshell, ANOVA, as analysis of variance is known, is a technique for comparing means of a quantitative variable between categories or groups formed by a categorical variable. Suppose, for example, that you have a variable (X) with three categories, A, B, and C, and a sample of observations within each of those categories. For each observation there is a measurement on a quantitative dependent variable (Y). Thus within every category or group you can find the mean of Y. ANOVA digs into such data to discover (1) if there are any differences among the means, (2) which specific means differ and by how much, and (3) assuming the observations are sampled from a population, whether the observed differences could have arisen by chance or whether they reflect real variation among the categories or groups in X.

We start by comparing two means, a situation that is a special case of the analysis of more than two means, which we take up in the next section.

Difference of Means or Effect Size

Imagine that you are running an experiment similar to that of Ansolabehere and Iyengar in their study of the effect of negative campaign ads on the likelihood of voting (as discussed in Chapter 5). In this experiment, you have two groups of randomly assigned participants: an experimental group that watches a television newscast with a campaign ad, and a control group that watches a television newscast without a campaign ad. Following our advice in Chapter 5, you have the participants in each group take a pre-test before watching the newscast and the pre-test includes a measure of your dependent variable, intention of voting in the next election. After each group has watched the newscast they are asked to answer questions on a post-test that once again measures the intention of voting. Having collected your data you are now ready to measure the effect of your independent variable, watching a campaign ad, on your dependent variable, intention of voting. To make conclusions on the effect of the campaign ad, you can compare the responses on the pre- and post-tests before and after watching the newscast with the embedded campaign ad. We will assume that the control group's answers to the pre- and post-tests were identical. We would expect this result because this group was not exposed to any political stimulus and therefore their intention of voting should not change. But the experimental group, however, did experience a change in their intention of voting. By examining the mean response to the question on intention of voting on the pre-test with the answers on the post-test, you can calculate the difference between the means. The difference between one mean and another is an **effect size,** one of the most basic measures of relationships in applied statistics. The question is whether the difference between the means is large enough for you to conclude that watching the campaign ad caused voters to change their intention of voting. We can interpret the **difference of means test** as follows: the larger the difference of means, the more likely that the difference is not due to chance and is instead due to a relationship between the independent and dependent variables. In other words, if the difference between the means is quite small, it is likely that the intervening treatment did not have an effect. It is essential, then, to establish a way to determine when the difference of means is large enough to conclude that there was a meaningful effect.

In the examples that follow, we demonstrate how difference of means tests allow you to make conclusions about the size of the difference.

Effect = Mean of treatment group – Mean of control group.

For those who appreciate more formality, a population effect can be written succinctly as

$$\Delta = \mu_{experimental} - \mu_{control},$$

where Δ is the effect size and the μ's are the means of the "population" treatment and experimental groups.[19] A logical sample estimator of μ is the difference in sample means:

$$D = \bar{Y}_{experimental} - \bar{Y}_{control},$$

where D is the symbol for the sample estimator of an effect size and the $\bar{Y}$'s are the sample means for the experimental and control groups.

If the research rests on a randomized experiment (see Chapter 5), the estimated effect can be used for making causal inferences. In studies using nonexperimental research designs, making causal claims is trickier but it follows the same logic. Estimating and testing the statistical significance of effect sizes is a valuable and common procedure in political and policy sciences. In the two examples that follow, we use difference of means tests in nonexperimental settings. These tests can be used in many different ways, depending on factors such as the sample size, whether you know the population standard deviation, whether population variances are equal or unequal, whether you are testing means between groups with equal or unequal sample sizes, and whether you are testing means drawn from the same or different groups. Regardless of the specific characteristics of your research design or data, however, the basic logic of all difference of means tests is the same. Although each test uses slightly different formulas, each has two identical properties: the numerator indicates the difference of means, and the denominator indicates the standard error of the differrence of means. As described earlier, the difference of means is calculated by subtracting one mean from the other. Each test, however, uses a slightly different formula to calculate the standard error of the mean.

The standard error of the difference of means captures in one number how likely the difference of sample means reflects differences in the population. This standard error is partly a function of the variability of the sample means and the sample sizes. Both variability and sample sizes provide information about how much confidence we can have that observed difference is representative of the population difference, Δ. In the first place, variability in a sample always affects confidence in inferences. Take a single mean, for example. If, in a sample every observation equals 5, the mean will also be 5, and we will be pretty confident that the population mean is close to 5. If, on the other hand, a sample has a wide spread of observed values, we will be less certain that the sample mean is representative of the population mean. To capture this in a formula, every calculation of a standard error includes measures of the sample variability. Second, as pointed out numerous time before, we have greater confidence in larger than in smaller samples. Suppose, for instance, our sample has only three observations; then a sample mean based on it is not likely to be a reliable estimator of the population mean. But given a sample of 1,200, the sample mean is more likely to be "near" the true, population mean. What is true of individual means also hold for the difference of means. The larger the sample sizes and the less the variation of scores in them, the greater the confidence that the observed difference, D, will correctly estimate the population difference, Δ.

Suppose we want to compare the average degree of democracy in developed and less developed countries. One indicator of democracy is the voice and accountability (VA) scale presented in Table 11-1. VA, as it is abbreviated, measures how much capacity citizens have to hold their political leaders responsible and the level of free speech they enjoy. The index is constructed so that 0 is average, negative scores indicate less accountability, and positive scores indicate more accountability.

Figure 12-4 shows boxplots for developing and developed nations side by side to illustrate differences in the distributions of VA. As you may recall, this type of graph displays central tendency, dispersion, skewness, and extreme values. We see that the developing countries have on average lower VA ratings (compare the means indicated by the circle symbols). The line connecting the means highlights the difference. There is also considerably more variation in scores (compare interquartile ranges and maximum and minimum values).

This figure presents no surprises: all industrial and postindustrial democracies have pretty much reached the same plateau in governance. So their smaller variation and higher average VA value make sense. It reinforces our impression that, if anything, the mean of the more economically advanced nations will be larger than the mean of the developing countries.

We will assume that the data in Table 11-1 are independent random samples of $N_1 = 17$ developed countries and $N_2 = 19$ developing countries. The sample VA means are 1.225 and −.548, respectively, and the estimated or observed effect of development on voice and accountability—the difference of means—is

$$D = \bar{Y}_{developed} - \bar{Y}_{developing} = 1.225 - (-.548) = 1.773.$$

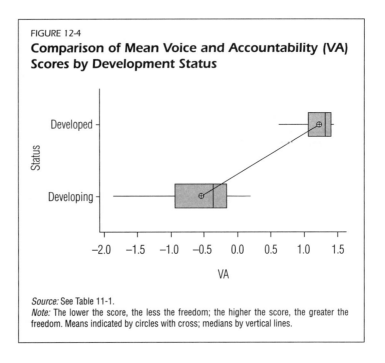

FIGURE 12-4

Comparison of Mean Voice and Accountability (VA) Scores by Development Status

Source: See Table 11-1.
Note: The lower the score, the less the freedom; the higher the score, the greater the freedom. Means indicated by circles with cross; medians by vertical lines.

To the degree that the VA scale is abstract, it is hard to give a substantive or practical meaning to the difference, 1.773. But the variable ranges from about –2 to 1.5 (see Figure 12-4). Therefore, the almost two-unit difference in means observed here represents quite a gap in political freedom.

The statistic D is a point estimator of the population difference of means. Whenever data come from random samples, you can reasonably ask if the observed effect represents a true difference or is the result of sampling error. This question leads to statistical hypothesis testing. The procedures for testing for the significance of a difference of means depend on (among other considerations) sample sizes. For reasons of simplicity we begin with so-called large-sample tests first and postpone the analysis of VA scores until the following section.

As mentioned earlier, there are many different versions of difference of means tests. These tests use different calculations, but each calculation uses essentially the same information (difference of means divided by the standard error of the difference of means). In our examples, we present a difference of means test based on the assumption that the population variances, σ^2_1 and σ^2_2, are equal. The tests differ only in that one is based on a large samples (both N's greater than 20), while the other is based on small samples (both N's less than 20).

LARGE-SAMPLE TEST OF THE DIFFERENCE OF MEANS. We are going to compare two sample means, $\bar{Y}_1$ and $\bar{Y}_2$, based on samples of size N_1 and N_2, respectively, with samples drawn from populations having means μ_1 and μ_2 and variances σ^2_1 and σ^2_2. In this section we assume that both N's are greater than or equal to 20. To illustrate a large-sample test, let's return to the gender-gap hypothesis about male-female differences on current social and political issues. One commonly heard argument is that women are on average a bit more suspicious and hostile toward the military. The 2004 NES survey contains a question that touches on that possibility; it asked respondents to place themselves on a "thermometer" of feelings toward the military with 0 degrees being coldest or most negative and 100 degrees being hottest or most favorable. Presumably, 50 degrees is the neutral position, neither hot or cold.[20] In the context of American politics the existence of a gender gap would imply that women have on average a slightly less favorable opinion of the military than men do.

The null hypotheses, that the thermometer means are equal, can be expressed in three equivalent ways:

$$H_0: \mu_{male} = \mu_{female} \text{ or } H_0: \mu_{male} - \mu_{female} = 0 \text{ or } H_0: \Delta_{\mu_{male} - \mu_{female}} = 0.$$

The alternative hypothesis is that women have on average lower thermometer means than men:

$$H_A: \mu_{male} > \mu_{female} \text{ or } H_A: \mu_{male} - \mu_{female} > 0 \text{ or } H_A: \Delta_{\mu_{male} - \mu_{female}} > 0.$$

In essence, we are testing whether the population difference of means is 0 or positive. Since only values much greater than 0 are of interest, this is a one-tailed test. We would use a two-tailed test if we specified a test that could be different from 0 in the positive or negative direction (representing both tails in the normal distribution). Let us choose the .01 level of significance, which means that if we should reject the null hypothesis, we may be making a type I error (falsely rejecting H_0), but the chances of doing so are 1 in 100 (.01).

A sample difference of means based on large samples has a normal distribution; thus, to find an appropriate critical value for testing the null hypothesis, we use the tabulated standard normal distribution or z distribution in Appendix A. The critical value has to be chosen in such a way as to make the probability of a type I error equal to .01. Recall that the table gives z scores that cut off the upper α proportion of the distribution. The critical value that cuts off .1 percent (.01) of the area under the normal curve is 2.325. (We interpolated between 2.32 and 2.33.) Any observed test statistic greater than or equal to this value will lead to the rejection of H_0. If the test statistic is less than 2.325, the null hypothesis still stands.

TABLE 12-21

Large-Sample Difference of Means Test (Gender by Attitudes toward the Military)

Gender	Sample Size	Mean	Standard Deviation	Standard Error of Mean
Male	518	79.56	20.59	.905
Female	531	80.07	21.99	.954

$D = -.517$.

$z_{obs} = -.393$, p-value = .614.

Confidence interval: -3.913, 2.878.

Source: 2004 National Election Study.

To test the effect, D is converted to an observed test statistic, z, by the general formula

$$z_{obs} = \frac{Estimated\ difference}{Estimated\ standard\ error}.$$

The absolute value of z_{obs} is compared to the critical z to reach a decision; thus, if $|z_{obs}| \geq z_{crit}$, reject H_0; otherwise, do not reject.

The estimated effect size, $\bar{Y}_{male} - \bar{Y}_{female} = 79.56 - 80.07 = -.517$, is negative, which is not at all what we hypothesized. So, at this point we could immediately reject our alternative hypothesis, but let's follow through with the test of statistical significance, just to illustrate the procedure. The test results appear in Table 12-21.

The observed test statistic, $|z_{obs}| = .393$, is considerably less than the critical value of 2.325. In view of these data, there is no reason to reject the null hypothesis. More important, the conclusion is that there appears to be no gender gap on this issue measured in this way. But bear in mind the timing of the survey, October to December 2004. The war in Iraq was just a year old and the shock of September 11 had not dissipated. A survey taken at another time might have produced different results.

CONFIDENCE INTERVALS. Chapter 11 underscored the value of confidence intervals. Most software routinely calculates these lower and upper boundaries that probably enclose the population value of an estimator. In Table 12-21 the 99 percent confidence intervals for the estimated gender effect are reported after the value of D. The intervals extend from -3.913 to 2.878. As discussed in Chapter 11, confidence intervals are constructed in such a fashion that (in this case) 99 times out of 100 the intervals will include the population difference of means. Crudely stated, we are 99 percent certain that the difference in men's and women's mean thermometer scores lies between –3.913 and 2.878. Since the intervals extend from about –4 to 3, we have reason to believe that 0 is a possible value of the population difference of means. Because these limits are based on the same α level used in the test of significance, we have in effect tested the null hypothesis a second way. By observing that the 99 percent confidence interval included the (null) hypothesized value of 0, we do not reject H_0. How sure are we? We are 99 percent certain. The substantive interpretation is that there is no difference between men and women on this issue.

SMALL-SAMPLE TEST OF THE DIFFERENCE OF MEANS. The preceding test assumed relatively large sample sizes. What about smaller ones? The cross-national data presented in

---

How It's Done

LARGE-SAMPLE TEST FOR THE DIFFERENCE OF THE MEANS

Assume N_1 and N_2, the size of samples 1 and 2, respectively, are each greater than 20. Assume the population variances $\sigma_1^2 =$ and σ_2^2 are equal.

The observed test statistic for the null hypothesis H_0: $\mu_1 = \mu_2$ is calculated as follows:

$$z_{obs} = \frac{(\bar{Y}_1 - \bar{Y}_2)}{\hat{\sigma}_{\bar{Y}_1 - \bar{Y}_2}} = \frac{(\bar{Y}_1 - \bar{Y}_2)}{\sqrt{\dfrac{\hat{\sigma}_1^2}{N_1} + \dfrac{\hat{\sigma}_2^2}{N_2}}},$$

where the $\hat{\sigma}^2$'s are the sample estimators of the population variances.

Reject H_0 if $|z_{obs}| \geq z_{crit}$; otherwise, do not reject.

Table 11-1 consisted of $N_1 = 17$ developed nations and $N_2 = 19$ developing countries. In this case, both sample sizes were smaller than 20. Let's test the hypothesis that socioeconomic development has little or nothing to do with a country's level of democracy. To operationalize the concept of democracy, we return to the voice and accountability scale. Given small samples, the t distribution is used instead of the z distribution.

Again, we will start our discussion with a recap of the test procedures. The null hypothesis is that mean voice and accountability (VA) scores are equal:

$$H_0\text{: } \mu_{developed} = \mu_{developing} \text{ or } H_0\text{: } \mu_{developed} - \mu_{developing} = 0$$
$$\text{or } H_0\text{: } \Delta_{\mu_{developed} - \mu_{developing}} = 0.$$

To demonstrate a two-sided test, we set the alternative hypothesis to be that the two population means are not equal, but we do not predict which is larger. The alternative hypothesis can thus be stated as follows:

$$H_A\text{: } \mu_{developed} \neq \mu_{developing} \text{ or } H_A\text{: } \Delta = 0.$$

We are testing whether or not the difference in the population means (the effect of development status) is 0. Consequently this is a two-sided test. Let's use the $\alpha = .05$ level of significance.

For small samples (that is, if N_1 or N_2 or both are less than 20), the sample estimator of the difference of means (D) follows a t distribution. To find the critical value in a t distribution, it is necessary to calculate the degrees of freedom. If the population standard deviations are assumed to be equal, as we do here, then the degrees of freedom equals $N_1 + N_2 - 2$, or in this case 17 + 19 − 2 = 34. (If you look at Figure 12-4 you will see that the variation of VA ratings is higher in the developing nations than in the developed countries. Presumably they are not the same in the "populations" of countries either. But treating the variances as if they are equal greatly simplifies the calculations, an advantage in an introductory work.[21]) You will also need to choose between the one-tailed and two-tailed tests, and the alpha level. In Appendix B the degrees of freedom are listed in the first column on the left, increasing from one to infinity. Across the top of the table you will see the alpha levels for a one-tailed test, and the alpha levels for the two-tailed test below that. You will need to specify the degrees of freedom, the directionality of the test, and the alpha level to select a critical t value for comparison with the computed t score to determine whether or not to accept the null hypotheses. For this example we use a two-tailed test with an alpha level of .05, and we have 34 df. Consulting the t table, you will see that 34 df does not appear on the table. In this case, we must approximate the critical t value by selecting the critical t value for the next smallest df, or 30 df. We choose the next smallest, rather than the next largest degrees of freedom, because it will provide a more stringent test of the hypothesis. So, we have a critical t value of 2.042. Any observed t value greater than or equal to this number will lead to the rejection of the null hypothesis in favor of the alternative.

The observed test statistic for comparison with the critical value closely resembles the one calculated for large samples, except that the t replaces the z.

$$t_{obs} = \frac{(\bar{Y}_{developed} - \bar{Y}_{developing}) - (\mu_{developed} - \mu_{developing})}{\text{standard error}} = \frac{D}{\text{standard error}}.$$

Simplification of the numerator is possible because $\mu_{developed} - \mu_{developing}$ is hypothesized to be 0 and $D = \bar{Y}_{developed} - \bar{Y}_{developing}$. Table 12-22 presents the test results.

The observed t (12.33) easily exceeds the critical value—in fact, it is greater than all t's with 30 df—so we reject the null hypothesis at the .05 level and indeed at the .001 level. As a reminder, the phrase "p-value = .000" at the bottom of Table 12-22 is the attained probability of the observed t. This means that if the null hypothesis of no difference of means is true, we have found an extremely unusual result, one with a probability of less than .001 or 1 in a thousand. Therefore, we conclude that in all likelihood the population means are *not* equal, or $\Delta \neq 0$. Observe in addition that the 95 percent confidence interval (1.481, 2.065) does not include 0. This result just restates the results of

TABLE 12-22
Small-Sample Difference of Means Test

Status	Sample Size	Mean	Standard Deviation	Standard Error of Mean
Developed	17	1.225	.232	.056
Developing	19	−.548	.550	.130

$D = 1.773$.
$t_{obs} = 12.33$, 34 df; p-value = .000.
Confidence interval: 1.481, 2.065.

the hypothesis test—we are reasonably certain that there is a difference between developing and developed nations in this respect—but now we have a range of possible values for the magnitude of the difference of VA means.

How much have we learned from this exercise? Developed countries certainly differ from other nations on this one indicator of democracy. In fact, quite a discrepancy exists. This finding perhaps comes as no surprise. At least, however, we have demonstrated how one can provide objective evidence for a hypothesis.

Difference of Proportions

Closely related to an analysis of the difference of means is a comparison of proportions. Testing for a difference of proportions follows exactly the same steps as the previous tests except for relatively minor adjustments in formulas. The steps involve (1) measuring the difference between two sample proportions, p_1 and p_2 (with the data from two independent samples), (2) placing a confidence interval around the estimated difference, and (3) testing the hypothesis that the population difference of proportions, $P_1 - P_2$, equals a specific value, usually 0. Let's stick with the gender-gap problem. The 2004 NES asked respondents to rate various groups on feeling thermometers. This time we take attitudes toward "feminists" as the dependent variable, but instead of comparing means, we investigate the differences in the proportions of men and women who rate feminists negatively. For this exposition, thermometer ratings of less than 50 are considered unfavorable ratings, whereas scores of 50 and higher are considered favorable. (We use this admittedly arbitrary dividing line merely to illustrate the method. As a rule of thumb, you should apply the statistical technique that preserves the level of measurement.)

Table 12-23 summarizes the data and analysis results. For large samples, a difference of proportions has a normal distribution with mean $P_1 - P_2$ and standard error $\sigma_{P_1-P_2}$. As a consequence, we compute an observed z to compare with the critical value. For the sake of argument, we choose a two-tailed test at the .05 level. The null hypothesis is therefore $H_0: P_1 = P_2$ or $H_0: P_1 - P_2 = 0$ and

471

How It's Done

SMALL-SAMPLE TEST FOR A DIFFERENCE OF THE MEANS AND CONFIDENCE INTERVALS

Assume N_1 and N_2, the size of samples 1 and 2, respectively, are less than 20. Assume population standard deviations σ_1 and σ_2 are equal.

Testing the null hypothesis, H_0: $\mu_1 = \mu_2$ requires three calculations: (1) a "pooled" estimator of the population standard deviation $(\hat{\sigma}_{pooled})$, (2) the standard error of the difference of means $(\hat{\sigma}_{\bar{Y}_1 - \bar{Y}_2})$, and (3) the observed t statistic (t_{obs}):

1. $\hat{\sigma}_{pooled} = \sqrt{\dfrac{(N_1 - 1)\hat{\sigma}_1^2 + (N_2 - 1)\hat{\sigma}_2^2}{N_1 + N_2 - 1}}$,

where $\hat{\sigma}_1^2$ and $\hat{\sigma}_2^2$ are the sample variances.

2. $\hat{\sigma}_{\bar{Y}_1 - \bar{Y}_2} = \hat{\sigma}_{pooled} \sqrt{\dfrac{1}{N_1} + \dfrac{1}{N_2}}$.

3. $t_{obs} = \dfrac{(\bar{Y}_1 - \bar{Y}_2)}{\hat{\sigma}_{\bar{Y}_1 - \bar{Y}_2}}$.

The t statistic has $N_1 + N_2 - 2$ degrees of freedom.

Reject H_0 if $|t_{obs}| \geq t_{crit}$; otherwise, do not reject.

Confidence intervals are calculated as follows:

$(\bar{Y}_1 - \bar{Y}_2) \pm t_{\alpha/2, N_1 + N_2 - 2}$.

where the $t_{\alpha/2, N_1 + N_2 - 2}$ comes from the t table with $N_1 + N_2 - 2$ degrees of freedom and $(1 - \alpha)$ is the desired level of confidence.

the alternative is simply H_A: $P_1 \neq P_2$. The sample sizes ($N_{male} = 496$, $N_{female} = 518$) are quite large, so the z distribution is appropriate and the critical value for the test is 1.96.

The estimated proportions are the cell frequencies divided by the totals (for example, $p_{male} = 139/496 = .280$). The test statistic has the same general form as the one for the difference of means:

$$z_{obs} = \frac{(p_1 - p_2) - (P_1 - P_2)}{\text{standard error}}$$

Usually (as in the present case) the null hypothesis is simply H_0: $P_1 - P_2 = 0$ and the last term in the numerator drops out. Because the observed test statistic (3.12; see Table 12-23) is considerably larger than 1.96, the hypothesis that the population proportions are the same is rejected at the $\alpha = .05$ level. In addition, we report that the attained probability of this z is less than .0014, further confirming the decision. Thus the difference in sample proportions is statistically significant. There appears to be a gap in attitudes toward feminists between men and women.

Before spending too much time interpreting the result, however, note the 95 percent confidence intervals shown in Table 12-23. They extend from about .03 to .14. In other words, the male-female split could be as small as .03, a difference that might have little if any practical importance. Moreover, the upper limit, .14, may not be large either. True, statistically speaking we have evidence of a small gender gap in these and previously reported data. Yet the difference does not seem to be earth shattering nor is it evidence of a major divide in American politics. The gender gap probably pales in comparison with other divisions in society such as racial, regional, class, or ethnic cleavages. A key lesson is the importance of examining data from several angles. Besides identifying the statistical significance of a result, measuring and thinking about its magnitude is important as well. Once again we see the value of confidence intervals. In this case, they tell us that there might be a modest gap in feelings, but a trivial gap is a possibility as well.

Analysis of Variance

ANOVA extends the previous method to the comparison of more than two means. As before, the dependent variable (Y) is quantitative. The independent or explanatory variable (X), sometimes referred to as a treatment or factor, consists of several categories. This procedure treats the observations in the categories as independent samples from populations. If the data constitute random samples (and certain other conditions are met), you apply ANOVA to test a null hypothesis such as H_0: $\mu_1 = \mu_2 = \mu_3$ . . . and so on, where the μ's are the population means of the groups formed by X.

TABLE 12-23
Difference of Proportions Test (Gender by Attitudes toward "Feminists")

Estimated proportion	Gender		Total
	Male	Female	
Proportion negative	.280	.197	.238
	(139)	(102)	(241)
Proportion positive	.720	.803	.762
	(357)	(416)	(773)
Total	1.0	1.0	1.0
	(496)	(518)	(1,014)

Estimated difference, $p_1 - p_2$: .280 − .197 = .083.
z_{obs} = 3.12; z_{crit} = 1.96.
Since z_{obs} > 1.96, reject H_0 at .05 level; p-value < .0014.
Confidence interval: .031, .136.

Source: 2004 National Election Study.

EXPLAINED AND UNEXPLAINED VARIATION. The inclusion of the word *variance* in "analysis of variance" may throw you. If the procedure deals with means, why not call it the "analysis of means?" As we said in Chapter 2 and elsewhere, the goal of empirical research is to explain differences. In statistics, the variation from all sources is frequently called the **total variation.** Identified and measured variables explain some of this overall variation; this is called **explained variation.** The remaining variation is called **unexplained variation.** Figure 12-5 may help clarify this point. It shows the data in Table 12-24 as a dot plot of individual values. Notice first the considerable variation among the points. By looking at the graph carefully, you may see two kinds of variation. For example, the members of group or category A differ from members of categories B and C. But these observations also vary among themselves. In all three groups, four out of five observations lie above or below their category means. (The mean in A, for instance, is 14, and two scores are above and below it.)

In ANOVA parlance, two types of variation add up to the overall variation. If we denote the overall variance as *total,* the within-category variance as *within* (or *unexplained*) and the between-group variance as *between* (or *explained*), then the fundamental ANOVA relationship is

Total variance = Within variance + Between variance.

The terms *between* or *explained* refer to the fact that some of the observed differences seems to be due to "membership in" or "having a property of" one

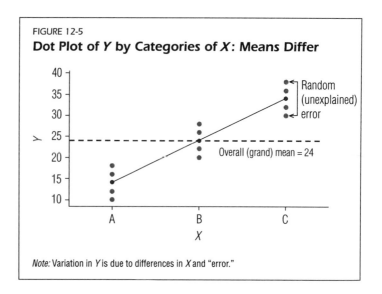

FIGURE 12-5
Dot Plot of Y by Categories of X: Means Differ

Note: Variation in *Y* is due to differences in *X* and "error."

TABLE 12-24
Measurements on Y within Three Categories of X

	Categorical Variable (X)		
	A	B	C
	10	20	30
	12	22	32
	14	24	34
	16	26	36
	18	28	38
Number of cases (N_i)	5	5	5
Mean ($\bar{Y}_i$)	14	24	34
Standard deviation ($\hat{\sigma}_i$)	3.16	3.16	3.16

Overall "grand" mean = 24.

category of *X*. That is, on average the A's differ from the B's. Knowing that a case has character-istic A tells us roughly what its value on *Y* will be. The prediction will not be perfect, however, because of the internal variation among the A's. Yet if we could numerically measure these dif-ferent sources of variability, we could determine what percentage of the total was explained:

Percent explained = (between/total) × 100.

If the portion of total variance explained by the independent variable is relatively large, there is reason to believe that at least two of the population means are not equal. You can figure out which are different only by examining graphs and calculating effect sizes.

Now look at Figure 12-6. It shows two things: the means of A, B, and C are all the same, and the observations differ among themselves but not because they belong to one or another group. Each level of *X* has the same mean. So the total variation in scores has nothing to do with levels of the factor. There is no difference among means and hence no explained or between variation. The analysis of the relationship is thus total variation = within variation + 0, and the percent explained variation is 0:

Percent explained = explained/total = 0/total = 0.

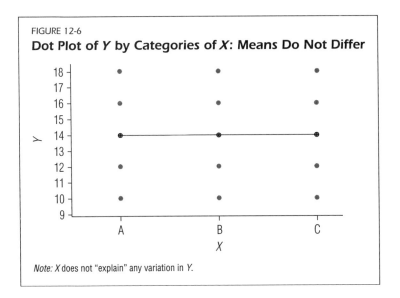

FIGURE 12-6
Dot Plot of Y by Categories of X: Means Do Not Differ

Note: X does not "explain" any variation in Y.

The mathematics of ANOVA simply involves quantifying the types of variation and using these numbers to make inferences. The standard measure of variation is the sum of squares, which is a total of squared deviations about a mean. Let's not get bogged down in computations and instead simply express the idea of explained variation with these terms: TSS stands for the total variability in the data (read this as "the total sum of squares"); BSS represents the between means variability ("between sum of squares"); and WSS ("within sum of squares") is the within groups or categories variation. With these definitions we have

$$\text{TSS} = \text{BSS} + \text{WSS},$$

and

$$\text{Percent explained} = (\text{BSS}/\text{TSS}) \times 100.$$

The percent of variation explained is a commonly used (and abused!) measure of the strength of a relationship. In the context of ANOVA, the percent variation explained is sometimes called **eta-squared** (η^2). One of the properties of eta-squared may be obvious: it varies between 0, which means the independent variable (statistically speaking) explains nothing about Y, to 1, which means it accounts for all of the variation. You frequently read something to the effect that "X explains 60 percent of the variation in Y and hence is an important explanatory factor." Whether or not the data justify a statement of this sort depends on a host of considerations that we take up later.

TABLE 12-25
Typical ANOVA Table Format

Source of Variation	Sum of Squares	Degrees of Freedom	Mean Square	Observed F
Between				
(name of variable)	BSS	$df_{between} = K - 1$	$BMS = BSS/df_{between}$	$F_{obs} = \dfrac{BSS/(K-1)}{WSS/(N-K)} = \dfrac{BMS}{WMS}$
Within				
(unexplained or error)	WSS	$df_{within} = N - K$	$WMS = WSS/df_{within}$	
Total	TSS	$df_{total} = N - 1$		

Note: Assume X has K categories.

SIGNIFICANCE TEST FOR ANALYSIS OF VARIANCE. A test of the hypothesis that K subpopulation means are equal (H_0: $\mu_1 = \mu_2 = \mu_3 = \ldots = \mu_K$) rests on several assumptions, especially that the observations in one group are independent of those in the other groups. In addition, we assume large N_K's, and equal population variances (that is, $\sigma^2_1 = \sigma^2_2 = \sigma^2_3 = \ldots$). Test results are most often summarized in an ANOVA table (Table 12-25 provides an example).

The terms inside the table may seem intimidating at first, but the numbers are straightforward. The sums of squares are calculated from formulas described elsewhere (these days, computers do most of the work). Each sum of squares has an associated degrees of freedom. They are easy to calculate: the between df is the number of categories (K) of the independent variable minus one, or $K - 1$; and the within df is N, the total sample size, minus the number of categories, or $N - K$. Together they sum to the degrees of freedom for the total sum of squares or $(K - 1) + (N - K) = N - 1$.

Whenever a sum of squares is divided by its degrees of freedom, the quotient is called a mean square. The between mean square is divided by the within mean square to obtain an observed test statistic called the F statistic. Like the other statistics we have discussed (for example, chi-square, t, and z), the observed F has a distribution called (sensibly enough) the F distribution. (As in the case of the t distribution, the F distribution is a family, each member of which depends on *two* degrees of freedom, one for the between component and one for the within component.) A decision about the null hypothesis of equal population means is made by comparing F_{obs} to F_{crit}.

The general idea should be familiar. Suppose we use the hypothetical data in Table 12-24 to test the hypothesis that $\mu_A = \mu_B = \mu_C$ against the alternative that at least two of them differ. (Technically, we should have larger samples, but this is just an illustration.) For this test we choose the .001 level of significance. Figure 12-7 displays the results from a computer program, along with a few comments. The observed F is 50, considerably larger than the critical F (with 12 and 2 df) of 12.97. (The critical value is found in Appendix D.) First decide on a level of significance, then the degrees of freedom for the within sum of squares (12) and for the between sum of squares (2). The needed value will be in the third F table. Since F_{obs} exceeds F_{crit}, the null hypothesis is rejected at the .001 level. (Indeed, if the null hypothesis were true, the p-value (.000) tells us we have obtained a very improbable result.)

What does all this mean? Eta-squared is BSS/TSS = 1000/1120 = .89, which suggests that X explains 89 percent of the variation in Y. Yet this information plus the fact that the relationship is statistically significant still leaves you with the task of providing a substantive interpretation to the results. These are hypothetical data, so not much can be said. But in a real-world context you would be expected to offer some reasons why X has such a tight relationship to Y.

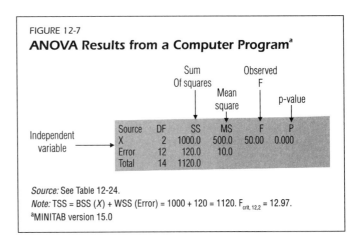

FIGURE 12-7
ANOVA Results from a Computer Program[a]

Source	DF	SS	MS	F	P
X	2	1000.0	500.0	50.00	0.000
Error	12	120.0	10.0		
Total	14	1120.0			

Source: See Table 12-24.
Note: TSS = BSS (*X*) + WSS (Error) = 1000 + 120 = 1120. $F_{crit, 12,2}$ = 12.97.
[a]MINITAB version 15.0

Regression Analysis

The procedure for exploring the relationship between two quantitative variables, **regression analysis,** is essentially a toolbox of methods for describing how, how strongly, and under what conditions an independent and dependent variable are associated. The regression framework is at the heart of empirical social and political research because of its versatility. Regression techniques can be used to analyze all sorts of variables having all kinds of relationships. But its value goes even further. In Chapter 5 we discussed the difficulty of making causal inferences on the basis of nonexperimental observations. Many scholars believe regression analysis and related approaches act as reasonable surrogates for controlled experiments and can, if applied carefully, lead to causal inferences.

Scatterplots

As mentioned before, the real work of empirical political science involves answering "why" questions: Why do some countries develop democratic forms of government? Why do nations go to war? Why are some people Republicans, others Democrats, and still others independent or apolitical? To answer these questions, researchers translate verbal theories, hypotheses, even hunches into models. A model shows how and under what conditions two (or more) variables are related. The beginning point of this process is identifying

477

associations or correlations between pairs of variables, and graphs provide the best first step.

One common graph is the scatterplot. Intended for quantitative data, a **scatterplot** contains a horizontal axis representing one variable and a vertical axis (drawn at a right angle) for the other variable. The usual practice is to place the values of the independent variable along the x-axis and values of the dependent variable along the y-axis. The scales of the axes are in units of the particular variables, such as percents or thousands of dollars. The X and Y values for each observation are plotted using this coordinate system. The measurements for each case are placed at the point on the graph corresponding to the case's values on the variables.

As an example, Figure 12-8 shows how five Y and X values are plotted on a scatterplot. Each case is located or marked at the intersection of the line extending from the x- and y-axes. The first pair of observations, 5 and 10, appears at the point $Y = 5$ and $X = 10$. Scatterplots are handy because they show at a glance the form and strength of relationships.

In this example, increases in X tend to be associated with increases in Y. Indeed, we have drawn a straight line on the graph in such a fashion that most observations fall on it or very near to it. In the language introduced at the beginning of the chapter, this pattern of points indicates a strong positive linear correlation.

For a more realistic illustration, refer to the data first described in Chapter 11. One hypothesis asserts that democratic institutions will take root and flourish only when a society has achieved a sufficiently high standard of living. "Sufficiently high" is, of course, an ambiguous criterion, but we can still investigate the proposition by examining the relationship between indicators of well-being and democracy. Figure 12-9 plots the relationship between gross national product (GNP) per capita and the voice and accountability (VA) variable, an index ranging between –2 and 2 that supposedly quanti-

A Reminder about Confidence Intervals and Tests of Statistical Significance

In this text we describe both hypothesis tests and confidence intervals. The former are common in scholarly and popular reports of statistical findings, but the latter, we believe, give you all the information a hypothesis test does and then some.

Suppose you hypothesize that in a population A – B = 0. You conduct a t or z test on a sample and reject this hypothesis at the .05 level. This means you are pretty sure A does not equal B. You are not positive, of course, but if you carried out the analysis correctly, there is only a 1 in 20 (.05) chance of rejecting a hypothesis that should not be rejected. That is the meaning of a level of significance.

What about finding 95 percent confidence limits for the A – B difference? Notice that 1 – .95 = .05. The connection between .95 and .05 is not coincidental. A confidence interval puts a positive spin on your inference: you are 95 percent sure that the true difference lies somewhere in the interval. By contrast, a significance test at the .05 level is in a sense a negative statement: you think there's only a 5 percent chance (.05 probability) that you are wrong. But as far as probability theory is concerned, one procedure is as good as another.

So why do we prefer confidence intervals? Because besides enabling you to discard (or not) a hypothesis, confidence intervals explicitly show you possible values of the effect. Better still, they use the same scale as the variables in the problem at hand. If you reject the hypothesis that A equals B, you only know that A probably does not equal B. But if the confidence interval for the difference runs from, say, 1 to 2, you may conclude that there is not a meaningful theoretical or practical difference. However, if the interval extends from 1 to 50, that might suggest that the independent variable could have a large effect on the value of the dependent variable and is something worthy of further investigation.

fies how much freedom citizens have to control their governments.[22]

The dots represent the joint GNP-VA scores for the 39 nations in the data matrix (see Table 11-1). A clear pattern exists (one, incidentally, that can't be detected simply by looking at two columns of raw numbers). For additional clarity, the names of a few countries lie juxtaposed to their *X-Y* scale positions. Luxembourg, Norway, and the United States typify wealthy democratic societies, whereas Cuba and Egypt typify two poorer countries. (If the data set is not too large [less than 20, say], all the observation names can be displayed.)

Although there is not a strict one-to-one relationship between the variables, notice that as income increases, scores on VA increase. Also, if you study the "flow" of dots, they seem to form an arc that points up and to the right. This configuration suggests a positive monotonic correlation, a relationship characterized by the pattern "an increase in *X* is generally accompanied by an increase in *Y*." But the data do not fall on or near a straight line or, in regression terms, there is not a linear correlation.

What does the graph tell us? Consider the lower end of the income scale. If you go up, small jumps in GNP appear to lead to relatively large increases in the democracy scores. The difference between Egypt's and Cuba's income levels is not huge, but their VA scores differ by almost .72

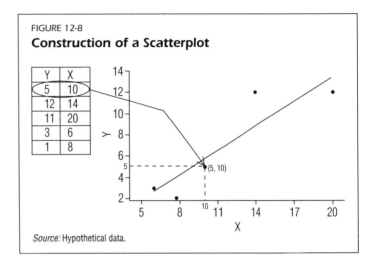

FIGURE 12-8
Construction of a Scatterplot

Y	X
5	10
12	14
11	20
3	6
1	8

Source: Hypothetical data.

A TIP AND A WARNING
If you have a large batch of quantitative data (say, more than 500 cases), you can obtain clearer, more interpretable results if you have your software program first select a sample of the data (25–75 cases) and then plot those numbers. If the sample is truly representative, the plot will reveal the important features of the relationship. Creating a scatterplot from an entire data set may produce a graphic filled with so many dots that nothing intelligible can be detected. Furthermore, scatterplots are suitable only for quantitative variables. They are not intended for categorical (nominal and ordinal) data. If you tried, for instance, to have your software plot party identification by gender, the result would be two parallel lines that tell you nothing.

units on a scale that varies only between –2 and 2. Now look at Norway and Luxembourg or at the United States and Luxembourg: there is quite a difference in their GNPs but not in their democracy scores. Apparently once a nation passes a certain economic threshold, the marginal returns to democracy from growth diminish. Countries do not have to be as rich as the United States or Luxembourg to be democratic, but they can't be dirt poor either.

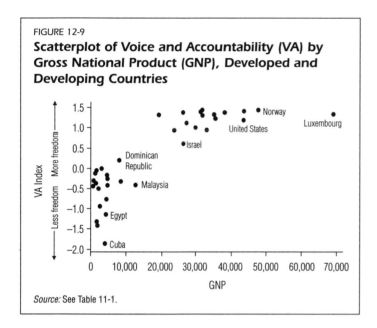

FIGURE 12-9

Scatterplot of Voice and Accountability (VA) by Gross National Product (GNP), Developed and Developing Countries

Source: See Table 11-1.

That the rate of change in Y per unit change in X is not constant also has statistical implications. What we see in Figure 12-9 is a quintessential case of a curvilinear relationship. Because regression analysis works best when variables have a linear or straight line correlation, statisticians have developed numerous techniques to transform data so that they better conform to the assumption of linearity. (We will mention a couple as we go along.)

Matrix Plots

Let us continue with the idea that higher levels of socioeconomic development are conducive to democracy. Given the multifaceted nature of the research hypothesis, it will be advantageous to measure development several ways. That is why we recorded (in Table 11-1) several economic and social indicators. At the outset of a study, it is always beneficial to look at the interrelationships among all pairs of independent variables before trying to assess if and how they individually or collectively affect the dependent variables. A powerful tool for doing so is the **matrix plot,** a graphic that simultaneously displays the scatterplots of each pair of variables. Conceptually simple, matrix plots are not easily constructed by hand; fortunately most statistical software does the job. The user simply lists the variables, and the software plots each item in the list against every other one.

Figure 12-10 is an example. Observe that each diagonal cell in the figure contains the name of a variable. The graphs to the right and left show scatterplots of that variable against others. You will also see that each relationship is displayed twice, once on each side of the diagonal. (As an example, the relationship between physicians and infant mortality appears in the last cell of the first row and again in the last cell of the first column.) The first row, for example, plots infant mortality rates against "phones," agricultural labor, GNP, and physicians per 100,000 population. (The definitions of these variables appear in Table 11-2.) Although it is hard, if not impossible, to discern precise details, the important features of the relationships—their form and strength—are clear.

Examine the first row in more detail. The first panel reveals a strong curvilinear relationship between infant mortality rates per 1,000 births (IMR) and telephone availability. The scale of phones runs along the top while the values

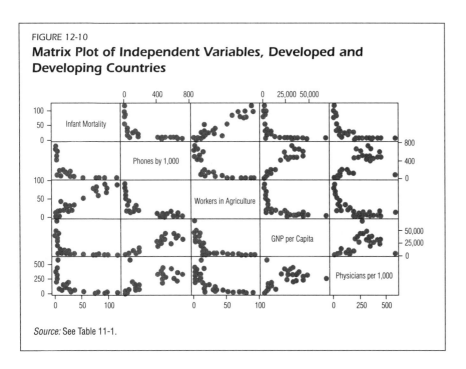

FIGURE 12-10

Matrix Plot of Independent Variables, Developed and Developing Countries

Source: See Table 11-1.

of IMR appear on the left vertical axis. As the number of phones increases slightly, the infant mortality rate plunges sharply. Remember that these are indicators of development and well-being. A country with a rudimentary communications system is also likely to be one with poor nutrition and neonatal care. Enhancements in physical infrastructure such as communications are likely to be accompanied by improvements in human services. Hence, more phones does not cause infant mortality to decline, but because of development the two go hand in hand. At some point, though, the decrease in infant deaths reaches nearly zero. Now, no matter how much the telephone system improves, it cannot be connected with further declines in IMR, which has come close to a lower bound. We can summarize the relationship with two observations. First, if you imagine or even draw a line through the middle of the points, it will form a curve indicative of a negative monotonic association. (Common sense tells you how to interpret the data: "the more phones, the lower infant mortality" or, going in the other direction, "the fewer phones, the higher death rates.") Second, notice that the association appears to be quite strong; most of the points lie close to the imaginary downward-sloping curved line.

The other plots are interpreted in the same way: find the scale scores for each pair of variables and see if the points follow a pattern. Notice, for instance, the strongly linear correlation between infant mortality and the percent of agricultural workers. (Look in the third panel of the top row or, conversely, in the third panel in the first column. Remember, each relation or plot is displayed twice.)

481

We conclude by offering a heads-up. Chapter 13 addresses multiple correlation, a technique that allows you to investigate the impact of more than one independent variable. It provides a sophisticated way to investigate multifaceted hypotheses. Yet it can be thrown off track by many factors, not the least of which is multicollinearity or correlations among explanatory variables. Suppose, for example, that you want to investigate the joint contribution of infant mortality and telephone usage on an indicator of democracy. The matrix plot reveals the tight connection between those two independent variables; as a result, measuring the separate impacts of these variables is going to be difficult in the absence of remedial steps. This book does not cover the multicollinearity problem, but it is important that you be aware of it.

Modeling Linear Relationships

The examination of scatterplots is a good first step in describing statistically or modeling a two-variable relationship. Figure 12-11 shows the relationship between telephone availability (main lines per 1,000 population) and infant mortality (the number of deaths per 1,000 live births) in just the 13 developing nations. It reveals a pattern: large values of infant mortality are associated with small values of phone availability, and vice versa. The relationship is by no means perfect, but there does seem to be a negative correlation. (We have added a line that clarifies the correlation.) The slope of this line is a negative number, which means that as we go up the scale on the x-axis, we move down the y-axis.

These ideas can be clarified further by recalling high school algebra. The equation for the graph of a straight line has the general form

$$Y = a + bX.$$

In the equation, X and Y are variables. The first letter (a) is called the constant and equals the value of Y when X equals zero (just substitute 0 for X). The equation has a geometric interpretation as well. If the graph of the equation is plotted, a is the point where the line crosses the y-axis. The letter b stands for the slope of the line, which indicates how much Y changes for each one-unit increase in X. For a positive b (that is, b > 0), if we move up the X scale one unit, b indicates how

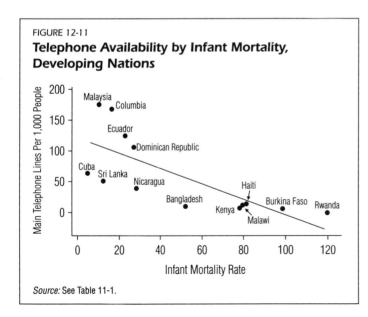

FIGURE 12-11

Telephone Availability by Infant Mortality, Developing Nations

Source: See Table 11-1.

much Y changes. (If we applied this type of equation to the data shown in Figure 12-11, b would be *negative* and would indicate how much Y, phone availability, decreases for every unit increase in X, infant deaths.)

If the relationship were positive—that is, if the slope were positive—the line would slant upward, and increases in X would be associated with increases in Y. If there were no (linear) relationship between the two variables, the slope of the line would be 0 and its graph would be horizontal and parallel to the x-axis.

An important feature of a line's slope is that its numerical value rests partly on the measurement scale of X. So, if you inadvertently used Y as the independent variable, a slope could be calculated, but its magnitude would in general *not* be the same as if X were treated as independent. Recall from the terminology introduced at the beginning of the chapter that the slope is an asymmetric parameter.

The Regression Model

Regression analysis can be thought of as applying these ideas to two-variable relationships, where both variables are numeric or quantitative. (Actually, regression analysis is general enough to include categorical variables but only in special ways. We discuss this possibility in Chapter 13.) The goal here is to find an equation, the graph of which "best fits" the data. If the points are approximately positively linearly correlated, then a straight line with a positive slope (b > 0) will pass through most of the points. If, however, a line with a negative slope (b < 0) goes through most of the data, then the correlation is negative. Finally, if the line has a slope of zero (b = 0), then X and Y are not linearly correlated.[23]

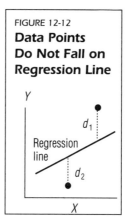

FIGURE 12-12
Data Points Do Not Fall on Regression Line

What exactly does "fit" mean in this context? In regression, an equation is found in such a way that its graph is a line that minimizes the squared vertical distances between the data points and the line drawn. In Figure 12-12, for example, d_1 and d_2 represent the distances of observed data points from an estimated regression line. Regression analysis uses a mathematical procedure that finds the single line that minimizes the squared distances from the line (for instance, the line that minimizes the sum d_1^2 and d_2^2). This procedure is called least squares and is often called "ordinary least squares," or OLS for short.

It is customary to describe the regression model or equation using Greek letters for the population parameters, the constant and the slope, and to add an error term, or "disturbance":

$$Y = \alpha + \beta X + \varepsilon$$

The two constants are called regression parameters. The first, α, is the constant and is interpreted exactly as indicated before: it is the value of Y when X equals zero. The second, β, is the **regression coefficient** and tells how much Y changes if X changes by one unit. The regression coefficient is always measured in units of the dependent variable.

The error (ε) indicates that observed data do not follow a neat pattern that can be summarized with a straight line. It suggests instead that an observation's score on Y can be broken into two parts: one that is "due to" the independent variable and is represented by the linear part of the equation, $\alpha + \beta X$, and another that is "due to" error or chance, ε. In other words, if we know an observation's score on X and also know the equation that describes the relationship, we can substitute the number into the equation to obtain a *predicted* value of Y. This predicted value will differ from the observed value by the error:

$$\text{Observed value} = \text{Predicted value} + \text{Error.}$$

If there are few errors, that is, if all the data lie near the regression line, then the predicted and observed values will be very close. In that case, we would say the equation adequately explains, or fits, the data. In contrast, if the observed data differ from the predicted values, then there will be considerable error and the fit will not be as good.

Figure 12-13 ties these ideas together. Suppose we consider a particular case. Its scores on X and Y (X_i and Y_i) are represented by a dot ($\bullet$). Its score on X is denoted as X_i. If we draw a line straight up from X_i to the regression line and then draw another line to the y-axis, we find the point that represents the predicted value of Y, denoted $\hat{Y}_i$. The difference between the predicted value, $\hat{Y}_i$, and the observed value, Y_i, is called the error or **residual.** Stated somewhat differently, ε represents the difference between the predicted score based on the regression equation, which is the mathematical equation describing the relationship between the variables—and the observed score, Y_i. (As we see in a moment, it stands for that part of a Y score that is unexplained.) Regression-computing formulas pick values of α and β that minimize the sum of all these squared errors. Because the regression equation gives predicted values, it is

How It's Done

CALCULATION OF ESTIMATED REGRESSION COEFFICIENTS

The regression coefficient is calculated as follows:

$$\hat{\beta}_{YX} = \frac{N \sum_{i=1}^{N} X_i Y_i - \left(\sum_{i=1}^{N} X_i \right)\left(\sum_{i=1}^{N} Y_i \right)}{N \sum_{i=1}^{N} X_i^2 - \left(\sum_{i=1}^{N} X_i \right)},$$

where N is the number of cases and X_i and Y_i are the X-Y values of the ith case.

The regression constant is calculated as follows:

$$\hat{\alpha} = \bar{Y} - \hat{\beta}_{YX} \bar{X},$$

where $\bar{Y}$ and $\bar{X}$ are the means of Y and X, respectively, and $\hat{\beta}_{YX}$ is the regression coefficient as calculated above.

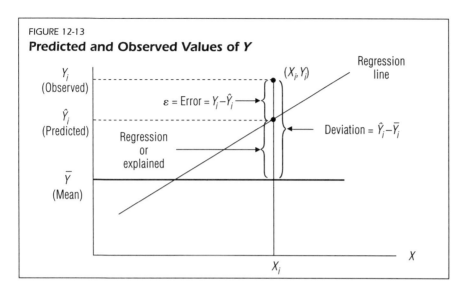

FIGURE 12-13

Predicted and Observed Values of Y

sometimes written without an error term but with a "hat" over Y to indicate that the dependent variable is not an exact function of X:

$$\hat{Y} = \hat{\alpha} + \hat{\beta} X_i.$$

Although the minimizing procedure may sound complicated, computer software and many handheld calculators make finding regression equations relatively easy. The tricky part is understanding the results.

For the data in Figure 12-11 (telephone availability [Y] and infant mortality [X]), the estimated regression coefficient (slope) is –1.2464 and the constant (intercept) is 120.62. Hence, the regression line is

$$\hat{Y} = 120.62 - 1.2464 \text{ (Infant mortality rate)}.$$

Interpretation of Parameters

The slope of the line and the Y-intercept describe the nature and strength of the connection between the two variables. Again, the Y-intercept is the value of the dependent variable when X (the independent variable) = 0, or, stated differently, it is the place where the regression line crosses the y-axis when $X = 0$.

Here the Y-intercept of 121 means that when a country's infant mortality rate is 0, its predicted telephone usage (measured in available phone lines per 1,000 population) is about 121 lines. This value is, of course, a prediction for a case with "no" infant mortality. In many instances the actual value is not of much substantive interest because a zero value on the independent variable usually does not make much theoretical sense. The slope or regression coefficient is a different matter, however.

The slope measures the amount the dependent variable changes when the independent variable, X, changes one unit. In this case, the slope of -1.25 tells us that for every increase in X of one infant death, there is a predicted or estimated decrease of 1.25 in telephone lines. These measures of development are tied together in a commonsense way. We noted earlier that lower infant mortality implies better health care and higher standards of living. Both of these factors would be associated with more advanced communications systems. If you think about the substance and context of the variables, interpreting regression coefficients is not so difficult.

The value of the regression coefficient (-1.25) may seem small and abstract. But putting different values of X in the regression equation can help you to gauge the impact of the regression coefficient:

- No infant deaths (that is, $X = 0$) implies $\hat{Y} = 121 - 1.25(0) = 121$ phone lines per 1,000 population.
- 1 infant death implies $\hat{Y} = 121 - 1.25(1) = 119.75$ phone lines per 1,000 population.
- 10 infant deaths implies $\hat{Y} = 121 - 1.25(10) = 108.5$ phone lines per 1,000 population.
- 100 infant deaths implies $\hat{Y} = 121 - 1.25(50) = 58.5$ lines per 1,000 population.

An increase from 1 to 10 deaths per 1,000 live births is associated with a decline of $119.75 - 108.5 = 11.25$ telephone lines. Similarly, an increase in the infant mortality rate from 10 to 100 "leads to" a decrease in phone lines of about 50 ($108.5 - 58.5 = 50$) lines, a decline of about 46 percent. Of course there is no causal connection between the variables. The examples do, however, show how strongly the variables are (negatively) correlated. The relationship is spurious partly due to a third factor, economic-social development. Also remember that the regression parameter is asymmetric. Had we used infant mortality rates as the dependent variable (thereby considering telephone lines per 1,000 people as independent), we would get a completely different equation:

$$\hat{Y}_{(IMR)} = 75.6 - .449\, X_{(phones)}.$$

The same interpretation applies—"a one telephone line increase is accompanied by a decrease of about a half an infant death"—and in this context results in the same theoretical conclusion, namely that improvements in communications are associated with lower infant mortality. But mixing the order of variables in other situations may lead to nonsensical results, especially if the analysis includes a hint of causality. Suppose, for example, you were studying the interaction between age and annual income. If you treated age as the dependent variable and regressed it on income, the resulting regression coefficient could be interpreted as "a one dollar increase in income leads to a specified change in aging." This reading is not what you intended, but it illustrates the importance of choosing the appropriate dependent and independent variables.

Measuring the Fit of a Regression Line

Look again at Figure 12-13. Earlier in the chapter we introduced the total sum of squares (TSS). This value is the sum of squared deviations from the mean. Now, by examining Figure 12-13 you can see that an observation's total deviation from the mean, denoted $Y_i - \bar{Y}$, can be divided into two additive parts. The first is the difference between the mean and the predicted value of Y. Let's label that portion as the "regression," or "explained," part (RegSS). It is explained in the sense that a piece of the deviation from the overall mean is attributed to X, the independent variable. The second component of the total deviation is called the residual sum of squares (ResSS), which measures prediction errors. This term is frequently labeled the "unexplained sum of squares" because it represents the differences between our predictions—that is, $\hat{Y}$—and what is observed. If all the predictions were perfect, there would be no errors and the ResSS would be zero. The ResSS provides the numerator of the conditional standard deviation of Y, a statistic used later on to test hypotheses and construct confidence intervals.

These three quantities are identical to the ones presented earlier in connection with the analysis of variance, except two of them have slightly different names, for they are now called the regression and error sums of squares. (Their computing formulas also differ slightly.) Yet the same fundamental relationship still holds:

$$TSS = RegSS + ResSS$$

The TSS represents all the variation in the data, explained or not, whereas the RegSS corresponds to the part of this total that is explained (in a statistical sense) by the independent variable via the regression equation. So, as in ANOVA, we can calculate the "proportion of total variation explained" by X as follows:

$$R^2 = RegSS/TSS$$

This measure (R^2) is known as **R-squared.** Despite its name, it is identical to eta-squared and is one of the most commonly reported statistics in the social sciences. Its popularity derives partly from its simplicity and partly from the belief that it indicates how well a regression model fits data.[24] For

TABLE 12-26

Regression Sums of Square and _R_-Squared and _r_

Source	Value
Regression (RegSS)	27,032
Residual (ResSS)	21,270
Total (TSS)	48,302

$R^2 = 27{,}032/48{,}302 = .56\ (56\%);\ r = -.75.$

example, if R-squared is multiplied by 100, the result is often interpreted as the percentage of (total) variation in Y that X explains.

Table 12-26 shows the sums of squares from the regression of telephone availability on infant mortality. The explained variation is .56, which, statistically speaking, means that X explains somewhat more than half of the variation in Y.

An important property of R-squared is that it is symmetrical, meaning that it has the same value no matter which variable is treated as dependent. This is a key difference from the regression coefficient, which does change depending on the choice of dependent variable. Also, R-squared must be at least zero; it cannot be negative because it is the quotient of two squared terms.

Figure 12-14 offers some additional insights into the properties and interpretation of R-squared. In the first set of graphs (a), we see that if all the data points lie on a straight line, there will be no residual or unexplained deviations, and consequently X explains 100 percent of the variation in Y. This is true for both perfect positive ($\hat{\beta} > 0$) or negative ($\hat{\beta} < 0$) relationships. Hence, R^2 equals 1. However, if the points have a general tendency to lie on a positively or negatively sloping line, R^2 will be less than 1 but will indicate that some portion of the variation in Y can be attributed to X. (See section b of Figure 12-14.) Finally, if no linear relationship exists between X and Y, R^2 will be 0. A value of 0 means only that there is no relationship describable by a straight line. It does not mean statistical independence. The variables may have no association at all, or they may be strongly curvilinearly related or connected in some other fashion (see Figure 12-14, section c). In both situations, R^2 will be 0 or close to 0, but the meaning will differ. A good way to spot the difference is to examine a plot of Y versus X. A scatterplot can help you determine the pattern your data come closest to.

In regression analysis, the term _explained_ has a different meaning than it does in day-to-day conversation. In statistics it means that the variation in one variable can be mathematically divided into two quantities. One, the so-called explained part, is the squared sum of differences between predicted values and the overall mean. These are strictly statistical terms. In ordinary discourse, _explain_ usually implies substantive understanding or knowledge of functional relationships of how things work. Thus we might find a large R-squared between two variables (for example, income and contacting political leaders), which in statistical terms implies that a lot of variation has been explained. But this finding does not necessarily indicate that we really understand why and how rich people make more contacts than the poor do. In fact, as we explained in Chapter 5, a relationship can be spurious, meaning that a false connection is caused by other factors. Always be cautious when confronted with seemingly large values of R-squared.

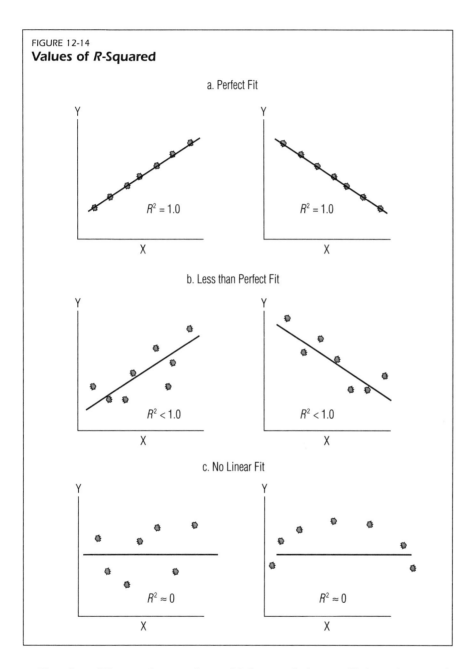

FIGURE 12-14
Values of *R*-Squared

a. Perfect Fit

$R^2 = 1.0$

$R^2 = 1.0$

b. Less than Perfect Fit

$R^2 < 1.0$

$R^2 < 1.0$

c. No Linear Fit

$R^2 \approx 0$

$R^2 \approx 0$

Here is a different slant on the multiple correlation coefficient, R-squared. If you think of the regression equation as a method for making predictions, then R-squared can be understood as a proportional-reduction-in-error (PRE) measure for assessing the strength of a relationship. Faced with a quantitative variable, *Y*, and lacking any other information, your best guess of a case's value on that variable would be the mean, $\bar{Y}$. When *X* is known, however, a better guess is achieved from the predicted value produced by the regression of Y on X. It can be shown that the reduction in errors is exactly R-squared. For instance, knowing the infant mortality rate of a nation reduces errors in predicting telephone availability by about half (56 percent), as you can see in Table 12-26.

The Correlation Coefficient

A close kin of R-squared is the **correlation coefficient,** a measure of the strength and direction of the linear correlation between two quantitative variables. The definition and computation of the correlation coefficient, denoted r, depend on standardizing the covariation between Y and X by dividing by their standard deviations. To find the covariation between two variables, you multiply the "deviations from the mean"—as shown in Chapter 11, a deviation is a value minus the mean—and add them together. This total is then divided by the variables' standard deviations. Since you will in all likelihood have software do the calculations, we focus here on the meaning of the coefficient.

We listed many general properties of measures of association at the beginning of this chapter, and you may wish to refresh your memory because the correlation coefficient exhibits many of those properties. Known under a variety of labels—the product-moment correlation, Pearson's r, or, most plainly, r—this coefficient reveals the direction of a regression line (positive or negative) and how closely observations lie near it. Its properties include the following:

How It's Done

THE CORRELATION COEFFICIENT

The (Pearson) correlation coefficient is calculated as follows:

$$r = \frac{\sum_i (Y_i - \bar{Y})(X_i - \bar{X})}{(N-1)\hat{\sigma}_Y \hat{\sigma}_X},$$

where Y_i and X_i are the values on Y and X of the ith observation; $\bar{Y}$ and $\bar{X}$ are the means of Y and X, respectively; $\hat{\sigma}_Y$ and $\hat{\sigma}_X$ are the sample standard deviations of Y and X, respectively; N is the sample size; and the summation is over all pairs of data points.

- It is most appropriate if the relationship is approximately linear.

- Its value lies between –1 and 1. The coefficient reaches the lower limit when Y and X are perfectly negatively correlated, which is to say all the data lie on a straight line sloping downward from left to right. Its maximum value (1) is achieved when the variables are perfectly positively correlated.

- It will equal 0 if there is no linear correlation or, to be more exact, when the slope is 0.

- The closer r is to either of its maximum values, the stronger the correlation. The nearer to 0, the weaker the correlation. (Consequently, $r = .8$ or $-.8$ implies a stronger relationship than $r = .2$ or $-.2$.)

- It has the same sign as the regression coefficient. (For example, if $\hat{\beta}$ is negative, r will be, too.)

- Unlike the regression parameter, it is symmetric in that its numerical value does not depend on which variable is considered dependent or independent.

- The correlation coefficient is scale independent in that its magnitude does not depend on either variable's measurement scale. It does not matter if, say, X is measured in dollars or thousands of dollars; the value r stays the same. This is not true of the regression coefficient.

Because of the last property, the correlation coefficient can be regarded as a kind of regression coefficient that does not depend on the units of Y or X. As a matter of fact, r has this association with the slope:

$$r = \left(\frac{\hat{\sigma}_X}{\hat{\sigma}_Y} \right) \hat{\beta},$$

where the $\hat{\sigma}$'s are the sample standard deviations of X and Y. (As an aside, notice that r is partly a function of the size of the standard deviations. Given two samples with identical $\hat{\beta}$'s between Y and X, the one with the larger standard deviation, σ_X, will *appear* to have the larger r and hence the larger linear correlation. But the magnitude of the relationship may be simply a function of the variability of X, not any intrinsic strength of the relationship.) We discuss the pros and cons of this feature of r in the next section.

If we look at our cross-national data analysis, we see that the correlation between infant mortality and telephone lines is about –.75 (see Table 12-26). This indicates a strong negative (linear) correlation.

To grasp the meaning of the numerical values of the correlation, try studying the patterns in Figure 12-15. In graphs (a) and (b) most of the data lie near a straight line, and r is close to either its maximum value of 1 or its minimum value of –1. By contrast, graphs (c) and (d) show what you are likely to encounter in data analysis, moderate to weak relationships. Observe that the sign of r reflects the direction of the correlation. In each of these graphs the data "behave" (this is a term statisticians commonly use) very well: either there is a correlation or there isn't. Look, however, at Figure 12-16. It illustrates a highly curvilinear relationship. A strong connection exists between Y and X—knowing the value of one would help you accurately predict values of the other—but the correlation coefficient ($r = -.15$) might suggest a weak relationship, until you remember that it measures the fit to a line. Once again, we stress the importance of examining scatterplots along with numbers. You are not likely to encounter such a strongly curved relationship in typical social science data, but it

FIGURE 12-15

Degree and Direction of Correlation

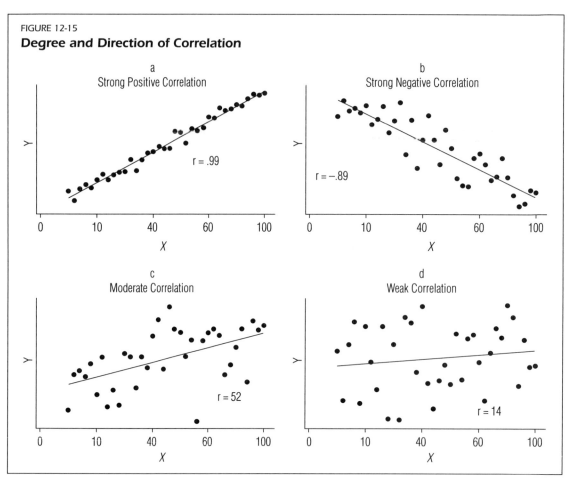

a
Strong Positive Correlation

r = .99

b
Strong Negative Correlation

r = −.89

c
Moderate Correlation

r = 52

d
Weak Correlation

r = 14

FIGURE 12-16

Curvilinear Relationship

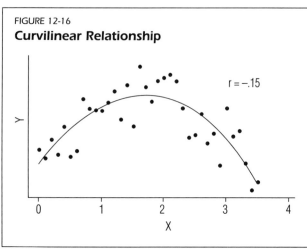

r = −.15

is important to remember that at this level of study, regression performs "best" when associations are linear or can be transformed to linearity.

Standardized Regression Coefficients

The regression coefficient indicates how much Y changes—in values of the dependent variable—for a one-unit change in X. If you were relating years of education (X) to annual wages (Y), the regression parameter would be expressed in, say, dollars. In the previous example, the estimated coefficient, 1.25, should be thought of and interpreted as "1.25 telephone lines per a one-unit increase in IMR (that is, one infant death)." In many, if not most cases, this measure of association is exactly what is needed. Yet social scientists sometimes prefer a "scale-free" statistic. For example, if independent variables had a common scale, their effect or impact could be compared unequivocally. (This statement rests on a major caveat discussed below.) Let's say a research team wants to explain variation in infant mortality rates. The problem with the regression coefficient is that a "one-unit change in X" means different things depending on the measurement scale. If X is income measured in dollars (a unit is $1), the coefficient may have a very small numerical value; if it is measured in thousands of dollars (a unit is $1,000), the coefficient will be larger.

Researchers often rescale all variables so that a one-unit change has common meaning. To do so, you subtract the mean from each value and divide the remainder by the standard deviation. Consider a variable, X. Its sample mean is $\bar{X}$ and its standard deviation is $\hat{\sigma}_X$. Denoting the corresponding standardized value of X with lowercase x, the standardization is

$$x_i = \frac{(X_i - \bar{X})}{\hat{\sigma}_X}.$$

Keep a close eye on the symbols: lowercase letters denote standardized variables, whereas uppercase letters represent the raw data.

Standardized variables have several interesting and (for statisticians) important features. First, the mean and standard deviation of the recalibrated variables are always 0 and 1.0, respectively. (Standardization is based on the fact that deviations from a mean add to zero.) Table 12-27 illustrates the technique for and properties of standardizing a variable.

More important, the advantage of standardization lies in the seeming simplicity of regression results. For example, when both X and Y are standardized and y is regressed on x, the resulting equation simplifies to

$$\hat{y} = \hat{\beta}^* x.$$

Notice that there is no constant (α): whenever Y and X have been standardized, the regression constant drops out. Also pay attention to $\hat{\beta}^*$. Called the **standardized regression coefficient,** this number is interpreted as in the usual way except that now a one-unit change in x is a *one-standard-deviation change.* In other words, the independent variable's effect is measured in standard deviations of y, not the scale of the original dependent variable. A standardized coefficient has the same sign as its unstandardized cousin. The only difference is in their numerical values and interpretations.

TABLE 12-27
Raw and Standardized Variables

Observation i	Raw Score (X_i)	Raw Deviation $(X_i - \bar{X})$	Squared Deviation $(X_i - \bar{X})^2$	Standardized Score (x_i)	Squared Deviations $(x_i - \bar{x})^2 = x_i^2$
1	10	$10 - 15 = -5$	25	−1.3361	1.78251
2	12	$12 - 15 = -3$	9	−.80178	.64296
3	14	$14 - 15 = -1$	1	−.26726	.07143
4	16	$16 - 15 = 1$	1	.26726	.07143
5	18	$18 - 15 = 3$	9	.80178	.64286
6	20	$20 - 15 = 5$	25	1.3361	1.7251
Sum	90	0	70	0	5

Raw scores:

Mean of X: $\bar{X} = 90/6 = 15$.

Standard deviation of X: $\hat{\sigma}_X = \sqrt{70/5} = 1.7852$.

Standardized scores:

Mean of x: $\bar{x} = \% = 0$.

Standard deviation of x: $\hat{\sigma}_X = \sqrt{5/5} = 1.0$.

Incidentally, in two-variable regression the standardized slope equals the correlation coefficient. Thus it is not surprising that after standardizing IMR and phone lines (the data in Figure 12-11), we find that $\hat{\beta}^*$ is −.75, the same value reported in the section on correlation.

The standardized regression coefficient of phone availability on infant mortality is −.75, which tells us that a one-standard-deviation increase in infant mortality corresponds to a .75 standard-deviation decrease in telephone availability. At the end of the day, though, the coefficients $\hat{\beta}$ and $\hat{\beta}^*$ differ numerically, but the picture they convey is the same: telephone availability and infant mortality are negatively correlated.

So, aside from simplifying the equation, why bother with standardized variables? Some social scientists believe that standardized regression coefficients enable you to rank the "relative importance" of independent variables on a dependent variable.[25] Imagine that you are trying to explain political participation. Your study includes (1) education in years of schooling, (2) annual family income in dollars, and (3) degree of partisanship measured on a five-point scale (1 for "least partisan" to 5 for "highly partisan"). The literature tells you to expect a positive relationship between all three variables and the indicator of participation. But some authors say socioeconomic factors are more important explanations of political behavior than are political leanings; others disagree completely. So you perform an analysis and find the regression coefficients for education, income, and partisanship to be, respectively, .0001, .5, and 1. Numerically, $\hat{\beta}_{participation\ partisan}$ is larger than either of the other two;

hence, psychology seems more important than economic class. The problem is that since the variables have different measurement scales, the coefficients cannot be compared directly. If, however, you standardize all of the variables, the standardized coefficients might turn out to be .8, .5, and .2, in which case the socioeconomic variables seem to have the strongest relationship.

Although calculating standardized variables may be a good idea, their use in the preceding example works only if the independent variables are independent of one another, a situation that rarely arises in observational studies. In addition, even if you feel compelled to take advantage of the standardized version, you should calculate the nonstandardized coefficient as well.

Inference for Regression Parameters

This section builds on the ideas presented in Chapter 11 and in previous sections regarding hypothesis testing. You may wish to review those topics briefly before proceeding.

Like any other procedure, regression can be applied to sample or population data. In the present context, we assume that in a specified population a relationship exists between X and Y and that one way of describing it is with the regression coefficient β. This unknown quantity must be estimated with $\hat{\beta}$, the sample regression coefficient. The test for its significance can go in two directions, both of which end up in the same place.

The estimated regression coefficient has a t distribution with $N - 2$ df. (When N becomes large—roughly 30 or more cases—the t blends into the standard normal distribution for which a z statistic is appropriate.) The null hypothesis is usually simply H_0: $\beta = 0$.[26] This possibility is tested against a two-sided ($\mu \neq 0$) or one-sided ($\mu < 0$ or $\mu > 0$) alternative. The test statistic has the typical form: the estimated coefficient divided by the estimated standard error:

$$t_{obs} = \frac{\hat{\beta}}{\hat{\sigma}_{\hat{\beta}}},$$

where $\hat{\sigma}_{\hat{\beta}}$ is the estimated standard error of the regression coefficient. Remember that if an estimator is calculated from many, many independent samples, these sample statistics will have a distribution, called the sampling distribution, with its own mean and standard deviation. When $\hat{\beta}$ is the statistic, its sampling distribution is the t distribution (the standard normal for a large N), which has mean β and standard error or deviation $\sigma_{\hat{\beta}}$. Confidence intervals can be constructed in the usual way:

$$\hat{\beta} \pm t_{\alpha/2, N-2} \hat{\sigma}_{\hat{\beta}},$$

where $t_{\alpha/2, N-2}$ is the value in Appendix B that cuts off the upper $\alpha/2$ proportion of the distribution.

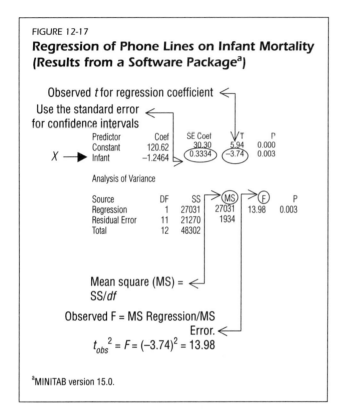

FIGURE 12-17

Regression of Phone Lines on Infant Mortality (Results from a Software Package[a])

Observed *t* for regression coefficient

Use the standard error for confidence intervals

Predictor	Coef	SE Coef	T	P
Constant	120.62	30.30	5.94	0.000
X → Infant	−1.2464	0.3334	−3.74	0.003

Analysis of Variance

Source	DF	SS	MS	F	P
Regression	1	27031	27031	13.98	0.003
Residual Error	11	21270	1934		
Total	12	48302			

Mean square (MS) = SS/*df*

Observed F = MS Regression/MS Error.

$$t_{obs}^2 = F = (-3.74)^2 = 13.98$$

[a]MINITAB version 15.0.

The second method for testing significance is to compute sums of squares, divide them by their degrees of freedom to obtain mean squares, and form the ratio of the mean square for regression to the mean square for error. This results in an observed F with 1 and $N - 2$ *df*. (Remember that the F distribution is a collection of distributions and their shapes and areas are partially functions of the number of variables and the sample sizes.) The tests are equivalent in this situation, where we have just one independent variable $t_{obs}^2 = F_{1,N-1}$.

Figure 12-17 gives an example of significance test results. The number of cases in this study is $N = 13$. The observed t (abbreviated "T" in this program) is −3.74. (Since the assumed null hypothesis in this test is 0, the t statistic has the same sign as the regression coefficient. When comparing it to a t derived from a table, ignore the sign or use the absolute value.) The attained probability (the p-value) is .003, which indicates significance not only at the .05 level but also at the .01 level. Remember that the p-value states, in effect, that if the null hypothesis ($\beta = 0$) is true, the observed result ($t_{obs} = -3.74$) or a larger one is less than (in this case) .003. And, you can verify that squaring −3.74 gives 13.98, the value of the F statistic. It too indicates that the null hypothesis is untenable and so might be rejected at the .01 level. Finally, you can substitute the standard error for the regression coefficient (.3334; denoted "SE Coef") in the formula for confidence intervals. Say you want 99 percent limits. The critical t (with $N - 2 = 13 - 2 = 11$ *df*; that is, $t_{.01/2,11\ df}$) from the table is 3.106 and the confidence intervals are

$$-1.2464 - (3.106)(.3334) \le \beta \le -1.2464 + (3.106)(.3334)$$
$$-1.2464 - 1.0355 \le \beta \le 1.2464 + 1.0355$$
$$-2.28194 \le \beta \le -.21086.$$

The interval does not contain zero, although it comes close. The overall interpretation is that a negative relationship exists between telephone lines and infant mortality, which lends credence to the general hypothesis that one form of development (growth in communications) improves another (lowering of neonatal infant deaths).

Regression Is Sensitive to Large Values

We wrap up the discussion of two-variable regression with a caution. Just as the sample mean and standard deviation are sensitive to outlying or deviant values, so too are the regression parameter ($\hat{\beta}$) and correlation coefficient (r). The point is best demonstrated with an example. The media sometimes report that health care costs continue to rise partly because Americans may be "over treated." The argument runs as follows: To the extent that medical facilities and physicians are available they will be utilized. Consequently, areas densely populated with, say, MRI devices will experience higher rates (per capita) of use than will places where they are scarce. Figure 12-18 takes a slightly different view of the problem. It shows the relationship between the number of surgical procedures carried out in the 50 states plus the District of Columbia by the number of surgical specialists. (Both variables have been converted to per-capita indices.) The plot in Figure 12-18a seem to back up the notion that more surgeons are accompanied by more surgeries. Naturally, it is possible that specialists migrate to places with a lot of sick people. However, it seems more probable that availability drives usage. The correlation coefficient (.53) supports a claim of a relatively strong positive correlation between operations and surgeons.

Yet even a cursory glance at the distribution might raise suspicions because one point lies far from all the others. It is the value for the District of Columbia, which has a high density of surgical specialists *and* operations. (The city is known for its many and famous health centers.) You may recall from Chapter 11 that a distant point is called an outlier. Sitting far from all the other states, D.C. is a prime example of an outlier. Correlation and regression are functions of the sums of squared deviations from means, and if one or two observations differ greatly from the others, their deviations will contribute a disproportionate amount to the totals. In technical terms, an outlier can exert great leverage on the numerical values of coefficients. That this is the case here can be seen in Figure 12-18b. It displays the procedures by specialists for all the data *except* D.C. With that outlier removed—a valid, even necessary step in data analysis—the linear correlation disappears; the correlation coefficient $r = -.18$ has changed directions and moved into the "weak" range.

Statisticians know full well that standard regression models are not robust against problems such as leverage points and recommend paying attention to and adjusting for them. Luckily, most regression software offers the option of flagging these kinds of data.

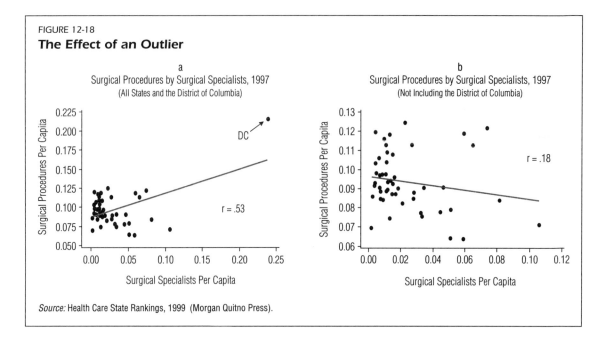

FIGURE 12-18

The Effect of an Outlier

Source: Health Care State Rankings, 1999 (Morgan Quitno Press).

Conclusion

In this chapter we have shown how to measure the existence, direction, strength, and statistical significance of relationships between two variables. We have also noted the difference between association and causation. The particular techniques used—cross-tabulation analysis, difference of means test, analysis of variance, regression analysis, and correlation—depend in part on the levels of measurement—nominal, ordinal, interval, or ratio—of the independent and dependent variables. Just because a relationship is present, however, does not mean that a cause for the dependent variable has been discovered. We explain this concept in more detail in Chapter 13.

Notes

1. Many techniques are available for analyzing relationships at a given level of measurement. The ones presented in this chapter are the most common and least complicated.

2. These labels represent an interpretation we have imposed on the question responses. It would be perfectly legitimate, for instance, to redefine the ideology scale as the "degree of liberalism." What matters is that you keep straight how the variables are treated and make your explanations consistent with those definitions.

3. You might think of ties as a "penalty" for the imprecise measurement that classification involves. But however they are interpreted, tied pairs count against all the measures except gamma in the sense that the more ties, the smaller the numerical value of the coefficient.

4. For further information about the calculation of each of these statistics, see Alan Agresti and Barbara Finlay, *Statistical Methods for the Social Sciences,* 3d ed. (Upper Saddle River, N.J.: Prentice-Hall, 1997), 272–282.

5. Partly because these coefficients do not generally describe the complexities of relationships between categorical variables, they have fallen out of favor with many social scientists. Sociologists and statisticians have developed methods for modeling the frequent multiplicity of interactions between categories in a table. We touch on a few techniques later in the chapter but leave the bulk of them to more advanced texts. A good introduction is Alan Agresti, *Analysis of Ordinal Categorical Data* (New York: Wiley, 1984).

6. Changing levels of measurement, although common, discards a certain amount of information and is not generally recommended. We do so only for illustrative purposes.

7. The situation is an instance of the ecological inference problem introduced in Chapter 3.

8. "Strongly" favor and oppose responses have been blended into two categories shown here.

9. The odds ratio is not the same as a related measure, relative risk. The relative risk measures the probability of one response divided by the probability of another response.

10. A good but slightly challenging introduction is Daniel A. Powers and Yu Xie, *Statistical Methods for Categorical Data Analysis* (San Diego: Academic Press, 2000).

11. For example, a lambda of .134 can be interpreted as a 13 percent reduction in errors in prediction responses to the gun control issue when gender is known.

12. Note, however, that small effects can in some circumstances have theoretical or substantive importance.

13. See Agresti and Finlay, *Statistical Methods for the Social Sciences,* 264–265 for more information and ideas about how to proceed.

14. Its formula is $\chi_{LR} = 2 \sum_i \sum_j f_{ij} \log\left(\dfrac{f_{ij}}{e_{ij}}\right)$, where f_{ij} is the observed frequency in the ijth cell,

   e_{ij} is the corresponding expected frequency under the hypothesis of statistical independence, and "log" is the natural (to base e) logarithm.

15. Agresti, *The Analysis of Ordinal Categorical Data,* 13.

16. *Ibid.*

17. Additional odds ratios can be formed, but they will be functions of the basic set.

18. A good place to begin is Stephen Fienberg, *The Analysis of Cross-Classified Categorical Data* (Cambridge: MIT Press, 1980); David Knoke and Peter J. Burke, *Log-Linear Models,* Sage University Paper Series on Quantitative Applications in the Social Sciences, series no. 20 (Beverly Hills, Calif.: Sage Publications, 1980), and Graeme Hutcheson and Nick Sofroniou, *The Multivariate Social Scientist* (London: Sage Publications, 1999).

19. The quotes around "population" are necessary because, technically speaking, there are no population experimental and treatment group means. These are hypothetical or theoretical quantities. They could exist only if a researcher could somehow conduct an experiment on an entire population and at the same time treat it as a control. This is a subtle point but one that has far-reaching consequences for how the results of experimental and observational studies are interpreted. An excellent and accessible introduction to this topic is Christopher Winship and Stephen L. Morgan, "The Estimation of Causal Effects from Observational Data," *Annual Review of Sociology* 25 (1999): 659–706.

20. The full question is "I'll read the name of a [group or person] and I'd like you to rate that [group or person] using something we call the feeling thermometer. Ratings between 50 degrees and 100 degrees mean that you feel favorable and warm toward the [group or person]. Ratings between 0 degrees and 50 degrees mean that you don't feel favorable toward the [group or person] and that you don't care too much for that [group or person]. You would rate the [group or person] at the 50 degree mark if you don't feel particularly warm or cold toward the [group or person]. . . . Using the thermometer, how would you rate the following. . . ." American National Election Study (ANES) 2004, HTML Codebook produced July 14, 2006. The American National Election Studies (www.electionstudies.org). The 2004 National Election Study [data set]. Ann Arbor, MI: University of Michigan, Center for Political Studies [producer and distributor].

21. Agresti and Finlay have this to say about the situation: "Though it seems disagreeable to make an additional assumption [that the two population variances are equal], confidence intervals

and two-sided tests are fairly robust against violations of this ... assumption, particularly when the sample sizes are similar and not extremely small." Agresti and Finlay, *Statistical Methods for the Social Sciences,* 221.

22. See Table 11-2 for the variable definitions.

23. Two variables may be related—that is, they may not be statistically independent—but still have no linear correlation between them. In other words, it might be possible to find a line that passes through most of the data, but it will not be a straight line.

24. R-squared is also called the multiple correlation coefficient or multiple *R* or the coefficient of determination. These terms usually come into play when analyzing the effect of several independent variables.

25. Invoking "relative importance" in a conversation among social scientists is akin to waving a red cape in a bull fighting ring. Considerable controversy exists about the very meaning of interpretation in statistics, and no one seems to agree if it can or cannot be measured. A gentle introduction to the topic can be found in Johan Bring, "How to Standardize Regression Coefficients," *The American Statistician* 48 (August 1994): 209–213. See also, among countless other excellent discussions, Christopher Achen, *Interpreting and Using Regression,* Sage University Paper Series on Quantitative Applications in the Social Sciences, series no. 29 (Beverly Hills, Calif.: Sage Publications, 1982); Gary King, "How Not to Lie in Statistics: Avoiding Common Mistakes in Quantitative Political Science," *American Journal of Political Science* 30 (August 1986): 666–687; Robert C. Luskin, "Abusus Non Tollit Usum: Standardized Coefficients, Correlations, and R^2's," *American Journal of Political Science* 35 (November 1991): 1032–1046; and Erik A. Hanusek and John E. Jackson, *Statistical Methods for Social Sciences* (New York: Academic Press, 1977): 78–79.

26. If you had reason to believe that the population regression coefficient equaled a particular value, you could make it the subject of the null hypothesis. Most of the time investigators simply test the hypothesis that it is 0. If you did not know if *Y* and *X* were (linearly) correlated, that is a reasonable procedure. Frequently, however, you know ahead of time that the variables are related (only their form or strength of association remains unknown), so checking the hypothesis $\beta = 0$ probably will not lead to much new information, unless the hypothesis is accepted.

Terms Introduced

ANALYSIS OF VARIANCE. A technique for measuring the relationship between one nominal- or ordinal-level variable and one interval- or ratio-level variable.

CHI-SQUARE. A statistic used to test the statistical significance of a relationship in a cross-tabulation table.

CORRELATION COEFFICIENT. A statistic that indicates the strength and direction of the relationship between two interval- or ratio-level variables. It is a standardized measure, meaning that its numerical value does not change with linear scale changes in the variables.

CROSS-TABULATION. Also called a contingency table, this array displays the joint frequencies and relative frequencies of two categorical (nominal or ordinal) variables.

DEGREES OF FREEDOM. The number of independent observations used to calculate a statistic.

DIFFERENCE OF MEANS TEST. A statistical procedure for testing the hypothesis that two population means have the same value.

EFFECT SIZE. The effect of a treatment or an experimental variable on a response. Most commonly the size of the effect is measured by obtaining the difference between two means or two proportions.

ETA-SQUARED. A measure of association used with analysis of variance that indicates the proportion of the variance in the dependent variable that is explained by the variance in the independent variable.

EXPLAINED VARIATION. That portion of the total variation in a dependent variable that is accounted for by the variation in the independent variable(s).

GOODMAN AND KRUSKAL'S GAMMA. A measure of association between ordinal-level variables.

GOODMAN AND KRUSKAL'S LAMBDA. A measure of association between one nominal- or ordinal-level variable and one nominal-level variable.

KENDALL'S TAU B OR C. A measure of association between ordinal-level variables.

MATRIX PLOT. A graph that shows several two-variable scatterplots on one page in order to present a simultaneous view of several bivariate relationships.

MEASURES OF ASSOCIATION. Statistics that summarize the relationship between two variables.

ODDS RATIO. A measure of association between two dichotomous variables in which the odds of having a trait or property in one group are divided by the corresponding odds of the other group.

PROPORTIONAL-REDUCTION-IN-ERROR (PRE) MEASURE. A measure of association that indicates how much the knowledge of the value of the independent variable of a case improves prediction of the dependent variable compared to the prediction of the dependent variable based on no knowledge of the case's value on the independent variable. Examples are Goodman and Kruskal's lambda, Goodman and Kruskal's gamma, eta-squared, and R-squared.

R-SQUARED. A statistic, sometimes called the coefficient of determination, defined as the explained sum of squares divided by the total sum of squares. It purportedly measures the strength of the relationship between a dependent variable and one or more independent variables.

REGRESSION ANALYSIS. A technique for measuring the relationship between two interval- or ratio-level variables.

REGRESSION COEFFICIENT. A statistic that indicates the strength and direction of the relationship between two quantitative variables.

RESIDUAL. For a given observation, the difference between its observed value on a variable and its corresponding predicted value based on some model.

SCATTERPLOT. A plot of Y-X values on a graph consisting of an x-axis drawn perpendicular to a y-axis. The axes represent the values of the variables. Usually the x-axis represents the independent variable if there is one and is drawn horizontally; the y-axis is vertical. Observed pairs of Y-X values are plotted on this coordinate system.

SOMER'S D. A measure of association between ordinal-level variables.

STANDARDIZED REGRESSION COEFFICIENT. A regression coefficient based on two standardized variables. A standardized variable is derived from a transformed raw variable and has a mean of 0 and standard deviation of 1.

STATISTICAL INDEPENDENCE. A property of two variables in which the probability that an observation is in a particular category of one variable *and* a particular category of the other variable equals the simple or marginal probability of being in those categories.

STRENGTH OF A RELATIONSHIP. An indication of how consistently the values of a dependent variable are associated with the values of an independent variable.

TOTAL VARIATION. A quantitative measure of the variation in a variable, determined by summing the squared deviation of each observation from the mean.

UNEXPLAINED VARIATION. That portion of the total variation in a dependent variable that is *not* accounted for by the variation in the independent variable(s).

Suggested Readings

Achen, Christopher. *Interpreting and Using Regression,* Sage University Paper Series on Quantitative Applications in the Social Sciences, series no. 29. Beverly Hills, Calif.: Sage Publications, Inc., 1982.

Agresti, Alan. *An Introduction to Categorical Data Analysis.* New York: Wiley, 1996.

Agresti, Alan, and Barbara Finlay. *Statistical Methods for the Social Sciences,* 3d ed. Upper Saddle River, N.J.: Prentice-Hall, 1997.

Lewis-Beck, Michael S., ed. *Basic Statistics,* Vol. 1. Newbury Park, Calif.: Sage Publications, 1993.

Pollock, Philip H. III. *The Essentials of Political Analysis,* 2d ed. Washington, D.C.: CQ Press, 2005.

Ryan, Thomas P. *Modern Regression Methods.* New York: Wiley, 1997.

Velleman, Paul, and David Hoaglin. *The ABC's of EDA: Applications, Basics, and Computing of Exploratory Data Analysis.* Duxbury, Mass.: Duxbury Press, 1981.

CHAPTER 13

Multivariate Analysis

Probably one of the most vexing problems facing both political scholars and practitioners is establishing causal relationships. It is no easy matter to make a causal inference such as "the presence of a death penalty deters crime in a state" or "increasing literacy in a nation will reduce its level of domestic violence." Both assertions contain causal claims in the words *deters* and *will reduce*. How does one prove arguments like these, which abound in the discourse of politics?

We suggested in Chapter 5 that the controlled randomized experiment offers a logically sound procedure for establishing causal linkages between variables. The difficulty with experimentation, however, is that it is not often practical or ethical. Researchers then have to rely on nonexperimental methods. One general approach, which is the subject of this chapter, is **multivariate analysis.** Generally speaking, multivariate analysis investigates the interrelationships of more than two variables. The technique uses a wide variety of statistics to find and measure associations between variables when additional sets of factors have been introduced. It is especially appropriate for causal analysis, although this is by no means its only purpose.

The idea is that if one can identify a connection between, say, X and Y that persists even after other variables (for example, W and Z) have been held constant, then there may be a basis for making a causal inference. We know by now that simply because a factor exhibits a strong relationship with a dependent variable, it does not follow that the former caused the latter. Both the independent and dependent variables might be caused by a third variable, which could create the appearance of a relationship between the first two and lead to an erroneous conclusion about the effects of the independent variable. The possibility that a third variable is the real cause of both the independent and dependent variables must be considered before one makes causal claims. Only by eliminating this possibility can a researcher achieve some confidence that a relationship between an independent and a dependent variable is a causal one. Figure 13-1 illustrates the problem of distinguishing possible causal explanations.

The case studies presented in Chapter 1 illustrate the problem of making causal inferences. Recall that Ansolabehere, Iyengar, and Simon wondered

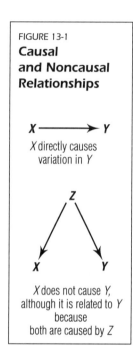

FIGURE 13-1
Causal and Noncausal Relationships

$X \longrightarrow Y$

X directly causes variation in *Y*

X does not cause *Y*, although it is related to *Y* because both are caused by *Z*

what effect negative political advertising (so-called attack ads) would have on political participation.[1] They conducted a laboratory experiment that offered convincing evidence about the deleterious or damaging (to voter turnout) effects of exposure to hostile political commercials. But their research has been questioned on several grounds, one of which is that the results cannot be generalized to broader, more "realistic" populations. Whether this criticism is fair or not, it raises the question: Could the causal effects be demonstrated in another way? Multivariate analysis provides an answer.

We also introduced in Chapter 11 cross-national data that could be used to explore the prerequisites of democracy. Until now, however, we conducted just preliminary analyses. Now we provide tools you can use to develop and test full-fledged models. We saw, in Chapter 12, that one indicator of democracy was related to gross national product (GNP) per capita. Before concluding that democracy flourishes best in wealthy countries, however, it might be advisable to explore the simultaneous effects of other factors. Or, it is possible that GNP simply acts as a proxy for other variables.

This chapter explains how political scientists use multivariate techniques to *control for* the effects of a third variable. This means that the impact of other variables is removed or taken into account when measuring the strength and direction of the relationship between an independent and a dependent variable. Generally, the impact of a third variable may be controlled either experimentally or statistically. **Experimental control** is introduced by randomly assigning the subjects in a study to experimental and control groups to control for other factors and then limiting exposure to the experimental stimulus to just the experimental group. **Statistical control,** the procedure used more frequently by political scientists, involves measuring each observation's values on the control variables and using these measures to make comparisons between observations.

Multivariate Analysis of Categorical Data

Suppose that we have hypothesized a relationship between attitudes toward government spending and presidential voting. Our hypothesis is that "the more a person favors a decrease in government spending, the more likely he or she is to vote Republican." Table 13-1 seems to confirm the hypothesis, since 64 percent of those who favor decreased spending voted Republican, whereas only 46 percent of those who favored keeping spending the same or increasing it voted Republican. This difference of 18 percentage points among a sample of 1,000 suggests that a relationship exists between attitudes toward government spending and presidential voting behavior.

At this point, you might ask, "Is there a causal relationship between opinion and vote [see the upper arrow diagram in Figure 13-1] or is there another factor, such as wealth, that creates the apparent relationship?" Or, even if you are not interested in causality, the question arises, "Can the explanation of presidential voting be increased by including another variable?" After all, 36 percent of those who favored decreased spending voted contrary to the hypothesis, as did 54 percent of those in favor of maintaining or increasing spending levels. Perhaps it would be possible to provide an explanation for those voters' behavior and hence improve the understanding of presidential voting behavior.

A second independent variable that might affect presidential voting is income. People with higher incomes might favor decreased government spending because they feel they gain little from most government programs. Those with higher incomes might also be more likely to vote Republican because they perceive the GOP to be the party that favors government policies that benefit the affluent. By the same token, people having lower incomes might feel both that increased government spending would help them *and* that Democrats generally support their interests. Therefore, income might influence both attitudes toward government spending and presidential voting and thus could create the appearance of a relationship between the two.

To consider the effect of income we need to bring it explicitly into the analysis and observe the resulting relationship between attitudes and voting.

TABLE 13-1

Relationship between Attitudes toward Government Spending and Presidential Vote

Dependent Variable: Presidential Vote	Independent Variable: Attitudes toward Government Spending		(N)
	Decrease Spending	Keep Spending the Same or Increase It	
Republican	64%	46%	(555)
Democratic	36	54	(445)
Total	100	100	
(N)	(550)	(450)	(1,000)

Note: Hypothetical data.

505

TABLE 13-2

Spurious Relationship between Attitudes and Presidential Voting When Income Is Controlled

Control Variable: Income; Dependent Variable: Presidential Vote	Independent Variable: Attitudes toward Government Spending		
	Decrease Spending	Keep Spending the Same or Increase It	(*N*)
High income			
Republican	80%	80%	(240)
Democratic	20	20	(60)
Total	100	100	
(*N*)	(250)	(50)	(300)
Medium income			
Republican	60%	60%	(210)
Democratic	40	40	(140)
Total	100	100	
(*N*)	(200)	(150)	(350)
Low income			
Republican	30%	30%	(105)
Democratic	70	70	(245)
Total	100	100	
(*N*)	(100)	(250)	(350)

Note: Hypothetical data.

In a **multivariate cross-tabulation,** we control for a third variable by holding it constant. In effect, we **control by grouping,** that is, we group the observations according to their values on the third variable and then observe the original relationship within each of these groups. In our example, each group consists of people with more or less the same income. If a relationship between opinions on spending and voting in these groups remains, it cannot be due to income.

Table 13-2 shows what might happen were we to control for income by grouping respondents into three income levels: high, medium, and low. Notice that it contains three contingency tables: one for each category of income, the control variable. Within each of the categories of income there is now *no* relationship between spending attitudes and presidential voting. Regardless of their attitudes on spending, 80 percent of respondents with high incomes voted Republican, 60 percent with medium incomes voted Republican, and 30 percent with low incomes voted Republican. Once the variation in income was removed by grouping those with similar incomes, the attitude-vote relationship disappeared. Consequently, income is a possible alternative explanation for the variation in presidential voting.

The original relationship, then, was spurious. Remember that a spurious relationship is one in which the association between two variables is caused by a third. In this case, attitudes cannot be a direct cause of presidential voting, because they had no effect on voting once income had been taken into account. Respondents did not vote the way they did because of their different attitudes toward government spending.

Note, however, that these remarks do not mean that there is *no* relationship between spending attitudes and presidential voting, for there is such a relationship, as Table 13-1 shows. But this original relationship occurred only because of the variables' relationships with a third factor, income. This relationship also means that the original relationship was not a causal one; spending attitudes cannot possibly be a cause of presidential voting because within income groups they make no difference whatever. The only reason for the relationship between spending attitudes and presidential voting is the effect of income on both variables. (See the lower arrow diagram in Figure 13-1.)

Because we have been using hypothetical data, we can easily illustrate other outcomes. Suppose, for instance, the control variable had absolutely no effect on the relationship between attitudes and vote. The result might look like the outcomes in Table 13-3. We now see that the strength and direction of the relationship between attitudes and voting is the same at all levels of income. In this particular situation members of the upper-income group behave just like those in the lower levels. Given these data we might be tempted to support the argument that attitudes toward government spending are causally related to candidate choice. But of course a critic could always say, "But you didn't control for Z." That would be a valid statement provided the skeptic was willing to provide a plausible reason why Z would have an effect on the original relationship. A randomized controlled experiment, by contrast, theoretically eliminates all alternative explanatory variables at one fell swoop.

In any event, these hypothetical data illustrate ideal situations. Consider, then, an actual multivariate cross-tabulation analysis. In Chapter 1 we presented the "Burnham problem," or the argument that voter turnout in American national elections is decreasing much more among the lower classes than among higher-status individuals. Although political scientists disagree about this supposition, which is often called "selective class demobilization," nearly everyone concurs that turnout is related to social and economic factors. Let us investigate the simple hypothesis that occupational status, one possible indicator of class, is related to voting in the way the literature suggests, namely, the higher the status, the greater the propensity to vote.

The data come from the 2000 American National Election Study, part of an ongoing effort to study the characteristics and behavior of the electorate.[2] The questionnaire asked respondents if they voted and, if not, why not. To simplify matters we simply lump people into "yes" and "no" categories. Sorting people

TABLE 13-3

Relationship between Attitudes and Presidential Voting after Income Is Controlled

Control Variable: Income	Independent Variable: Attitudes toward Government Spending		
Dependent Variable: Presidential Vote	Decrease Spending	Keep Spending the Same or Increase It	(N)
High income			
Republican	64%	46%	(183)
Democratic	36	54	(117)
Total	100	100	
(N)	(250)	(50)	(300)
Medium income			
Republican	64%	46%	(197)
Democratic	36	54	(153)
Total	100	100	
(N)	(200)	(150)	(350)
Low income			
Republican	64%	46%	(179)
Democratic	36	54	(171)
Total	100	100	
(N)	(100)	(250)	(350)

Note: Hypothetical data.

by occupation is more challenging, however. The original study recorded "actual" occupations and assigned them codes according to a Census Bureau classification. It then recoded these detailed classifications into a dozen or so broader groups. For our analysis we further reduce these categories by assigning them to one of four statuses: high, medium, skilled, and low.[3] This scheme permits us to test the hypothesis about voter turnout and socioeconomic status.

Table 13-4 reveals a strong relationship between occupation and voting. The results support the proposition. First, note the row marginal totals. Overall about 75 percent of the respondents claimed to have voted. Official election statistics tell us that this proportion is way too high; in fact slightly more than half of the eligible electorate participated in 2000. These figures illustrate the overreporting problem discussed in Chapter 1. Nevertheless, if we assume that overreporting occurs fairly consistently across occupations—admittedly a questionable assumption—we can still observe patterns in turnout among the status categories. And Table 13-4 shows that voting and nonvoting follow the predictions: there is almost a 28 percentage point difference in turnout between the high- and low-status groups. In fact, reading

TABLE 13-4
Cross-tabulation of Turnout by Occupation

Did Respondent Vote in 2000?	Low Status	Skilled Status	Medium Status	High Status	Totals
No, for whatever reason	42.1%	31.3%	27.1%	14.4%	25.8%
Yes, voted in 2000	57.9	68.7	72.9	85.6	74.2
Marginal column totals	100.0%	100.0%	100.0%	100.0%	100.0%
	(178)	(297)	(424)	(451)	(1350)

Source: 2000 National Election Study (ICPSR 3131).

Note: Only non-farm workers included; Chi-square = 60.5 with 3 degrees of freedom (significant at .001 level); tau b = .19; gamma = .35.

across the first row of numbers you can see that the proportion of those not voting drops steadily. This trend suggests a positive correlation, which is bolstered further by the two measures of association, tau b and gamma, that we report at the bottom of the table. Finally, the large chi-square statistic (χ^2 = 60.5) suggests that this relationship did not arise by chance or sampling error but probably reflects an association in the total population.

So far, so good. But in a sense the data are not especially informative. Or, more specifically, they do not tell us *why* the variables are connected. What is it about one occupation rather than another that induces its members to vote more regularly? Is it workplace experiences? Interworker relationships? Is this, in fact, a causal link or is it explained by some third factor? A likely explanation is that education accounts for the original relationship. That is, if we control for or hold constant the amount of schooling people have, we may find that occupational status has little independent effect on their political behavior. Or, and this possibility may be even more probable, we may find that with *some* levels of education there is little or no connection between occupation and voting, while with other levels such a relationship exists. Looking at a problem this way has been called **explication,** or the specification of the conditions under which X and Y are and are not related.

Table 13-5 provides some answers. It may look daunting at first. But just think of it as four cross-tabulations, or contingency tables, one for each of the educational groups. Thus section a of Table 13-5 shows the relationship between turnout and occupational status for just those respondents with *less than* a high school diploma. In this subtable there is a slight association between the variables. The connection, however, is much weaker than in Table 13-4. A quick way to see this is to compare the chi-square, or measures of association, in the two tables. The chi-square in the full table (that is, Table 13-4), for example, is 60.5 (with 3 degrees of freedom), which indicates a strong departure from statistical independence. The corresponding chi-square for those with the least education is 11.2, again with 3 degrees of freedom; the

TABLE 13-5

Cross-tabulation of Turnout by Occupation, Controlling for Education

Did Respondent Vote in 2000	Low Status	Skilled Status	Medium Status	High Status	Marginal Row Totals
a. Education: Less than High School[a]					
No, for whatever reason	68.5%	46.3%	50.0%	23.1%	52.4%
Yes, voted in 2000	31.5	53.7	50.0	76.9	47.6
Marginal column totals	100.0%	100.0%	100.0%	100.0%	100.0%
	(54)	(67)	(32)	(13)	(166)
b. Education: High School Graduates[b]					
No, for whatever reason	31.2%	32.1%	35.3%	20.0%	31.9%
Yes, voted in 2000	68.8	67.9	64.7	80.0	68.1
Marginal column totals	100.0%	100.0%	100.0%	100.0%	100.0%
	(80)	(134)	(170)	(45)	(429)
c. Education: High School Graduate Plus Non-College Training[c]					
No, for whatever reason	32.0%	25.0%	26.3%	20.0%	24.5%
Yes, voted in 2000	68.0	75.0	73.7	80.0	75.5
Marginal column totals	100.0%	100.0%	100.0%	100.0%	100.0%
	(25)	(64)	(99)	(90)	(278)
d. Education: College and/or Advanced Degree[d]					
No, for whatever reasons	27.8%	10.0%	9.8%	11.3%	11.4%
Yes, voted in 2000	72.2	90.0	90.2	88.7	88.6
Marginal column totals	100.0%	100.0%	100.0%	100.0%	100.0%
	(18)	(30)	(122)	(302)	(472)

[a] Chi-square = 11.2 with 3 degrees of freedom (significant at .01 level); tau b = .21; gamma = .35.
[b] Chi-square = 3.85 with 3 degrees of freedom (not significant at .05 level); tau b = .02; gamma = .03.
[c] Chi-square = 1.92 with 3 degrees of freedom (not significant at .05 level); tau b = .06; gamma = .13.
[d] Chi-square = 5.12 with 3 degrees of freedom (not significant at .05 level); tau b = .02; gamma = .06.

number of degrees of freedom in a table is *always* $(R-1)(C-1)$, where R and C are the numbers of rows and columns, respectively). This smaller value may be partly due to the number of cases in each table, but mostly it points to a much weaker relationship.

Next, examine the section b of Table 13-5, for high school graduates. If you compare, say, the first row of this table with the first row of Table 13-4, you will see that among the high school graduates there is virtually no connection between turnout and occupational status. Comparison of the chi-square statistics and measures of association provides further evidence of what happens when we control for education. Gamma in Table 13-4 is .35 but here it is only .03. These results suggest that controlling for education greatly reduces the original relationship. Examination of the other two parts of Table 13-5 reinforces that conclusion. Look, for example, at the tau b measure of association in Table 13-4 (.19) and in the four parts of Table 13-5 (.21, .02, .06, and .02). Only the tau b in section a of Table 13-5 matches the original tau b; the values in the other sections are about zero. The other summary statistic (gamma) and the test of independence follow the same pattern. After a little reflection, we might regard these results as sensible. Education is probably related to *both* occupation and voter turnout. People with college degrees tend to have high-paying, prestigious jobs. At the same time, they have the cognitive skills and free time to follow and participate in elections. Conversely, those with less education go into less-well-regarded jobs and at the same time perhaps have less time and inclination for politics. So of course there is an interrelationship among these variables.

What does all this mean for a practical understanding of voter turnout? We can say in the first instance that there is a link between class (as measured here) and voting. And this finding agrees with what previous research has found. But it is equally important to note that education may partly explain or specify this association. For when we control for education, the original association between voter turnout and occupation largely disappears. The disappearance of a relationship often indicates spuriousness. Finally, section a of Table 13-5 points to the possibility that job status among those with less than a high school diploma is for some reason connected with voting behavior because those with the most prestigious occupations are more apt to vote than others.[4]

These ideas can be extended to more complicated situations. In complex contingency tables containing more than one control factor and having variables with more than two or three response categories, it is often difficult to discern changes in the magnitude and nature of the original relationships. Moreover, even in a relatively simple situation in which a researcher controls for a third variable, five general results can occur. First, the original relationship may disappear entirely, indicating that it was spurious. This was the case with the hypothetical government-spending attitudes and presidential-voting example. Second, the original relationship may decline somewhat but not disappear completely, indicating that it was partly spurious. A **partly spurious relationship** is one in which some but not all of the original relationship may be accounted for by the control variable. Third, the original relationship may remain unchanged, indicating that the third variable was not responsible for

TABLE 13-6
Five Results When Controlling for a Third Variable

Uncontrolled Relationship	Measure of Association	Significance Level
a.	+.35	.001
Category 1	+.05	.07
Category 2	−.03	.14
Category 3	+.01	.10
b.	+.42	.001
Category 1	+.20	.01
Category 2	+.17	.05
Category 3	+.26	.001
c.	−.28	.01
Category 1	−.30	.05
Category 2	−.35	.01
Category 3	−.27	.05
d.	−.12	.08
Category 1	−.42	.001
Category 2	−.30	.01
Category 3	−.35	.001
e.	+.31	.001
Category 1	+.55	.001
Category 2	+.37	.01
Category 3	+.16	.05

Note: Hypothetical data.

the original relationship (see Table 13-3). Fourth, the original relationship may increase, indicating that the control variable was disguising or deflating the true relationship between the independent and dependent variables. Fifth, the controlled relationship may be different for different categories of the control variable. This is called a **specified relationship.** It indicates that the relationship between two variables is dependent on the values of a third. For example, the relationship between attitudes and voting may be strong only among medium-income people. If the relationship between two variables differs significantly for different categories of a third variable, then we say that the third variable has specified the relationship, or that an "interaction" is present.

As we saw in the previous example, it is frequently helpful to use a measure of association in addition to percentages to see the effects of a control variable. Table 13-6 presents a summary of the possible results when a third variable is held constant. The top row of each section shows a hypothetical measure of association for the uncontrolled relationship between the independent and dependent variables. The next three lines display the same information for each *category* of the control variable. Comparing the uncontrolled results with the controlled results indicates whether the relationship is spurious (a), partially spurious (b), unchanged (c), has increased (d), or has been specified (e). Note, for example, that in section b the measure of association decreases in each level of the control variable but does not approach zero. This pattern suggests that the control factor partly but not totally explains the original relationship.

Multivariate cross-tabulation analysis, then, may be used to assess the effect of one or more control variables when those variables are measured at the nominal or ordinal level. Although this is a relatively straightforward and frequently used procedure, it has several disadvantages.

First, it is difficult to interpret the numerous cross-tabulations required when a researcher wishes to control for a large number of variables at once and all of them have a large number of categories. Suppose, for example, that you wanted to observe the relationship between television news viewing and political knowledge while controlling for education (five categories), newspaper exposure (four categories), and political interest (three categories). You would need to construct sixty different groups and look at the relationship in each one!

Second, controlling by grouping similar cases together can rapidly deplete the sample size for each of the control situations, producing less accurate and statistically insignificant estimates of the relationships. Suppose that we had started out with a standard sample size of 1,500 respondents in our example of the relationship between television news exposure and political knowledge. By the time we had divided the sample into the sixty discrete control groups, each subtable measuring the relationship between news exposure and political knowledge would have, on average, only 25 people in each. In practice, many would have many fewer cases and some might not have any at all. All these potentially sparse tables would make it virtually impossible to observe a significant relationship between news exposure and political knowledge.

A third problem is that control groups in multivariate cross-tabulation analysis often disregard some of the variation in one or more of the variables. For example, to control for income in our government-spending attitude/presidential-voting example, we put all those with low incomes into one group. This grouping ignored what might be important variations in actual income levels.

For these reasons, social scientists are largely moving away from the analysis of multivariate cross-tabulations with percentages and measures of association. A variety of sophisticated and powerful techniques have been developed to describe complex contingency tables with "parsimonious" models.[5] There are two general approaches. Many sociologists, biometricians, demographers, and economists have developed methods designed explicitly to tease out of cross-tabulations as much information as possible. Political scientists along with other social scientists prefer to generalize and adapt regression methods to include categorical data. Unfortunately, the mathematics required in the new procedures is beyond the scope of this book. (We do, however, look at the analysis of dichotomous dependent variables in a later section.) In any event, the goal of these more complicated procedures is partly the same as that discussed here: to analyze the effects of two or more independent variables on a dependent variable.

We conclude this section by comparing the analysis of cross-tabulations with the randomized experimental design as discussed in Chapter 5. The goal of the latter is to see if one factor causes another. By randomly assigning individuals to treatment (experimental) and control groups, the investigator (in theory at least) can scrutinize a relationship between X and Y uncontaminated by other variables such as Z. In most research settings, however, randomization is simply not possible. Given a hypothesis about voter turnout and social class, for instance, how can a researcher randomly place someone in a particular occupation and then wait to see what effect this placement has on the person's behavior? Therefore, instead of using randomization to get rid of potentially contaminating variables, it is necessary to try to control for them manually. That is, the investigator has to explicitly identify variables (for example, Z) that might be influencing the X-Y relationship, measure them, and then statistically control for them just as we did in Table 13-5. In that case we looked at the association between the variables *within* levels of the third factor. This approach is possible if the control factor is categorical and the total number of cases is large. Other techniques are needed for different circumstances. In the next section we discuss the cases of one continuous dependent variable and two or more categorical test factors.

Multiple Regression

So far we have described how to control for a third variable by grouping cases that have identical or similar values on the variable. Although this procedure works when the values of the control variable are discrete, control by grouping causes major problems, as we discussed earlier, because of the potential proliferation of control groups when analyzing several variables and the reduction in the sample size within each control group.

This problem is not a concern with quantitative variables. When all the variables are measured at the interval or ratio level (and even sometimes when they are not), we can use **multiple regression analysis** to control for other variables. Recall from Chapter 12 that two-variable, or bivariate, regression analysis involves finding an equation that best fits or approximates the data and thus describes the nature of the relationship between the independent and dependent variables. Recall in addition that a regression coefficient, which lies at the core of the regression equation, tells how much the dependent variable, *Y,* changes for a one-unit change in the independent variable, *X.* Regression analysis also allows us to test various statistical hypotheses such as β = 0, which means that, if true, there is no linear relationship between an independent and a dependent variable. A regression equation moreover may be used to calculate the predicted value of *Y* for any given value of *X.* And the residuals or distances between the predicted and observed values of *Y* lead to a measure (R^2) of how well the equation fits the data.

As the name implies, multiple regression simply extends these procedures to include more than one independent variable. The main difference—and what needs to be stressed—is that a **multiple regression coefficient** indicates how much a one-unit change in an independent variable changes the dependent variable *when all other variables in the model have been held constant or controlled.* The controlling is done by mathematical manipulation, not by literally grouping subjects together. **Control by adjustment** is a form of statistical control in which a mathematical adjustment is made to assess the impact of a third variable. That is, the values of each case are adjusted to take into account the effect of all the other variables rather than by grouping similar cases.

The general form of a linear multiple regression equation is

$$Y = \alpha + \beta_1 X_1 + \beta_2 X_2 + ... + \beta_K X_K + \epsilon.$$

Let's examine this equation to make sure its terms are understood. In general, it says that the values of a dependent variable, *Y,* are a *linear* function of (or perhaps caused by) the values of a set of independent variables. The function is linear because the effects of the variables are additive. How the independent variables influence *Y* depends on the numerical values of the βs.

As in previous chapters, parameters are denoted by lowercase Greek letters. The first, alpha (α), is a **regression constant.**[6] It can be interpreted in many ways, the simplest being α is the value of *Y* when all the independent variables have scores or values of zero. (Just substitute 0 for each *X* and note

THE MEANING OF PARTIAL REGRESSION COEFFICIENTS
In the interest of simplicity, the text uses a sparse system of notation. The Greek lowercase letter β with a numerical subscript stands for a particular regression coefficient. Hence β_1 is the coefficient for the first *X.* This practice is fine as long as you realize that these are *partial* regression coefficients. Their magnitude depends on the dependent variable and any other independent variables in the equation. To be precise, our notation should allow you to see immediately what particular variables are involved. For instance, the symbol $\beta_{YX_1X_2X_3}$ is the (partial) regression coefficient between *Y* and X_1 when X_2 and X_3 have been held constant. It is not the same as β_{YX_1} for the regression of *Y* on *X* alone. The context should make clear what is a partial regression coefficient and what is a total regression coefficient. But be on the lookout for this notation in other texts and scholarly presentations.

that all the terms except the first drop out, leaving $Y = \alpha + \varepsilon$.) As in simple regression, the βs indicate how much Y changes for a one-unit change in the independent variables when the other variables are held constant. Each β is called a **partial regression coefficient** because it indicates the relationship between a particular X and the Y after all the other independent variables have been "partialed out" or simultaneously controlled. The presence of ε (epsilon), which stands for error, means that Y is not a perfect function of the Xs. In other words, even if we knew the values of X, we could not completely or fully predict Y; there will be errors. (In the symbols used in Chapter 12, we denoted this idea by $Y - \hat{Y}$.) But regression proceeds on the assumption that the errors are random or cancel out and that their average value is zero. Hence we can rewrite the regression equation as

$$E(Y) = \alpha + \beta_1 X_1 + \beta_2 X_2 + ... + \beta_K X_K.$$

Read this last equation as "the expected value of Y is a linear (or additive) function of the Xs."

Finally, predicted values of Y (denoted $\hat{Y}$) may be calculated by substituting any values of the various independent variables into the equation.

Interpretation of Parameters

So important are the partial regression coefficients that we should examine their meaning carefully in the context of a specific example. Let's consider Ansolabehere and colleagues' study of the effects of campaign advertising on the decision to vote. Recall from Chapter 1 that the investigators relied primarily on an experimental design to assess the causal impact of negative television commercials on voter turnout. After finding a strong, presumably causal relationship, they wanted to increase the external validity of the findings further by replicating the study in a broader context. To do so they collected data on thirty-four Senate elections in 1992. The main dependent variable was voter turnout measured as the number of votes cast for senator divided by the state's voting-age population. Naturally they could not randomly assign states to different types of campaigns. Instead, they read numerous newspapers and magazine reports about each of the Senate contests and classified the "tone" of each campaign into "positive" (scored 1), "mixed" (scored 0), or "negative" (scored –1).[7] Tone is the main explanatory or independent variable. The basic idea is that negative campaigning decreases interest in and satisfaction with politics and hence discourages participation. So they expected a negative correlation between turnout and campaign tone: the more negative the tone, the lower the turnout.

Had they just analyzed the two variables, however, any claims of causality would have been highly suspect. By now we know why. To demonstrate a

TABLE 13-7

Multiple Regression Coefficients, Standard Errors, and *t* Statistics for Senate Turnout Data

Independent Variable	Estimated Coefficient	Standard Error	*t* statistic
Constant	−.295	.124	2.38
Campaign tone[a]	.021	.006	3.50
Turnout in 1988	.571	.090	6.34
Mail-back rate	.340	.125	2.72
South/non-South[b]	.047	.013	3.62
College education (%)	.172	.076	2.26
Incumbent's campaign spending[c]	.011	.006	1.83
Closeness of race	−.068	.039	1.74

Source: Based on Stephen D. Ansolabehere, Shanto Iyengar, Adam Simon, and Nicholas Valentino, "Does Attack Advertising Demobilize the Electorate," *American Political Science Review* 88 (December 1994): 829–838, Table 2.

Note: $N = 34$; $R^2 = .94$.

[a] Campaign tone: 1 = positive tone, 0 = mixed tone, −1 = negative tone.
[b] South/non-South: 0 = southern state, 1 = non-southern state.
[c] Logarithm of incumbent candidates campaign expenditures.

causal connection between X and Y, you have to eliminate possible alternative explanations for any observed relationship. The experiment accomplishes this by randomizing cases into treatment and control groups. Nonexperimental research requires that potential confounding factors be identified, measured, and *statistically* held constant. Consequently, the Ansolabehere team collected data on several additional variables that it believed might be relevant to turnout.

A portion of their results appears in Table 13-7. This arrangement represents a standard format for presenting multiple regression results. The first column lists the variables; the next two give the estimated partial regression coefficients and the standard errors; and the third supplies a corresponding observed t statistic. The bottom row provides the multiple R-squared and other information. Consider each component in turn. Note that because the authors' dependent variable is measured as a proportion, as were presumably some of the independent variables, we will have to multiply by 100 to convert the results to percentages.

Look first at the coefficient for tone, the principal independent variable, "Campaign tone." It indicates the direction and strength of the relationship between mood of the campaign (negative or positive) and voter turnout, when all the other factors in the equation have been held constant. Here is one way to understand the estimated value of .021: If somehow we could shift a Senate race one unit up the tone index—that is, make it one unit more positive (or less negative)—then participation would increase by about 2 percent. It is important to note that this increase would occur if the other factors in the model did not change. Since the investigators used a rough classification system to place the campaigns in a tone category, it is not intuitively or substantively clear what a one-unit change means. But we know that negative campaigns were assigned a score of –1, whereas neutral ones were coded 0. Thus, a one-unit increase represents quite an alteration in campaign style and content, and that amount of change would "cause" a predicted 2 percent increase in voter turnout. That may not be a lot. But many factors besides tone affect voting, so perhaps getting candidates to be more positive and less hostile toward one another would increase electoral participation a bit.

Other coefficients can be interpreted the same way. Look, for instance, at "Turnout in 1988." The researchers believed a state's civic climate would affect voter turnout beyond the effects of other factors. So they tried to adjust for level of "civic duty" and used 1988 voter turnout as a surrogate indicator. (Presumably states with civic-minded cultures will exhibit higher voting rates than those with less political interest and involvement.) The estimated partial coefficient is .571. One way to interpret this number is to set all the other independent variables to zero. Then the predicted value for the dependent variable is given by

$$\hat{Y}_i = -.295 + .571(\text{Turnout in 1988}),$$

where $\hat{Y}_i$ is the predicted value of turnout when all the independent variables except 1988 turnout have been set to zero. (The regression constant –.295 indicates a "negative" turnout rate when all variables including turnout in 1988 have been set to zero. This seemingly nonsensical result raises two points. First, as we indicated in Chapter 12, the regression constant does not always have a meaningful substantive interpretation. How can voter turnout be less than zero? Second, linear multiple regression has to be interpreted cautiously when the dependent variable is constrained or must lie within a specified range of values such as 0 to 1. The reason is that predicted values may fall outside the allowable range, which can cloud interpretations.) In any event, when the other independent variables are set to zero, the predicted turnout proportion in the 1992 Senate race is –.295 when there was no turnout in 1988 and increases to .276 when the independent variable goes up one unit (that is, when turnout climbs from no one voting to 100 percent participation). Admittedly, this is a nonsensical situation because all the other variables, such as spending, have been set to zero. But it illustrates the basic idea. To see how this result is obtained, just write the equations with the actual values of X being considered:

$$\hat{Y}_i = -.295 + .571(0) = -.295$$
$$\hat{Y}_i = -.295 + .571(1) = .276.$$

In short, if $X_{turnout1988} = 1$ (that is, if participation in 1988 were 100 percent) instead of $X_{turnout1988} = 0$, then the predicted turnout in the Senate races would jump by .571 (or 57.1 percentage) points. This relatively large rise suggests that the participation level in one election will track or follow the rate in previous elections, and in this research context it is clearly appropriate to control for this fact.

By writing down the equations with values of the independent variables included, you can find similar predicted rates and substantive interpretations. Sometimes it helps to write out the entire estimated equation and substitute values for independent variables one at a time. It just takes time and a bit of hypothetical reasoning and common sense to figure out what the numbers mean. The "South/non-South" indicator is very instructive in this regard.

Dummy Variables

How can a variable like "region" be measured? The Ansolabehere study used a common method, dummy variable coding. A **dummy variable** is a hypothetical index that has just two values: 0 for the presence (or absence) of a factor and 1 for its absence (or presence). Dummy variables are frequently used to convert categorical data into a form suitable for numerical analysis.[8] In the current situation the investigators assigned southern states the number 0 and the remainder 1. Since 0 and 1 are perfectly legitimate numbers, they can be used in a quantitative analysis. More important, this type of code leads to an especially simple interpretation. Once again a good way to interpret parameters, even dummy variables, is to write the equation explicitly. This time let us write the equation with all the independent variables set to specific values. Then we can "move" a state from South to non-South to see what happens to voter turnout when everything else has been held constant. We will give the independent variables these values:

- Campaign tone is "positive" = 1

- Turnout in 1988 = .5 (that is, half of eligible electorate voted)

- Mail-back rate = .8 (that is, 80 percent of the citizens returned their Census Bureau questionnaire)

- Proportion of population with a college education = .4

- Incumbent's campaign expenditures = $100,000, which the authors convert to logarithm or 5.[9]

- Closeness of race = 0. This is as competitive an election as is possible because the variable used by Ansolabehere and his associates is the squared difference between the proportions voting Democratic and Republican. If that squared difference is 0, then presumably the parties split the (two-party) vote evenly. (Numbers greater than 0 mean less competitive races.)

■ Region = 0 if South; 1 otherwise (non-South)

For the South (that is, X_{Region} = 0), the estimated equation with these values leads to a predicted proportion (the coefficients appear in column two of Table 13-7):

$$\hat{Y}_i = -.295 + .021(1) + .571(.5) + .340(.8) + .172(.4) + .011(5) - .068(0) + .047(0) = .9023.$$

For the non-South (X_{Region} = 1) it is (look at the last term)

$$\hat{Y}_i = -.295 + .021(1) + .571(.5) + .340(.8) + .172(.4) + 0.11(5) - .068(0) + .047(1) = .9493.$$

Assuming everything else is held constant, we see that if we could shift a Senate election from the South to elsewhere, we would increase turnout by about 5 percent. We should stress that coding a categorical variable such as "region" with a dummy variable is perfectly reasonable from a statistical standpoint and leads to an intuitively appealing interpretation.

Moreover, if we had a categorical variable with more than two categories, we could create separate dummy variables for each category. Suppose, for instance, the states had been divided into four regions: North, South, Midwest, and West. The four dummy variables that represent this "meta" variable would be as follows:

■ X_{East} = 1 if East;
  0 otherwise (i.e., not eastern state)

■ X_{South} = 1 if South;
  0 otherwise

■ $X_{Midwest}$ = 1 if Midwest;
  0 otherwise

■ X_{West} = 1 if West;
  0 otherwise.

Take New York, for example. Its scores on the four variables would be 1, 0, 0, 0. Alabama would be coded 0, 1, 0, 0. California's codes would be 0, 0, 0, 1. Other states are assigned scores in the same way according to the definitions of the dummy variables. For mathematical reasons we have to drop one of the dummy variables when using them in multiple regression. The omitted variable, which really stands for a category, becomes a reference point. Suppose we wanted to estimate voter turnout based on region (as coded above), campaign tone, and turnout in 1988. The regression equation could be written as follows:

$$\hat{Y}_i = \alpha + \beta_{1t} X_{East} + \beta_2 X_{South} + \beta_3 X_{Midwest} + \beta_4 X_{\text{``Tone''}} + \beta_5 X_{turnout88}.$$

In this formulation the West is a reference category, and the partial regression coefficients β_1 through β_3 measure the expected change in Y if a state were to move from the West to another region. The other regression coefficients would be interpreted in the usual fashion.

Estimation and Calculation of a Regression Equation

Where do the numerical values of the regression coefficients come from? Just as in bivariate regression, the α and βs are calculated according to the principle of least squares: a mathematical procedure that selects the (unique) set of coefficients that minimizes the squared distances between each data point and its predicted Y value. Nearly every statistical software system includes a computer program for performing multiple regression analysis.

Standardized Regression Coefficients

As discussed in Chapter 12, a regression coefficient calculated from standardized variables is called a standardized regression coefficient, or, sometimes, a beta weight, and under certain, restricted circumstances might indicate the relative importance of each independent variable in explaining the variation in the dependent variable when controlling for all the other variables. Standardizing a variable, you may remember, means subtracting its mean from each individual value and dividing by the standard deviation. To obtain the standardized coefficient, you do this for all the variables, including Y, and then regress the standardized Y on the standardized Xs. It is the same procedure demonstrated in Chapter 12 except that now there are more than two variables. A standardized coefficient shows the partial effects of an X on Y in standard deviation units. The larger the absolute value, the greater the effect of a one-standard-deviation change in X on the mean of Y, controlling for or holding other variables constant. Most software offers the option of calculating unstandardized or standardized coefficients.

Table 13-8 presents a comparison of the regression using standardized and unstandardized variables. The data are a sample of twenty congressional districts. The goal of this example analysis is to see if President Clinton's 1996 vote in these districts was related to their level of urbanization, X_1, and concentration of African Americans, X_2. The hypothesis is that Democrats traditionally attract support from urban areas and from minority groups. Although it would be problematic to do so, we might also try to answer the question,

How It's Done

Calculating a Partial Standardized Regression Coefficient

$Y, X_1, X_2, \ldots X_k$ are the dependent and independent variables, respectively. Let $\bar{Y}$ be the mean of Y and $\bar{X}_k$ be the mean of the kth independent variable. Let $\hat{\sigma}_Y$ be the standard deviation of Y and $\hat{\sigma}_{Y_k}$ be the sample standard deviation of the kth independent variable.

Method 1:
Standardize all variables. For example for the ith case and kth X variable:

1. $\quad y_i = \dfrac{(Y_i - \bar{Y})}{\hat{\sigma}_Y} \text{ and } x_{ik} \dfrac{(X_{ik} - \bar{X}_k)}{\hat{\sigma}_{X_k}}.$

2. Regress Y on all Xs. The result is the partial standardized coefficient, .

$\hat{\beta}_{YX_k|X_1,\ldots,X_{k-1},X_{k+1},\ldots,X_k}.$

Method 2:

$\hat{\beta}^*_{YX_k|X_1,\ldots,X_{k-1},X_{k+1},\ldots,X_k} = \hat{\beta}_{YX_k|X_1,\ldots,X_{k-1},X_{k+1},\ldots,X_k} \dfrac{\hat{\sigma}_{X_k}}{\hat{\sigma}_Y},$

where $\hat{\beta}_{YX_k|X_1,\ldots,X_{k-1},X_{k+1},\ldots,X_k}.$ is the unstandardized partial regression coefficient.

Which is a "more important" explanation of Clinton's 1996 vote: urbanization or ethnicity?

Columns 2 through 4 of Table 13-8 give the raw scores, and columns 5 through 7 show the standardized versions of the variables. The standardized values are calculated with the method introduced in Chapter 12. For instance, to convert or transform Y, a raw score, to a standardized score, y, use the following formula:

$$y_i = \frac{(Y_i - \bar{Y})}{\hat{\sigma}_Y}.$$

where $\bar{Y}$ is the mean of Y and $\hat{\sigma}_Y$ is its standard deviation.

Each of the regression coefficients can be interpreted in the usual fashion, and we leave it to the reader to provide a substantive explanation of the data. We only note that in the second equation the measurement scales are deviation units, so, for instance, a one-unit increase in X_1 means a one-standard-deviation change in urbanization. This scale may not have much intuitive appeal, but using standardized scores allows us to assess the relative importance of the variables in explaining variation in voting. We see that the black

TABLE 13-8

Regression for Sample of Congressional Districts (Percentage of Vote for Clinton in 1996 Regressed on Percentage of Urban and Percentage Black)

Case Number	Percentage Clinton 1996	Percentage Urban	Percentage Black	Standard Score Clinton	Standard Score Urban	Standard Score Black
1	46	31.5	5.6	−0.357	−1.010	−0.390
2	47	94.8	1.3	−0.278	0.992	−0.656
3	71	100.0	3.5	1.608	1.156	−0.520
4	42	19.0	22.0	−0.671	−1.406	0.626
5	31	69.2	5.6	−1.536	0.182	−0.390
6	47	72.1	4.0	−0.278	0.274	−0.489
7	36	13.7	1.2	−1.143	−1.573	−0.662
8	47	15.8	3.9	−0.278	−1.507	−0.495
9	47	74.6	1.6	−0.278	0.353	−0.638
10	47	0.0	1.0	−0.278	−2.006	−0.675
11	81	99.2	71.0	2.394	1.131	3.660
12	28	64.1	1.1	−1.772	0.021	−0.669
13	51	23.6	3.3	0.036	−1.260	−0.532
14	48	78.1	4.8	−0.200	0.463	−0.440
15	84	100.0	42.7	2.630	1.156	1.907
16	86	100.0	62.3	2.787	1.156	3.121
17	59	95.5	15.8	0.665	1.014	0.242
18	58	5.9	17.9	0.586	−1.820	0.372
19	64	92.0	11.6	1.058	0.903	−0.018
20	39	52.3	2.7	−0.907	−0.352	−0.570

Source: Bureau of the Census data. These numbers are a small subset of data drawn from Kenneth Janda, "Statistics for Political Research, CD2000." Retrieved May 24, 2004, from www.janda.org/c10/data%20sets/menu.html.

Note: Regression equations:

Unstandardized variables: $\hat{Y} = 38.1 + 0.116X_{Urban} + 0.555X_{Black}$ $R^2 = .674$.

Standardized variables: $\hat{Y} = .228X_{Urban} + .705X_{Black}$ $R^2 = .674$.

coefficient (.705) is about three and a half times larger than the one for urban population (.228). We might conclude then that ethnicity is a more important explanation of Clinton's success than urban size. If so, Democrats might concentrate their mobilization efforts on that source of support.[10]

We note finally two points. First, transforming variables by standardization just changes their measurement scales. It does not alter their interrelationships. Therefore, tests of significance and measures of fit are the same for both sets of data. This is apparent from the two equal R^2 values (that is, R^2 = .674 in both instances). This will always be the case. And the regression constant drops out of the equation when standardized variables are used.

Many computer programs routinely report standardized regression coefficients, and they are commonly found in scholarly articles and books. The seeming comparability of the standardized coefficients tempts some scholars into thinking that the explanatory power of, say, X, can be compared with that of another independent variable, say, Z. It would be easy to conclude, for example, that if b_{YX} is larger in absolute value than b_{YZ}, the former might be a more important or powerful predictor of Y than the latter. (Remember we are talking about the standardized coefficients, which now presumably have the same measurement scale.) Yet you should be extremely careful about inferring significance from the numerical magnitudes of these coefficients. Such comparisons of the "strength of relationship" are possible only to the extent that *all the original independent variables have a common scale or unit of measurement.* The standardization process just changes the variables to standard deviation scales. It does not change or enhance their substantive interpretation. Also, standardization is affected by the variability in the sample, as can be seen by noting the presence of the standard deviations in the above formula. So if one independent variable exhibits quite of bit of variation while another has hardly any at all, it may be wrong to say the first is a more important explanation than the second, even if its standardized coefficient is larger.[11]

Measuring the Goodness of Fit

The overall success of a multiple regression equation in accounting for the variation in the dependent variable is partly indicated with the **multiple correlation coefficient,** often called the multiple R, or coefficient of determination, R^2. As we explained in Chapter 12, R^2 is the ratio of the explained variation in the dependent variable to the total variation in the dependent variable; hence, it equals the proportion of the variance in the dependent variable that may be explained by the set of independent variables:

$$R^2 = \frac{TSS - ResSS}{TSS} = \frac{RegSS}{TSS},$$

where

TSS = the **total** sum of squares;
$ResSS$ = the **residual** sum of squares; and
$RegSS$ = the **regression** sum of squares.

R^2 itself can vary from 0 to 1, and it has a corresponding significance test that indicates whether the entire regression equation permits a statistically significant explanation of the dependent variable. R^2 never decreases as independent variables are added. But just throwing more variables into a model will not help you understand Y. Each independent variable added must be carefully considered. The researchers studying negative campaigning did not report the sums of squares, but they noted that the R^2 for the Senate election data is .94 (see Table 13-7). This number might suggest that the set of independent variables explains a large portion of the variance in 1992 voter turnout.

Yet keep in mind that the model includes voter turnout in 1988, and, at the state level, participation in one election no doubt tracks or follows what happened in the previous contest. It is possible that most of the total explained variation is due to that one variable, 1988 turnout. Indeed, this explains why the researchers included 1988 turnout: they wanted to see what effect negative campaigning had *net of* or controlling for those other factors that influence voting. (Hence, we see how multiple regression provides an analogue to the controlled experiment. By statistically holding variables constant, we try to approximate the power of randomly assigning units to various treatments.)

Tests of Significance

Because population parameters such as regression coefficients are unknown, investigators use sampling and estimation methods to find statistically sound guesses for them. But the question always remains: do the observed results support various hypotheses? For example, the Ansolabehere research developed a model that supposedly shows how negative campaigns and other factors affect voter turnout. But since they are dealing with a sample,[12] we can ask if there are any such effects in the "population of Senate elections."

Most social scientists respond by making one or both of two tests. The first, which we do not explain in detail, assesses the overall model. In particular, the null hypothesis is

$$H_0: \alpha = \beta_1 = \beta_2 = \beta_3 = \ldots = \beta_K = 0.$$

That is, the test is of the hypothesis that all the coefficients (that is, α and the βs) equal 0. The rival or alternative hypothesis is that at least one of them is nonzero, but the particular ones are unspecified. This test, which usually comes first, is called an F test. It involves the F distribution that we came across earlier. Most computer regression programs automatically churn out the necessary statistics, and the results are fairly easy to evaluate. The interpretation follows the test procedures outlined in Chapter 12 and elsewhere: compare an observed F—one calculated from different sums of squares in the data—with a critical F, which is selected from a tabulated or computed F distribution.

The authors of the data in Table 13-7 do not report any F values or sums of squares, so let us say hypothetically that the observed F is 10. This value can be compared with a critical F, which in this particular instance might be 2.39 for the .05 level of significance and 3.42 for the .01 level. These critical values are found in Appendix D. To use that appendix, you have to know two degrees of freedom, one for the regression model (called "between groups" in the table) and one for the error or residual (called "within groups"). For multiple regression these degrees of freedom are K = number of independent variables and $N - K - 1$, respectively. Table 13-7 contains seven independent variables, so $K = 7$; there are $N = 34$ observations and thus the error degrees of freedom is $34 - 7 - 1 = 26$. Comparing the hypothetical observed F with the critical values, we see that it exceeds both and therefore we reject the null hypothesis at the .01 level. This result means that we have reason to believe that at least one of the regression coefficients is nonzero.

But which ones? The general practice is to compute t statistics for each coefficient and compare the observed t with a critical t based on $N - K - 1$ degrees of freedom.[13] The observed t values are calculated, as shown in Chapter 11, from the formula

$$t_{observed} = \frac{(\hat{\beta} - 0)}{\hat{\sigma}_{\hat{\beta}}} = \frac{(\hat{\beta})}{\hat{\sigma}_{\hat{\beta}}},$$

where $\hat{\beta}$ is the estimated coefficient and $\hat{\sigma}_{\hat{\beta}}$ is the estimated standard error or standard deviation of the regression coefficient. We use 0 in the numerator because in most published research the null hypothesis is that the population coefficient, β, is zero. This t can be checked against a critical value obtained from a table like the one in Appendix B. Most computer programs carry out the calculations and tests automatically, so the user simply has to interpret the results.

Refer back to Table 13-7. The last column gives the observed t for each coefficient. Because this analysis is based on $N = 34$ cases and there are 7 independent variables and 1 constant, the appropriate number of degrees of freedom is $34 - 7 - 1 = 26$. The corresponding critical t for a two-tailed test (for now we do not hypothesize that a β is positive or negative if it is not 0) at the .05 level of significance is 1.706. All the ts in Table 13-7 exceed this value. This finding can be interpreted as evidence that in the population of Senate elections, all these variables have an effect on voter turnout.

Logistic Regression

Suppose we want to explain why people in the United States do or do not vote. As we have suggested many times before, such a study should start from a theory or at least a tentative idea of political participation. We might suppose, for example, that demographic factors such as education and race are related to voter turnout: well-educated whites vote more frequently than do less-

TABLE 13-9
Raw Data

Respondent	Turnout	Race	Years of Education
1	1	1	11
2	1	0	16
3	1	1	16
4	1	1	15
5	1	1	17
6	1	1	11
7	1	1	12
8	0	1	12
9	1	1	18
10	1	1	18
11	0	1	18
12	1	1	15
13	NA	1	12
14	1	0	4
15	0	0	10

Source: James A. Davis, Tom W. Smith, and Peter V. Marsden, General Social Surveys, 1972–2002: Cumulative File.

Note: Data are part of a larger data set that includes 4,417 respondents interviewed in 1993 and 1994. Race coded 1 for "white" and 0 for "nonwhite"; turnout coded 1 for "voted in the 1992 presidential election" and 0 for "did not vote in the 1992 presidential election." NA indicates not available.

educated nonwhites. To test this proposition we could collect measures of education, race, and voting from a survey or poll.

Table 13-9 shows a very small portion of such data drawn from the General Social Surveys conducted by the Opinion Research Center at the University of Chicago. It contains indicators of voter turnout (coded 1 for "voted in the 1992 presidential election" and 0 for "did not vote in the 1992 presidential election"), race (coded 1 for "white" and 0 for "nonwhite"), and highest year of schooling completed. (In the example that follows, only the first 15 out of 4,417 respondents are shown, to save space.) The codes given to voting and race are admittedly arbitrary, but we will see that this scoring system has convenient properties.

One might wonder how we could use a method like multiple regression to analyze these data, since, strictly speaking, the dependent variable, voter turnout, is not numeric or quantitative. (Earlier we saw that categorical independent variables can be coded as dummy variables and entered into regression equations along with quantitative variables.) Indeed, a major problem for the social scientist is to explain variation in dependent variables of this type, which are often called dichotomies or binary responses. Consider, for instance, Figure 13-2, which shows the plot of turnout against number of

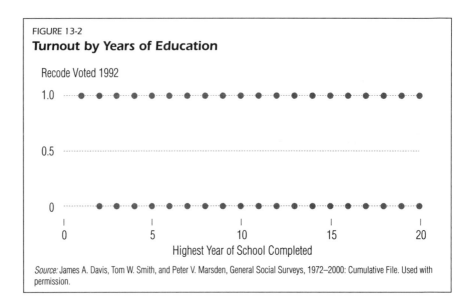

FIGURE 13-2

Turnout by Years of Education

Recode Voted 1992

Source: James A. Davis, Tom W. Smith, and Peter V. Marsden, General Social Surveys, 1972–2000: Cumulative File. Used with permission.

years of schooling. We see two parallel lines of dots that do not tell us much, if anything, about the relationship between voting and education.

Nevertheless, we might conceptualize the problem this way. Denote the two outcomes of the dependent variable, Y, as 1 for "voted" and 0 for "did not vote." Each person in the study, in other words, is assigned a score on the dependent variable of 1 or 0, depending on whether or not they voted. We can then interpret the expected value of Y as "the probability that Y equals 1" because

$$E(Y) = [1 \times P(Y = 1)] + [0 \times (P = 0)] = P(Y = 1).$$

Note that $P(Y = 1)$ means "the probability that Y equals 1," which in turn is the probability that a person voted. $P(Y = 0)$ is defined similarly.[14] As noted before, the expected value of a variable can be thought of roughly as the sum of its possible values times the probabilities of their occurrence.[15]

We can now construct a linear regression model for the probability that Y equals 1, which we will denote simply as P. That is, for two independent variables our desired model has the general form

$$E(Y) = P = \alpha + \beta_1 X_1 + \beta_2 X_2.$$

This is called a **linear probability model,** and it means that the expected value of the binary dependent variable or (what is the same thing) the probability that Y equals 1 is a linear function of the independent variables X_1 and X_2. The regression coefficients, the betas, simply indicate how much the predicted probability changes with a one-unit change in an independent variable, given that all the other variables in the model have been held constant.

The idea might be clarified by calculating a linear probability model for the General Social Survey data. The result is

$$\hat{Y} = .214 + .035 \; Education + .043 \; Race.$$

The parameters still have the usual interpretation: when education equals 0 and race is also 0 (that is, nonwhite), the predicted probability of voting is .214.[16] For each one-year increase in education, with race held constant, the probability of turnout increases .035. So, for example, an African American with one year of schooling completed would have a predicted vote score of

$$\hat{Y} = .214 + .035(1) + .043(0) = .249.$$

To make sure that you understand this point, substitute different education and race values to get the predicted chances that various types of people will vote.

Although all the coefficients are statistically significant by the usual standards,[17] the typical measure of goodness of fit, R^2, is quite low at .058 (as the plot of Figure 13-2 suggests). In view of this poor fit and the fact that the dependent variable is a dichotomy, it is reasonable to wonder if linear regression is in fact the right technique for analyzing dichotomous dependent variables.

The linear probability model works reasonably well when all predicted values lie between .2 and .8, but statisticians still believe that it should not generally be used. One reason is that the predicted probabilities can have strange values, since the linear part of the model can assume just about any value from minus to plus infinity, but a probability by definition must lie between 0 and 1. For example, a white person with twenty-two years of schooling would have a predicted probability of voting of

$$\hat{P} = .214 + .035(22) + .043(1) = 1.027,$$

which is greater than 1.

In addition, the linear probability model violates certain assumptions that are necessary for valid tests of hypotheses. For example, the results of a test of the hypothesis that a β is zero in a linear probability equation might be wrong. For these and other reasons, social scientists generally do not use a linear probability model to analyze dichotomous dependent variables.

So what can be done? We certainly do not want to give up because many dichotomies or binary dependent variables or responses are frequently worth investigating. A common solution is to use **logistic regression,** a nonlinear model in which the log odds of one response as opposed to another is the dependent variable.[18] (We explain odds and log odds a little later.) The logistic regression function, which for two independent variables, X_1 and X_2, and a dichotomous dependent variable, Y, has the form

$$Prob\,(Y = 1) = P = \frac{e^{(\alpha + \beta_1 X_1 + \beta_2 X_2)}}{1 + e^{(\alpha + \beta_1 X_1 + \beta_2 X_2)}},$$

is a rather mysterious-looking formula that can actually be easily understood simply by looking at some graphs and making a few calculations. First note that e, which is often written *exp*, stands for the exponentiation function. A function can be thought of as a machine: put a number in and another, usually different number comes out. In this case, since e is a number that equals approximately 2.718218, X enters as the exponent of e and emerges as another number, 2.71828^X. For instance, if X equals 1, then e^1 is (approximately) 2.7182, and if $X = 2$, e^2 is about 7.3891. (Many hand-held calculators have an exponentiation key, usually labeled e^X or $exp\,(X)$. To use it just enter a number and press the key.) Although this function may seem rather abstract, it appears frequently in statistics and mathematics and is well known as the inverse function of the natural logarithm; that is, $\log\,(e^x) = x$. For our purposes it has many useful properties.

The logistic function, which uses e, can be interpreted as follows: the probability that Y equals 1 is a nonlinear function of X, as shown in Figure 13-3. Curve a shows that as X increases, the probability that Y equals 1 (the probability that a person votes, say) increases. But the amount or rate of the increase is not constant across the different values of X. At the lower end of the scale, a one-unit change in X leads to only a small increase in the probability. For X values near the middle, however, the probability goes up quite sharply. Then, after a while, changes in X again seem to have less and less effect on the probability, since a one-unit change is associated with just small increases. Depending on the substantive context, this interpretation might make a great deal of sense. Suppose, for instance, that X measures family income and Y is a dichotomous variable that represents ownership or nonownership of a beach house. (That is, $Y = 1$ if a person owns a beach house and 0 otherwise.) Then for people who are already rich (that is, have high incomes) the probability of ownership would not be expected to change much, even if they increased their income considerably. Similarly, people at the lower end of the scale are not likely to buy a vacation cottage even if their income does increase substantially. It is only when someone reaches a threshold that a one-unit change might lead to a large change in the probability.

Curve b in Figure 13-3 can be interpreted the same way. As X increases, the probability that Y equals 1 decreases, but the amount of decrease depends on the magnitude of the independent variable.

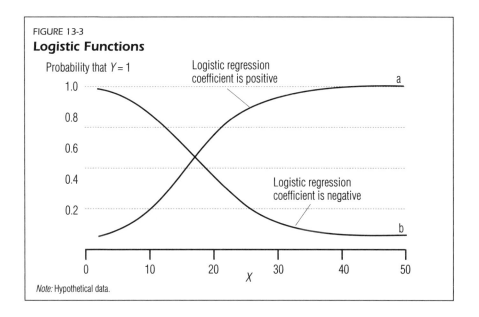

FIGURE 13-3
Logistic Functions

Probability that $Y = 1$

Logistic regression coefficient is positive

1.0
0.8
0.6
0.4
0.2

Logistic regression coefficient is negative

a

b

0   10   20   30   40   50

X

Note: Hypothetical data.

The essence of nonlinear models is that the effects of independent variables are not constant but depend on particular values. So the logistic regression function has a reasonable interpretation. It also meets the objections mentioned earlier, namely, that predicted values will lie between 0 and 1, which are the minimums and maximums for probabilities, and that the assumptions of hypothesis testing will be met.[19]

The logistic regression can be further understood with a numerical example. Using a procedure to be described shortly, the estimated logistic regression equation for the General Social Survey voting data is

$$\hat{P} = \frac{e^{-1.593 + .182\ Education + .210\ Race}}{1 + e^{-1.593 + .182\ Education + .210\ Race}}.$$

In this particular equation, α equals -1.593, β_1 equals $.182$, and β_2 equals $.210$. These numbers are called **logistic regression coefficients,** which are related to multiple regression coefficients in that they show how the probability of voting changes with changes in the independent variable.

Although an explanation of how the coefficients were calculated goes beyond the scope of this book (and computer programs for doing the work are widely available), we can start to examine their meaning by substituting some values for the independent variables into the equation. Keep in mind, however, that logistic regression coefficients (the βs) are similar to regular regression coefficients: they indicate the effect that a change in a particular independent variable produces when the other independent factors in the model have been held constant. In this sense they are like partial coefficients of multiple regression because each isolates the impact of a specific X net of all the other Xs in the equation. But remember that a β does not directly measure the change in Y for a given change in X. It instead shows the net effect on a quantity called the log odds or logit, which is connected to the probability that Y equals 1. This interpretation becomes clearer as we go on.

TYPES OF MODELS FOR A DICHOTOMOUS VARIABLE

In bivariate and multiple regression analysis the dependent variable (Y) is quantitative or numerical, and one statistical goal is to explain its variation. Because the conceptualization of Y seems so natural, understanding regression coefficients is relatively straightforward. In the case where Y has just two categories (such as 1 and 0), however, there are a couple of ways of setting up and interpreting models. One approach is to examine Y directly by modeling the *probability* that Y equals 1 or 0. These models have regression-like coefficients for the Xs, but they appear in the exponents of somewhat-complicated-looking equations for the probabilities and cannot be understood in the simple "a-one-unit-change-in-X-produces-a . . ." framework of ordinary regression. So understanding the meaning of logistic coefficients is not intuitive. It is possible, however, to model, not the probability that Y equals 1, but the *odds* that Y equals 1 as opposed to 0. (The odds that Y equals 1 is *not* the same as the probability that Y equals 1, as we emphasize later in the chapter.) In this formulation the odds become a kind of dependent variable, and the analyst's objective is to study what affects it. Furthermore, it is frequently convenient to transform the odds by taking their natural logarithm to get "logits." So logits, too, can be considered as a sort of dependent variable. The use of logits is popular because models for them are linear in the explanatory factors, and a (partial) logistic regression coefficient does have the interpretation that a one-unit change in X is associated with (partial) beta-unit change in the *logit* or log odds when other Xs have been controlled. The difficulty, of course, is that now the meaning of the dependent variable—a logit—is not obvious. Fortunately, all these formulations are equivalent, and it is possible to move back and forth among them. The first part of this logistic regression section develops and explains models for probabilities, and a latter part looks at models for the log odds.

An example helps clarify the meaning. Consider a person who reports zero years of schooling ($X_1 = 0$) and who is nonwhite ($X_2 = 0$). Then the equation becomes

$$\hat{P} = \frac{e^{-1.593+.182\,(\,0\,)+.210\,(\,0\,)}}{1 + e^{-1.593+.182\,(\,0\,)+.210\,(\,0\,)}}$$
$$= \frac{e^{-1.593}}{(1 + e^{-1.593})}$$
$$= .169.$$

This expression means that the estimated probability that a nonwhite person with zero years of education will vote is .169. (Notice that the level of education has been effectively controlled.) Consider next a white person ($X_2 = 1$) with the same amount of education ($X_1 = 0$). The predicted probability of voting is now

$$\hat{P} = \frac{e^{(-1.593+.182\,(\,0\,)+.210\,(\,1\,))}}{1 + e^{(-1.593+.182\,(\,0\,)+.210\,(\,1\,))}}$$
$$= \frac{e^{(-1.593+.210\,)}}{1 + e^{(-1.593+.210\,)}} =$$
$$= \frac{e^{(-1.488\,)}}{1 + e^{(-1.488\,)}}$$
$$= .201.$$

The results show that whites with zero years of education have a slightly higher predicted probability of voting than nonwhites with zero years of education, .201 versus .169. The "effect

TABLE 13-10
Estimated Log Odds, Odds, and Probabilities for Voting in 1992 Presidential Election

Years of Education	Race	Log Odds	Odds	Probability
0	Nonwhite (0)	−1.593	.203	.169
0	White (1)	−1.488	.226	.184
10	Nonwhite (0)	.227	1.255	.557
10	White (1)	.437	1.548	.608
20	Nonwhite (0)	2.074	7.745	.886
20	White (1)	2.152	8.602	.896

of being white" for this level of education is to modestly increase the chances of voting. (At different levels of education the effect will be different.)

A similar substitution shows that the probability that a nonwhite with nine years of education will vote is

$$\hat{P} = \frac{e^{-1.593 + .182\,(\,9\,) + .210\,(\,0\,)}}{1 + e^{-1.593 + .182\,(\,9\,) + .210\,(\,0\,)}}$$
$$= \frac{e^{+.045}}{(1 + e^{+.045})}$$
$$= .511.$$

Table 13-10 shows the predicted probabilities of voting for a few other combinations of education and race. Look at the entries in the first two and the last columns. (Ignore the "Odds" columns for the moment.) They show, for instance, that a nonwhite person with twenty years of schooling has an (estimated) probability of voting in the 1992 presidential election of .886, versus .896 for a white person with the same educational attainment. So, at the high end of the education scale, nearly everyone votes no matter what his or her race. Now look at the middle two rows, which compare nonwhite and white people who have had ten years of schooling. The estimated probabilities of voting are .557 and .608, respectively.

It may not be apparent, but the magnitude difference between whites' and nonwhites' probabilities of voting depends on the level of education. At the lower end of the scale, for example, the differences in probabilities are noticeable, if small. But at the upper end they almost vanish. This pattern reflects the nonlinear relationship between the independent variables and the dichotomous dependent variable. (Look at curve a of Figure 13-3 again.) So we conclude that unlike ordinary regression, a one-unit change in one of the independent variables has a *variable* impact on the dependent variable. This property actually makes sense substantively. After all, a person may have so much of a property like income or education that an additional increase in it will not really affect his or her chances of voting.

Estimating the Model's Coefficients

It is natural to wonder how the coefficient estimates are derived, and it would certainly simplify things if we could provide straightforward formulas for calculating them. Unfortunately, there are no such easy equations. Instead, logistic regression analysis is best performed with special computer programs. Logistic regression has become so widely used that the appropriate tools can be found in many statistical program packages such as SPSS, MINITAB, R, and SAS. Your instructor or computer consultant can help you find and use these programs. We recommend that if you have a dependent variable with two categories and want to perform regression, ask for a logistic regression program.[20]

Although the details are beyond the scope of the book, the method used to estimate unknown coefficients relies on a simple idea: pick those estimates that maximize the likelihood of observing the data that have in fact been observed. In effect, we propose a model that contains certain independent variables and hence unknown coefficients. Associated with the model is a likelihood function, L. The parameters in the function L give the probability of the observed data. That is, the data points are treated as fixed or constant, and the likelihood is a function of unknown parameters. Using the principles of differential calculus, a numerical algorithm selects values of the parameters that maximize the likelihood function. Logically enough they are called maximum-likelihood estimators. Therefore, the aim of the behind-the-scenes number crunching is to find those values for the parameters that maximize the probability of obtaining the observed data that we did. For computational purposes the logarithm of the likelihood function is calculated to give the *log likelihood function,* or LL. Log likelihood functions, which are somewhat analogous to sums of squares in regression, appear in many model fitting and testing procedures, as we see below.

If the estimated coefficients are calculated correctly and certain assumptions are met, they have desirable statistical properties. They are, for instance, unbiased estimators of corresponding population parameters and can be tested for statistical significance.

Measures of Fit

As in the case of simple and multiple regression, researchers want to know how well a proposed model fits the data. The same is true of logistic regression. After estimating a model we want to know how well it describes or fits the observed data. Several measures of goodness of fit exist, although they have to be interpreted cautiously. In fact, considerable disagreement exists about which measure is best, and none of the alternatives has the seemingly straightforward interpretation that the multiple regression coefficient, R^2, has.

With ordinary regression analysis, one way to calculate the fit is to compare predicted and observed values of the dependent variable and measure the number or magnitude of the errors. Alternatively (and equivalently), we can determine what proportion of the variation in Y is statistically explained by the independent variables. We can also observe the resulting indicators, such as the multiple R or coefficient of determination, R^2, which provide some useful information but can in certain circumstances be misleading.

Logistic regression involves roughly similar steps. But the procedures are more complicated and cumbersome, and so we simply sketch the general ideas. Our main objective is to provide a working understanding of substantive research articles and computer output.

Most logistic regression software programs routinely report the values of log likelihood functions, LL. (They will be negative numbers.) Occasionally, as with the popular program package SPSS, the result given is –2 times the log likelihood, but you can switch back and forth easily by the appropriate multiplication or division. As an example, the log likelihood for the logistic regression of education and race on reported turnout is LL = –2,508.82. This number looks large, but what exactly does it mean? Unfortunately, the number is not terribly informative by itself. But it can be compared with the LLs obtained for other models. And these comparisons can be used to gauge the overall fit and test hypotheses about sets of coefficients.

A simple strategy for assessing fit is to contrast the log likelihood of a model with just one constant term, LL_0, with a model that contains, say, two independent variables, X_1 and X_2. This log likelihood we denote LL_C, for "current" model. A measure of "improved" fit, then, is

$$R^2_{pseudo} = \frac{LL_0 - LL_C}{LL_0}.$$

The denominator plays the role of the total sum of squares that we have seen on numerous occasions. The numerator shows the difference in the fit when independent variables have been added and might be loosely considered the "explained" portion. The "pseudo" in the resulting R-squared indicates that this statistic is not the same as the R^2 of ordinary regression, and it certainly does not represent explained variation. But the basic idea is the same: pseudo-R^2 roughly suggests the relevance of a set of independent variables in understanding the probability that $Y = 1$.

For the turnout example, LL_0 for the model with no independent variables (only a constant term) is $-2,639.80$, and LL_C for the model with education and race included is $-2,508.82$. Thus, the pseudo R^2 is

$$R^2_{pseudo} = \frac{[-2639.80 - (-2508.82)]}{-2639.80} = \frac{-130.98}{-2639.80} = .05.$$

The difference, -130.98, is often reported as the "improvement of goodness of fit," or the "omnibus test of model coefficients" in some software packages. It can be easily calculated because the log likelihoods are routinely reported. (Keep in mind also that some programs report -2 times the log likelihood. The same conclusions will result because the -2 factors cancel out.) This number suggests that the addition of two independent variables did not really improve the fit very much, for the proportional improvement is only .05 (or about 5 percent). But before rejecting the model, keep in mind that the pseudo R^2 is not an infallible indicator of fit and that others have been proposed.[21]

In addition, there is a different approach to assess goodness of fit. If our model describes the data well, then it ought to lead to accurate predictions. That is, we should be able to plug in values of the independent variables for each observation, obtain predicted probabilities for every one, and use these predictions to predict whether or not a person has a score of 1 on Y. (For instance, given an individual's scores on the independent variables, we should be able to predict if a person has voted or not.) We can then count the number of observations correctly and incorrectly classified to obtain a "correct classification rate" (CCR). If a model has any explanatory power, the CCR should be relatively high, say, more than 75 percent and certainly more than 50 percent. For the General Social Survey data, as an example, we use the estimated model to predict whether or not each person will vote and then compare those predictions with what the respondents actually did. Table 13-11 shows the results.

We see that the model made $3,060 + 75 = 3,135$ correct predictions and $1,184 + 98 = 1,282$ incorrect ones. Since there were 4,417 individuals in the study, the CCR is $3,135/4,417 = .7098$, or about 71 percent. (Again, many logistic regression software programs report the CCR as part of their output.) By this standard the model seems to fit modestly well.

Measuring goodness of fit is not as straightforward in logistic regression as it is in ordinary regression, and all the proposed methods have shortcomings as well as strengths. Moreover, perhaps because logistic regression has been incorporated into standard political analysis relatively recently, there is no widely accepted and used list of measures. Some authors provide several indicators, whereas others give few. Thus, when reading articles and papers

TABLE 13-11
Cross-tabulation of Predicted and Observed Votes

	Model Prediction	
Actual Observation	Respondent Did Not Vote	Respondent Voted
Respondent did not vote	75	1,184
Respondent voted	98	3,060

that use dichotomous dependent variables and logistic regression, you may have to reserve judgment about the evidence in favor of a particular model.[22]

Significance Tests

To start, we can perform a test analogous to the F test in multiple regression to investigate the statistical significance of a set of coefficients.[23] This procedure follows the steps in the previous section. Let LL_C be the log likelihood for a current or "complete" model—the one with all the explanatory variables of interest included—and let LL_0 be the log likelihood for the "reduced" model—the one with one or more independent variables eliminated. Then the difference between the two likelihoods forms a basis for a test of a test statistic:

$$G = -2(LL_0 - LL_C)$$

G, which tests the null hypothesis that a β or a set of βs is zero, has a chi-square distribution with k degrees of freedom, where k is the number of variables dropped from the complete model to obtain the reduced one. It can be used to test one coefficient at a time, in which case the number of degrees of freedom is $k = 1$. A small G (that is, near zero) means the "tested" coefficients are not statistically significant and perhaps should not be included, whereas a large one suggests that they may be (statistically) important.

For the education, race, and turnout example the null hypothesis of interest is $\beta_{race} = \beta_{education} = 0$. The alternative hypothesis is that at least one of the parameters is not zero. The computed or observed G is equal to 260.36. Given that there are 2 degrees of freedom—that is, the comparison is between a model with just a constant and one with *two* independent variables—we find that this difference is highly significant. We know this by checking the chi-square table in Appendix C, where at the .001 level the critical value of chi-square with 2 degrees of freedom under the null hypothesis is 13.82. Since the observed chi-square greatly exceeds the critical value, the null hypothesis can be rejected.[24]

Articles in the scholarly literature frequently report significance tests for the individual coefficients using a different statistic. Recall that a statistical test of significance is a test of a null hypothesis that a population parameter

TABLE 13-12

Estimated Coefficients, Standard Errors, and Tests of Significance for Voting Data

Variable	Estimated Coefficient	Standard Error	Wald Statistic	Degrees of Freedom	Probability
Constant	−1.593	.169	89.044	1	.000
Education	.182	.012	223.545	1	.000
Race	.210	.091	5.359	1	.021

equals some specific value, often zero. In the case of logistic regression we usually want to test the hypothesis that in the population a β equals zero. As an example, we might want to test the null proposition that the partial logistic coefficient relating education to turnout is zero. The form of this kind of test is roughly similar to the others we have described throughout the book: divide an estimated coefficient by its standard error. In this case if the sample size is large (say, greater than 200), the result gives a z statistic that has a normal distribution with a mean of zero and standard deviation. That is,

$$z = \left(\frac{\hat{\beta}}{s.e.} \right),$$

where $\hat{\beta}$ is the estimated coefficient and $s.e.$ is its estimated standard error. This quotient, often labeled a "Wald" statistic, can be investigated with a usual z test procedure: establish a critical value under the null hypothesis that a β equals some value, compare the observed z to the critical value, and make a decision. (Recall that critical values for a z can be found in Appendix A. Decide on a level of significance for a one- or two-tailed test and consult the appropriate cell of the table.)

Software invariably reports the coefficients and their standard errors and usually the z or Wald statistic as well, so we need not worry about computing them by hand. Table 13-12 shows the result for the General Social Survey data, which suggests that we can reject the hypotheses that $\beta_{education}$ and β_{race} are zero. We can thus conclude that the variables have a statistically significant effect on the probability of voting.

A slight nomenclature problem arises with the Wald or z statistic. Some authors and software define the *square* of the z as the Wald statistic. (SPSS does this, for example.) In this version the Wald statistic (that is, z^2) has a chi-square distribution with 1 degree of freedom and can be analyzed using the methods presented previously and in Chapter 12.

We conclude this section by pointing out that the accuracy of the Wald (or z) statistic depends on many factors, such as the sample size. As a result, some statisticians advise using the G statistic applied to one coefficient at a time. That is, test a model with K independent variables and hence coefficients against one with $K - 1$ parameters. (The former would be the "current" model, the latter the "reduced" model.) If G is significant, the variable left out should perhaps be included. Otherwise, we might not reject the hypothesis that its coefficient is zero. But since the z or z^2 appears so frequently, it is important to be aware of its purpose.

An Alternative Interpretation of Logistic Regression Coefficients

We might summarize to this point by saying that logistic regression analysis involves developing and estimating models such that the probability that Y equals 1 (or 0) is a *nonlinear* function of the independent variable(s):

$$P(Y = 1) = \text{Nonlinear function of } X.$$

It is possible, though, to rewrite the logistic regression equation to create a linear relationship between the Xs and Y. Doing so provides an alternative way to interpret logistic regression results. Instead of explaining variation in Y with a linear probability model or P with a logistic regression, we can work with odds, which is the probability of one response or value of a variable over the probability of another response or value of a variable, and use it as a dependent variable (see Chapter 11).

Suppose we sampled a person at random from a group of eligible voters. We could ask, "What is the probability (P) that this individual actually voted?" or, a related question, "What are the *odds* that this individual voted?" Probability and odds are not the same, for the odds are the ratio of *two* probabilities, the probability of voting compared with the probability of not voting:

$$Odds = O = \frac{P_{vote}}{(1 - P_{vote})},$$

where P_{vote} is the probability of voting.

Some examples will help to illustrate the difference. Suppose the probability that a randomly selected citizen votes is .8. Then the *odds* of her doing so are $.8/(1 - .8) = .8/.2 = 4$, or, as is commonly said, 4 to 1. The person, in other words, is four times as likely to vote as not. As another example, suppose the probability of turning out is .4, then the *odds* are $.4/(1 - .4) = .4/.6 = .6667$, or about .667 to 1. In this case, the citizen is less likely to vote than not to vote. In both examples, the terms in the denominator of the fraction are just the $1 - P$, which is the probability of not voting. (Since probabilities must add to 1—either a person did or did not vote—the probability of not voting is $1 - P$.)[25] It

TABLE 13-13
Probabilities and Odds

Probabilities	Odds
1	∞
.7	2.333
.5	1
.4	.667
.1	.111
0	0

Note: Read the odds as "*X* to 1."

is important not to confuse probabilities and odds; they are related, but not the same.

More generally, consider a variable, Y, that takes just two possible values, 0 and 1. Let P be the probability that $Y = 1$ and $Q = 1 - P$ be the probability that $Y = 0$. Then the odds that Y is 1 as opposed to 0 are

$$O = \frac{P}{(1 - P)} = \frac{P}{Q}.$$

The term O has intuitive appeal, since it accords with common parlance. The odds, O, can vary from 0 to ∞, or infinity. If $O = 1$, then the "chances" that $Y = 1$ or 0 are the same, namely 1 to 1. If O is greater than 1, the probability that $Y = 1$ is greater than 1/2, and conversely if O is less than 1, the probability is less than 1/2. Table 13-13 shows a few more examples of probabilities and odds in a case in which a random process can lead to just one of two possible outcomes.

Why bother with odds? Take a look at the logistic model. It is really a formula that relates P to some Xs, so we ought to be able to rewrite it by putting 1 in front to obtain $1 - P$. Then we could put the two equations together to get an expression for P over $1 - P$. Here is how. To simplify, let $Z = \alpha + \beta_1 X_1 + \beta_2 X_2$. Now the expression for P can be written

$$P = \frac{e^z}{1 + e^z}.$$

In the same fashion we can write $1 - P$ as

$$1 - P = 1 - \frac{e^z}{1 + e^z}.$$

This latter expression can be simplified to

$$1 - P = \frac{1}{1 + e^z}.$$

Now we can put the two equations for P and $1 - P$ together to obtain an expression for the odds, $O = P/(1 - P)$:

$$O = \frac{P}{1 - P} = \frac{\dfrac{e^z}{1 + e^z}}{\dfrac{1}{1 + e^z}}.$$

This expression in turn simplifies to

$$O = e^z.$$

Remember that we let $Z = \alpha + \beta_1 X_1 + \beta_2 X_2$, so this expression is really

$$O = e^{\alpha + \beta_1 X_1 + \beta_2 X_2}.$$

We have thus found a simple expression for the odds. But it is still nonlinear because of the exponentiation, e. But a property of the exponentiation function is that $\log(e^Z) = Z$, where log means the natural logarithm. So we find that the logarithm of the odds—called the log odds, or logit—can be written as a linear function of the explanatory variables:

$$Logit = \log O = \alpha + \beta_1 X_1 + \beta_2 X_2.$$

This model can be interpreted in the same terms as multiple linear regression if we keep in mind that the dependent variable is the logit, or log odds, not Y or probabilities. Refer, for instance, back to Table 13-11. The middle two columns show the predicted log odds and the odds for voting for various combinations of race and education. As an example, a nonwhite ($X_2 = 0$) with no schooling ($X_1 = 0$) has an estimated .203 to 1 chance of voting. This compares with, say, a highly educated white ($X_1 = 20$, $X_2 = 1$), whose odds of voting are about 8.6 to 1.

Also note that if we exponentiate the linear logit model, we obtain an equation for the plain odds:

$$O = e^{\alpha + \beta_1 X_1 + \beta_2 X_2}.$$

This equation in turn can be rewritten

$$O = e^{\alpha} e^{\beta_1 X_1} e^{\beta_2 X_2}.$$

This formulation shows that the logistic regression coefficients can be interpreted as the multiplicative effect of a one-unit increase in an X on the odds when other variables are constant.

We should stress that these remarks are simply an alternative but equivalent way of interpreting logistic regression coefficients. Moreover, we can move from one view to the other by simply manipulating the results with a pocket calculator. Most computer programs and articles report the coefficients, along with other statistical information. To make sense of them often requires substituting actual data values into the equations and seeing what the probabilities or odds turn out to be.

PROBABILITY VERSUS ODDS

Keep terms straight. A probability is not the same as odds, at least in statistical analysis. A probability refers to the chances of something happening such as a person voting. Odds compare two probabilities, as the probability of voting to the probability of not voting. If N_Y is the number of people out of a sample of N who reply yes, for example, the estimated probability of a yes response is

$$\hat{P} = \frac{N_Y}{N}.$$

The estimated probability of a no is

$$\hat{Q} = 1 - \hat{P} = \frac{N - N_Y}{N}.$$

The estimated odds of observing a yes as opposed to a no, however, are

$$\hat{Q} = \frac{\hat{P}}{\hat{Q}} = \frac{\dfrac{N_Y}{N}}{\dfrac{N - N_Y}{N}} = \frac{N_Y}{N - N_Y}.$$

If the probability of yes is .6, then the probability of no is $1 - .6 = .4$, and the corresponding odds of yes are $.6/.4 = 1.5$ or 1.5 to 1.

A Substantive Example

To further the understanding of logistic regression parameters, we use an example from *The Bell Curve: Intelligence and Class Structure in American Life,* by Richard Herrnstein and Charles Murray. Herrnstein and Murray's work became extremely controversial because it argued that intelligence plays a larger role in economic success than does socioeconomic background and that many socially undesirable behaviors stem more from low cognitive ability than from prejudice or lack of opportunity. Many observers have tried to interpret their results as saying genes or nature are more important in explaining success and achievement than are family background and other environmental variables. If true, such findings would have enormous implications for affirmative action, Head Start, special education, and a host of other public policies.

We of course cannot address the correctness of Herrnstein and Murray's argument. But using this rather contentious book as an example allows us to kill two birds with one stone. Most important, the authors' use of logistic regression analysis to bolster their positions provides an interesting example of the method. But

TABLE 13-14
Logistic Regression Coefficients for Log Odds of Being on Welfare

Variable	Estimate	Probability
Constant (α)	−1.036	.000
Intelligence (β_1)	−.580	.002
Socioeconomic status (β_2)	−.061	.726[c]
Age (β_3)	−.113	.439[c]
Poverty status (β_4)[a]	−.900	.000
Marital status (β_5)[b]	1.053	.000

Source: Adapted from Richard Herrnstein and Charles Murray, *The Bell Curve* (New York: Free Press, 1994), 607.

Note: R_{pseudo} = .312.

[a] Was woman living in poverty at the time of birth of her first child?
[b] Was woman married or not at time of birth of child?
[c] Coefficient is not significantly different from zero.

the book also shows how statistics influence policy analysis and how an understanding of statistical techniques can help you evaluate the strengths and weaknesses of substantive claims.

Table 13-14 presents a typical example of Herrnstein and Murray's results. The authors wanted to know how intelligence, which is measured by a standardized test,[26] and various demographic factors affect the probability of welfare dependency. The variables include a socioeconomic status index, age, and two additional indicators, poverty status (on or off welfare) prior to the birth of a child, and marital situation at the time of the child's birth (married or unmarried). The data in Table 13-14 are based on a sample of "women with at least one child born prior to January 1, 1989."[27] The "Estimate" column of Table 13-14 contains the components of a model that predicts the probability or odds that a woman is on welfare.

The model for the estimated log odds can be read from the middle column:

$$\text{Logit} = -1.036 - .580\,Intelligence - .061\,Social - .113\,Age - .900\,Poverty + 1.053\,Marital,$$

while the equation for the predicted probability takes the form

$$\hat{P} = \frac{e^{-1.036 - .580\,Intelligence - .061\,Social - .113\,Age - .900\,Poverty + 1.053\,Marital}}{1 + e^{-1.036 - .580\,Intelligence - .061\,Social - .113\,Age - .900\,Poverty + 1.053\,Marital}}.$$

If one substitutes values for the independent variable in the equations, it is easy to find predicted log odds, odds, and probabilities for various combinations of attributes. The way Herrnstein and Murray measured the independent variables makes interpretations easy because each factor is scored in standard deviation units. If, for instance, a score on intelligence is 0, then the person has the mean or average level of intelligence for the group being studied. If the score is 1, the individual is one standard deviation above the mean. Similarly, a score on intelligence of –1 indicates below-average intelligence. The socioeconomic index and age variable are interpreted in the same way. The other two factors, which need not concern us, are just dichotomous variables indicating the presence or absence of a condition. (A woman who gives birth out of wedlock, for example, receives a score of 0; otherwise the score is 1.) Putting these facts together allows us to compare women with different combinations of social background and intelligence and supposedly draw conclusions about which is the most important explanatory factor.

Consider first a woman of average intelligence, socioeconomic standing, and age ($X_1 = 0$, $X_2 = 0$, $X_3 = 0$), and who has scores of zero on the two indicator variables, poverty and marital status ($X_4 = 0$, $X_5 = 0$). Then her predicted log odds of being on welfare are[28]

$$\text{Estimated Logit} = -1.036 - .580(0) - .061(0) - .113\,Age - .900(0) \\ +1.053(0) = -1.036.$$

The log odds translate into estimated odds:

$$\hat{O} = e^{-1.036} = .355.$$

This number tells us that a woman with mean or average characteristics has a less than even chance—.335 to 1—of being on welfare. And her estimated probability is

$$\hat{P} = \frac{e^{-1.036}}{1 + e^{-1.036}}$$

$$= \frac{.355}{(1 + .355)}$$

$$= .262.$$

Now let's look at a woman who is one standard deviation above the mean on intelligence but who has the same characteristics on the other variables. The coefficient in Table 13-14 or in the equation indicates that the log odds will be decreased by .580. This effect translates into a decline in the odds and probability of being in poverty:

$$\hat{O} = e^{-1.036 - .580(1)}$$

$$= .199.$$

and

$$\hat{P} = \frac{e^{-1.036-.580\,(1)}}{1 + e^{-1.036-.580\,(1)}}$$

$$= \frac{.199}{(1 + .199)}$$

$$= .166.$$

We can see that the data suggest that being above average in intelligence lowers one's chances of going on public assistance. By substituting zeros and ones for the values of other variables we can see how they affect welfare status. (Remember, the coefficients are based on Herrnstein and Murray's particular standard deviation measurement scales. In this context, letting the age be 20 would not make much sense because it would mean "20 standard deviations above the average," which would perhaps be a person well over 200 years old.)

Notice that the table shows the level of significance of each variable and the pseudo R^2. The latter is .312, a value suggesting that the model fits the data reasonably well. Moreover, we see from Table 13-14 that intelligence, poverty, and marital status are significant, but the other variables are not. The authors take this fact as further evidence that IQ, which they believe is largely inherited, has a greater effect on the chances of being on welfare than does family background.

A last comment. The results in Table 13-14 seem to say that intelligence has a much greater impact on the chances or probability of being on welfare than does socioeconomic background because (1) the magnitude of the coefficient for IQ is about nine times as large as the one for social and economic position and (2) it is statistically significant, whereas the coefficient for socioeconomic status is not. It is, however, highly debatable whether we can draw any firm conclusions about the relative importance of variables from just these data. The size of the coefficients depends on more than just the strength of their relationships with other variables. And statistical significance, as we indicated in Chapter 12, is not always a good guide to substantive significance.

Conclusion

As we have seen, multivariate data analysis helps researchers provide more complete explanations of political phenomena and produce causal knowledge. Observing the relationship between an independent and a dependent variable while controlling for one or more control variables allows researchers to assess more precisely the effect attributable to each independent variable and to accumulate evidence in support of a causal claim. Being able to observe simultaneously the relationship between many independent variables and a dependent variable also helps researchers construct more parsimonious and complete explanations for political phenomena.

Multivariate data analysis techniques control for variables in different ways. Multivariate cross-tabulations control by grouping similar observations; partial correlation and multiple and logistic regression control by adjustment. Both types of procedures have their advantages and limitations. Control by grouping can result in the proliferation of analysis tables, the reduction of the number of cases within categories to a hazardous level, and the elimination of some of the variance in the control variables. Control by adjustment, in contrast, can disguise important specified relationships, that is, relationships that are not identical across the range of values observed in the control variables.

Notes

1. Stephen D. Ansolabehere, Shanto Iyengar, and Adam Simon, "Replicating Experiments Using Aggregate and Survey Data: The Case of Negative Advertising and Turnout," *American Political Science Review* 93 (December 1999): 901–910.

2. Nancy Burns, Donald R. Kinder, Steven J. Rosenstone, Virginia Sapiro, and the National Election Studies. "National Election Studies, 2000: Pre-/Post-election Study," ICPSR Study No. 3131 (Ann Arbor: University of Michigan, Center for Political Studies, 2001).

3. The "high" category consists of executives, managers, and professionals; "skilled" contains traditional blue-collar workers, including transportation, construction, and a few protective service employees; the "medium" level holds sales, clerical, technicians, and most service personnel; "low" is reserved for operatives, manual laborers, and household workers. We exclude the relatively few agricultural workers and members of the armed forces. The reference population, therefore, is a slightly truncated portion of the total American workforce. But these data should be more than satisfactory for our purposes.

4. Note, however, that the sample size with this combination of education and job status in section a of Table 13-5 is relatively small.

5. An excellent introduction is Alan Agresti, *Categorical Data Analysis,* 2d ed. (New York: Wiley, 2002). Another very useful book is Bayo Lawal, *Categorical Data Analysis with SAS and SPSS Applications* (Mahwah, N.J.: Erlbaum, 2003).

6. In Chapter 12 we also called this term the intercept because it has a simple geometric interpretation.

7. Stephen D. Ansolabehere, Shanto Iyengar, Adam Simon, and Nicholas Valentino, "Does Attack Advertising Demobilize the Electorate," *American Political Science Review* 88 (December 1994): 833.

8. This is not the only way to treat categorical data. Another common procedure is "effect coding" or "deviation coding," which uses scores −1, 0, and 1 as measurement units. See Graeme Hutcheson and Nick Sofroniou, *The Multivariate Social Scientist* (Thousand Oaks, Calif.: Sage Publications, 1999), 85–94.

9. It is not clear from the text, but the investigators apparently used common logarithms and so the log to the base 10 of 100,000 is 5.

10. Of course, urbanization and percentage black are themselves correlated. Hence, the best strategy might be to mobilize urban areas with large numbers of African Americans.

11. For essentially the same reasons, you might not want to compare standardized regression co-efficients based on samples from two different populations. See John Fox, *Applied Regression Analysis, Linear Models, and Related Methods* (Thousand Oaks, Calif.: Sage Publications, 1997), 105–108.

12. Well, actually, they do not have a random sample of anything. They have simply collected data on thirty-four Senate elections that took place in 1992. These data are obviously not a sample from a population of such elections. Nevertheless, as we noted in Chapter 12, researchers often proceed as if they had a sample. This is what Ansolabehere and his colleagues did, and it is what we do for explanatory purposes.

13. The usual explanation for this formula for degrees of freedom is that to estimate the necessary standard deviations, we "lose" one degree of freedom for each regression coefficient plus one for the constant. A more precise explanation can be found in most statistics texts such as Alan Agresti and Barbara Finlay, *Statistics for the Social Sciences,* 3d ed. (Englewood Cliffs, N.J.: Prentice-Hall, 1997).

14. If you want to clarify expressions like these, simply replace the variable's symbols and codes with substantive names. Thus, for example, $P(Y = 0)$ can in the present context be read literally as "the probability that 'turnout' equals 'did not vote.'"

15. More precisely, the expected value of a probability distribution is called the mean of the distribution.

16. Note that these estimates are calculated on the basis of the full sample, not just the data in Table 13-10.

17. The t statistic for education and race are 16.043 and 2.391, respectively, both of which are significant at the .05 level.

18. Logistic regression is just one technique for analyzing categorical dependent variables. Another common approach is probit analysis. Although you run across this approach frequently in social science research, and it rests on many of the same ideas as the logistic approach, space considerations prevent us from covering the topic. A very good reference is J. Scott Long, *Regression Models for Categorical and Limited Dependent Variables* (Thousand Oaks, Calif.: Sage Publications, 1997).

19. Of course, like any statistical technique, logistic regression analysis assumes certain conditions are true and will not lead to valid inferences if these conditions are not met.

20. Quite a few different methods can be used to analyze these kinds of data. A related procedure, called probit analysis, is widely used, and if the data are all categorical, log-linear analysis is available.

21. See, for example, Long, *Regression Models for Categorical and Limited Dependent Variables,* 104–113. Note also that some statisticians recommend against using most R^2-type measures in logistic regression work. See, for example, David W. Hosemer and Stanley Lemeshow, *Applied Logistic Regression Analysis* (New York: Wiley, 1989), 148.

22. The same comment applies to any data analysis technique: empirical results have to be interpreted and accepted with caution.

23. Consider, for example, a model that contains two types of variables—one group measuring demographic factors and another measuring attitudes and beliefs. The investigator might want to know if the demographic variables can be dropped without significant loss of information.

24. Computer programs usually report the calculated or obtained probability of the observed chi-square, so we do not even have to look up a critical value in a table.

25. That is, $P + (1 - P) = 1$.

26. Their measure of intelligence was the Armed Forces Qualification Test, "a paper and pencil test designed for teens who have reached their late teens." See Richard Herrnstein and Charles Murray, *The Bell Curve* (New York: Free Press, 1994), 579.

27. Ibid., 607.

28. You might not get exactly these results if you repeat the calculations on your own because they were carried out to more decimal places than indicated.

Terms Introduced

CONTROL BY ADJUSTMENT. A form of statistical control in which a mathematical adjustment is made to assess the impact of a third variable.

CONTROL BY GROUPING. A form of statistical control in which observations identical or similar to the control variable are grouped together.

DUMMY VARIABLE. A hypothetical index that has just two values: 0 for the presence (or absence) of a factor and 1 for its absence (or presence).

EXPERIMENTAL CONTROL. Manipulation of the exposure of experimental groups to experimental stimuli to assess the impact of a third variable.

EXPLICATION. The specification of the conditions under which X and Y are and are not related.

LINEAR PROBABILITY MODEL. Regression model in which a dichotomous variable is treated as the dependent variable.

LOGISTIC REGRESSION. A nonlinear regression model that relates a set of explanatory variables to a dichotomous dependent variable.

LOGISTIC REGRESSION COEFFICIENT. A multiple regression coefficient based on the logistic model.

MULTIPLE CORRELATION COEFFICIENT. A statistic varying between 0 and 1 that indicates the proportion of the total variation in Y, a dependent variable, that is statistically explained by the independent variables.

MULTIPLE REGRESSION ANALYSIS. A technique for measuring the mathematical relationships between more than one independent variable and a dependent variable while controlling for all other independent variables in the equation.

MULTIPLE REGRESSION COEFFICIENT. A number that tells how much Y will change for a one-unit change in a particular independent variable, if all the other variables in the model have been held constant.

MULTIVARIATE ANALYSIS. Data analysis techniques designed to test hypotheses involving more than two variables.

MULTIVARIATE CROSS-TABULATION. A procedure by which cross-tabulation is used to control for a third variable.

PARTIAL REGRESSION COEFFICIENT. A number that indicates how much a dependent variable would change if an independent variable changed one unit and all other variables in the equation or model were held constant.

PARTLY SPURIOUS RELATIONSHIP. A relationship between two variables caused partially by a third.

REGRESSION CONSTANT. Value of the dependent variable when all the values of the independent variables in the equation equal zero.

SPECIFIED RELATIONSHIP. A relationship between two variables that varies with the values of a third.

STATISTICAL CONTROL. Assessing the impact of a third variable by comparing observations across the values of a control variable.

Suggested Readings

Anderson, T. W. *Introduction to Multivariate Statistical Analysis.* New York: Wiley, 1958.

Blalock, Hubert M. *Causal Inference in Non-Experimental Research.* Chapel Hill: University of North Carolina Press, 1964.

———. *Social Statistics,* 2d ed. New York: McGraw-Hill, 1972.

Draper, N. R., and H. Smith. *Applied Regression Analysis.* New York: Wiley, 1966.

Kendall, M. G., and A. Stuart. *The Advanced Theory of Statistics,* Vol. 2. London: Griffin, 1961. Chapter 27.

Kerlinger, F. N., and E. Pedhazer. *Multiple Regression in Behavioral Research.* New York: Holt, Rinehart and Winston, 1973.

Long, J. Scott. *Regression Models for Categorical and Limited Dependent Variables.* Thousand Oaks, Calif.: Sage Publications, 1997.

Overall, J. E., and C. Klett. *Applied Multivariate Analysis.* New York: McGraw-Hill, 1973.

Pampel, Fred C. *Logistic Regression: A Primer.* Thousand Oaks, Calif.: Sage Publications, 2000.

Scheffe, Henry A. *The Analysis of Variance.* New York: Wiley, 1959.

Chapter 14

The Research Report:

An Annotated Example

In the preceding chapters we described important stages in the process of conducting a scientific investigation of political phenomena. In this chapter we discuss the culmination of a research project: writing a research report. A complete and well-written research report that covers each component of the research process will contribute to the researcher's goal of creating transmissible, scientific knowledge.

This chapter examines how two researchers conducted and reported their research. We evaluate how well the authors performed each component of the research process and how adequately they described and explained the choices they made during the investigation. To help you evaluate the report, the major components of the research process and some of the criteria by which they should be analyzed are presented here as a series of numbered questions. Refer to these questions while you read the article and jot the number of the question in the margin next to the section of the article in which the question is addressed. For easier reference, the sections of the article have been assigned letters and numbers.

1. Do the researchers clearly specify the main research question or problem? What is the "why" question?
2. Have the researchers demonstrated the value and significance of their research question and indicated how their research findings will contribute to scientific knowledge about their topic?
3. Have the researchers proposed clear explanations for the political phenomena that interest them? What types of relationships are hypothesized? Do they discuss any alternative explanations?
4. Are the independent and dependent variables identified? If so, what are they? Have the authors considered any alternative or control variables? If so, identify them. Can you think of any variables that the researchers did not mention?
5. Are the hypotheses empirical, general, and plausible?
6. Are the concepts in the hypotheses clearly defined? Are the operational

definitions given for the variables valid and reasonable? What is the level of measurement for each of the variables?

7. What method of data collection is used to make the necessary observations? Are the observations valid and the measures reliable?

8. Have the researchers made empirical observations about the units of analysis specified in the hypotheses?

9. If a sample is used, what type of sample is it? Does the type of sample seriously affect the conclusions that can be drawn from the research? Do the researchers discuss this?

10. What type of research design is used? Does the research design adequately test the hypothesized relationships?

11. Are the statistics that are used appropriate for the level of measurement of the variables?

12. Are the research findings presented and discussed clearly? Is the basis for deciding whether a hypothesis is supported or refuted clearly specified?

Research Report Example

A   BIAS IN NEWSPAPER PHOTOGRAPH SELECTION

1   On the morning of April 30, 2002, the *Milwaukee Journal Sentinel* published at the top of the front page two large, side-by-side photographs of the candidates vying for the office of Milwaukee County Executive in a special election that day. In the first photograph, a confident-looking Jim Ryan looks a potential voter in the eye as he shakes his hand. In the second picture, Scott Walker is seen giving an awkward thumbs-up sign to the same potential voter with his eyes closed and his chin on his chest. To any impartial observer, the Ryan photograph is much more flattering than the picture of Walker.

2   By themselves, these two photographs on the morning of the election might not be cause for concern. However, these contrasting pictures followed a month in which the *Milwaukee Journal Sentinel* ran numerous editorials about the race which either had a strongly pro-Ryan slant or outright endorsed him.[1] This incident raises important questions about the selection of candidate photographs by newspapers. Was the front-page placement of a favorable Ryan photograph next to an unfavorable picture of his opponent an extension of the pro-Ryan editorial stance over the previous month? If so, do these two pictures reflect a broader, underlying—and grossly under-examined—tendency for candidate photographs to reinforce newspaper editorial and endorsement positions, or do they represent an isolated incident?

3   This study seeks answers to these questions by examining whether the "political atmosphere" of various newspapers leads to discernible differences in the favorableness of photographs of political candidates. We test a hypothesis of bias at the level of the media outlet by examining 435 candidate photographs from a sample of U.S. Senate, gubernatorial, and local races from seven newspapers during the 1998 and 2002 general election seasons. We conclude that the favorableness of candidate photographs differs, often markedly, and that these differences are related to a given newspaper's political atmosphere.

B   Relevant Literature

1 Over the past few decades, scholars have produced substantial evidence that the media can help create and/or shape perceptions of political figures (Dalton, Beck, and Huckfeldt 1998; Covington et al. 1993; Patterson 1993; Graber 1972). McCombs et al. (1997), for example, found a link between newspaper and television news coverage and Spanish voters' images of candidates during the 1995 regional and municipal elections in Spain. Likewise, King (1997: 40) discovered that "the press significantly contributed to the construction of candidate images in the heads of the voters" during the 1994 Taipei mayoral election.

2 Most studies concerned with the media's impact on mass attitudes and voter perceptions focus on text and/or verbal messages (e.g., Bystrom, Robertson, and Banwart 2001; Niven 1999; Dalton, Beck, and Huckfeldt 1998; Domke et al. 1997; Mutz 1992). But beyond these written and verbal messages, newspapers, newsmagazines, television news broadcasts, and the internet also provide news consumers with visual images. These visuals can be influential. Studies from a variety of disciplines have uncovered strong evidence that visual images influence people's attitudes and ability to learn about individuals, events, and issues (Gilliam and Iyengar 2000; Palvio 1991; Graber 1987; Nisbett and Ross 1980), and that nonverbal cues can be much more important than verbal ones in affecting assessments of other individuals (Argyle, Alkema, and Gilmour 1971).

3 Visual images also can influence the impressions of a candidate. Rosenberg et al. (1986) and Rosenberg and McCafferty (1987) provide evidence that the quality of candidate photographs in campaign fliers impacts perceptions of candidates and, ultimately, votes. Likewise, Martin (1978), who examines voting patterns in Australian local elections, discovered that vote decisions are strongly related to the first impressions of the candidates that voters form on the basis of candidate photographs. Finally, based on their experiment using videotaped excerpts of all candidates in the 1984 U.S. presidential election, Sullivan and Masters (1988: 363) found that "viewers' attitudes toward candidates can be influenced by the experience of watching brief excerpts on television.[2]

4 Even if images can influence voter perceptions, the favorableness of candidate photographs is largely a moot issue unless the media demonstrate a tendency to use these images to favor one candidate over another in their campaign coverage. This possibility raises the broader issue of whether a political and/or ideological bias exists in news coverage. This topic has received a great deal of attention from scholars, journalists, and pundits (Alterman 2003; Coulter 2002; Goldberg 2001; Domke et al. 1999; Niven 1999; Domke et al. 1997; Lee and Solomon 1990).

Despite the large number of studies that have examined this issue, there is little consensus on whether any media bias exists, however.

5 We believe that a major flaw in previous studies regarding the existence of a political and/or ideological media bias is the tendency to develop or apply theories designed for the level of analysis of the media as a whole instead of individual media outlets. In other words, there has been too much focus on the forest and not enough attention given to the trees. Since most studies in this area have focused on an overall media bias, the results of these works have often been inconclusive. By not considering the political leanings of individual media outlets, previous studies have likely under-estimated the extent and importance of biased coverage. Therefore, our study is concerned not with an overall political and/or ideological media bias, but whether the political atmosphere of individual newspapers (consciously or unconsciously) affects their selection of candidate photographs.[3]

6 Although such a media outlet-based theory has been inconspicuous, when scholars have considered the connection between coverage and the political atmosphere of particular media outlets, their findings have generally supported such a relationship.[4] In his analysis of coverage of the Nixon and Stevenson fund stories during the 1952 presidential election, Rowse (1957: 127–28), for example, found that almost all of the 31 newspapers he examined "showed evidence of favoritism in their news columns." In addition, "almost every example of favoritism in the news columns coincided with the paper's editorial feelings." Likewise, Page (1996: 112) argues that "in many newspapers and other media outlets, political points of view are not confined to editorial and op-ed pages but pervade news stories as well." Finally, examining 60 senatorial campaigns across three election cycles, Kahn and Kenney (2002) discovered that information about these campaigns on news pages is biased in favor of the candidate endorsed on the editorial page.

C   Political Atmosphere Theory

1 Our hypothesis is based on what we call Political Atmosphere Theory ("PAT"). Rather than addressing bias across the media as a whole, PAT explains bias within individual media outlets. PAT posits that media outlets develop an organizational political culture that influences the nature of their coverage. We do not argue that news organizations deliberately slant their political coverage toward their preferred candidate, although we do not dismiss this possibility either.[5] Instead, supported by the aforementioned evidence that political points of view in newspapers are not limited to editorial pages (Kahn and Kenney 2002; Page 1996; Rowse 1957), PAT is based on the following two contentions. First, as described by Page (1996: 129):

2 *Correspondence between news and editorial points of view can occur without any conspiracy, without any contact, without any breach of the "wall of separation" between the news and editorial departments. All that is needed is a tendency (even an unconscious tendency) for owners and publishers to hire and retain like-minded editors in both realms, with or without subsequently intervening in what they do.*

3 Simply put, identifiable political atmospheres are perpetuated over time at particular media outlets—in part at least—by hiring practices and the self-selection of job applicants.

4 Second, we contend that those employees who do not hold political opinions consistent with the prevailing political atmosphere at a media outlet will—intentionally or unintentionally—shape their reporting or editing to "fit in."[6] In other words, we maintain that those employees of a given media outlet responsible for the content of that outlet's news product will tend either to share core values and political leanings or act as if they do. This contention is based on Schein's (1985) ideas regarding organizational culture. Schein explains how organizations develop and perpetuate "basic assumptions and beliefs" that lead to observed behavioral regularities, norms, dominant values, a guiding philosophy, the rules of the game for surviving in the organization, and a climate of interaction with outsiders.

5 The result of these two practices is that the day-to-day coverage of politics by newspapers—including the selection of candidate photographs—will reflect an overall political atmosphere, more than likely consistent with the political slant of the editorial page. Our resultant hypothesis is that *newspaper photographs of political candidates will favor the candidate who best reflects the political atmosphere of a given newspaper.* Testing this hypothesis requires establishing both the political atmosphere of particular newspapers and their pattern of pictorial coverage of political candidates.

D   Why Newspapers

1 Although one could examine bias in the selection of candidate images in a variety of media, we focus on newspaper coverage of political campaigns for a number of reasons. First, most research on the pattern and impact of visual news images has already focused on television (e.g., Gilliam and Iyengar 2000; Graber 1990). This stands to reason as a majority of Americans today receive their news from television. However, millions still read newspapers daily and local papers are a particularly important source of news about statewide and local political campaigns.

2 Second, unlike television where viewers are bombarded with a continuous stream of visuals, fewer images accompany newspaper articles, which potentially increases the importance of any single image.[7] Moreover, since images presented in newspapers are fixed, every time readers glance at a newspaper photograph, they are presented with the same (and now reinforced) visual message. This reinforcement can even take place over time, because newspapers will often use the same photograph (such as a headshot of a candidate) repeatedly during a campaign.

3 Lastly, testing the relationship between news coverage and a particular media outlet's political atmosphere is easier for newspapers than it is for television. One could certainly claim that slanted political coverage exists on television news, but such broadcasts do not generally contain an official statement of the media outlet's position on issues (i.e., no editorials and/or endorsements). Although some have claimed that newspapers are less likely to endorse political candidates than in the past (Stanley and Niemi 1994) and that newspapers are less partisan now than in earlier periods (Rubin 1981), our examination indicates that endorsements remain common for major races and that partisan leanings in endorsement and editorial practices generally remain clear.

E   Research Design

1 Photographs of candidates were collected from seven different newspapers covering seven different statewide and local races during both the 1998 and 2002 election cycles. These included U.S. Senate races in Illinois, Missouri, Ohio, and Wisconsin; gubernatorial races in Illinois and Wisconsin; and the Milwaukee County (Wisconsin) Executive election. For three of the elections, photographs came from multiple newspapers covering the same race, resulting overall in eleven separate sets of newspaper photographs. Criteria for selecting the races examined included the availability of the newspaper and the closeness of the race.[8] We limited our analysis to newspapers from the Midwest for reasons of practicality. There is no reason to expect, however, that bias in photograph favorableness would exist in the Midwest but not elsewhere in the United States.[9]

2 We collected 435 photographs for this study[10] and coded each of these pictures in terms of their favorableness. In particular, the photographs were coded separately by three coders[11]—who were unaware of which newspaper generated a particular photograph at the time of its coding[12]—on a five point scale, ranging from –2 (highly unfavorable) to 2 (highly favorable). Photographs coded as favorable presented a smiling or confident-looking candidate, the presence of cheering supporters, the existence of a dramatic backdrop, or actions such as the hugging of a small child. Unfavorable photographs showed the candidate with a bewildered or angry look on his/her face, speaking to a mostly empty room, or caught in an awkward physical position.[13]

3 Although the characteristics of favorable and unfavorable photographs were discussed prior to coding, we chose not to have each picture coded through the use of a checklist of attributes (facial expressions, backgrounds, camera angles, etc.) as others have done (Waldman and Devitt 1998; Moriarty and Popovich 1991; Moriarty and Garramone 1986). We find the use of such a checklist troubling, particularly because studies using this approach quantify and weight each measure of various attributes equally; i.e., the candidate's arms above his/her head, a formal setting for the picture, and the camera angle looking up at the candidate are all considered to be equally positive attributes. We contend that a checklist of equally-weighted attributes adds an unnecessary degree of artificiality to the evaluation of the photographs. Instead, we based individual coding decisions on quick impressions of each photograph, as we wanted to replicate as much as possible the average readers' reaction to pictures in the newspaper.[14] Granted, one might criticize this coding of the favorableness of candidate photographs as too subjective. Yet, the independent coding decisions of the pictures included in our sample produced a high intercoder reliability statistic (Alpha = 0.85), indicating that these quick impressions were quite consistent across the three coders.

4 Each picture also was examined to determine its size. Photograph size was coded as 1 (very small head shot), 2 (small), 3 (medium), or 4 (large). Taking stock of photograph size is based on the assumption—reasonable, we believe—that four favorable small head shots of a candidate could be countered by one very large and unfavorable photograph of the same candidate. Plus, there is some empirical evidence that the size of photographs in newspapers can have an influence on readers' perceptions (Wanta 1986). Therefore, we created an additional variable by multiplying the favorableness scale by the size of each photograph. The potential range of values for this variable was –8 (for a large, highly unfavorable photograph) to 8 (for a large, highly favorable one).

5 We measured the concept of political atmosphere in two ways. The first approach considers which candidate the newspaper endorsed at the end of the general election period. The advantage of this measure is its objectivity (and thus ease of replication). Some might question such an approach. After all, the actual endorsement takes place after most of the photographs have appeared, making a direct causal claim nonsensical. This would be a valid criticism if we asserted that the endorsement itself affects the coverage of the campaign. This is not our claim. Instead, we use the endorsement as an indicator of the political character of the newspaper. Furthermore, the criticism of this measure of political atmosphere is weakened by the presence of a second approach that looked at endorsements and editorials over time. This approach is less objective but more theoretically powerful, as it places the election in question within the context of endorsements and editorials of a given newspaper over several election cycles.

6 To assess the political atmosphere of newspapers over time, we first examined the endorsements of the major statewide or national races (presidential, senate, gubernatorial) for the election year of the race in question and for the previous two even-year elections.[15] This typically yielded seven endorsements for each paper.[16] If all of these endorsements were for candidates of the same political party, we coded the paper's political atmosphere as consistent with that party's positions. This was the case for a majority of the newspapers. The *St. Louis Post-Dispatch* and *Capital Times* (Madison, Wisconsin) endorsed all Democratic candidates. They were coded as newspapers with "Democratic" political atmospheres. *The Chicago Tribune, Columbus Dispatch,* and *Cincinnati Enquirer* favored all Republican candidates. They were coded as having "Republican" political atmospheres.[17]

7 For two of the newspapers, there was a mixture of Republican and Democratic endorsements over the three election cycles examined for each race. In these cases, further analysis was undertaken to determine the political atmosphere of each newspaper. Editorials were collected from the first week of every odd month in 1998 and 2002. Those dealing with major, ideologically divisive issues were rated by coders unfamiliar with the other elements of this study. The individual editorials were coded as strongly Republican (1), leaning Republican (2), neutral (3), leaning Democrat (4), or strongly Democrat (5). The intercoder reliability statistic was high (Alpha = 0.80).

8 One of the newspapers, the *Milwaukee Journal Sentinel,* had consistently strong "Democratic" editorials in both 1998 and 2002. Using the 1–5 scale, the mean coder score for the sample of editorials in 1998 was 4.31, while in 2002 it was 4.23. This was consistent with our expectations, since the *Milwaukee Journal Sentinel* has a reputation in the Milwaukee community of faithfully producing editorials supportive of Democratic Party positions. As a result of the analysis of the editorials, the *Milwaukee Journal Sentinel* was coded as having a Democratic political character for both the 1998 and 2002 races examined. The other newspaper with a mixed endorsement pattern, the *Wisconsin State Journal* (Madison, Wisconsin), was less easily put into a party camp. Analysis of the samples of its editorials from 1998 and 2002 yielded mean coder scores of 2.22 in 1998 and 3.45 in 2002. As a result, its overall political atmosphere was coded as "mixed" in 1998 and 2002, the only one of the seven newspapers examined for which this coding was employed.

F   Findings

1 The relationship between the political atmosphere of the newspapers and the favorableness of candidate photographs was investigated using both of the alternative approaches to measuring political atmosphere. The first approach—employing the newspaper's endorsement in the race in question—produced evidence of a strong relationship between the political character of the newspaper and the favorableness of candidate photographs as presented in Table 1. Taking the measure of photograph favorableness to be an interval level measurement,[18] a difference of means test was performed on the favorableness of photographs of endorsed (0.75) and non-endorsed (0.59) candidates, not weighting for photograph size; a Pearson's correlation statistic was also calculated. Both the mean difference and the correlation were significant at the $p \leq .05$ level, indicating significantly more favorable photographs for endorsed candidates compared to non-endorsed candidates.[19]

2 As presented in Table 1, taking into account the size of each photograph confirms the relationship between the favorableness of pictures and newspaper endorsements. The difference in average favorableness between endorsed and non-endorsed candidates (1.392 and 0.990, respectively) was statistically significant at $p \leq .05$, as was the Pearson correlation. Therefore, we can conclude that utilizing the more objective of the two approaches to measuring political atmosphere—actual endorsement of the candidate in the race in question—indicates a strong relationship between the political leanings of the newspaper and how candidates are treated pictorially.

3 The second approach to capturing political atmosphere, examining endorsements and editorials over time, also supports the link between a newspaper's political atmosphere and its selection of candidate photographs. Once again, we investigated photograph favorableness regardless of size as well as weighted by size. To use this approach to measure political atmosphere, photographs were placed into two categories: candidate photographs of the same political party as the newspaper's political atmosphere and pictures of candidates from the opposite party as the political atmosphere of the newspaper. Photographs of candidates in a newspaper with a "mixed" political atmosphere (specifically, photographs from

TABLE 1

Measures of "Political Atmosphere" and Favorableness of Candidate Photographs, Unweighted and Weighted for Size of the Photograph

Candidate	Unweighted for Size of Photographs			Weighted for Size of Photographs		
	# of Photos[a]	Mean Favorableness (SE of Mean)	Difference[b]	# of Photos[a]	Mean Favorableness (SE of Mean)	Difference[b]
Endorsement of the Candidate in the Given Election						
Endorsed	227	0.747 (0.048)	+0.154*	227	1.392 (0.120)	+0.402*
Not endorsed	208	0.593 (0.063)		208	0.990 (0.147)	
Correlation Statistic						
Pearson *R* Value		.095*			.102*	
Endorsement/Editorial Pattern over Time						
Candidate party same as pattern	163	0.699 (0.055)	+0.262**	163	1.546 (0.155)	+0.452**
Candidate party opposite of pattern	145	0.437 (0.070)		145	0.998 (0.193)	
Correlation Statistic						
Pearson *R* Value		.129**			.119**	

Note: **Statistically significant at $p \leq .01$ level, one-tailed; *significant at $p \leq .05$, one-tailed.

[a] The number of photographs in the difference of means analysis of newspaper endorsement in the particular race differs from that of the analysis of endorsement/editorial pattern over time. This is due to the exclusion of the *Wisconsin State Journal*, judged to have a mixed endorsement/editorial pattern, from the latter analysis.

[b] Significance levels for the difference of means test based on independent samples *t*-tests with assumption of unequal variances.

the *Wisconsin State Journal*) were excluded from this part of the analysis. The results of this analysis are also found in Table 1.

4   The difference in means of photograph favorableness for candidates of the same party (0.699) and opposite party (0.437) as the political atmosphere of each newspaper was highly significant statistically.[20] The Pearson correlation between photograph favorableness and editorial pattern was significant at the $p \leq .01$ level. These findings remained constant when the size of the photographs was taken into account. Candidates of the same political party as the editorial leanings of the newspaper had a mean score of 1.546 on the "favorableness times size" scale, while those of the party opposite to the political atmosphere of the newspaper had a mean score of 0.998. This difference was significant at the $p \leq .01$ level. A statistically significant relationship was also indicated by the correlation statistic.

5   In addition to demonstrating that, on the whole, there was a statistically significant relationship between the favorableness of candidate photographs and each newspaper's political atmosphere, we examined this relationship for each particular race covered by a given newspaper in the sample. Table 2 contains the summary statistics of the favorableness of the candidate photographs in each of the eleven sets of pictures examined, not weighting for photograph size, as well as that newspaper's endorsement in the race and its political atmosphere over time.

TABLE 2
Favorableness of Candidate Photographs in the Eleven Races Examined

Race	Newspaper	"Political Atmosphere"	Party of Candidate	# of Pict.	Mean Favorableness (SE)	Difference[a]	Same as Endorsed	Same as Political Atmosphere
1998 IL Senate	Chicago Tribune	Rep	Dem	21	0.75 (0.16)	+0.42*	Yes	Yes
			Rep	14	1.17 (0.11)			
1998 OH Senate	Cincinnati Enquirer	Rep	Dem	5	−0.27 (0.19)	+0.90**	Yes	Yes
			Rep	9	0.63 (0.24)			
1998 OH Senate	Columbus Dispatch	Rep	Dem	10	−0.07 (0.13)	+0.51*	Yes	Yes
			Rep	12	0.44 (0.22)			
1998 WI Senate	Milwaukee Journal Sentinel	Dem	Dem	23	0.65 (0.13)	+0.61**	Yes	Yes
			Rep	16	0.04 (0.20)			
1998 WI Senate	Wisconsin State Journal (Madison)	Mixed	Dem	14	0.67 (0.23)	+0.76**	Yes	(NA)
			Rep	9	−0.19 (0.22)			
2002 MO Senate	St. Louis Post-Dispatch	Dem	Dem	23	0.43 (0.13)	+0.05	No	No
			Rep	22	0.48 (0.15)			
2002 IL Governor	Chicago Tribune	Rep	Dem	17	0.35 (0.19)	+0.12	Yes	Yes
			Rep	19	0.47 (0.13)			
2002 WI Governor	Capital Times (Madison)	Dem	Dem	25	1.01 (0.13)	+0.19	Yes	Yes
			Rep	20	0.85 (0.20)			
2002 WI Governor	Milwaukee Journal Sentinel	Dem	Dem	26	0.56 (0.17)	+0.06	Yes	Yes
			Rep	22	0.50 (0.22)			
2002 WI Governor	Wisconsin State Journal (Madison)	Mixed	Dem	52	0.95 (0.10)	+0.18	No	(NA)
			Rep	52	1.13 (0.12)			
2002 Milw. Cty. Exec.	Milwaukee Journal Sentinel	Dem	Dem	12	1.06 (0.16)	+0.70*	Yes	Yes
			Rep	12	0.36 (0.24)			

Note: **Statistically significant at p ≤ .01 level, one tailed; * significant at p ≤ .05, one-tailed.
[a] Significance levels for the difference of means tests based on independent-samples *t*-tests, with assumption of unequal variances.

6 Setting aside for a moment the issue of whether the differences are statistically significant, the most conspicuous result from Table 2 is that nine out of eleven candidates endorsed in a given race received more favorable photographs from that newspaper during the general election period than did his or her opponent. Likewise, of the nine races in which the newspaper in question had a clearly Democratic or Republican political atmosphere, eight out of nine candidates of the party consistent with that atmosphere received more favorable photographs.

7 As one might expect, however, not all of the individual races produced statistically significant differences in favorableness consistent with the political leanings of the newspaper. In some cases, this was due to the overall difference between the two candidates being rather small (e.g., the photos published by the *St. Louis Post-Dispatch* in the 2002 U.S. Senate race in Missouri). Nonetheless, six out of eleven sets of photographs generated pictures with statistically significant differences in favorableness consistent with their endorsement in the campaign. In addition, five of the nine races covered by a newspaper with a Democratic or Republican political character over time had statistically significant differences in favorableness consistent with their political atmosphere. Finally, there was never a statistically significant case of a candidate whose opponent was endorsed by a given newspaper—or of a candidate whose political leanings were contrary to a paper's political atmosphere—receiving more favorable photographs.

8 Weighting the photographs by size yielded similar—albeit somewhat weaker—differences in favorableness relative to the endorsement in the particular race and the political atmosphere of the newspapers over time. These results are presented in Table 3. Nine of the eleven endorsed candidates received more favorable photographs, while eight of the nine candidates of the same party as each paper's political atmosphere over time did as well. In five out of eleven sets of photographs, these differences were significant at the $p \leq .01$ or $p \leq .05$ level. These findings again demonstrate a tendency for newspapers to publish more favorable photographs of candidates they endorse or who reflect that paper's political atmosphere.

G   Conclusion

1 This study offers strong evidence that the newspaper photograph selection process is biased, even if this bias may not be as intentional as it is often portrayed. Specifically, we uncovered strong and statistically significant results supporting the proposition that newspapers offer slanted pictorial presentations of political candidates. Rather than an overall liberal or conservative bias, however, these findings support Political Atmosphere Theory and its expectation that bias in visual images will be consistent with the political character of individual newspapers.

2 The use of multiple measures and analytical techniques increases our confidence in our findings. In particular, we came at our central hypothesis from a number of different angles and continually found the same results. We measured the political atmosphere of a newspaper in two separate ways: the endorsement in the race in question and the endorsement/editorial pattern over time. We also

TABLE 3

Favorableness of Candidate Photographs (Weighted by Photograph Size) in the Eleven Races Examined

Race	Newspaper	"Political Atmosphere"	Party of Candidate	# of Pict.	Mean Favorableness (SE)	Difference[a]	Same as Endorsed	Same as Political Atmosphere
1998 IL Senate	Chicago Tribune	Rep	Dem	21	1.86 (0.45)	+0.54	Yes	Yes
			Rep	14	2.40 (0.43)			
1998 OH Senate	Cincinnati Enquirer	Rep	Dem	5	−0.27 (0.27)	+1.42*	Yes	Yes
			Rep	9	1.15 (0.54)			
1998 OH Senate	Columbus Dispatch	Rep	Dem	10	0.17 (0.33)	+0.08	Yes	Yes
			Rep	12	0.25 (0.39)			
1998 WI Senate	Milwaukee Journal Sentinel	Dem	Dem	23	1.20 (0.29)	+1.33**	Yes	Yes
			Rep	16	−0.13 (0.35)			
1998 WI Senate	Wisconsin State Journal (Madison)	Mixed	Dem	14	0.76 (0.23)	+1.28**	Yes	(NA)
			Rep	9	−0.52 (0.39)			
2002 MO Senate	St. Louis Post-Dispatch	Dem	Dem	23	0.71 (0.34)	+0.24	No	No
			Rep	22	0.95 (0.43)			
2002 IL Governor	Chicago Tribune	Rep	Dem	17	1.06 (0.65)	+0.24	Yes	Yes
			Rep	19	1.30 (0.43)			
2002 WI Governor	Capital Times (Madison)	Dem	Dem	25	2.27 (0.47)	+1.37*	Yes	Yes
			Rep	20	0.90 (0.33)			
2002 WI Governor	Milwaukee Journal Sentinel	Dem	Dem	26	1.86 (0.48)	+0.07	Yes	Yes
			Rep	22	1.79 (0.79)			
2002 WI Governor	Wisconsin State Journal (Madison)	Mixed	Dem	52	1.15 (0.19)	+0.00	No	(NA)
			Rep	52	1.15 (0.19)			
2002 Milw. Cty. Exec.	Milwaukee Journal Sentinel	Dem	Dem	12	2.61 (0.49)	+1.69*	Yes	Yes
			Rep	12	0.92 (0.66)			

Note: **Statistically significant at p ≤ .01 level, one tailed; * significant at p ≤ .05, one-tailed.
[a] Significance levels for the difference of means tests based on independent-samples t-tests, with assumption of unequal variances.

established differences between candidate photographs by looking both at individual races and at all the photos together. Finally, when considering favorableness of the photographs, we both weighted for photograph size and did not. Based on these various approaches, we repeatedly came to the same conclusion. There was a significant correlation between photograph favorableness and the political atmosphere of each newspaper.

3 Nonetheless, when we disaggregated the data by newspaper and by race, not all of the races examined were characterized by a statistically significant imbalance in the favorableness of candidate pictures. In other words, not every newspaper was guilty of allowing its political atmosphere to influence the photograph selection process to a statistically significant degree. In fact, newspapers such as the *St. Louis Post-Dispatch* should be applauded for their apparent objectivity despite an easily identifiable political slant on the editorial page of the newspaper. Nonetheless, six of the seven newspapers we examined demonstrated statistically significant levels of bias in their selection of photographs corresponding to at least one of the two measures of political atmosphere.

4 An additional point worth noting is that the favorableness of candidate photographs may at least partly be due to the physical traits of the candidates themselves. Some candidates are quite handsome or beautiful, making it difficult to take a bad picture, while other candidates are—to put it kindly—"less physically gifted." The more attractive a candidate is physically the more likely it is that newspapers will have good pictures of that candidate to publish and the more favorable the impressions of that candidate will be among voters (Ottati and Deiger 2002). Yet, if they set their minds to it, media outlets should be able to publish favorable pictures of the more unattractive candidates and unfavorable pictures of the more attractive ones. Thus, candidate attractiveness is another factor that needs to be considered, but it does not eliminate the existence or importance of the political-atmosphere, photograph-favorableness relationship we uncover.

5 We cannot state for certain that differing degrees of favorableness in newspaper photographs affect election results in the United States. Voters are flooded with hundreds of visual images—as well as verbal messages—about candidates from newspapers, television news, and campaign ads during the course of a campaign. Nonetheless, the existing literature does indicate that visual images can influence voter attitudes about political candidates. An experiment we conducted involving more than 250 participants also supports this claim. We found that attitudes regarding traits associated with and the general impression of a particular political candidate, as well as the likelihood of voting for that candidate, were all affected by varying the favorableness of newspaper photographs (Barrett and Barrington 2005). It is also important to emphasize that the bias we uncovered was consistent over the course of a campaign. Therefore, readers of a particular newspaper are generally exposed to a series of favorable or unfavorable candidate photographs that can help both to create a particular impression of a candidate and to reinforce that impression.

6 Finally, if there is bias in the newspaper photograph selection process, there may easily be a similar bias in television news departments, Internet news

sites, and other sources of visual images.[21] There is also no reason to assume that biases from various news sources would necessarily cancel each other out in a particular election. The various forms of media bias may indeed be complementary, and unflattering newspaper photographs of a candidate could certainly complement negative visual images from non-media sources, such as mailers or television ads targeting that candidate. Questions about the relationship between our findings and these other sources of visual messages warrant further study.

Notes

1. The newspaper ran ten editorials during the month of April 2002. Four of these endorsed Ryan with explanations why he was the newspaper's choice, three others mentioned the endorsement without much detail as to why, and one did not mention the newspaper's official position but had a strongly pro-Ryan slant. Only two of these editorials were fairly neutral—discussing the issues but not focusing on the candidates—while none of them indicated Walker was the better candidate.

2. Masters and Sullivan (1989) again discovered evidence that attitudes toward political candidates are influenced by televised images when they replicated their earlier experiment in France shortly before the legislative elections of March 1986. Conversely, Riggle (1992) argues that candidate image matters less than partisanship and ideology.

3. For examinations of whether various newsmagazines, newspapers, and television networks were biased in their selection of photographs or images of presidential candidates, see Moriarty and Garramone (1986), Moriarty and Popovich (1991), Windhauser and Evarts (1991), and Waldman and Devitt (1998).

4. See D'Alessio and Allen (2000); but see also Carter, Fico, and McCabe (2002) on local television coverage of state campaigns.

5. Goldman and Beeker (1985: 351) seem more willing to suggest that the selection of newspaper photographs represents a conscious bias. As they put it, "The newsphoto is thus by its very nature a form of recontextualizing, and hence, by its very nature an ideologically loaded process."

6. Davis (2001: 139) discusses claims that the physical proximity of media outlet employees facilitates socialization. He cites Paletz and Entman (1981: 202), who argue that because reporters are "physically concentrated in the same places," they interact and "share the perceptions and gossip that shape stories."

7. As Lang, Newhagen, and Reeves (1996: 461) point out, recipients of media messages are information processors with a "limited capacity of processing." Photographs provide information to the media consumer at less cost than the flood of visual images television viewers experience.

8. Close races were selected for two reasons. First, these are precisely the kind of races in which a pattern of biased coverage could play a role in the outcome. Establishing that such coverage takes place in close races, therefore, is much more important than establishing that it happens in races in which the outcome is never in doubt. Second, one potential criticism of our study is that newspapers endorse candidates that are likely to win to ensure access to the candidate after the election. By selecting close races, we strengthen the claim that the endorsement captures the underlying political atmosphere at each newspaper rather than a rational calculation on the part of the editorial board.

9. One could claim that the selection of newspapers only from the Midwest introduces a regional bias to our study. Although it is true that regions of the United States differ in their political cultures, political atmosphere theory focuses on biased coverage at the level of the individual media outlet. Such a claim, therefore, mistakes our study of biased coverage in individual media outlets for a study of bias across the media as a whole (see the discussion above on the media outlet as our level of analysis). Generalizing from one region to another a finding about the *overall* scope and direction of bias *across the media* is indeed shaky. But it is unclear why one would expect findings of a relationship between political atmosphere and photograph favorableness at the level of the media outlet in one region not to hold up in other regions as well.

10. The number of photographs actually collected was 438. Copies of the original photographs were gathered during the campaigns from the print versions of each newspaper for the 2002 races, beginning two months prior to the day of the general election. In cases where a party's primary fell within this two-month window, we began collecting photographs on the day after that party's candidate was selected. Photos were obtained from microfilm for the 1998 races. Three photographs were discarded from the analyses because the quality of the microfilm image was too poor to allow proper coding. Photographs in which both candidates were pictured together were included in the coding set twice, once for each candidate.

11. The three coders included the two authors as well as a research assistant who was not involved in the collection of the photographs.

12. The candidate photographs were each attached to a blank sheet of paper with the date and newspaper name recorded on the back so that the coders would not know which paper the photograph came from (in three of the races, more than one newspaper's photos were analyzed).

13. To give readers a better idea about our coding decisions, here are a few examples of pictures that were coded as having different levels of favorableness. In one photograph, judged to be highly favorable by all three coders, an Illinois senatorial candidate—Peter Fitzgerald—is seen smiling, holding a small child, and surrounded by a large group of supporters holding Fitzgerald campaign signs. Another photograph, which received a neutral code from all three coders, was simply a head shot of Wisconsin senatorial candidate Russ Feingold. In this picture, Feingold has a blank expression on his face—neither smiling nor frowning—and the tight cropping of the photograph provides a minimal (and plain) background. A third photograph, coded as highly unfavorable by two of the coders and unfavorable by the third, shows Wisconsin gubernatorial candidate Scott McCallum in a debate. McCallum is behind a podium, looking at a competitor with his head cocked to one side, his brow wrinkled, and one eye closed.

14. Newspaper readers undoubtedly take into account—consciously or unconsciously—things such as facial expressions and backgrounds; we did as well. But newspaper readers do not evaluate photographs with a checklist, and it is certainly questionable whether they weight equally the various attributes of the photograph in their assessment of it.

15. Because the *Milwaukee Journal Sentinel* was formed in 1996 through the merger of the *Milwaukee Journal* and the *Milwaukee Sentinel,* we could only look at the 1996 and 1998 endorsements for the 1998 race.

16. In some cases, no endorsement was given for a particular race. For example, the *Chicago Tribune* did not endorse either major party candidate for governor in 1992. Because every other major race endorsement by this newspaper from 1992 to 2002 was for a Republican candidate, however, the *Tribune* was coded as having a "Republican" political atmosphere.

17. One could potentially criticize our use of endorsements alone as a measure of political atmosphere in these cases. However, since existing research has found that endorsements are generally consistent with editorials in the United States (Gaziano 1989), and since every major endorsement was for a candidate of the same political party in these five cases, we feel confident that the measure accurately captures the overall editorial climate at the newspapers examined.

18. One could argue that how favorable a photograph is can be relatively easily ordered ("unfavorable" compared to "neutral") but not so easily assigned numeric values (–2 compared to 2) that achieve interval consistency. At the same time, the interval level assumptions do not seem as great a leap of faith in this case as they are in many other studies in which interval level measurement approaches are employed. It is not unreasonable to say that the difference between "highly unfavorable" and "favorable" is approximately the same as the difference between "neutral" and "unfavorable." Therefore, we only present interval level statistics in this study. We also, however, conducted ordinal level analyses—not reported in this article—that produced similar findings in practically all cases.

19. Because directional expectations existed prior to the analysis, all statistical significance levels in this study are for one-tailed tests.

20. For the difference of means tests, only favorableness of photographs in newspapers at the two endpoints of the three-category political atmosphere variable were examined. However, in calculating the Pearson correlation statistic, the "mixed" political atmosphere newspaper photographs were included.

21. In fact Kepplinger (1982) already has found evidence of visual bias in the television coverage of the 1976 campaign for the office of the Chancellor of the Federal Republic of Germany.

References

Alterman, Eric. 2003. *What Liberal Media?: The Truth about Bias and the News.* New York: Basic Books.

Argyle, Michael, Florisse Alkema, and Robin Gilmour. 1971. "The Communication of Friendly and Hostile Attitudes by Verbal and Non-Verbal Signals." *European Journal of Social Psychology* 1: 385–402.

Barrett, Andrew W., and Lowell W. Barrington. 2005. "Is a Picture Worth a Thousand Words? Newspaper Photographs and Voter Evaluations of Political Candidates." *Harvard International Journal of Press/Politics* 10: 98–113.

Bystrom, Dianne G., Terry A. Robertson, and Mary Christine Banwart. 2001. "Framing the Fight: An Analysis of Media Coverage of Male and Female Candidates in Primary Races for Governor and U. S. Senate in 2000." *American Behavioral Scientist* 44: 1999–2013.

Carter, Sue, Frederick Fico, and Jocelyn A. McCabe. 2002. "Partisan and Structural Balance in Local Television Election Coverage." *Journalism and Mass Communication Quarterly* 79: 41–53.

Coulter, Ann H. 2002. *Slander: Liberal Lies about the American Right.* New York: Crown.

Covington, Cary R., Kent Kroeger, Glenn Richardson, and J. David Woodard. 1993. "Shaping a Candidate's Image in the Press: Ronald Reagan and the 1980 Presidential Election." *Political Research Quarterly* 46: 783–98.

D'Alessio, Dave, and Mike Allen. 2000. "Media Bias in Presidential Elections: A Meta-Analysis." *Journal of Communication* 50: 133–56.

Dalton, Russell J., Paul A. Beck, and Robert Huckfeldt. 1998. "Partisan Cues and the Media: Information Flows in the 1992 Presidential Election." *American Political Science Review* 92: 111–26.

Davis, Richard. 2001. *The Press and American Politics: The New Mediator.* Upper Saddle River, NJ: Prentice Hall.

Domke, David, David P. Fan, Michael Fibison, Dhavan V. Shah, Steven S. Smith, and Mark D. Watts. 1997. "News Media, Candidates and Issues, and Public Opinion in the 1996 Presidential Election." *Journalism and Mass Communication Quarterly* 74: 718–37.

Domke, David, Mark D. Watts, Dhavan Shah, and David P. Fan. 1999. "The Politics of Conservative Elites and the 'Liberal Media' Argument." *Journal of Communication* 49: 35–58.

Gaziano, Cecelie. 1989. "Chain Newspaper Homogeneity and Presidential Endorsements." *Journalism Quarterly* 66: 836–45.

Gilliam, Franklin D., and Shanto Iyengar. 2000. "Prime Suspects: The Influence of Local Television News on the Viewing Public." *American Journal of Political Science* 44: 560–73.

Goldberg, Bernard. 2001. *Bias: A CBS Insider Exposes How the Media Distort the News.* Washington, DC: Regnery.

Goldman, Robert, and Gloria L. Beeker. 1985. "Decoding Newsphotos: An Analysis of Embedded Ideological Values." *Humanity and Society* 9: 351–63.

Graber, Doris A. 1972. "Personal Qualities in Presidential Images: The Contribution of the Press." *Midwest Journal of Political Science* 16: 46–76.

_____. 1987. "Kind Words and Harsh Pictures: How Television Presents the Candidates." In Kay Lehman Schlozman, ed., *Elections in America.* Boston, MA: Allen & Unwin.

_____. 1990. "Seeing Is Remembering: How Visuals Contribute to Learning from Television News." *Journal of Communication* 40: 134–55.

Kahn, Kim Fridkin, and Patrick J. Kenney 2002. "The Slant of the News: How Editorial Endorsements Influence Campaign Coverage and Citizens' Views of Candidates." *American Political Science Review* 96: 381–94.

Kepplinger, Hans Mathias. 1982. "Visual Biases in Television Campaign Coverage." *Communication Research* 9: 432–46.

King, Pu-tsung. 1997. "The Press, Candidate Images, and Voter Perceptions." In Maxwell McCombs, Donald L. Shaw, and David Weaver, eds., *Communication and Democracy:*

Exploring the Intellectual Frontiers in Agenda-Setting Theory. Mahwah, NJ: Lawrence Erlbaum.

Lang, Annie, John Newhagen, and Byron Reeves. 1996. "Negative Video as Structure: Emotion, Attention, Capacity, and Memory." *Journal of Broadcasting and Electronic Media* 40: 460–77.

Lee, Martin A., and Norman Solomon. 1990. *Unreliable Sources: A Guide to Detecting Bias in News Media.* Secaucus, NJ: Lyle Stuart.

Martin, D. S. 1978. "Person Perception and Real-Life Electoral Behavior." *Australian Journal of Psychology* 30: 255–62.

Masters, Roger D., and Denis G. Sullivan. 1989. "Nonverbal Displays and Political Leadership in France and the United States." *Political Behavior* 11: 123–56.

McCombs, Maxwell, Juan Pablo Llamas, Esteban Lopez-Escobar, and Federico Rey 1997. "Candidate Images in Spanish Elections: Second-Level Agenda-Setting Effects." *Journalism and Mass Communication Quarterly* 74: 703–17.

Moriarty, Sandra E., and Gina M. Garramone. 1986. "A Study of Newsmagazine Photographs of the 1984 Presidential Campaign." *Journalism Quarterly* 63: 728–34.

Moriarty, Sandra E., and Mark N. Popovich. 1991. "Newsmagazine Visuals and the 1988 Presidential Election." *Journalism Quarterly* 68: 371-80.

Mutz, Diana C. 1992. "Mass Media and the Depoliticization of Personal Experience." *American Journal of Political Science* 36: 483–508.

Nisbett, Richard E., and Lee Ross. 1980. *Human Inference: Strategies and Shortcomings of Social Judgment.* Englewood Cliffs, NJ: Prentice-Hall.

Niven, David. 1999. "Partisan Bias in the Media?: A New Test." *Social Science Quarterly* 80: 847–57.

Ottati, Victor C., and Megan Deiger. 2002. "Visual Cues and the Candidate Evaluation Process." In Victor C. Ottati, R. Scott Tindale, John Edwards, Fred B. Bryant, Linda Heath, Daniel C. O'Connell, Yolanda Suarez-Balcazar, and Emil J. Posavac, eds., *Social Psychology of Politics: Social Psychological Applications to Social Issues,* Vol. 6. New York: Kluwer Academic/Plenum.

Page, Benjamin I. 1996. *Who Deliberates? Mass Media in Modern Democracy.* Chicago: University of Chicago Press.

Paletz, David L., and Robert M. Entman. 1981. *Media Power Politics.* New York: Free Press.

Palvio, Allan. 1991. *Images in Mind: The Evolution of a Theory.* New York: Harvester Wheatsheaf.

Patterson, Thomas. 1993. *Out of Order.* New York: Knopf.

Riggle, Ellen D. 1992. "Cognitive Strategies and Candidate Evaluations." *American Politics Quarterly* 20: 227–46.

Rosenberg, Shawn W., Lisa Bohan, Patrick McCafferty, and Kevin Harris. 1986. "The Image and the Vote: The Effect of Candidate Presentation on Voter Preference." *American Journal of Political Science* 30: 108–27.

Rosenberg, Shawn W., and Patrick McCafferty. 1987. "The Image and the Vote: Manipulating Voters' Preferences." *Public Opinion Quarterly* 51: 31–47.

Rowse, Arthur Edwards. 1957. *Slanted News: A Case Study of the Nixon and Stevenson Fund Stories.* Boston, MA: Beacon.

Rubin, Richard. 1981. *Press, Party and Presidency.* New York: Norton.

Schein, Edgar H. 1985. *Organizational Culture and Leadership.* San Francisco, CA: Jossey-Bass.

Stanley, Harold, and Richard Niemi. 1994. *Vital Statistics on American Politics.* Washington, DC: Congressional Quarterly Press.

Sullivan, Denis G., and Roger D. Masters. 1988. "'Happy Warriors': Leaders' Facial Displays, Viewers' Emotions, and Political Support." *American Journal of Political Science* 32: 345–68.

Waldman, Paul, and James Devitt. 1998. "Newspaper Photographs and the 1996 Presidential Election: The Question of Bias." *Journalism and Mass Communication Quarterly* 75: 302–11.

Wanta, Wayne. 1986. "The Effects of Dominant Photographs: An Agenda Setting Experiment." *Journalism Quarterly* 63: 728–34.

Windhauser, John W., and Dru Riley Evarts. 1991. "Watching the Campaigns on Network Television." In Guido H. Stempel III and John W. Windhauser, eds., *The Media in the 1984 and 1988 Presidential Campaigns.* Westport. CT: Greenwood.

Analysis of the Research Report Example

Now that you have read this example of a research report and noted whether and where the authors have addressed each of the twelve research questions, compare your findings with ours. The letters and numbers after each question refer to where in the article the question under discussion is addressed.

1. *Do the researchers clearly specify the main research question or problem? What is the "why" question?* (A-3)

The research question in this article is clearly stated. The authors, Andrew W. Barrett and Lowell W. Barrington, ask, "whether the 'political atmosphere' of various newspapers leads to discernable differences in the favorableness of photographs of political candidates."

2. *Have the researchers demonstrated the value and significance of their research question and indicated how their research findings will contribute to scientific knowledge about their topic?* (B, D)

The authors note that previous research indicates that the media help shape voters' views of candidates and that voters respond to visual images of candidates. They also note that a large number of studies have examined whether media bias exists, but that there is little consensus. However, much of this research has asked whether the media as a whole are biased, whereas few researchers have looked at specific media outlets and whether there is a connection between the political atmosphere of particular media outlets and the nature of the coverage given to candidates. Furthermore, most research has focused on the pattern and impact of visual news coverage in television, even though many people read newspapers, which are an important source of state and local campaign news. The visual images in newspaper photographs are more enduring because they are seen every time someone glances at the newspaper and because newspaper photographs tend to be used repeatedly. Because newspapers typically contain editorials and endorsements of candidates, it is easier to establish the political atmosphere of newspapers than it is for television. Therefore, the authors' research on whether the political atmosphere of individual newspapers affects their selection of candidate photographs is a direct test of whether media bias exists and will contribute to the debate over this topic.

3. *Have the researchers proposed clear explanations for the political phenomena that interest them?* (C) *What types of relationships are hypothesized?* (C-5) *Do they discuss any alternative explanations?* (G-4)

The authors use "Political Atmosphere Theory" (PAT) to explain bias in individual media outlets. According to PAT, "media outlets develop an organizational political culture that influences the nature of their coverage." Political atmospheres are maintained through hiring practices, self-selection of job applicants, and the tendency of "employees who do not hold political opinions consistent with the prevailing political atmosphere at a media outlet" to change their reporting and editing work to fit in with the dominant values of the news organization. The result, the authors contend, is that the coverage of politics, including the selection of photographs, will be consistent with the political slant of the editorial page. Specifically, they hypothesize that "newspaper photographs of political candidates will favor the candidate who best reflects the political atmosphere of a given newspaper." They do not suggest any alternative explanations for why the favorableness of photographs of political candidates would vary systematically until the conclusion, when they consider that the favorableness of candidate photographs may be due to the physical attractiveness of the candidates.

4. *Are the independent and dependent variables identified?* (D-2, D-5) *If so, what are they? Have the authors considered any alternative or control variables? If so, identify them. Can you think of any variable that the researchers did not mention?*

The authors do not specify which is the independent variable and which is dependent variable, but these can be inferred from their discussion. The dependent variable is favorability of candidates' photographs. The independent variable is the political atmosphere of the newspaper from which the photograph is drawn. There are no control or alternative variables used in the analysis and no obvious ones seem to have been overlooked.

5. *Are the hypotheses empirical, general, and plausible?* (C-5)

The hypothesis is empirical. It is also general in that it relates characteristics of newspapers and newspaper photographs in general. The hypothesis is plausible given previous research that shows favorability of news coverage of candidates that coincided with editorial points of view.

6. *Are the concepts in the hypotheses clearly defined? Are the operational definitions given for the variables valid and reasonable? What is the level of measurement for each of the variables?* (E)

Each of the variables is clearly defined, and the operational definitions appear to be valid and reasonable. Favorability of photographs was measured using a five-point scale. The authors did not provide a checklist of features to distinguish favorable from unfavorable photographs. Instead they relied on the "quick impressions" of the photographs by three coders. The test they used to check intercoder reliability showed the coders' impressions were quite consistent. The authors argue that quick impressions are a more valid measure of the favorability of photographs, because readers also form quick impressions and do not read the newspaper and look at the photographs with checklists. Barrett and Barrington created a second favorability rating by coding a photograph's size (1 through 4) and multiplying the first favorability rating by the code for the photograph's size so that the second measure ranged from –8 for a large, very unfavorable photograph to 8 for a large, very favorable one.

The concept of political atmosphere was operationalized in two ways. The first measure was based on which candidate the newspaper endorsed in the election. In the second, the authors examined newspaper endorsements of the major statewide or national races for the election year of the race in question and for the previous two years. If all of a paper's endorsements were for Republican candidates, the newspaper's political atmosphere was coded as Republican. Similarly, if all of a paper's endorsements were for Democratic candidates, the newspaper's political atmosphere was coded as Democratic. Where the pattern of endorsements was mixed, the newspaper's editorials were analyzed further to see whether the editorials were consistently in one partisan camp or the other. If this was the case, the newspaper was coded as Democratic or Republican; if not, the newspaper was coded as "mixed" and was not used for analysis. The pattern of newspaper endorsements over time was used to determine whether a candidate's party was the same as the endorsement pattern or opposite to the pattern. The first measure of political atmosphere is objective and clearly replicable. The second is harder to replicate as it depends on coding editorials on a five-point scale ranging from strongly Republican to strongly Democratic. Nevertheless, the measure is reasonable and appears to be valid.

The level of measurement of the independent variable is nominal. Favorability of photographs is certainly ordinal, but if the size of the intervals between the scores is assumed to be equal, then the measure is interval level.

7. *What method of data collection is used to make the necessary observations?* (E) *Are the observations valid and the measures reliable?* (E)

The researchers use document analysis to measure the variables. They base their measures on pictures, endorsements, and editorials in newspapers. They check the reliability of their measures using intercoder reliability testing.

8. *Have the researchers made empirical observations about the units of analysis specified in the hypotheses?* (E)

The units of analysis in this study are newspaper photographs of political candidates. Each variable is a characteristic of a photograph or the newspaper in which the photograph was printed, which can be observed empirically.

9. *If a sample is used, what type of sample is it? Does the type of sample seriously affect the conclusions that can be drawn from the research? Do the researchers discuss this?* (E-1)

Barrett and Barrington did not use a probability sample. Rather they selected seven newspapers covering seven different statewide and local races during the 1998 and 2002 election cycles in four Midwestern states. They selected close races (see note 8) to avoid a potential criticism that newspapers endorse candidates that are likely to win to ensure access to the candidate after the election. Once the races were chosen, they selected all of the newspapers' photographs of the candidates.

The authors argue that choosing newspapers from the Midwest should not limit the generalizability of their findings, because there is no reason to expect that a relationship between political atmosphere and photograph favorableness for individual media outlets in the Midwest would not be found in other regions (see note 9).

10. *What type of research design is used? Does the research design adequately test the hypothesized relationships?* (E-1, E-2, E-5, E-6)

The researchers use a nonexperimental research design. The researchers do not exercise any control over the independent variable, political atmosphere of newspapers. They use naturally occurring differences in political atmosphere. They measure political atmosphere using newspaper endorsements, which occur *after* most of the photographs have appeared; they argue it is not a problem that the independent variable (endorsements) occurred after the dependent variable because it is not the endorsement itself that affects coverage of the campaign. Rather, endorsements reflect the political atmosphere that prevailed during the campaign. The second method they use to operationalize political atmosphere includes a time period prior to the campaigns analyzed in the current study.

Barrett and Barrington use this research design to conduct two sets of analyses. In the first, they use all of the photographs. In the second, they look at the relationship between political atmosphere and favorableness of photographs for each of the newspapers included in their study. The research design is adequate to test the hypothesized relationships.

11. *Are the statistics that are used appropriate for the level of measurement of the variables?* (F-1)

The authors treat the measure of photograph favorableness as an interval-level measure. This is acceptable practice. Therefore, because they have nominal-level independent variables, they use a difference of means test.

They use a one-tailed t test of statistical significance assuming unequal variances, which is appropriate. They also report (in note 18) that they conducted ordinal-level analyses, which they did not report in the article, and obtained similar findings "in practically all cases." It is not clear what this means.

They use Pearson's r as a correlation coefficient to measure the strength of the relationship, which is also acceptable, although eta-squared, a proportionate-reduction-in-error measure of association could also be used.

12. *Are the research findings presented and discussed clearly? Is the basis for deciding whether a hypothesis is supported or refuted clearly specified?* (F, G)

The authors find there are statistically significant differences in the favorableness of photographs in the direction they hypothesized for both measures of political atmosphere for the photographs taken together. They clearly present their results in the tables. They do not report the actual p value or level of significance; they follow a typically used convention of reporting whether their findings were significant at the .01 or .05 level.

While their results clearly are statistically significant, the authors also claim that they are strong and "significant." For example, they conclude: "The first approach—employing the newspaper's endorsement in the race in question—produced evidence of a strong relationship. . . ." (F-1) Later they write that this finding "indicates a strong relationship between the political leanings of the newspaper and how candidates are treated pictorially." (F-2) But, are these relationships strong and significant in the everyday sense of the word? The Pearson's r values range from .095 to .129 for the first set of analyses reported in Table 1. These values indicate weak relationships since they are quite close to zero. Furthermore, the authors do not discuss how to interpret the size of the difference of the mean favorableness. How much of a difference in the favorableness of photographic treatment does there have to be to have a significant impact on voters? Is a .154 difference out of a scale that runs from –2 to 2 or a .452 difference in a measure that can vary from –8 to 8 a significant difference?

Conclusion

A research report rarely answers all the questions that can be raised about a topic. But a well-written report, because it carefully explains how the researcher conducted each stage in the research process, makes it easier for other researchers to evaluate the work. Other investigators may build on that work by varying the method of data collection, the operationalization of variables, or the research design.

By now you should understand how scientific knowledge about politics is acquired. You should know how to formulate a testable hypothesis, choose valid and reliable measures for the concepts that you relate in a hypothesis, develop a research design, conduct a literature review, and make empirical observations. You should also be able to analyze data using appropriate univariate, bivariate, and multivariate statistics. Finally, you should be able to evaluate most research reports as well as write a research report yourself.

We encourage you to think up research questions of your own. Some of these may be feasible projects for a one- or two-semester course. You will learn much more about the research process by doing research than by just reading about it. We wish you success.

Appendixes

Normal Curve Tail Probabilities. Standard normal probability in right-hand tail (for negative values of z, probabilities are found by symmetry)

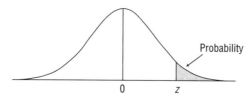

Second Decimal Place of z

z	.00	.01	.02	.03	.04	.05	.06	.07	.08	.09
0.0	.5000	.4960	.4920	.4880	.4840	.4801	.4761	.4721	.4681	.4641
0.1	.4602	.4562	.4522	.4483	.4443	.4404	.4364	.4325	.4286	.4247
0.2	.4207	.4168	.4129	.4090	.4052	.4013	.3974	.3936	.3897	.3859
0.3	.3821	.3783	.3745	.3707	.3669	.3632	.3594	.3557	.3520	.3483
0.4	.3446	.3409	.3372	.3336	.3300	.3264	.3228	.3192	.3156	.3121
0.5	.3085	.3050	.3015	.2981	.2946	.2912	.2877	.2843	.2810	.2776
0.6	.2743	.2709	.2676	.2643	.2611	.2578	.2546	.2514	.2483	.2451
0.7	.2420	.2389	.2358	.2327	.2296	.2266	.2236	.2206	.2177	.2148
0.8	.2119	.2090	.2061	.2033	.2005	.1977	.1949	.1922	.1894	.1867
0.9	.1841	.1814	.1788	.1762	.1736	.1711	.1685	.1660	.1635	.1611
1.0	.1587	.1562	.1539	.1515	.1492	.1469	.1446	.1423	.1401	.1379
1.1	.1357	.1335	.1314	.1292	.1271	.1251	.1230	.1210	.1190	.1170
1.2	.1151	.1131	.1112	.1093	.1075	.1056	.1038	.1020	.1003	.0985
1.3	.0968	.0951	.0934	.0918	.0901	.0885	.0869	.0853	.0838	.0823
1.4	.0808	.0793	.0778	.0764	.0749	.0735	.0722	.0708	.0694	.0681
1.5	.0668	.0655	.0643	.0630	.0618	.0606	.0594	.0582	.0571	.0559
1.6	.0548	.0537	.0526	.0516	.0505	.0495	.0485	.0475	.0465	.0455
1.7	.0446	.0436	.0427	.0418	.0409	.0401	.0392	.0384	.0375	.0367
1.8	.0359	.0352	.0344	.0336	.0329	.0322	.0314	.0307	.0301	.0294
1.9	.0287	.0281	.0274	.0268	.0262	.0256	.0250	.0244	.0239	.0233
2.0	.0228	.0222	.0217	.0212	.0207	.0202	.0197	.0192	.0188	.0183
2.1	.0179	.0174	.0170	.0166	.0162	.0158	.0154	.0150	.0146	.0143
2.2	.0139	.0136	.0132	.0129	.0125	.0122	.0119	.0116	.0113	.0110
2.3	.0107	.0104	.0102	.0099	.0096	.0094	.0091	.0089	.0087	.0084
2.4	.0082	.0080	.0078	.0075	.0073	.0071	.0069	.0068	.0066	.0064
2.5	.0062	.0060	.0059	.0057	.0055	.0054	.0052	.0051	.0049	.0048
2.6	.0047	.0045	.0044	.0043	.0041	.0040	.0039	.0038	.0037	.0036
2.7	.0035	.0034	.0033	.0032	.0031	.0030	.0029	.0028	.0027	.0026
2.8	.0026	.0025	.0024	.0023	.0023	.0022	.0021	.0021	.0020	.0019
2.9	.0019	.0018	.0017	.0017	.0016	.0016	.0015	.0015	.0014	.0014
3.0	.00135									
3.5	.000233									
4.0	.0000317									
4.5	.00000340									
5.0	.000000287									

Source: R. E. Walpole, *Introduction to Statistics* (New York: Macmillan, 1968). Used with permission.

Critical Values from *t* Distribution

	Alpha Level for One-Tailed Test						
	.05	.025	.01	.005	.0025	.001	.0005
Degree of Freedom (df)	Alpha Level for Two-Tailed Test						
	.10	.05	.02	.01	.005	.002	.001
1	6.314	12.706	31.821	63.657	127.32	318.31	636.62
2	2.920	4.303	6.965	9.925	14.089	22.327	31.598
3	2.353	3.182	4.541	5.841	7.453	10.214	12.924
4	2.132	2.776	3.747	4.604	5.598	7.173	8.610
5	2.015	2.571	3.365	4.032	4.773	5.893	6.869
6	1.943	2.447	3.143	3.707	4.317	5.208	5.959
7	1.895	2.365	2.998	3.499	4.029	4.785	5.408
8	1.869	2.306	2.896	3.355	3.833	4.501	5.041
9	1.833	2.262	2.821	3.250	3.690	4.297	4.781
10	1.812	2.228	2.764	3.169	3.581	4.144	4.587
11	1.796	2.201	2.718	3.106	3.497	4.025	4.437
12	1.782	2.179	2.681	3.055	3.428	3.930	4.318
13	1.771	2.160	2.650	3.012	3.372	3.852	4.221
14	1.761	2.145	2.624	2.977	3.326	3.787	4.140
15	1.753	2.131	2.602	2.947	3.286	3.733	4.073
16	1.746	2.120	2.583	2.921	3.252	3.686	4.015
17	1.740	2.110	2.567	2.898	3.222	3.646	3.965
18	1.734	2.101	2.552	2.878	3.197	3.610	3.922
19	1.729	2.093	2.539	2.861	3.174	3.579	3.883
20	1.725	2.086	2.528	2.845	3.153	3.552	3.850
21	1.721	2.080	2.518	2.831	3.135	3.527	3.819
22	1.717	2.074	2.508	2.819	3.119	3.505	3.792
23	1.714	2.069	2.500	2.807	3.104	3.485	3.767
24	1.711	2.064	2.492	2.797	3.091	3.467	3.745
25	1.708	2.060	2.485	2.787	3.078	3.450	3.725
26	1.706	2.056	2.479	2.779	3.067	3.435	3.707
27	1.703	2.052	2.473	2.771	3.057	3.421	3.690
28	1.701	2.048	2.467	2.763	3.047	3.408	3.674
29	1.699	2.045	2.462	2.756	3.038	3.396	3.659
30	1.697	2.042	2.457	2.750	3.030	3.385	3.646
40	1.684	2.021	2.423	2.704	2.971	3.307	3.551
60	1.671	2.000	2.390	2.660	2.915	3.232	3.460
120	1.658	1.980	2.358	2.617	2.860	3.160	3.373
∞	1.645	1.960	2.326	2.576	2.807	3.090	3.291

Source: James V. Couch, *Fundamentals of Statistics for the Behavioral Sciences* (St. Paul: West, 1987), 327. Used with permission.

Chi-Squared Distribution Values for Various Right-tail Probabilities

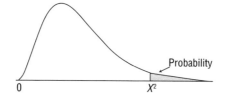

Right-Tail Probability

df	0.250	0.100	0.050	0.025	0.010	0.005	0.001
1	1.32	2.71	3.84	5.02	6.63	7.88	10.83
2	2.77	4.61	5.99	7.38	9.21	10.60	13.82
3	4.11	6.25	7.81	9.35	11.34	12.84	16.27
4	5.39	7.78	9.49	11.14	13.28	14.86	18.47
5	6.63	9.24	11.07	12.83	15.09	16.75	20.52
6	7.84	10.64	12.59	14.45	16.81	18.55	22.46
7	9.04	12.02	14.07	16.01	18.48	20.28	24.32
8	10.22	13.36	15.51	17.53	20.09	21.96	26.12
9	11.39	14.68	16.92	19.02	21.67	23.59	27.88
10	12.55	15.99	18.31	20.48	23.21	25.19	29.59
11	13.70	17.28	19.68	21.92	24.72	26.76	31.26
12	14.85	18.55	21.03	23.34	26.22	28.30	32.91
13	15.98	19.81	22.36	24.74	27.69	29.82	34.53
14	17.12	21.06	23.68	26.12	29.14	31.32	36.12
15	18.25	22.31	25.00	27.49	30.58	32.80	37.70
16	19.37	23.54	26.30	28.85	32.00	34.27	39.25
17	20.49	24.77	27.59	30.19	33.41	35.72	40.79
18	21.60	25.99	28.87	31.53	34.81	37.16	42.31
19	22.72	27.20	30.14	32.85	36.19	38.58	43.82
20	23.83	28.41	31.41	34.17	37.57	40.00	45.32
25	29.34	34.38	37.65	40.65	44.31	46.93	52.62
30	34.80	40.26	43.77	46.98	50.89	53.67	59.70
40	45.62	51.80	55.76	59.34	63.69	66.77	73.40
50	56.33	63.17	67.50	71.42	76.15	79.49	86.66
60	66.98	74.40	79.08	83.30	88.38	91.95	99.61
70	77.58	85.53	90.53	95.02	100.4	104.2	112.3
80	88.13	96.58	101.8	106.6	112.3	116.3	124.8
90	98.65	107.6	113.1	118.1	124.1	128.3	137.2
100	109.1	118.5	124.3	129.6	135.8	140.2	149.5

Source: Alan Agresti and Barbara Finlay, *Statistical Methods for the Social Sciences,* 3d edition (Upper Saddle River, NJ: Prentice Hall: 1997) p. 670. Used with permission.

F Distribution

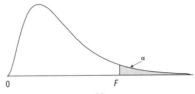

$\alpha = .05$

df_2	df_1									
	1	2	3	4	5	6	8	12	24	∞
1	161.4	199.5	215.7	224.6	230.2	234.0	238.9	243.9	249.0	254.3
2	18.51	19.00	19.16	19.25	19.30	19.33	19.37	19.41	19.45	19.50
3	10.13	9.55	9.28	9.12	9.01	8.94	8.84	8.74	8.64	8.53
4	7.71	6.94	6.59	6.39	6.26	6.16	6.04	5.91	5.77	5.63
5	6.61	5.79	5.41	5.19	5.05	4.95	4.82	4.68	4.53	4.36
6	5.99	5.14	4.76	4.53	4.39	4.28	4.15	4.00	3.84	3.67
7	5.59	4.74	4.35	4.12	3.97	3.87	3.73	3.57	3.41	3.23
8	5.32	4.46	4.07	3.84	3.69	3.58	3.44	3.28	3.12	2.93
9	5.12	4.26	3.86	3.63	3.48	3.37	3.23	3.07	2.90	2.71
10	4.96	4.10	3.71	3.48	3.33	3.22	3.07	2.91	2.74	2.54
11	4.84	3.98	3.59	3.36	3.20	3.09	2.95	2.79	2.61	2.40
12	4.75	3.88	3.49	3.26	3.11	3.00	2.85	2.69	2.50	2.30
13	4.67	3.80	3.41	3.18	3.02	2.92	2.77	2.60	2.42	2.21
14	4.60	3.74	3.34	3.11	2.96	2.85	2.70	2.53	2.35	2.13
15	4.54	3.68	3.29	3.06	2.90	2.79	2.64	2.48	2.29	2.07
16	4.49	3.63	3.24	3.01	2.85	2.74	2.59	2.42	2.24	2.01
17	4.45	3.59	3.20	2.96	2.81	2.70	2.55	2.38	2.19	1.96
18	4.41	3.55	3.16	2.93	2.77	2.66	2.51	2.34	2.15	1.92
19	4.38	3.52	3.13	2.90	2.74	2.63	2.48	2.31	2.11	1.88
20	4.35	3.49	3.10	2.87	2.71	2.60	2.45	2.28	2.08	1.84
21	4.32	3.47	3.07	2.84	2.68	2.57	2.42	2.25	2.05	1.81
22	4.30	3.44	3.05	2.82	2.66	2.55	2.40	2.23	2.03	1.78
23	4.28	3.42	3.03	2.80	2.64	2.53	2.38	2.20	2.00	1.76
24	4.26	3.40	3.01	2.78	2.62	2.51	2.36	2.18	1.98	1.73
25	4.24	3.38	2.99	2.76	2.60	2.49	2.34	2.16	1.96	1.71
26	4.22	3.37	2.98	2.74	2.59	2.47	2.32	2.15	1.95	1.69
27	4.21	3.35	2.96	2.73	2.57	2.46	2.30	2.13	1.93	1.67
28	4.20	3.34	2.95	2.71	2.56	2.44	2.29	2.12	1.91	1.65
29	4.18	3.33	2.93	2.70	2.54	2.43	2.28	2.10	1.90	1.64
30	4.17	3.32	2.92	2.69	2.53	2.42	2.27	2.09	1.89	1.62
40	4.08	3.23	2.84	2.61	2.45	2.34	2.18	2.00	1.79	1.51
60	4.00	3.15	2.76	2.52	2.37	2.25	2.10	1.92	1.70	1.39
120	3.92	3.07	2.68	2.45	2.29	2.17	2.02	1.83	1.61	1.25
∞	3.84	2.99	2.60	2.37	2.21	2.09	1.94	1.75	1.52	1.00

Source: From Table V of R. A. Fisher and F. Yates, *Statistical Tables for Biological, Agricultural and Medical Research*, published by Longman Group Ltd., London, 1974. (Previously published by Oliver & Boyd, Edinburgh.) Reprinted by permission of the authors and publishers.

$\alpha = .01$

df_2	df_1									
	1	2	3	4	5	6	8	12	24	∞
1	4052	4999	5403	5625	5764	5859	5981	6106	6234	6366
2	98.49	99.01	99.17	99.25	99.30	99.33	99.36	99.42	99.46	99.50
3	34.12	30.18	29.46	28.71	28.24	27.91	27.49	27.05	26.60	26.12
4	21.20	18.00	16.69	15.98	15.52	15.21	14.80	14.37	13.93	13.46
5	16.26	13.27	12.06	11.39	10.97	10.67	10.27	9.89	9.47	9.02
6	13.74	10.92	9.78	9.15	8.75	8.47	8.10	7.72	7.31	6.88
7	12.25	9.55	8.45	7.85	7.46	7.19	6.84	6.47	6.07	5.65
8	11.26	8.65	7.59	7.01	6.63	6.37	6.03	5.67	5.28	4.86
9	10.56	8.02	6.99	6.42	6.06	5.80	5.47	5.11	4.73	4.31
10	10.04	7.56	6.55	5.99	5.64	5.39	5.06	4.71	4.33	3.91
11	9.65	7.20	6.22	5.67	5.32	5.07	4.74	4.40	4.02	3.60
12	9.33	6.93	5.95	5.41	5.06	4.82	4.50	4.16	3.78	3.36
13	9.07	6.70	5.74	5.20	4.86	4.62	4.30	3.96	3.59	3.16
14	8.86	6.51	5.56	5.03	4.69	4.46	4.14	3.80	3.43	3.00
15	8.68	6.36	5.42	4.89	4.56	4.32	4.00	3.67	3.29	2.87
16	8.53	6.23	5.29	4.77	4.44	4.20	3.89	3.55	3.18	2.75
17	8.40	6.11	5.18	4.67	4.34	4.10	3.79	3.45	3.08	2.65
18	8.28	6.01	5.09	4.58	4.25	4.01	3.71	3.37	3.00	2.57
19	8.18	5.93	5.01	4.50	4.17	3.94	3.63	3.30	2.92	2.49
20	8.10	5.85	4.94	4.43	4.10	3.87	3.56	3.23	2.86	2.42
21	8.02	5.78	4.87	4.37	4.04	3.81	3.51	3.17	2.80	2.36
22	7.94	5.72	4.82	4.31	3.99	3.76	3.45	3.12	2.75	2.31
23	7.88	5.66	4.76	4.26	3.94	3.71	3.41	3.07	2.70	2.26
24	7.82	5.61	4.72	4.22	3.90	3.67	3.36	3.03	2.66	2.21
25	7.77	5.57	4.68	4.18	3.86	3.63	3.32	2.99	2.62	2.17
26	7.72	5.53	4.64	4.14	3.82	3.59	3.29	2.96	2.58	2.13
27	7.68	5.49	4.60	4.11	3.78	3.56	3.26	2.93	2.55	2.10
28	7.64	5.45	4.57	4.07	3.75	3.53	3.23	2.90	2.52	2.06
29	7.60	5.42	4.54	4.04	3.73	3.50	3.20	2.87	2.49	2.03
30	7.56	5.39	4.51	4.02	3.70	3.47	3.17	2.84	2.47	2.01
40	7.31	5.18	4.31	3.83	3.51	3.29	2.99	2.66	2.29	1.80
60	7.08	4.98	4.13	3.65	3.34	3.12	2.82	2.50	2.12	1.60
120	6.85	4.79	3.95	3.48	3.17	2.96	2.66	2.34	1.95	1.38
∞	6.64	4.60	3.78	3.32	3.02	2.80	2.51	2.18	1.79	1.00

APPENDIX D (continued)

$\alpha = .001$

df_2	1	2	3	4	df_1 5	6	8	12	24	∞
1	405284	500000	540379	562500	576405	585937	598144	610667	623497	636619
2	998.5	999.0	999.2	999.2	999.3	999.3	999.4	999.4	999.5	999.5
3	167.5	148.5	141.1	137.1	134.6	132.8	130.6	128.3	125.9	123.5
4	74.14	61.25	56.18	53.44	51.71	50.53	49.00	47.41	45.77	44.05
5	47.04	36.61	33.20	31.09	29.75	28.84	27.64	26.42	25.14	23.78
6	35.51	27.00	23.70	21.90	20.81	20.03	19.03	17.99	16.89	15.75
7	29.22	21.69	18.77	17.19	16.21	15.52	14.63	13.71	12.73	11.69
8	25.42	18.49	15.83	14.39	13.49	12.86	12.04	11.19	10.30	9.34
9	22.86	16.39	13.90	12.56	11.71	11.13	10.37	9.57	8.72	7.81
10	21.04	14.91	12.55	11.28	10.48	9.92	9.20	8.45	7.64	6.76
11	19.69	13.81	11.56	10.35	9.58	9.05	8.35	7.63	6.85	6.00
12	18.64	12.97	10.80	9.63	8.89	8.38	7.71	7.00	6.25	5.42
13	17.81	12.31	10.21	9.07	8.35	7.86	7.21	6.52	5.78	4.97
14	17.14	11.78	9.73	8.62	7.92	7.43	6.80	6.13	5.41	4.60
15	16.59	11.34	9.34	8.25	7.57	7.09	6.47	5.81	5.10	4.31
16	16.12	10.97	9.00	7.94	7.27	6.81	6.19	5.55	4.85	4.06
17	15.72	10.66	8.73	7.68	7.02	6.56	5.96	5.32	4.63	3.85
18	15.38	10.39	8.49	7.46	6.81	6.35	5.76	5.13	4.45	3.67
19	15.08	10.16	8.28	7.26	6.61	6.18	5.59	4.97	4.29	3.52
20	14.82	9.95	8.10	7.10	6.46	6.02	5.44	4.82	4.15	3.38
21	14.59	9.77	7.94	6.95	6.32	5.88	5.31	4.70	4.03	3.26
22	14.38	9.61	7.80	6.81	6.19	5.76	5.19	4.58	3.92	3.15
23	14.19	9.47	7.67	6.69	6.08	5.65	5.09	4.48	3.82	3.05
24	14.03	9.34	7.55	6.59	5.98	5.55	4.99	4.39	3.74	2.97
25	13.88	9.22	7.45	6.49	5.88	5.46	4.91	4.31	3.66	2.89
26	13.74	9.12	7.36	6.41	5.80	5.38	4.83	4.24	3.59	2.82
27	13.61	9.02	7.27	6.33	5.73	5.31	4.76	4.17	3.52	2.75
28	13.50	8.93	7.19	6.25	5.66	5.24	4.69	4.11	3.46	2.70
29	13.39	8.85	7.12	6.19	5.59	5.18	4.64	4.05	3.41	2.64
30	13.29	8.77	7.05	6.12	5.53	5.12	4.58	4.00	3.36	2.59
40	12.61	8.25	6.60	5.70	5.13	4.73	4.21	3.64	3.01	2.23
60	11.97	7.76	6.17	5.31	4.76	4.37	3.87	3.31	2.69	1.90
120	11.38	7.31	5.79	4.95	4.42	4.04	3.55	3.02	2.40	1.56
∞	10.83	6.91	5.42	4.62	4.10	3.74	3.27	2.74	2.13	1.00

Glossary

Accretion measures. Measures of phenomena through indirect observation of the accumulation of materials.

Actions. Physical human movement or behavior done for a reason.

Alternative hypothesis. A statement about the value or values of a population parameter. A hypothesis proposed as an alternative to the null hypothesis.

Alternative-form method. A method of calculating reliability by repeating different but equivalent measures at two or more points in time.

Ambiguous question. A question containing a concept that is not defined clearly.

Analysis of variance. A technique for measuring the relationship between one nominal- or ordinal-level variable and one interval- or ratio-level variable.

Antecedent variable. An independent variable that precedes other independent variables in time.

Applied research. Research designed to produce knowledge useful in altering a real-world condition or situation.

Arrow diagram. A pictorial representation of a researcher's explanatory scheme.

Assignment at random. Random assignment of subjects to experimental and control groups.

Bar graph. A graphic display of the data in a frequency or percentage distribution.

Branching question. A question that sorts respondents into subgroups and directs these subgroups to different parts of the questionnaire.

Causal relationship. A connection between two entities that occurs because one produces, or brings about, the other with complete or great regularity.

Central tendency. The most frequent, middle, or central value in a frequency distribution.

Chi-square. A statistic used to test the statistical significance of a relationship in a cross-tabulation table.

Classical experimental design. An experiment with the random assignment of subjects to experimental and control groups with a pre-test and post-test for both groups.

Closed-ended question. A question with response alternatives provided.

Cluster sample. A probability sample that is used when no list of elements exists. The sampling frame initially consists of clusters of elements.

Cohort. A group of people who all experience a significant event in roughly the same time frame.

Confidence interval. The range of values into which a population parameter is likely to fall for a given level of confidence.

Confidence level. The degree of belief that an estimated range of values—more specifically, a high or low value—includes or covers the population parameter.

Construct validity. Validity demonstrated for a measure by showing that it is related to the measure of another concept.

Constructionism. An approach to knowledge that asserts humans actually construct—through their social interactions and cultural and historical practices—many of the facts they take for granted as having an independent, objective, or material reality.

Content analysis. A procedure by which verbal, nonquantitative records are transformed into quantitative data.

Content validity. Validity demonstrated by ensuring that the full domain of a concept is measured.

Control by adjustment. A form of statistical control in which a mathematical adjustment is made to assess the impact of a third variable.

Control by grouping. A form of statistical control in which observations identical or similar to the control variable are grouped together.

Control group. A group of subjects who do not receive the experimental treatment or test stimulus.

Convenience sample. A nonprobability sample in which the selection of elements is determined by the researcher's convenience.

Correlation coefficient. A statistic that indicates the strength and direction of the relationship between two interval- or ratio-level variables. It is a standardized measure, meaning that its numerical value does not change with linear scale changes in the variables.

Correlation matrix. A table showing the relationships among discrete measures.

Covert observation. Observation in which the observer's presence or purpose is kept secret from those being observed.

Critical theory. The philosophical stance that disciplines such as political science should assess critically and change society, not merely study it objectively.

Cross-level analysis. The use of data at one level of aggregation to make inferences at another level of aggregation.

Cross-sectional design. A research design in which measurements of independent and dependent variables are taken at the same time; naturally occurring differences in the independent variable are used to create quasi-experimental and quasi-control groups; extraneous factors are controlled for by statistical means.

Cross-tabulation. Also called a contingency table, this array displays the joint frequencies and relative frequencies of two categorical (nominal or ordinal) variables.

Cumulative. Characteristic of scientific knowledge; new substantive findings and research techniques are built upon those of previous studies.

Cumulative proportion. The total proportion of observations at or below a value in a frequency distribution.

Data matrix. An array of rows and columns that stores the values of a set of variables for all the cases in a data set.

Deduction. A type of reasoning in which if the premises of an argument are true, the conclusion is necessarily true.

Degrees of freedom. The number of independent observations used to calculate a statistic.

Demand characteristics. Aspects of the research situation that cause participants to guess the purpose or rationale of the study and adjust their behavior or opinions accordingly.

Dependent variable. The phenomenon thought to be influenced, affected, or caused by some other phenomenon.

Descriptive statistic. The mathematical summary of measurements for one variable.

Dichotomous variable. A variable having only two categories that for certain analytical purposes can be treated as a quantitative variable.

Difference of means test. A statistical procedure for testing the hypothesis that two population means have the same value.

Direct observation. Actual observation of behavior.

Direction of a relationship. An indication of which values of the dependent variable are associated with which values of the independent variable.

Directional hypothesis. A hypothesis that specifies the expected relationship between two or more variables.

Dispersion. The variability or differences in a set of numbers.

Disproportionate sample. A stratified sample in which elements sharing a characteristic are underrepresented or overrepresented in the sample.

Double-barreled question. A question that is really two questions in one.

Dummy variable. An index that has just two values: 0 for the presence (or absence) of a factor and 1 for its absence (or presence).

Ecological fallacy. The fallacy of deducing a false relationship between the attributes or behavior of individuals based on observing that relationship for groups to which the individuals belong.

Ecological inference. The process of inferring a relationship between characteristics of individuals based on group or aggregate data.

Effect size. The effect of a treatment or an experimental variable on a response. Most commonly the size of the effect is measured by obtaining the difference between two means or two proportions.

Electronic database. A collection of information (of any type) stored on an electromagnetic medium that can be accessed and examined by certain computer programs.

Element. A particular case or entity about which information is collected; the unit of analysis.

Empirical generalization. A statement that summarizes the relationship between individual facts and that communicates general knowledge.

Empirical research. Research based on actual, "objective" observation of phenomena.

Empirical verification. Characteristic of scientific knowledge; demonstration by means of objective observation that a statement is true.

Episodic record. The portion of the written record that is not part of a regular, ongoing record-keeping enterprise.

Erosion measures. Measures of phenomena through indirect observation of selective wear of some material.

Estimator. A statistic based on sample observations that is used to estimate the numerical value of an unknown population parameter.

Eta-squared. A measure of association used with the analysis of variance that indicates the proportion of the variance in the dependent variable that is explained by the variance in the independent variable.

Expected value. The mean or average value of a sample statistic based on repeated samples from a population.

Experimental control. Manipulation of the exposure of experimental groups to experimental stimuli to assess the impact of a third variable.

Experimental effect. Effect, usually measured numerically, of the independent variable on the dependent variable.

Experimental group. A group of subjects who receive the experimental treatment or test stimulus.

Experimental mortality. A differential loss of subjects from experimental and control groups that affects the equivalency of groups; threat to internal validity.

Experimentation. A research design in which the researcher controls exposure to the test factor or independent variable, the assignment of subjects to groups, and the measurement of responses.

Explained variation. That portion of the total variation in a dependent variable that is accounted for by the variation in the independent variable(s).

Explanatory. Characteristic of scientific knowledge; signifying that a conclusion can be derived from a set of general propositions and specific initial considerations; providing a systematic, empirically verified understanding of why a phenomenon occurs as it does.

Explication. The specification of the conditions under which X and Y are and are not related.

External validity. The ability to generalize from one set of research findings to other situations.

Extraneous factors. Factors besides the independent variable that may cause change in the dependent variable.

Face validity. Validity asserted by arguing that a measure corresponds closely to the concept it is designed to measure.

Factor analysis. A statistical technique useful in the construction of multi-item scales to measure abstract concepts.

Factorial design. Experimental design used to measure the effect of two or more independent variables singly and in combination.

Falsifiability. A property of a statement or hypothesis such that it can (in principle, at least) be rejected in the face of contravening evidence.

Field experiments. Experimental designs applied in a natural setting.

Field study. Observation in a natural setting.

Filter question. A question used to screen respondents so that subsequent questions will be asked only of certain respondents for whom the questions are appropriate.

Focused interview. A semistructured or flexible interview schedule used when interviewing elites.

Formal model. A simplified and abstract representation of reality that can be expressed verbally, mathematically, or in some other symbolic system, and that purports to show how variables or parts of a system are interconnected.

Frequency distribution. The number of observations per value or category of a variable.

General. Characteristic of scientific knowledge; applicable to many rather than to a few cases.

Goodman and Kruskal's gamma. A measure of association between ordinal-level variables.

Goodman and Kruskal's lambda. A measure of association between one nominal- or ordinal-level variable and one nominal-level variable.

Guttman scale. A multi-item measure in which respondents are presented with increasingly difficult measures of approval for an attitude.

Histogram. A type of bar graph in which the height and area of the bars are proportional to the frequencies in each category of a categorical variable or intervals of a continuous variable.

History. A change in the dependent variable due to changes in the environment over time; threat to internal validity.

Hypothesis. A tentative or provisional or unconfirmed statement that can (in principle) be verified.

Independent variable. The phenomenon thought to influence, affect, or cause some other phenomenon.

Index of diversity. A measure of variation for categorical data that can be interpreted as the probability that two individuals selected at random would be in different categories.

Index of qualitative variation. A measure of variation for categorical data that is the index of diversity adjusted by a correction factor based on the number of categories of the variable.

Indirect observation. Observation of physical traces of behavior.

Induction. Induction is the process of drawing an inference from a set of premises and observations. The premises of an inductive argument support its conclusion but do not prove it.

Informant. Person who helps a researcher employing participant observation methods interpret the activities and behavior of the informant and the group to which the informant belongs.

Informed consent. Procedures that inform potential research subjects about the proposed research in which they are being asked to participate. The principle that researchers must obtain the freely given consent of human subjects before participation in a research project.

Institutional review board. Panel to which researchers must submit descriptions of proposed research involving human subjects for the purpose of ethics review.

Instrument decay. A change in the measurement device used to measure the dependent variable, producing change in measurements; threat to internal validity.

Instrument reactivity. Reaction of subjects to a pre-test measurement process.

Intercoder reliability. Demonstration that multiple analysts, following the same content analysis procedure, agree and obtain the same measurements.

Interitem association. A test of the extent to which the scores of several items, each thought to measure the same concept, are the same. Results are displayed in a correlation matrix.

Internal validity. The ability to show that manipulation or variation of the independent variable actually causes the dependent variable to change.

Interpretation. Philosophical approach to the study of human behavior that claims that one must understand the way individuals see their world in order to understand truly their behavior or actions; philosophical objection to the empirical approach to political science.

Interquartile range. The middle 50 percent of observations.

Interval measurement. A measure for which a one-unit difference in scores is the same throughout the range of the measure.

Intervening variable. A variable coming between an independent variable and a dependent variable in an explanatory scheme.

Intervention analysis. Also known as interrupted time series design. A nonexperimental design in which the impact of a naturally occurring event (intervention) on a dependent variable is measured over time.

Interviewer bias. The interviewer's influence on the respondent's answers; an example of reactivity.

Interviewing. Interviewing respondents in a nonstandardized, individualized manner.

Kendall's tau b or c. A measure of association between ordinal-level variables.

Leading question. A question that encourages the respondent to choose a particular response.

Level of measurement. The extent or degree to which the values of variables can be compared and mathematically manipulated.

Likert scale. A multi-item measure in which the items are selected based on their ability to discriminate between those scoring high and those scoring low on the measure.

Linear probability model. Regression model in which a dichotomous variable is treated as the dependent variable.

Literature review. A systematic examination and interpretation of the literature for the purpose of informing further work on a topic.

Logistic regression. A nonlinear regression model that relates a set of explanatory variables to a dichotomous dependent variable.

Logistic regression coefficient. A regression coefficient based on the logistic model that measures how the log-odds of a response change for a one-unit change in an independent variable when the values of all other variables (if any) in the equation are held constant.

Matrix plot. A graph that shows several two-variable scatterplots on one page in order to present a simultaneous view of several bivariate relationships.

Maturation. A change in subjects over time that affects the dependent variable; threat to internal validity.

Mean. A measure of central tendency found by the summing the values of the variable and dividing the total by the number of observations, N.

Mean absolute deviation. A measure of dispersion of data points for interval- and ratio-level data.

Measurement. The process by which phenomena are observed systematically and represented by scores or numerals.

Measures of association. Statistics that summarize the relationship between two variables.

Median. The category or value above and below which one-half of the observations lie.

Mode. The category with the greatest frequency of observations.

Multigroup design. Experimental design with more than one control and experimental group.

Multiple correlation coefficient. A statistic varying between 0 and 1 that indicates the proportion of the total variation in Y, a dependent variable, that is statistically explained by the independent variables.

Multiple regression analysis. A technique for measuring the mathematical relationships between more than one independent variable and a dependent variable while controlling for all other independent variables in the equation.

Multiple regression coefficient. A number that tells how much Y will change for a one-unit change in a particular independent variable, if all of the other variables in the model have been held constant.

Multivariate analysis. Data analysis techniques designed to test hypotheses involving more than two variables.

Multivariate cross-tabulation. A table that displays the relationship between three or more categorical (nominal, ordinal, or grouped quantitative) variables.

Negative relationship. A relationship in which the values of one variable increase as the values of another variable decrease.

Negatively skewed. A distribution of values in which fewer observations lie to the left of the middle value and those observations are fairly distant from the mean.

Nominal measurement. A measure for which different scores represent different, but not ordered, categories.

Nonexperimental design. A research design characterized by at least one of the following: presence of a single group, lack of researcher control over the assignment of subjects to control and experimental groups, lack of researcher control over application of the independent variable, or inability of the researcher to measure the dependent variable before and after exposure to the independent variable occurs.

Nonnormative knowledge. Knowledge concerned not with evaluation or prescription but with factual or objective determinations.

Nonprobability sample. A sample for which each element in the total population has an unknown probability of being selected.

Normal distribution. A distribution defined by a mathematical formula and the graph of which has a symmetrical, bell shape; in which the mean, mode, and median coincide; and in which a fixed proportion of observations lies between the mean and any distance from the mean measured in terms of the standard deviation.

Normative knowledge. Knowledge that is evaluative, value-laden, and concerned with prescribing what ought to be.

Null hypothesis. A statement that a population parameter equals a single or specific value. Often a statement that the difference between two populations is zero.

Odds. The ratio of the frequency of one event to the frequency of a second event.

Odds ratio. A measure of association between two dichotomous variables in which the odds of having a trait or property in one group are divided by the corresponding odds of the other group.

Open-ended question. A question with no response alternatives provided for the respondent.

Operational definition. The rules by which a concept is measured and scores assigned.

Ordinal measurement. A measure for which the scores represent ordered categories that are not necessarily equidistant from each other.

Overt observation. Observation in which those being observed are informed of the observer's presence and purpose.

Panel mortality. Loss of participants from a panel study.

Panel study. A cross-sectional study in which measurements of variables are taken on the same units of analysis at multiple points in time.

Parsimony. The principle that among explanations or theories with equal degrees of confirmation, the simplest—the one based on the fewest assumptions and explanatory factors—is to be preferred. (Sometimes known as Ockham's razor.)

Partial regression coefficient. A number that indicates how much a dependent variable would change if an independent variable changed one unit and all other variables in the equation or model were held constant.

Participant observation. Observation in which the observer becomes a regular participant in the activities of those being observed.

Partly spurious relationship. A relationship between two variables caused partially by a third.

Period effect. An indicator or measure of history effects on a dependent variable during a specified time period.

Pie diagram. A circular graphic display of a frequency distribution.

Political science. The application of the methods of acquiring scientific knowledge to the study of political phenomena.

Population. Term used in sampling and statistics to mean a clearly defined collection of objects.

Population parameter. A characteristic of a population.

Positive relationship. A relationship in which the values of one variable increase (or decrease) as the values of another variable increase (or decrease).

Positively skewed. A distribution of values in which fewer observations lie to the right of the middle value, and those observations are fairly distant from the mean.

Post-test. Measurement of the dependent variable after manipulation of the independent variable.

Precision matching. Matching of pairs of subjects with one of the pair assigned to the experimental group and the other to the control group.

Pre-test. Measurement of the dependent variable prior to the administration of the experimental treatment or manipulation of the independent variable.

Primary data. Data recorded and used by the researcher who is making the observations.

Probabilistic explanation. An explanation that does not explain or predict events with 100-percent accuracy.

Probability sample. A sample for which each element in the total population has a known probability of being selected.

Proportional-reduction-in-error (PRE) measure. A measure of association that indicates how much knowledge of the value of the independent variable of a case improves prediction of the dependent variable compared to the prediction of the dependent variable based on no knowledge of the case's value on the independent variable. Examples are Goodman and Kruskal's lambda, Goodman and Kruskal's gamma, eta-squared, and R-squared.

Proportionate sample. A probability sample that draws elements from a stratified population at a rate proportional to size of the samples.

Pure, theoretical, or recreational research. Research designed to satisfy one's intellectual curiosity about some phenomenon.

Purposive sample. A nonprobability sample in which a researcher uses discretion in selecting elements for observation.

Push poll. A poll intended not to collect information but to feed respondents (often) false and damaging information about a candidate or cause.

Question-order effect. The effect on responses of question placement within a questionnaire.

Questionnaire design. The physical layout and packaging of a questionnaire.

Quota sample. A nonprobability sample in which elements are sampled in proportion to their representation in the population.

***R*-squared.** A statistic, sometimes called the coefficient of determination, defined as the explained sum of squares divided by the total sum of squares. It purportedly measures the strength of the relationship between a dependent variable and one or more independent variables.

Random digit dialing. A procedure used to improve the representativeness of telephone samples by giving both listed and unlisted numbers a chance of selection.

Random start. Selection of a number at random to determine where to start selecting elements in a systematic sample.

Randomization. The random assignment of subjects to experimental and control groups.

Randomized response technique (RRT). A method of obtaining accurate answers to sensitive questions that protects the respondent's privacy.

Range. The distance between the highest and lowest values or the range of categories into which observations fall.

Ratio measurement. A measure for which the scores possess the full mathematical properties of the numbers assigned.

Regression analysis. A technique for measuring the relationship between two interval- or ratio-level variables.

Regression coefficient. A statistic that indicates the strength and direction of the relationship between two quantitative variables.

Regression constant. Value of the dependent variable when all of the values of the independent variables in the equation equal zero.

Relationship. The association, dependence, or covariance of the values of one variable with the values of another variable.

Relative frequency. Percentage or proportion of total number of observations in a frequency distribution that have a particular value.

Reliability. The extent to which a measure yields the same results on repeated trials.

Repeated measurement design. An experimental design in which the dependent variable is measured at multiple times after the treatment is administered.

Research design. A plan specifying how the researcher intends to fulfill the goals of the study; a logical plan for testing hypotheses.

Residual. For a given observation, the difference between an observed value of a variable and its corresponding predicted value based on some model.

Resistant measure. A measure of central tendency that is not sensitive to one or a few extreme values in a distribution.

Response quality. The extent to which responses provide accurate and complete information.

Response rate. The proportion of respondents selected for participation in a survey who actually participate.

Response set. The pattern of responding to a series of questions in a similar fashion without careful reading of each question.

Running record. The portion of the written record that is enduring and covers an extensive period of time.

Sample. A subset of observations or cases drawn from a specified population.

Sample bias. The bias that occurs whenever some elements of a population are systematically excluded from a sample. It is usually due to an incomplete sampling frame or a nonprobability method of selecting elements.

Sample statistic. The estimator of a population characteristic or attribute that is calculated from sample data.

Sample-population congruence. The degree to which sample subjects represent the population from which they are drawn.

Sampling distribution. A theoretical (nonobserved) distribution of sample statistics calculated on samples of size N that, if known, permits the calculation of confidence intervals and the test of statistical hypotheses.

Sampling error. The difference between a sample estimate and a corresponding population parameter that arises because only a portion of a population is observed.

Sampling fraction. The proportion of the population included in a sample.

Sampling frame. The population from which a sample is drawn. Ideally it is the same as the total population of interest to a study.

Sampling interval. The number of elements in a sampling frame divided by the desired sample size.

Sampling unit. The entity listed in a sampling frame. It may be the same as an element, or it may be a group or cluster of elements.

Scatterplot. A plot of Y-X values on a graph consisting of an x-axis drawn perpendicular to a y-axis. The axes represent the values of the variables. Usually the x-axis represents the independent variable if there is one and is drawn horizontally; the y-axis is vertical. Observed pairs of Y-X values are plotted on this coordinate system.

Search engine. A computer program that visits Web pages on the Internet and looks for those containing particular directories or words.

Search term. A word or phrase entered into a computer program (a search engine) that looks through Web pages on the Internet for those that contain the word or phrase.

Secondary data. Data used by a researcher that was not personally collected by that researcher.

Selection bias. Bias in the assignment of subjects to experimental and control groups; threat to internal validity.

Simple post-test design. Weak type of experimental design with control and experimental groups but no pre-test.

Simple random sample. A probability sample in which each element has an equal chance of being selected.

Simulation. A simple representation of a system by a device in order to study its behavior.

Single-sided question. A question with only one substantive alternative provided for the respondent.

Small-N design. Type of experimental design in which one or a few cases of a phenomenon are examined in considerable detail, typically using several data collection methods, such as personal interviews, document analysis, and observation.

Snowball sample. A sample in which respondents are asked to identify additional members of a population.

Somer's d. A measure of association between ordinal-level variables.

Specified relationship. A relationship between two variables that varies with the values of a third.

Split-halves method. A method of calculating reliability by comparing the results of two equivalent measures made at the same time.

Spurious relationship. A relationship between two variables caused entirely by the impact of a third variable.

Standard deviation. A measure of dispersion of data points about the mean for interval- and ratio-level data.

Standard error. The standard deviation or measure of variability or dispersion of a sampling distribution.

Standard normal distribution. Normal distribution with a mean of 0 and a standard deviation and variance of 1.

Standardized regression coefficient. A regression coefficient based on two standardized variables. A standardized variable is derived from a transformed raw variable and has a mean of 0 and standard deviation of 1.

Statistical control. Assessing the impact of a third variable by comparing observations across the values of a control variable.

Statistical independence. A property of two variables in which the probability that an observation is in a particular category of one variable *and* a particular category of the other variable equals the simple or marginal probability of being in those categories.

Statistical inference. The mathematical theory and techniques for making conjectures about the unknown characteristics (parameters) of populations based on samples.

Statistical regression. Change in the dependent variable due to the temporary nature of extreme values; threat to internal validity.

Statistical significance. The probability of making a type I error.

Stratified sample. A probability sample in which elements sharing one or more characteristics are grouped, and elements are selected from each group in proportion to the group's representation in the total population.

Stratum. A subgroup of a population that shares one or more characteristics.

Strength of a relationship. An indication of how consistently the values of a dependent variable are associated with the values of an independent variable.

Structured observation. Systematic observation and recording of the incidence of specific behaviors.

Summation index. A multi-item measure in which individual scores on a set of items are combined to form a summary measure.

Survey instrument. The schedule of questions to be asked of the respondent.

Survey research. The direct or indirect solicitation of information from individuals by asking them questions, having them fill out forms, or using other means.

Systematic sample. A probability sample in which elements are selected from a list at predetermined intervals.

Tautology. A hypothesis in which the independent and dependent variables are identical, making it impossible to disconfirm.

Test stimulus or factor. The independent variable introduced an controlled by an investigator in order to assess its effects on a response or dependent variable.

Testing. Effect of a pre-test on the dependent variable; threat to internal validity.

Test-retest method. A method of calculating reliability by repeating the same measure at two or more points in time.

Theory. A statement or series of statements that organize, explain, and predict phenomena.

Total variation. A quantitative measure of the variation in a variable, determined by summing the squared deviation of each observation from the mean.

Transmissible. Characteristic of scientific knowledge; indicates that the methods used in making scientific discoveries are made explicit.

Trimmed mean. The mean of a set of numbers from which some percentage of the largest and smallest values has been dropped.

Two-sided question. A question with two substantive alternatives provided for the respondent.

Type I error. Error made by rejecting a null hypothesis when it is true.

Type II error. Error made by failing to reject a null hypothesis when it is not true.

Unexplained variation. That portion of the variation in a dependent variable that is *not* accounted for by the variation in the independent variable(s).

Unit of analysis. The type of actor (individual, group, institution, nation) specified in a researcher's hypothesis.

Unstructured observation. Observation in which all behavior and activities are recorded.

Validity. The correspondence between a measure and the concept it is supposed to measure.

Variance. A measure of dispersion of data points about the mean for interval- and ratio-level data.

Weighting factor. A mathematical factor used to make a disproportionate sample representative.

Written record. Documents, reports, statistics, manuscripts, and other recorded materials available and useful for empirical research.

z score. The number of standard deviations by which a score deviates from the mean score.

Index

Note: Tables, figures, boxes, and note(s) are indicated by *t, f, b,* and *n (nn)*, respectively.